Ecological Studies, Vol. 102

Analysis and Synthesis

Edited by

G. Heldmaier, Marburg, FRG
O.L. Lange, Würzburg, FRG
H.A. Mooney, Stanford, USA
U. Sommer, Kiel, FRG

Ecological Studies

Volumes published since 1990 are listed at the end of this book.

Springer

Berlin
Heidelberg
New York
Barcelona
Budapest
Hong Kong
London
Milan
Paris
Sanza Clara
Singapore
Tokyo

Yuri I. Sorokin

Coral Reef Ecology

1st edition 1993. 2nd printing 1995

With 102 Figures

Springer

Professor Dr. Yuri I. Sorokin
Institute of Oceanology
Academy of Sciences
Krasnodar distr.
353470 Gelendzhik
CIS

Zoology Department
University of Queensland
Santa Lucia 4067
Queensland
Australia

ISBN 3-540-60532-0 2nd. printing Springer-Verlag Berlin Heidelberg New York

ISBN 3-540-56427-6 1st ed. Springer-Verlag Berlin Heidelberg New York

Cataloging-in-Publication applied for

Die Deutsche Bibliothek – CIP – Einheitsaufnahme
Sorokin, Jurij I.:
Coral reef ecology/Yuri I. Sorokin. - Berlin ; Heidelberg ;
New York ; Barcelona ; Budapest ; Hong Kong ; London ;
Milan ; Paris ; Santa Clara ; Singapore ; Tokyo : Springer, 1995
(Ecological studies ; Vol. 102)
ISBN 3-540-60532-0
NE: GT

Typesetting: Best-set Typesetters Ltd., Hong Kong
SPIN 10522070 31/3137-5 4 3 2 1 0 - Printed on acid-free paper

This book is dedicated to the memory of Thomas Goreau.

The author

Contents

Introduction

The coral reef communities constitute one of the most important ecosystems on our Planet. The main feature of coral reef ecosystems is their extremely high productivity. The photo-synthetic production in reef communities attains 5–20 $g\,m^{-2}$ of organic carbon or 50–200 g of wet biomass, it being actually independent of the nutrient concentration in surrounding waters. Primary production in coral reef ecosystems remains permanently high, while in other productive marine ecosystems – be it in pelagic communities of upwelling regions or in coastal shelf communities of the temperate zone – a high productivity (2–5 $g\,C\,m^{-2}\,day^{-1}$) is attained only periodically. An extremely high autotrophic production of coral reefs, which is based on the three-dimensional structure of their bottom biotopes, supports the existence of unusually dense communities of zoobenthos and fish. Their biomass reaches on reefs hundreds of grams per 1 m^{-2}. The fish resources of reefs comprise a significant portion of their total stock in the World Ocean – about 9% – and fishery in reef areas is an important source of protein in the rations of the population in tropical coastal countries. The export of organic matter and combined nitrogen from coral reefs significantly enhances productivity of surrounding oceanic waters. Reefs are also the place of propagation and juvenile growth of many pelagic fish species important for fishery. The reefs should be recognized as a prospective zone for development of aquaculture and as a source of raw materials for the production of medicaments. The immense recreation potential of coral reefs is one of their most important resources, which should be protected for the benefit of mankind.

The core groups of organisms composing reef bottom communities are represented by calcareous forms such as corals, foraminiferans and calcareous algae, in which the function of calcification is coupled with photosynthesis. The coupling greatly enhances this process. As a result, coral reef ecosystems produce over half of the total lime material sedimenting in the World Ocean. The modern reefs are an important element of the geomorphological structure and bottom topography of the ocean floor. Ancient reef structures form one of the main facies of sedimental rocks, being widespread in the USA, Canada, CIS, and Saudi Arabia, where they are the basic oil collectors. In the CIS, large oil deposits in the reefogenic rocks were found in Siberia, Bashkiria, the Volga-Ural region, Turkmenia, Uzbekistan, and Ukraine. The study of geomorphology, growth and ecology of the modern reefs enables us to understand the conditions of formation of ancient reef facies, thus helping in the search for the oil deposits they contain.

Of the total area of modern reefs in the World Ocean ($6 \times 10^5\,m^2$) 15% are situated in the Atlantic Ocean, 53% in south-eastern Asia, 19% in the Pacific, and 9% in the Red Sea. The main reef massifs of the Atlantic Ocean are situated in the Caribbean Sea by the coasts of southern Florida, Mexico, Honduras, Nicaragua, Venezuela, and around the Bahamas, Virgin Islands, and the Antilles. In this region are situated 10 atolls out of their total number of 300, and 2 of the 30 large barrier reefs of the World Ocean. The barrier and fringing reefs situated at the shelf off the continents and high islands occupy more than half of the total area of modern reefs. Among the largest reef systems we can point out the barrier reef systems of Belize, Madagascar, Fiji, New Caledonia, and the Great Barrier Reef of Australia. Some of them extend over 1000–2000 km. The total area of the Great Barrier Reef of Australia between the continental shore-line and the outer edge of the barrier reefs is about $215 \times 10^3\,km^2$.

The first studies of coral reefs started at the end of the eighteenth century were concerned mostly with their cartographic description, their geomorphology and origin, and the taxonomic composition of reef flora and fauna, including corals. The reefs were studied by outstanding naturalists of the past: Chamisso, Laiell, Darwin. In 1923, the Royal Society of London organized an expedition to the Great Barrier Reef of Australia led by one of the future founders of reef biology and ecology, sir Maurice Yonge. The Records of this expedition contained basic data on plankton and benthos of reefs, and on feeding and metabolism of corals. They constituted the basis for the future development of reef ecology. Later on, the main centers of coral reef studies were represented by the Institute of Marine Biology on Hawaii, Moorea Station, Australian biological stations on the Heron, One Tree and Lizard Islands, the Australian Institute of Marine Sciences in Townsville, biological stations on Guam Island, in Eilat by the Red Sea, on Jamaica, the Belize Barrier Reef, and in the Nhatrang Institute of Marine Biology (Vietnam). Significant insight into coral reef ecology has been gained by studies at the Enivetok and Bikini atolls after the nuclear tests in the 1950s. Classical investigations were carried out here on the geomorphology, hydrodynamics, productivity and metabolism of the reefs (Devaney et al. 1987). Further impulse to the development of coral reef ecology was given in the early 1970s by the incipient dangerous degradation of coral reefs as a consequence of human activity.

Coral reef studies are now among the most important directions of modern marine biology, marine geology, geomorphology, and paleontology. The International Symposia on Coral Reefs have been held every 4 years since 1969. The Proceedings of these Symposia marked the main stages on the way to developing coral reef studies. The bibliography in this field contains over 15000 papers and books. In 1982, the international journal

"Coral Reefs" was founded. The results of reef studies have been reviewed by Wiens (1962), Yonge (1963), Stoddart (1969), Lewis (1977b, 1981), Sorokin (1977b), Rosen (1982), Kühlmann (1982, 1985), and Preobrazhensky (1986). Results of coral reef studies were also discussed in collective monographs edited by Jones and Endean (1973–1977), Barnes (1983), Sokolov (1987), and by Dubinsky (1990). The methodology of these studies has been presented in a monograph edited by Stoddart and Johannes (1978) under the patronage of UNESCO. Several monographic works reviewed observations on separate reef systems or separate reefs (Maxwell 1968; Endean 1982; Rützler and Macintyre 1982; Zlatarsky and Estalella 1980; Bennett 1984; Mather and Bennett 1984; Devaney et al. 1987). The main goal of this book is development of the holistic concept of the coral reef ecosystem. To attain this goal it was necessary to analyze composition and community structure of all components of reef ecosystems, the features of their ecological physiology and trophodynamics. The analysis should deal with the biological and physical factors and mechanisms providing their permanently high biological productivity in nutrient-poor environments. A general insight is needed into the trophical structure and function of the reef ecosystem as a whole, as well as the quantifications of energy flow between its components at different trophical levels. Drastic destabilization trends that appeared in reef systems over the last two decades make it necessary to analyze carefully factors causing this development which now endangers the very existence of coral reefs. All these problems are taken up in this book along with a discussion of such aspects of coral reef ecology, as did not find comprehensive analysis in previous reviews and monographs. Among them are the hydrology of the coral reef, the dynamics of nutrients and organic matter exchange in reef environments, composition and ecology of plankton in reef waters, microbial ecology in coral reefs, and physiology and feeding of corals. The results of our own investigations, which are included in the book, concern the problems of primary production and microbiology of reefs, feeding of corals, biogeochemical metabolism and nutrient dynamics in reef ecosystems. These studies have been accomplished during the cruises of the R/V *Vityaz* and *Dmitry Mendeleev* in 1968–1971, *Kallisto* in 1975–1981, and *Alexander Nesmeyanov* in 1985, and at biological stations on Coconut Island (Hawaii), Heron, Lizard and One Tree Islands (Australia).

The name of the Great Barrier Reef of Australia, often mentioned in this book, is abbreviated in the text to GBR.

Acknowledgments. In these studies I received financial support from the Great Barrier Reef Committee, the Government of Australia, and the University of Queensland. I am greatly indebted for the invaluable support I enjoyed in my research and in the preparation of this book to Drs. M. Vinogradov, A. Zhirmunsky, B. Sokolov, E. Titlyanov, B. Preobrazhensky, M. Doty, R. Endean, A. Cameron, J. Kikkawa, J. Greenwood, D. McColm, M. and J. Lawn, D. Moriarty, P. Pollard, J. Baker, T. and K. Tubbenhauer.

1 Reef Lime Constructions

1.1 Geological History, Features of Geomorphology

The lime construction forms the structural basis of the coral reef ecosystem. It provides optimal disposition for sessile phototrophic organisms in the field of light. It presents refuge for the vagile fauna saving it from being grazed. It also forms the fortress that defends the reef biota from the destructive action of hydraulic stress by waves and currents. This construction is created largely by the hermatypic calcareous organisms – animals and plants, which extract carbonates from surrounding sea waters for building up their exoskeletons, shells, spicules or other calcareous elements of their bodies. The form and shape of the reef constructions themselves are largely resulting from the interaction of communities of sedentary hermatypic organisms with the hydrodynamic stress, which in the tropical zone has usually some definite direction (Heckel 1974). Therefore, in spite of a great morphological variety of reefal constructions they have many common geomorphological elements, such as buttress system, reef-flat, lagoon, patch reef zone etc. It demonstrates the existence of some common regularity in their origin and growth on different structural foundations of shelf and island coasts, situated in different parts of the tropical zone. This regularity was first recognized by Charles Darwin (Darwin 1842). In his book he distinguished three main types: atolls, barrier reefs, and fringing reefs, and also put forward his famous hypothesis, which explained the existence of the above-mentioned reef constructions by their subsequent genesis from one another. He supposed that first the fringing reef grew up around an island. If this island was subjected to tectonic subsidence in relation to the sea level, the fringing reef surrounding it transformed into a barrier reef, and then into an atoll (Fig. 1.1). This mode of development is supported by the preferential growth of corals on outer fringes and slopes of reefs and its inhibition in their internal zones and lagoons.

Darwin's theory is still valid. In fact, modern scenaria of the origin of ancient and recent reefs accept both its basic points: the vertical movement of sea – land juxstapositions, and the preferable coral growth on outer reef zones. Deep drillings proved the predictions based on Darwin's theory about the existence of great thicknesses of reefal carbonate rocks under

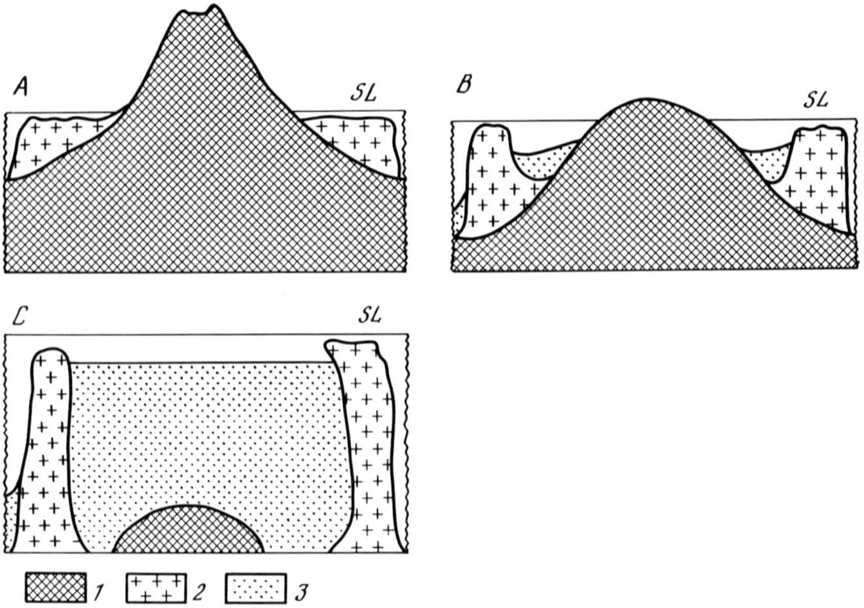

Fig. 1.1. Evolution of reefs during tectonic subsidence according to C. Darwin's hypothesis: *A* fringing reef; *B* barrier reef; *C* atoll. Facies: *1* basic rocks; *2* consolidated biogenic carbonate rocks; *3* unconsolidated carbonate rocks

Table 1.1. Thickness of identical reef constructions under modern atolls according to drilling results; data by Ladd (1969), Purdy (1963)

Basin	Atoll	Thickness of reefal lime columns, m	Geological age of carbonate rocks at the base of reef construction
Pacific Ocean	Funafuti	>400	Quaternary
	Bikini	>780	Tertiary
	Eniwetok	1400	Jertiary
	Mururoa	>400	Quaternary
	Midway		Quaternary
Indian Ocean	Maldives	2200	Eocene
Caribbean Sea	Tenerife	1300	Quaternary
	Glower reef	1000	Quaternary

atolls and barrier reefs (Table 1.1, Fig. 1.1). Modern geology relates their origin to the gradual tectonic subsidence of the whole sea floor together with the underwater mountains and reefs on their tops situated on it, which was caused by the movement of corresponding tectonic plates to the oceanic trenches (Ladd 1969; Scott and Dixon 1983). As we can see, the theory of Mobilism appeared to be fruitful also in understanding the reef's genesis. But as regards the origin of the main types of modern reef constructions

Darwin's theory was later completely revised, while his basic idea about the evolution of reefs is still present in modern theories on their morphogenesis developed by Maxwell (1968) and Hopley (1983). The revision of Darwin's theory in relation to modern reefs became necessary when proof appeared of drastic changes in sea level, unknown to Darwin, with amplitudes over 100 m during the Pleistocene-Holocene 150–10 × 10^3 years B.P., which were caused by periodic glaciations (Daly 1915).

The most ancient biogenic reefal constructions are known to have started in the Precambrian. They are known as "bioherms" and "biostromes" (Cuminga 1932) and have the shape of hillocks 2–20 m high. They consist of amorphous (bioherms) or skeleton remains structured by lime material of larger invertebrates (biostromes). During the Precambrian they were constructed by calcareous blue-green algae, which accumulated lime in their mucoid capsules. During diagenetic transformations these constructions gave origin to stromatolites – solid lime lumps which compose the main elements of the most ancient reef constructions (Copper 1974), which are known from Archean (2600 × 10^6 years B.P.). But the most widespread stromatolithic reefal constructions were formed during the Late Precambrian (Riphean) time 1700–670 × 10^6 years B.P. They were up to several tens of meters thick and several hundred meters long. Reef bioherms of this type were common in Proterozoic sediment rocks of North America, Eurasia, Africa (Cloud and Semikhatov 1969; Kuznetzov 1986). In the Early Cambrian more complex reef-like constructions appeared, which contained inside a kind of framework, composed either by skeletons of bryozoans, tubes of sedentary polychaetes – sabellarides or by colonies of archaeochaetes (Koroljuk and Mikhailova 1986). The latter were animals somewhat similar to calcareous sponges (Zhuravleva and Myagkova 1979), while some of them probably could have been also formed by ancient coralline algae. They have massive, firm funnel, columnar or cup-like lime colonies, which resemble the colonies of modern calcareous sponges or ancient rugose corals. The archaeocyathide biostromes looked like modern patch reefs and had a round shape with a diameter of 20–30 m (Stubblefield 1960). The ridges of such archaeocyathid reefs were wide-pread in Early – Middle Cambrian (520–540 × 10^6 years B.P.). At the shelf of warm seas they formed systems of constructions 5–10 m thick, which resembled atolls or barrier reefs (Zhuravleva 1960; Copper 1974). At that time large reef-like biostromes were constructed by large sedentary polychaetes having massive lime tubes. In the Late Cambrian-Early Ordovician bioherm-like reefs were also common, constructed by massive ancient bryozoans, Trepostomata and Cystoporata (Cuffey 1977). During the Ordovician biostrome reefal constructions were the most widespread created by calcareous sponges (Archaeoscyphia and Stromataporidea), by bryozoans and by the coralline algae Celathum (LeCompte 1956; Copper 1974). At the same time the first ancient hard corals appeared – tabulates and later rugoses (Sokolov 1955; Ivanovski 1975).

The earliest taxa of these had small colonies (10–30 cm crosswise). Nevertheless, they soon became basic elements of the frameworks in reefs which grew in that period (Hill and Stumm 1956; Koroljuk and Mikhailova 1986). Stromatoporide – tabulate reef rocks – are common in Silurian-Devonian sedimental facies (470–340 × 10^6 years B.P.). Their size runs to several kilometers with a thickness of 50 to 100 m (Lecompte 1956; Sokolov 1986). Rugose corals had firm colonies withstanding the wave stress (Sokolov 1955). With their appearance the processes of reef formation and growth expanded and large reef constructions appeared extending hundreds of kilometers. Where tectonic subsidence look place the growth of these Devonian rugose-based reefs resulted in the construction of 1000 m and more of thick reefal massifs (Nalivkin 1967). At the end of the Devonian corals became extinct due to the start of the cold period followed by glaciations and the drop of the sea level (Copper 1974). The growth of reefs then started anew only in the Triassic, but initially they were again structured like the biostromes with calcareous sponges as basis for their frameworks (Wilkinson 1983b), and with calcareous algae as main producers of carbonaceous filling material. Then in the Middle Triassic the hermatypic scleractinian corals appeared at the tropical coasts. Among them there were colonial as well as solitary forms. The most common of these were the families Thamnasteriidae, Astrocoeniidae, Conophyllidae, Montlivaltiidae, and Actinacididae (Sokolov 1955; Krasnov 1976; Sokolov 1986). It was supposed that they had descended from various taxa of rugoses. But the contention that all basic branches of hard corals, be it tabulates, rugoses or scleractinians, represented independent evolutionary lines of the ancient polypoid actinians (Wells 1956) seems to be more acceptable. They became extremely widespread among neritic fauna of the Thetis and then in all tropical zones (Potts 1983).

At mid-Late Jurassic scleractinians created large reef constructions on vast areas of tropical shelves. But in some places the thick reefs (up to 50 m) were built up also by sabellarid polychaetes and plecypodes. Reefs grew extensively also during the Cretaceous when the fast evolution of scleractinians simultaneously proceeded. At the end of this period many of their modern families appeared. A significant role in the formation of these reefs was played also by hydrocorals like milleporides and stylasterids, which flourished on the reef of the Late Cretaceous (Boshma 1956). In the Paleogene-Eocene many new modern families of scleractinians appeared. After some interruption, which took place during that period, the active growth of reefs started again and acquired the features of modern reefs. Their thickness in some places attained 600 m. In their formation, besides corals, also corallines, calcareous green algae and foraminiferans participated, forming the basic mass of carbonaceous material. The growth of reefs proceeded also in the Pleistocene, but with interruptions caused by large oscillations of the sea level in this period. In Fig. 1.2 the scheme of the geological history of the main reef-building is shown.

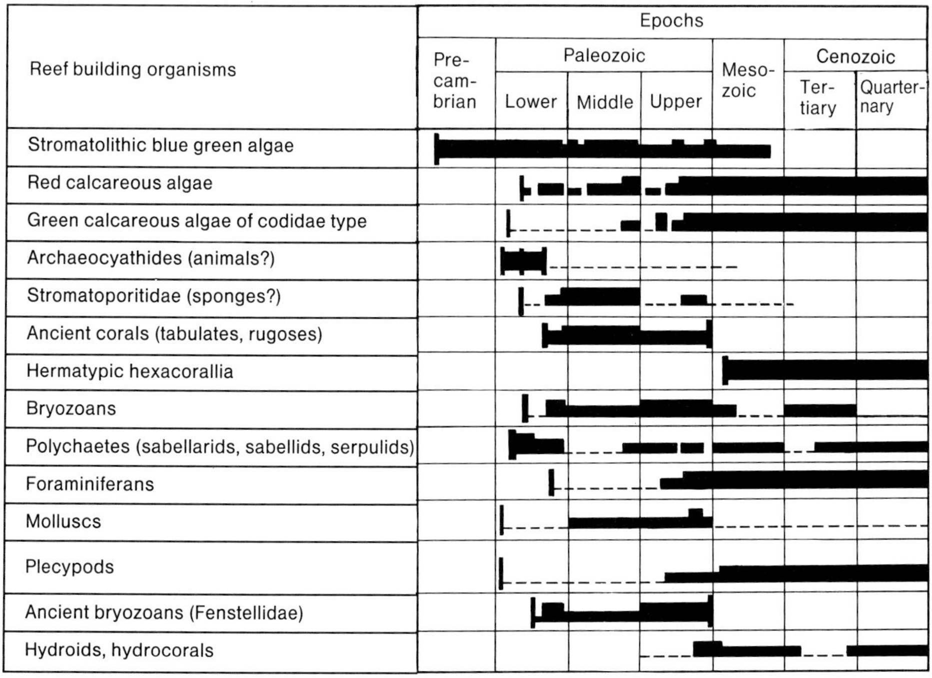

Fig. 1.2. Scheme of geological history of main groups of reef-building organisms: the *width of stripe* corresponds to their relative importance in reef genesis *vertical line* represents time of their appearance and extinction

The formation of large reef platforms, on which the reefs of the Pleistocene and then the modern reefs of the Holocene grew up, started at the Late Mesozoic-Early Cenozoic, 60–80 × 10^6 years B.P. Their formation was stimulated by a very active motion of the tectonic plates at that time. It stimulated vertical movements of the sea-floor surface mean to the sea surface. In the areas of subsidence of islands, on the tops of underwater mountains, and on the continental coasts the growth of Mesozoic-Cenozoic reefs proceeded very fast: 20–200 cm in 10^{-3} years, thus compensating the rate of submergence of their bases. By this process large massifs of the Late Tertiary reefs were formed (Ladd 1969). Their thickness attained 400 to 2000 m according to drilling data (Table 1.1). During the later 10–2 × 10^6 years B.P., when the shoreline of the World Ocean because established close to its modern state, the level of the latter underwent drastic oscillation. These oscillations, within 50–100 m, were caused by periodic drops in global temperature accompanied by glaciations. The second cause of these oscillations was the tectonic-eustatic oscillations of the ocean floor as a sequel to the irregularities in rotation of the Planet at this period (Shakleton and Opdyke 1973). Even during the later 350 × 10^3 years B.P. the sea-level changed several times (Fig. 1.3, Chappell and Veeh 1978). Reefs exposed

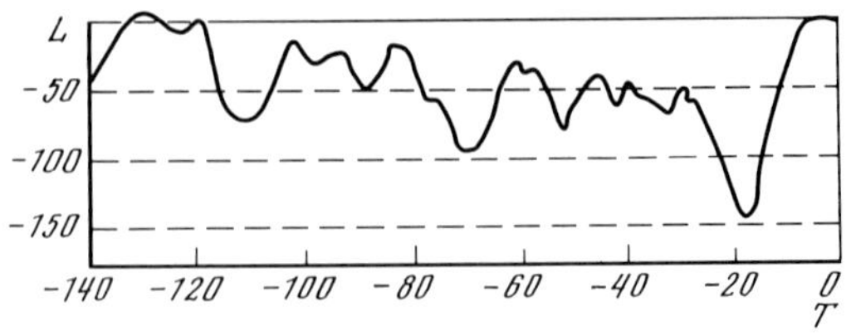

Fig. 1.3. Eustatic fluctuations of sea level (L, m) during the last 140 000 years; T time, 10^3 years. (After Chappell 1981; Potts 1983)

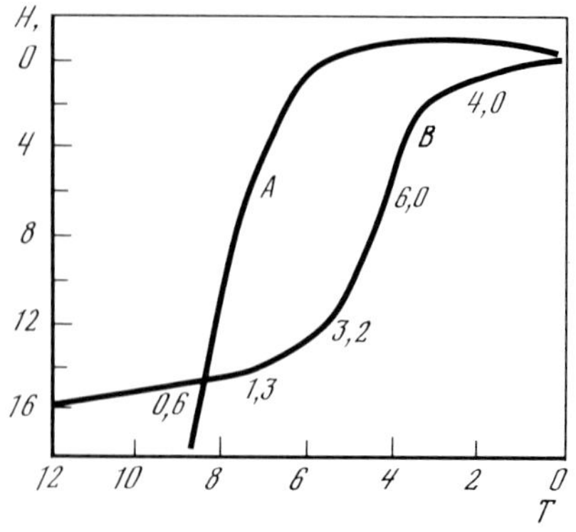

Fig. 1.4. Change in sea level (A) and reef growth in time (B) during the Holocene transgression. *Numbers* by *curve B* indicate rates of reef growth at various periods of transgression, m 10^3 years^{-1}; T age of reef, 10^3 years; H depth of reef column down from the modern surface of reef flat, m. (After Thom et al. 1978; McLean et al. 1978)

after the glaciation period next in turn died and were subjected to atmospheric erosion. After following elevations of the sea level the reefs extended to grow up at the base of these eroded ancient reef platforms. Therefore, intensive reef growth proceeded even under conditions of periodic exposures. Thus, the platforms of Pleistocene reefs were formed, which in some places attained a thickness of hundreds of meters. Those processes went on until 110–120 × 10^3 years B.P., when a large glaciation caused the descent of the sea level by over 50 m, which continued then with interruptions till the last Würm glaciation 25 000–16 000 years B.P., when the sea-level dropped by 100–130 m (Fig. 1.3). Then its quick rise started and went on until 5000 years B.P. This is known as the Holocene transgression (Chappell 1983). Its rate with reference to the coastal line was about 10 m in 10^{-3} years (Fig. 1.4). Around 7000 years B.P. the rate of transgression slowed down. At 5000–4000 years B.P. it stoped and then descended again 1–2 m till now (Thom et al. 1978). When the rising ocean level reached the surface of the exposed Pleistocene reef platforms between 12 000–8000 years B.P., the growth of reefs started on them again. This happened when the sea-level was 20–30 m below the present one. Therefore, all the modern Holocene reefs grown on the drowned ones during the transgression of the ancient platforms have a thickness of only 5–20 m as an average (Adey 1978). At the GBR (Great Barrier Reef), the longest discovered, their thickness was 23 m (Hopley 1983; Davies 1983), and in the Caribbean 33 m (reef Alcaran, Macintyre et al. 1977). Thus, the modern coral reefs form only a thin veneer over the hundreds of meters thickness of old Pleistocene platforms. The rocks at the base of modern reefs at a depth of 10–20 m under their surface have an age of $5-10 \times 10^3$ years shown by isotopic dating. Several meters below, the rocks of their Pleistocene base have an age of $1-2 \times 10^5$ years (Thurber et al. 1965).

The upper layer of these ancient reef platforms, having been exposed for about 1×10^5 years, was subjected to various kinds of aerial and karst erosion (Purdy 1974). The growth of Holocene reefs started and proceeded most actively at the elevated places of their relief. In depressions and funnels, where the sediments were accumulating, the growth of corals was inhibited by siltation. Later on, in depressions the lagoons were formed (Davies and Kinsey 1977; Hopley 1983). When the growing reefs were approaching the stabilized sea-level at $5-4 \times 10^3$ years B.P. their growth slowed down from 5–10 m in 10^{-3} years to 1–4 m in 10^{-3} years. Upon reaching the surface, growth stopped and the erosion of their surface, exposed during ebb tides, started. This resulted in the formation of consolidated reef flats. The rate of reef growth in the depressions was slow: 0.5–1.5 m in 10^3 years. Therefore, in the initial phase of growth in the Holocene in these places, usually deep lagoons were forming (Fig. 1.4). After the elevated edges of the reefs reached the sea surface and stopped growing, the lagoons gradually filled with sediment. At a later phase in the evolution of such reefs these shallow lagoons became covered by patch reefs and the edges of their reef flat gradually spread over the lagoon. This process is going on now on numerous modern reefs in the Indo-Pacific. Later, such a reef will reach the phase of an old platform or table reef (Fig. 1.5). Thus, among the main geological factors which controlled the growth rate of modern reefs and determined their future forms and shapes there

Fig. 1.5. Morphology of some Holocene reefs of the GBR region at different phases of their evolution; designations on figures are given as combinations of numbers (1 to 3) and letters (A, B, C); the *numbers* correspond to size or type of ancient Pleistocene platform bases of modern reefs: *1* medium-sized platform (1.75–3.25 km); *2* small platform (<1.75 km); *3* stretched ribbon-like platform; the *letters* correspond to the evolution phases of modern reefs: *A* Pleistocene platform at the end of the last decrease in sea level (*SL*); *B* phase of mature crescent or ring reef (atoll); *C* phase of old platform reef. Facies of sedimented rocks: *1* Pleistocene platform; *2* submerged part of reef or lagoon; *3* corals; *4* soft sediments; *5* algal flat (coralline algae); *6* sandy flat; *7* rubble; *8* ridges of gravel; *9* karst hillocks; *10* karst depressions on the Pleistocene platform; *11* sand cays. (After Hopley 1983)

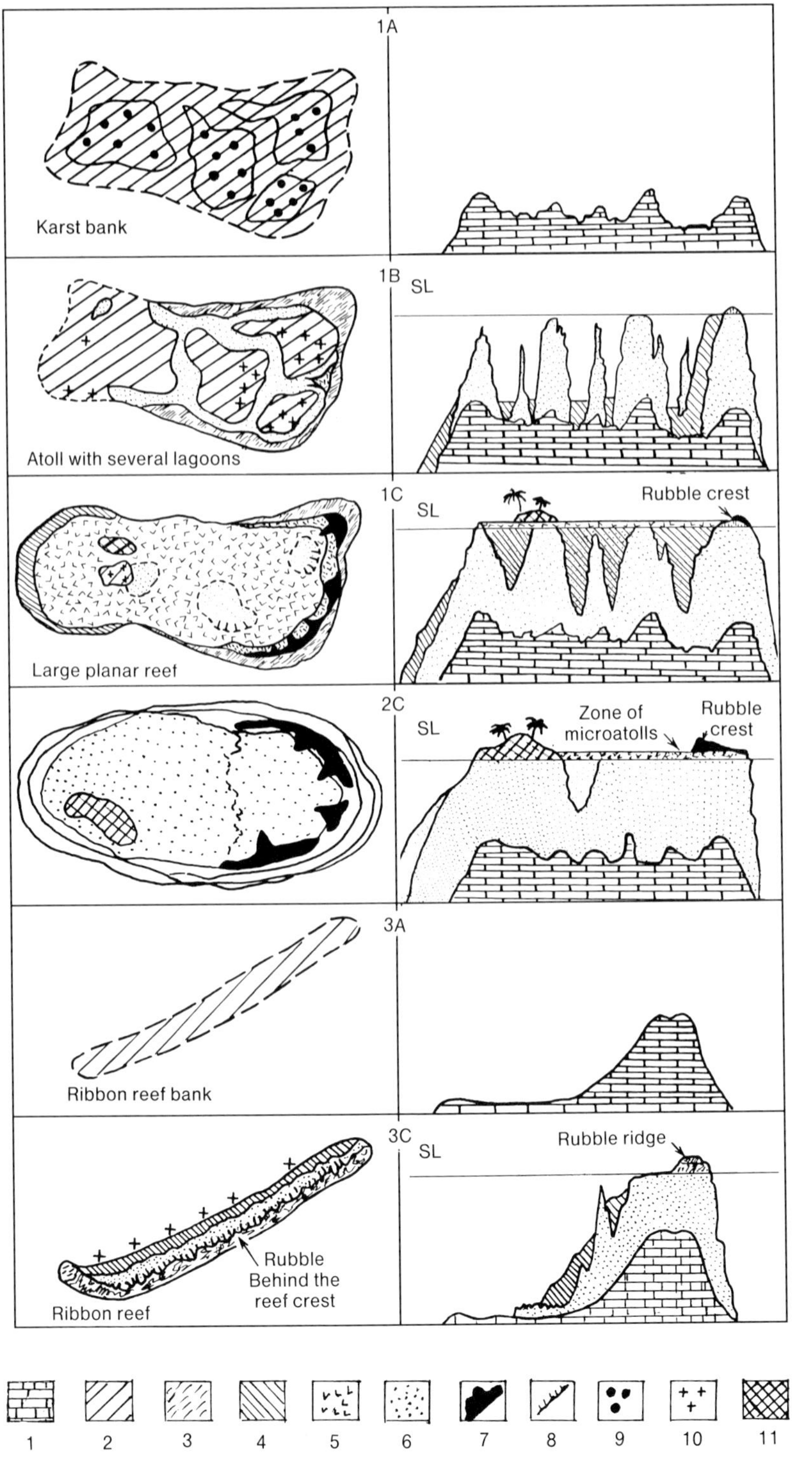
1A
Karst bank
1B
SL
Atoll with several lagoons
1C
SL
Rubble crest
Large planar reef
2C
SL
Zone of microatolls
Rubble crest
3A
Ribbon reef bank
3C
SL
Rubble ridge
Rubble Behind the reef crest
Ribbon reef
1
2
3
4
5
6
7
8
9
10
11

could be mentioned: (1) the size and the geomorphology of the Pleistocene reef platform or other bottom elements which started their growth in the middle phase of the Holocene transgression; (2) the rate of rise of the sea level in juxtaposition to the shoreline during the Holocene transgression; (3) the absolute time when the growing reef reached the stabilized sea surface at the end of the transgression, i.e., the time when their vertical growth stopped (Davies 1983). All these factors varied in different regions, thus resulting in a great variety of geomorphological features of modern reefs (Thom and Chappell 1975; Davies 1983; Guilcher 1988). They are represented by several types, such as atolls, fringing and barrier reefs, ribbon and planular reefs, sand cays and coral banks. Their genetical classification was developed by Maxwell (1968). But it has not been confirmed by subsequent geological data (Purdy 1974; Thom et al. 1978; Hopley et al. 1978). Later more adequate classification was presented by Hopley (1982, 1983). This classification accounts for the interaction of the main factors in the genesis of Holocene reefs, such as the geological (size and geomorphology of Pleistocene platform bases on which they grew, the rate of transgression), the biological (conditions of coral growth and its rate) and the physical (mechanical erosion, transport of sediments, etc.) factors. This classification concerns only the shelf reefs developing in the Great Barrier Reef region. But it contains the nesessary background for developing a universal genetic classification of modern coral reefs. By this classification all modern reefs are divided into four main groups depending on size and structure of the base platforms on which they have grown.

The first group includes the ring reefs, which have grown up on large Pleistocene platforms more than 3.25 km wide. Their surface had a complex relief which included several Karst depressions. On these platforms large crescent or ring reefs (atolls) grew up which had several lagoons mimicking the former depressions of their ancient bases (Fig. 1.5). Later their lagoons were filled with rubble and sand, and overgrown with patch reefs. Finally, at the latest stage of their evolution they transformed into flat planar reefs. The second group of average-sized ring reefs grew up on ancient platforms less than 3.25 km wide. They underwent actually the same evolution, but had only one central lagoon. To the third group belong the small round or ribbon planar reefs which have grown up on the small platforms less than 1.75 km wide or on narrow ridges left over from ancient reefs after their long exposure. Their peculiar feature is the absence of lagoons. As they were reaching the stabilized sea-level during their growth, the depressions on the surface of their bases were also filled with sediments. Finally, the fourth group of reefs had no Pleistocene base at all and grew up on hard rocky bottom elements in the tropical shelf. They usually do not form consolidated structures, often being represented by banks or cays. Some of these reefs resemble in their geomorphology and evolution the planer reefs of the third group. The reefs of this group are relatively rare in the Indo-Pacific, but are rather widespread in the Western Atlantic region. The barrier reefs near the high islands within this classification correspond to the mature ring reefs, and the fringing reefs to the late phase of their evolution – e.g., to the planar reefs – which is just the opposite of Darwin's theory. But in areas of active modern tectonic movements they are also an important factor in reef evolution and growth (Chappell 1983), as it was supposed by Darwin. The transgression interacted in such areas with the tectonic movements, and the direction of the latter was largely regulating the formation of the modern reefs which grew there (Thom and Chappell 1978). For example, because of transgression being simultaneous with tectonic subsidence the modern stabilized sea-level in the Carribbean reef region was reached at about 1000 years B.P., while in the Indo-Pacific it happened 5–4 $\times 10^3$ years B.P.

The reefs grown up on ancient reef platforms were definitely developing the topography of their base, which largely predicted their modern morphology. These reefs are classified as "structural." In fact, they have quite regular elements of structure, such as reef edge, algal ridge, reef flat, and patch-reef zone. These elements are disposed on structural reefs as stripes, which are situated normally in the direction of the dominating wind and wave stress (Stoddart 1969; Guilcher 1988). Each stripe or zone represents also a definite local ecological unity, characterized by a definite selection of key species in its biota. Actually, they are a kind of linear biocenoses or biotopes formed within a definite geomorphological structure of the reef. The zonal distribution of biocenoses on the structural reefs is created by physical stress and light distribution in their interaction with the reef biota (Graus and Macintyre 1989). The zonal structure makes an important background for describing the composition of reef biota and the distribution of its main components, as well as of their functional roles within the reef ecosystem (Yonge 1940; Stoddart 1969; Pichon and Morrissey 1981). The description and classification of zones were developed by Ladd (1977) for the Pacific atolls, by Maxwell (1968) for the GBR, by Pichon (1974) for the Barrier reef of Tulear, Madagascar, by Goreau and Goreau (1973) for Caribbean reefs. A general geomorphology of coral reefs was reviewed by Guilcher (1988). In Fig. 1.6 schemes are shown of the profiles across several main types of reef which demonstrate the position of zones and provide the terminology used for their designation (Leontjev et al. 1979). Among these profiles with an outer slope across the reef to the leeward of its side (or to landward on the Barrier reef) the following main zones should be mentioned:

1. *Zones of outer windward slope* (the ranges of depths are given as averages)

   a) F_s – *Reef slope zone*, depth 200–80 m. It is usually formed by the steep rocky slope of the ancient reef base platform, cut by grooves and ditches – signs of former aerial and then underwater erosion. The ridges between grooves are overgrown mostly by ahermatypic corals and sponges.

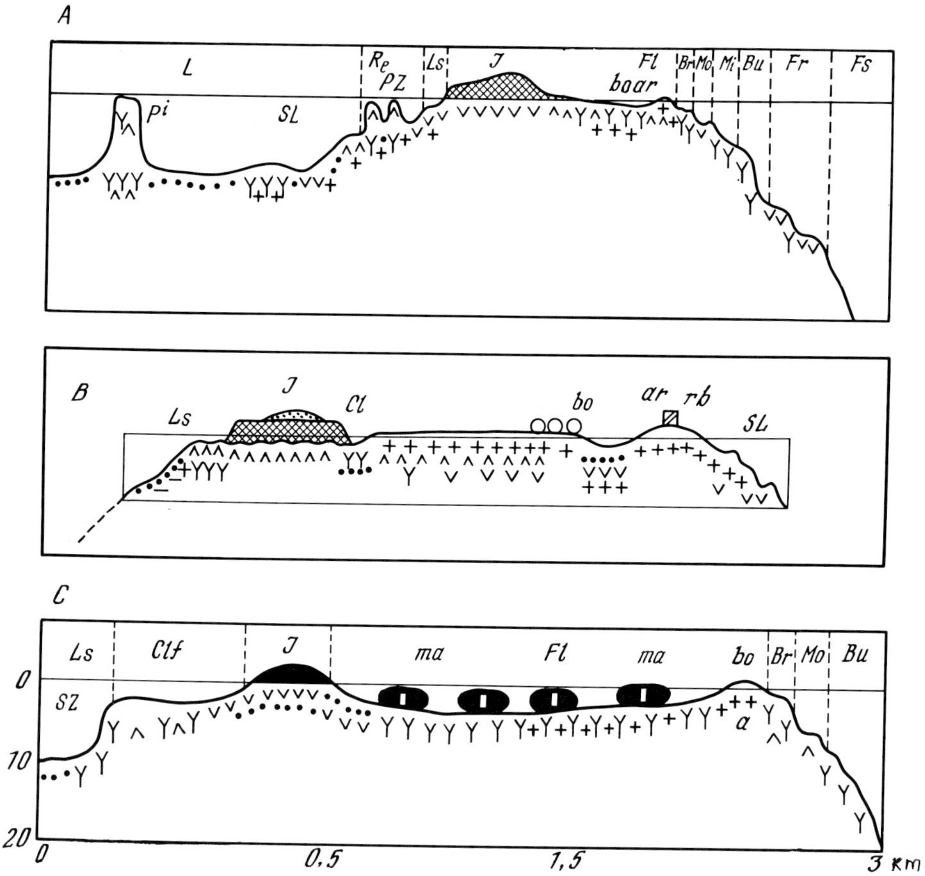

Fig. 1.6A–E. Schemes of zonal structure of some modern reefs. **A,B** windward reefs of Pacific atolls from Marshall and Lau groups; **C** windward flat of Indo-oceanic atoll Kargados-Karahos; **D** platform reef of Western Atlantic; **E** barrier reef off Jamaica; Zones and elements of reef structure: *I* island; *rc* reef crest; *ma* microatolls; *pr* patch reefs; *pi* pinnacles; *irf* inner reef flat; *ls* lagoonal slope; *L* lagoon; *CL* coastal lagoon ("boat channel"); for designations of other zones see text. Facies (biotopes): *1* living corals; *2* rubble; *3* consolidated reefogenic rock; *4* coralline algae and other macrophytes; *5* sand; *6* microatoll; *7* thickets of seagrass; *8* ancient reef base

b) F_s – *Reef zone*, depth 80–40 m. This zone is formed by the steep slope of the ancient reef platform on which the coral growth also started. Corals, calcareous sponges, sedentary polychaetes and coralline algae, which inhabit this lower boundary zone of depth distribution of hermatypic organisms, produce here lime constructions over the ancient reef base. This often results in the formation of steep precipices and impending peaks at depths of 80–50 m.

c) F_s – *Reef zone*, depth 20–40 m. This zone is also situated at the steep slope of the ancient platform. But through receiving more light it is

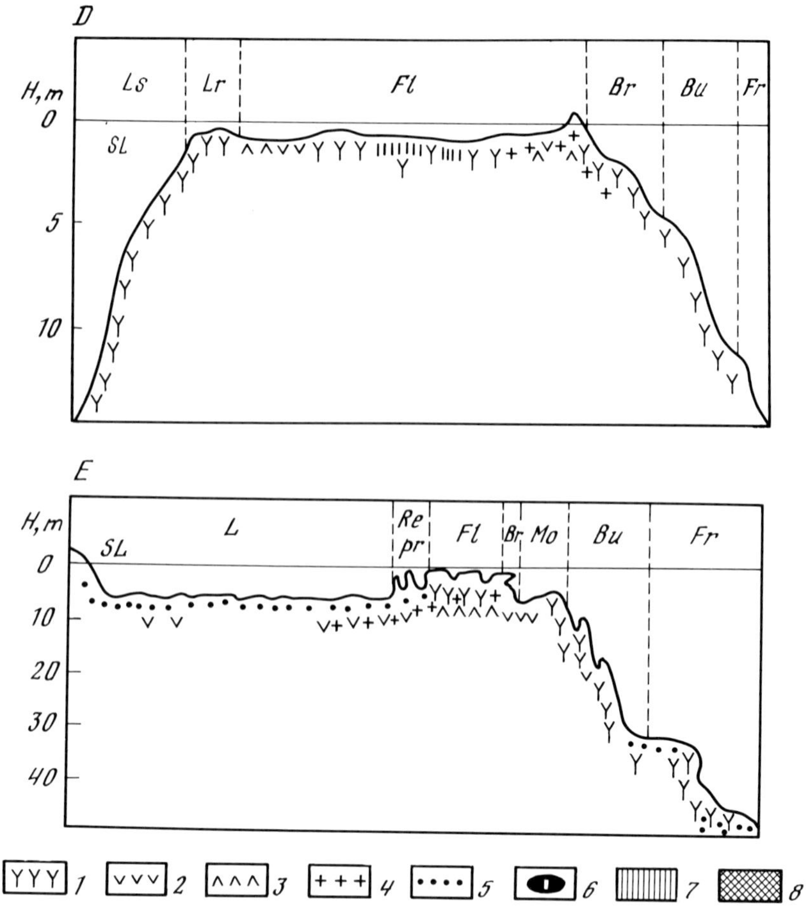

Fig. 1.6 *Continued*

characterized by abundant coral growth and fast calcification. This results in the formation of impending "sill reefs" – steep reef terraces formed by foliose or massive corals like Echinophyllia on Pacific atolls, or Montastrea, Agaricia and Madracis on Caribbean reefs. In this zone an abundant sessile fauna of sclerosponges, bryozoans, gorgonaceans, and hydroids flourishes.

d) Bu – *Buttress zone*, depth 20–8 m (see Fig. 2.1). The windward reef slope is formed in this zone by spoors with grooves between them. The spoors or buttresses are 3–8 m high ridges descending down the slope and are constructed by abundant coral populations, growing between the grooves, through which sediment and rubble material are moving down from the upper reef zones. The buttress zone is one of the most important elements in the coral reef ecosystem. Dissipating wave energy, the buttress systems preserve communities of the outer reef slope from destruction (Shinn 1963). The same function is accomplished also by knolls, which often grow up in this zone (Roberts et al. 1975). The spoors are places of luxurious coral growth. The coral communities here are of high species diversity with ramose scleractinians and various gorgonaceans dominating.

e) Mi – *Mixed zone*, depth 8–6 m. This zone was described by Goreau and Goreau (1973) as an intermediate between moat and buttress zones. The zone is characterized by rich coral growth in conditions of significant wave stress.

f) Mo – *Zone of frontal moat*, depth 6–4 m. The moat, 1–2 m deep, usually extends along the outer reef edge, and is formed through inhibition of coral growth under the influence of sediment and rubble shed from the reef flat by tidal currents and surf. Its bottom is usually covered with rough sand and rubble, overgrown with periphyton. It is populated by abundant vagile benthos and by solitary corals *Fungia*.

g) Br – *Braker zone*, depth 0–2 m. This represents the reef edge, subjected to maximal wave action. On Atlantic reefs this zone is usually colonized by massive corals, while on the Indo-Pacific by crustose coralline algae.

2. *Zones of windward reef-flat*

a) Ar – *Algal ridge zone*. The zone is represented by slightly sloping ridges 10–30 cm high, built up by coralline algae along the windward reef edges at elevated flats of some old atolls in the Indo-Pacific, and are situated in the Trade Wind Zone experiencing permanent wave stress. The same waves repel from this zone fish which graze on coralline algae, thus stimulating growth and formation of the ridge in this zone (Adey and Vassar 1975).

b) Bo – *Zone of boulder ramparts*. The ramparts are the ridges of calcareous boulders formed from coral colonies consolidated by coralline algae. The ridges are cut by deep trenches and channels through which the transport of sediments from the reef flat passes. Behind the boulders stripes of accumulated rubble and sand (the gravel tail) are situated, which form cays behind ramparts.

c) Fl – *Inner reef flat zone*. This is a planar consolidated reef plateau. Its construction and cover differ on different reefs depending on its disposition in relation to sea level and on the height of the tides. At elevated reefs, where flats are exposed during ebb-tides, cays of gravel are forming along their surface. It is cut by channels and covered mostly by coralline algae and by small rare coral colonies, which grew in channels and in depressions. The surface of the flat, exposed only partially during low water, is usually covered by thickets of macrophytes, opportunistic ramose corals which survive exposure, siltation and physical stress (Maxwell 1968), by zoantharians and by ramose coralline algae. On reefs with submerged flats or with depressions in their middle or rear parts abundant coral communities develop on the surface, consisting of ecomorphs which resist physical stress (Grassle 1973; Stoddart 1969). Ramose corals often form there microatolls (Scoffin and Stoddart 1978). A significant part of the submerged flats is covered by macrophytes, sea grasses and ramose Corallinaceae (Milliman 1969).

d) Re – *Rear or inner reef flat zone*, depth 1–4 m. This zone is stretching along the backward part of the reef flat or along the inner reef slopes facing the lagoon. The zone is structured by patch reefs – round planar lime hillocks 2–20 m large – which are constructed by corals and coralline algae in calm reef shallows on the sandy bottom. Their surface is often exposed during ebb-tides.

e) Ls – *Inner lagoonal reef slope zone*, depth 2–20 m. It is represented by steep sandy or rubble slopes from the inner reef edge at the "rear" zone towards the lagoon.

f) L – *Zone of lagoon*, depth 5–40 m. This is the depression fringed by the reef. Its depth is largely dependent upon the phase of reef evolution: the older is the reef, and shallower is its lagoon. In shallow lagoons over 50% of its sandy bottom is usually covered by patch reefs and by thickets of ramose corals. In deep lagoons at depths of 20–40 m knolls or pinnacles rise. Some of them reach the sea surface (Milliman 1969). In lagoon shallows microatolls are forming, and the bottom is covered by macrophytes and seagrasses. Lagoons of large ring and barrier reefs often occupy more than half of the total of their areas.

g) Lr – *Leeward reef zone*, depth 0–2 m. This zone is represented usually by stripes of patch reefs, micro-atolls, and by ridges of rubble and dead corals. Islands and sand cays often form there.

Some researchers in their studies of reef geomorphology make a distinction between specific biotopes (Pichon 1978) and typical landscapes (Preobrazhensky 1986), according to what seemed to be useful for their survey descriptions.

1.2 Biological Factors of Formation and Erosion

The sedimentation and accretion of biogenic carbonates is one of the most important functions of coral reef ecosystems. It is one of the most remarkable global biogeochemical processes. Even at its stationary phase of growth in conditions of stabilized sea level, the reef biota produces 4–5 kg m^{-2} of $CaCO_2$ per year, while at the phase of active reef growth during the Holocene transgression this value attained 20 kg m^{-2}. Annual production of $CaCO_2$ at modern reefs approximates 2.5×10^9 t. A large part of crumbly carbonate material thus produced is transferred from reefs to the bottom of the surrounding sea, thus composing a significant part of the lime component of carbonate oceanic sediments. The rocky lime reef constructions represent an important geomorphological element of the sea floor in tropical regions.

The basic mechanism of formation of consolidated reef rocks, either in modern coral reefs or in ancient reefs constructed by other organisms, is the same (Goreau and Goreau 1973; Hubbard 1985). On the shallow bottom of warm seas large sessile animals, which have external calcareous or chitin-phosphorite skeletons, build up the lime framework of the reef, which is then filled with friable carbonaceous material composed of rubble, coral sand, spicules, crumbs of coralline crusts, etc. This material is then consolidated during the processes of its lithification and dolomitization based upon the recrystallization of $CaCO_2$ (Marshall 1983). Finally, the solid reef rocky base is gradually constructed, which resists wave stress and composes the substrate for subsequent growth of sessile fauna (Land and Goreau 1970).

The main framework builders on Holocene reefs in the Pacific on their windward side were massive, cribose or corymbose corals, like *Goniopora*, *Goniastrea*, *Platygyra*, *Porites*, *Pocillopora*, *Symphillia*. On their leeward side, among the frameworkers the ramose coral was dominating of the genera *Acropora*, *Pocillopora*, *Millepora* (Hopley et al. 1978; Davies and Marshall 1979). On the Atlantic reefs, the basic frameworkers are two species of ramose acroporid corals: *A. palmata* and *A. Cervicornis*, which have large and firm branches (Macintyre et al. 1977; Adey 1978). On the reefs of Indian ocean (Reunion Island) the ramose corals from the genera *Acropora*, *Pocillopora* and *Montipora*, together with massive poritides, were recorded even in columns from the windward reef (Montaggini 1977). At the fore-reef zone, where coral growth is inhibited by definiency of light, the sclerospongia became a significant element of reef framework (Hartman 1977).

The ability to extract calcium carbonate from sea water and to deposit it either as calcite or aragonite in the skeletal structures is possessed by numerous reef algae and animals. This function of calcification is especially developed in photo-autotrophs – in plants such as calcareous macrophytes and those in symbiosis with algae animals, such as symbiotic foraminifera, corals and tridacnid bivalves. The function of calcification is possessed also by various asymbiotic animals like ahermatypic (asymbiotic) corals, sedentary polychaetes, bryozoans, tunicates, echinoderms, and asymbiotic forams. But in any case the main producers of lime material on reefs and the main reef builders remain the photoautotrophs (Kühlmann 1988a).

The calcification in animal symbiotic photoautotrophs results in the formation of crystals of lime in the form of aragonite. This process occurs in special calcyoblast cells of the epidermis (Vandermeuler and Watabe 1974). It is definitely coupled with photosynthesis of their algal symbionts the zooxanthellae (cf. Sect. 8.3). In experiments, the calcification rate in corals

Table 1.2. Rate of calcification (Rc) and its light dependence (after Goreau and Goreau 1960b)

Organisms	Taxa	Rc, mg $CaCO_3\,g^{-1}\,h^{-1}$	
		In the light	In the dark
Algae	*Galaxaura* sp.	143.0	92.0
	Halimeda opunta	8.9	6.4
	Lithothamnion sp.	2.4	2.1
Corals	*Acropora cervicornis* (apical branches)	3.3	0.4
	Acropora cervicornis (lateral branches)	0.9	0.1
	Porites furcata	0.28	0.08
	Millepora complanata	0.26	0.14

increased by an average of about one order of values in the light as compared with that in the dark (Goreau 1963; Barnes and Taylor 1973; Chalker 1983; Huston 1985; cf. Table 1.2). In some corals it increased in the light up to 40 times, but in the hydrocoral *Millepora* calcification appears to be less sensitive to light. The coupling of calcification with photosynthesis in scleractinian corals was also proved in experiments with a specific inhibitor of the latter – DCMU – which inhibited in them also the calcification (Chalker 1983). Concerning the mechanisms of this coupling, one thing seems to be obvious: the pumping out of CO_3^- ions (and probably other acid products, like the lower organic acids) by Zooxanthellae from the zones of calcification in the polyp's calcyoblasts (Muscatine 1971). They also consume NH_4^+ and PO_3^- ions which inhibit crystallization of $CaCO_3$ (Goreau 1963; Yonge 1973; Grossland and Barnes 1974). The zooxanthellae produce also organic polymers of chitine type which accomplish the function of nuclei for the aragonite crystals (Yonge 1973; Johnston 1979). Being the generators of chemically bound energy within the polyp's cells, they also provide the energy supply for the process of Ca^{2+}-ions transport through the membranes of polyp's cells (Barnes and Taylor 1973). Scleractinian hermatypic corals deposit $CaCO_3$ in state of aragonite crystals. Besides, in corals colonies the calcite crystals could be also found; they appear there as a result of secondary recrystallization which proceeds actively in holes made in them by borers (Macintyre and Towe 1976).

The rate of calcification (Rc) could be measured as the weight or as the volume accretion rates of colonies, by estimation of Ca^{2+}-ions consumption from water, via observations of CO_2-balance in water by means of recording pH changes in time or by using ratiotracers' ^{45}Ca and ^{14}C (Goreau 1963; Drew 1973; Stoddart and Johannes 1978; Jokiel et al. 1978; Barnes and Crossland 1978; Barnes and Deveraux 1984). The in situ measurements of this value showed that its diurnal variation in general coincided with the diurnal curve of the coral's gross photosynthesis (cf. Sect. 8.4; Chalker 1983). The light dependence curves of the coral's calcification appeared to be coincide with the corresponding light curves of photosynthesis (Barnes and Taylor 1973). The dependence of Rc upon the ambient light intensity (J) derived from these curves looks as follows: Rc = Rcm · tg J/Jk, where: Rcm is the maximal rate of calcification, J the observed and Jk the optimal illumination (Chalker 1983). Thus the values of Rc, and correspondingly the coral's growth rates, should decrease with depth due to the decrease of light. But, actually, because of light adaptation of the photosynthetic apparatus of their zooxanthellae (cf. Sect. 8.3), it does not significantly decrease down to a depth 20 m in transparent waters (Buddemeyer et al. 1974; Dustan 1975; Barnes and Taylor 1973). However in deeper waters the Rc and the coral growth rate usually slow down (Gareau 1963; Huston 1985).

Second in importance among animal symbiotic calcifiers in reef ecosystems are the foraminifera. Some of their species, which live in coral sands, are real giants in the protozoan's world, having a diameter of 5–8 mm ("sand dollars"). They grow up to a volume of 0.2–0.3 cm^3 during 2–3 years (Ross 1977), while the small also symbiotic benthic forams, living in shallow temperate environments, but having for symbionts cyanobacteria instead of Zooxanthellae, attain their maximal size only after 5–6 years. The rate of calcification in hermatypic forams is dependent upon light, as in corals (Duguay and Taylor 1978). Forams produce around 15% of the total lime material of coral sands (Maxwell 1968; cf. Sect. 5.5.8). Remarkable calcifiers on reefs are also the large hermatypic bivalves tridacnids (cf. Sect. 5.1.3). Their shells are an important element in lime rock of the Pleistocene as well as of modern reefs (Chapell and Polach 1976). A definite role in the production of lime on reefs is played also by numerous ahermatypic animals, which have carbonate elements in their skeletons such as carapaces, needles, spicules, radules, shells, globules. Among them there are urchins, asteroids, crinoids, holothurians, various molluscs, crabs, sedentary polychaetes, bryozoans, sponges, soft corals (Chave et al. 1972; Maxwell 1973; Cuffey 1977).

The function of calcification is peculiar also to numerous representatives of seaweeds inhabiting reefs. In total, they comprise about 100 genera of calcareous algae (Borowitzka 1977). Most of the thallomic calcareous macrophytes deposit $CaCO_3$ in the form of aragonite orthorhomboid crystals in their cell walls or in the intercellular spaces. Among these are the red algae from the family. Peyssonelliaceae, the green algae Dascillus, Caulerpa and the brown alga Pagina. The coralline algae deposit in their cell walls crystals of calcite. The blue-green algae from the family. Nostocales accumulate in their mucus small globules of calcite. In some algae calcium accumulates inside the cells in the form of Ca-oxalate (Acetabularia).

The most powerful calcifier among the reef macrophytes are the Halimeda and the red Corallinacea (Goreau 1963; Littler and Doty 1975; Borowitzka 1983). The contents of $CaCO_3$ in Halimeda is as high as 90–90% of dry weight of its thalloms, while in other green calcareous algae from the same family Udoteaceae, like Udotea, Penicillus, Riphoclavus it is within 40 to 60%. On most of the reefs Halimeda produces 10 to 50% of total carbonate material of soft sediments (coral sands). In the Caribbean this genus is represented mainly by *H. opunta*, *H. discoidea*, *H. incrassata*. In the GBR area of the Pacific the genera *H. tuna*, *H. copiosa*, *H. minima*, and also the same *H. opunta* are most common (Drew and Abel 1983).

The red coralline algae (fam. Corallinaceae) are represented on reefs by several common genera (cf. Sect. 4.3.1). They form either weakly branching articulose, knob-like or crustose solid lime colonies which well resist the hydrodynamic stress. Coralline algae play a key role in processes of reef growth and consolidation, especially at windward reef edges, where the crustose or knob-like forms dominate, such as Porolithon, Sporolithon, Lithophillum, Lithothamnium, Hydrolithon, and Neogoniolithon. Here they find optimal conditions for their intensive growth: strong light and the intensive surf which repells their grazers: parrot and sturgeon fish. They deposit in their cell walls a mixture of calcite plus 25% of $MgCO_3$ (Borowitzka 1983). The rate of their linear growth at the margins of colonies approximates 0.7–2 cm year, while their thickness growth ranges between 1–5 mm $year^{-1}$ (Adey and Vassar 1975). After the death of their cells the crystallization of both these carbonates continues in them. This process results in lithification of their colonies and thus in the building up of the flat rocks (Littler 1972). The branching corallines Jania and Amphiroa, which grow on the pieces of coral rubble, also finally cement them forming so the solid rock. The main function of branching coralline algae is the production of carbonaceous material of soft sediments, which was evaluated as being within 20–40% of its total production by reef biota. This same share belongs in coral sands to calcite, which is produced in reef environments almost exclusively by corallines. The calcification in Corallinacea is light-dependent (Borowitzka 1979) but is less pronounced in them as compared with corals: the ratio of Rc in the light and in the dark in these is ~1.5 (cf. Table 1.2). Calcification in them is inhibited by DCMU, as in corals. It was supposed that via photosynthesis some excess of ATP is produced in them, which takes part in the energy supply of Ca^{2+}-ions transport from seawater to the zones of calcification in algal tissues (Barnes and Taylor 1973).

The results of Rc estimates in various reef calcifiers, animals and plants are summarized in Table 1.3. In some branched coralline algae the rate of calcification could be even higher than in corals, attaining 2 mg $CaCO_3$ $g^{-1}h^{-1}$. In crustose corallines, such as Porolithon, it is about the same as in corals: 0.1–0.3 mg $CaCO_3\,g^{-1}h^{-1}$. A high rate of calcification have also the thallomic calcareous algae Halimeda, Penicillus, Udotea – up to 2.9 mg $g^{-1}h^{-1}$. Thus the thickets of these algae should produce 3–5 kg lime $m^{-2}year^{-1}$. In corals realistic Rc values are equal to 0.2–0.6 mg $CaCO_3$ $g^{-1}h^{-1}$. Their values given in Table 1.3, which Goreau (1963) obtained with the use of ^{45}Ca-label, were probably exaggerated due to taking no account of possible isotopic exchange (Borowitzka 1979). The above rates of cal-

Table 1.3. Rate of calcification (Rc, $mg\,g^{-1}h^{-1}$ of $CaCO_3$) in some reef calcareous animals and plants in daytime

Groups of calcifiers	Taxa	Rc	Reference
Coralline algae	*Amphiroa fragilissima*	2.09	Stark et al. (1969)
	Amphiroa foliacea	0.26	Borowitzka (1979)
	Porolithon oncoides	0.30	Littler (1976)
Green calcareous algae	*Halimeda opunta*	0.70	Stark et al. (1969)
	Halimeda tuna	2.90	Böhm (1978)
	Halimeda discoidea	0.35	Goreau (1963)
	Penicillus sp.	1.05	Goreau (1963)
	Udotea sp.	2.90	Goreau (1963)
Brown calcareous algae	*Padina sanctaecircus*	3.13	Goreau (1963)
Scleractinian corals	*Acropora cervicornis*	2.4–3.3	Goreau (1963)
	Porites furcata	0.28	Goreau (1963)
	Average for 11 species, Jamaica	1–2	Goreau (1963)
	Acropora cervicornis	0.20	Chalker (1983)
	Acropora sp.	0.20	Barnes et al. (1976)
	Acropora palmata	0.16–0.26	Gladfelter et al. (1978)
Hydrocorals	*Millepora complanata*	0.26	Goreau (1963)

cification in corals result in an annual production of 2–10 $kg\,m^{-2}$ of $CaCO_3$ by the coral cover of a bottom surface of 10–30% (Gladfelter et al. 1978).

The evaluation of calcification rates in whole coral reef communities may be based on: a) the experimental estimation of Rc in common species with taking into account the relative bottom area covered by them (Smith and Kinsey 1976); b) measurements of linear growth of the dominant corals (Jokiel et al. 1978); c) observations of changes in the topography of definite reefal constructions in time in conjunction with radiocarbon dating of the rock's age (Davies 1983), and estimations of the carbon balance of the whole reef (Sargent and Austin 1949). Linear and mass growth in corals could be measured by X-ray analysis of annual rings on cross-sections of colonies (Macintyre and Smith 1974). But the most widespread is the alizarine method. The colony of corals or its branch is covered in situ with a plastic bag, into which the stain is then injected. After several months of exposure on reef the colony is cut into sections and its growth beyond the boundary of the stained layer is measured (Stoddart and Johannes 1978; Oliver et al. 1983).

The linear growth of corals depends on the morphology of their colonies (Table 1.4). The highest values were recorded in common ramose acroporid corals, such as *Acropora cervicornis*, *A. palmata*, *A. formosa*, *A. prolifera*, *A. pulchra*: 4–20 cm $year^{-1}$. Average rates of linear growth is peculiar to the foliose, columnar, cribose corals and to corals with short branches like Agaricia, Turbinaria, Madracis, Pocillopora, Porites, Pavona: 0.5–4 cm $year^{-1}$. In the massive corals, in which the ratio of area to volume of

Table 1.4. Rates of linear growth of corals; data by Chave et al. (1972), Clausen and Roth (1975), Krasnov (1981), Isdale (1983), Davies (1983), Huston (1985), Logan and Tomascik T. (1991)

Basin	Taxa of coral	Growth rates, cm/year
Pacific	*Acropora pulchra*	10.0–17.0
	Pocillopora damicornis	0.9–3.9
	Montastrea sp.	1.4
	Pavona clavus	1.5–2.3
	P. gigantea	1.0–1.9
	Porites lobata	0.4–1.3
	P. compressa	0.3–2.4
	Favia pallida	0.5–0.7
	Fungia fungites	1.0–1.7
Atlantic, Caribbean	*Acropora cervicornis*	4.0–2.6
	A. palmata	4.0–9.9
	A. prolifera	3.7–8.2
	Agaricia agaricites	0.5–1.7
	Siderastrea siderea	0.2–0.5
	Porites porites	0.3–5.6
	P. furcata	0.9–2.3
	Diploria strigosa	0.3–1.0
	Montastrea annularis	0.9–1.1
	Madracis asperula	2.4

the colony is low, linear growth is also low: 0.2–0.5 cm $year^{-1}$ (Chave et al. 1972; Clausen and Roth 1975; Bak 1976; Glynn 1977; Gladfelter et al. 1978; Yap and Gomez 1984). The rate of coral growth is connected with the rate of calcification. But the environmental factors, which control calcification, influence not only their linear growth but also the morphometry of their colonies, because through the change in calcification rate the ratio between the surface of the skeleton parts, the mass of soft tissues and the number of polyps (Goreau 1963; Oliver et al. 1983) also changes. It finally results in the appearance of their ecomorphs. Among the main factors influencing the coral's linear growth are light intensity, temperature, turbidity, and wave stress (Dodge et al. 1974). The decrease in illumination and water temperature was supposed to be the main cause of an apparent decrease in the growth rate of corals down the slope of the outer reef at depths beyond 20–30 m (Huston 1985; Bablet 1985). In some corals (*Monastrea annularis*) the linear growth was maximal at some middle depth of 10–15 m. The linear growth at this depth could be faster even in ramose corals, while their mass growth remains maximal at the surface because of faster branching (Oliver et al. 1983). Turbidity, and in shallows also overheating and excess illumination are among the factors inhibiting linear growth and calcification in corals, which live in lagoons and patch-reef areas, where the calcification rate in them appears to be significantly less than at outer reef slopes (cf.

Table 1.5. Rates of lime biodeposition (Rc) and annual production (Rt) of carbonate material in reef bottom biotopes; designations of reef zones: *af* reef-flat covered by crustose coralline algae; *fl* reef-flat covered by living corals; *pr* patch reefs; *l* lagoon; *m* average value calculated per whole reef. Data by Stoddart (1969), Smith (1973), Smith and Pesret (1974), Stearn et al. (1977), Kinsey (1978), Hubbard et al. (1981b), Drew and Abel (1983), Johannes et al. (1983b), Barnes and Devereux (1984), Hubbard (1985)

Location	Reef zones	Rc, $g\,m^{-2}\,day^{-1}$	Rt, $kg\,m^{-2}$
One Tree Island ring reef,	af	11.6	4.0
GBR	fl	13.0	4.5
	l	4.3	1.0
	pr	1.5	0.3
Lizard Island,	fl	10.4	3.6
GBR	af	11.0	3.8
	m	5.8	1.8
Rib reef, GBR	fl	9.4	3.5
	l	6.1	2.5
Abrolhos I, Western Australia	fl	13.0–50.0	12.0
Oahu I., Hawaii	m	3.0	1.1
Enivetok atoll	fl	31.0	11.2
	af	11.7	4.7
	m	10.0	3.6
Rongelap atoll	m	24.0	9.2
Canton atoll	m	1.5	0.5
Barbados	fl	42.0	15.0
Saint Croix, Virgin I.	m	19.0	5.0–7.0
Bahamas	m	22.0	75

Table 1.5). The seasonal changes in temperature and illumination are also reflected in the seasonal variations of linear growth rates of corals (Shinn 1966; Buddemeyer and Kinzie 1976; Gladfelter et al. 1978; Bablet 1985). The inhibiting effect of eutrophication on coral growth is connected mainly with the increasing turbidity which weakens the illumination upon which the rate of calcification directly depends (Dodge and Vaisnys 1977; Dodge 1981). Another cause could be the inhibition of calcification by a high content of PO_4^{3-} in water, which also inhibits calcification (Simkiss 1964). The growth rate and calcification on reefs increase after storms, when new populations of young, fast-growing ramose corals recruited from broken branches start actively to grow (Highsmith et al. 1980).

The data on linear growth rate of corals as well as on the growth of their mass, when measured at ambient environments, are used as indicators of the in situ rate of their calcification.

The results of estimations of calcification rates in reef benthic communities are given in Table 1.5 (see also Tables 10.6, 10.8). They show that it progresses most actively in the reef-flat zone independently of domination of corals or of coralline algae in it. It comprises there 10–20 $g\,m^{-2}\,day^{-1}$, which is comparable to the primary production of organic matter or sometimes even exceeds. it. Maximum values were recorded at the reef-flat of the high-latitude Abrolhos reefs off West Australia – close to $50\,g\,m^{-2}$ day (Johannes et al. 1983a). An average annual production of $CaCO_3$ on these reefs was evaluated as being within $12\,g\,m^{-2}$ which corresponded to an annual reef growth of 2 cm. As an average, the annual production of lime in communities of reef-flat and outer reef slope was approximately within $3–5\,kg\,m^{-2}$ (Smith 1973, 1983; Table 1.5). The rate of calcification, the same as other processes of biogeochemical metabolism in reef communities, is subjected to seasonal fluctuations (Barnes 1988).

In lagoons the rate of calcification is usually lower: $4–6\,g\,m^{-2}\,day^{-1}$. The carbonate material is produced there mostly by calcareous algae and foraminifera. The annual production of $CaCO_3$ in lagoons ranges within 1 to $3\,kg\,m^{-2}$. Its average production in various reef or atoll communities was estimated to be within 1–7 kg $CaCO_3\,m^{-2}$ (Smith and Kinsey 1976; cf. Table 10.3). Per $1\,km^2$ of their area reefs are producing about 2–4 thousand tonnes of carbonaceous material. This quantity demonstrates, that coral reef ecosystems are extremely powerful generators of lime in marine environments. Reef biogenic carbonates thus sedimented contain besides $CaCO_3$ also many other elements and minerals, like $MgCO_3$, Sr, CO_3, $Mg(NH_4)PO_4$, PO_4^{3-}, Fe^{3+}, Cu^{2+}, Ni^{2+}, Zn^{2+}, Pb^{2+} (John 1974; Dodge et al. 1984). Global annual carbonate production on coral reefs was evaluated to be close to 2.5×10^9 tons.

Vertical linear growth rate of the reefs themselves was estimated by geologists with the aid of radiocarbon dating of cores. Before reaching stabilized sea level during the Holocene transgression, the rate of reef growth ranged within 2 to $12\,mm\,year^{-1}$. This growth rate corresponds to the calcification rate (without taking erosion into account) of 3 to $20\,g\,m^{-2}$ year, which was higher than the calcification rates on modern reefs that also reached stabilized sea level (cf. Table 1.5). Now it could be evaluated to be within 1–4 m 10^{-3} years (Hubbard 1985). The rate of vertical reef growth appears to be dependent upon the level of wave stress. Under conditions of permanent strong wave stress consolidated reef plateaus grow less fast, they being more solid than in calm environments because the reef body in the first case is constructed largely by slow-growing coralline algae which resist better the surf than corals, and in the second case mostly by fast-growing branching corals. As examples of slow-growing wave-stressed reefs the Pacific atolls could be mentioned, and of fast-growing less-stressed reefs, those of the Caribbean region (Adey 1978).

The consolidated reef rocks, which reef-building organisms construct on the plateau of modern reef-flat and patch reefs, are simultaneously subjected to intensive erosion (Highsmith 1981; Trudgill 1983). Its products, reworked by surf into soft sediments, are transported to the reef edge and moving down its slopes reach the bottom of the surrounding ocean. Thus they participate in the formation of carbonate oozes on island and continental slopes and on adjacent deep oceanic floor. The sedimentation rates in the lagoons of the ring and barrier reefs was estimated to be within 1–

3 mm year^{-1} (Bardach 1961; P.J. Davies 1977; Frankel 1977). The rate of erosion of the surface of consolidated lime rocks at flats in modern reefs, which also reached stabilized sea, should be balanced against the vertical growth of the consolidated reef-flat. The latter may be evaluated within 1–2 mm or 2–4 kg m^{-2} of lime material per year. The rate of erosion should be within the same range. Its direct estimate by using various menthods resulted in the above-shown range (Table 1.5). This process is most active in the intertidal zone. Deeper down it decreases more than ten times (Trudgill 1976).

The reef rocks show numerous signs of erosion in a multitude of various channels, moats, trenches, holes and caves which cut their surfaces and pierce their body (Ginsburg 1983). This erosion of the reef-flat rocks proceeds mainly biologically, various boring animals and plants participating in it. But also abiotic factors take a definite part in this process. The mechanical erosion caused by them on reef surfaces is largely enhanced by the abrasive impact of a mass of rubble moved by water. The reef rocks are subjected also to an intensive weathering process connected with the crystallization of marine salt inside their porous lime material during the drying out of sea water left behind in it between the tides. The weathering is intensified also by an excess of agressive CO_2 formed by the decay of the mass of organic matter produced by reef bottom communities. The production of CO_2 being accompanied by a drop in pH proceeds most actively at night in pools remaining on reef-flats at low water. The waters of lagoons are also enriched with free CO_2, especially at night. Passing over the reef flat it dissolves the lime material on its surface, while microzones enriched with free CO_2 are also forming at night under the thalli of sea weeds, rubble and detritus, as well as in numerous holes inhabited by their owners – various benthic invertebrates and fish, which produce CO_2 during respiration. The reef rocks are perforated by numerous boring molds, filamentous algae, worms, molluscs, and sponges (cf. Sect. 4.2). Their surfaces are permanently scraped and grazed by benthic animals and fish, which feed on periphyton and corals (Bak 1976; Trudgill 1983). The boring molds belong to the genera *Aspergillus*, *Cladosporum*, *Penicillum*, *Phialophora* (Kendrick et al. 1982). Among the boring algae penetrating into the colonies of corals and into solid reef rocks, the most numerous are the filamentous green algae Ostreobium and blue-green Hormatonema (Alexandersson 1975; Kobluk and Risk 1977). Their filaments dissolve the crystals of carbonate and penetrate into the rock material and in the colonies of dead and living corals.

The boring polychaete sponges and bivalves make a network of holes in rock material as well as in coral colonies (Rützler 1975; Hutchings 1985, cf. Sects. 5.1.4, 5.2.3, 5.3). They undermine the mechanical durability of corals and solid reef rocks exposed to the destroying action of surf. In regions of cyclones, the activity of reef borers is the main cause of destruction of the branching coral thickets on outer reef slopes and reef-flats (Warme 1977).

Table 1.6. Annual rates and depths of erosion of reef-flat rock surfaces in the intertidal zone; data by Davies (1983); Rützler (1975), Trudgill (1976), Hudson (1977), Bromley (1978)

Location	Rate ($kg\,m^{-2}$)	Depth (mm)
Florida	3–13	1.5–6.0
Bikini atoll	1	0.5
Bermuda	2.4–7.0	1.2
Australia	1–2	0.5–1
GBR of Australia	0.4–6.0	0.2–2.9
Aldabra atoll (upper part of intertidal zone)	1.3–9.0	0.6–4.0
Same, medium part	2.2–2.6	1–3
Same, lower part	1.1–4.4	0.5–2.0

Among the most common benthic reef borers there should be mentioned the sponges Cliona, Anthosigmella, Siphonodyction, the bivalves Lithophaga, Gastrochaena, Petricola, the vermetid gastropods Dendropoma, the sipunculid worms Phascolosoma, the polychaetes Eunice (miticulata), Polydora, Fabricinus, Hypiscomus (elegans), the barnacles Lithophrya (dorsalis), and the urchins Echinometra. On reef flats off the Virgin Islands, in a zone covered by coralline algae, they are active over 30% of their total surface, thus furthering their destruction by surf (Adey 1975; Bak 1989).

An approximate estimate of the rate of erosion of reef-flat rock gives its annual value as up to several $kg\,m^{-2}$ (Table 1.6). Alone the boring sponges in places of infestation may destroy 7–20 $kg\,m^{-2}\,year^{-1}$ of solid reef rock, which could be much more than the possible calcification rate. Average rates of bioerosion were evaluated to be within 0.4 $kg\,m^{-2}$ (Stearn et al. 1977). The share of bioerosion comprises about two thirds of the total (Trudgill 1976).

The mass of carbonate material produced by the erosion of reef into rubble and sand is transported by surf and currents and also by migrating animals (mainly by fish) ingesting the lime material to other zones of the reef. So the bioerosion of solid reef rocks is one of the main sources of formation of soft reef sediments (Stearn et al. 1977; Davies 1983). The reef-flat of modern reefs, whose vertical growth was stopped by the recent stabilization of the sea-level, became the zone of intensive production of friable lime material. Transferred out to the leeward side of the reef it is contributing to the further lateral growth of the reef and to the build-up of its submerged parts.

1.3 Bottom Sediments

Reef bottom sediments contain a larger part (ca. 60–80%) of total lime material of modern and also of the Pleistocene reefs (Thomassin 1978; Riddle 1988). The benthic biocenoses of the soft bottom are important components of the coral reef ecosystems. Studies of composition, origin and distribution of bottom sediments give a remarkable insight into the problems of reef geology and biology (Purdy 1963; Goreau and Goreau 1973; Maxwell 1968, 1973a; Stoddart 1969). The main part of the lime material composing the reef bottom sediments is of biological origin (Flood 1983; cf. Table 1.7). The terrigenous clastic material, consisting of clay, sand, pebbles, and ferrous oxides, comprises a significant part of sediments (10–15%) only in lagoons and on the flat of some fringing and barrier reefs (Ginsburg 1956; Morelock and Koenig 1967; Milliman 1967; Stoddart et al. 1969). As an example, the reefs off the coasts of Florida, Cuba and GBR could be mentioned (Stoddart 1969; Maxwell 1968; Riddle 1988). From Table 1.7, it can be seen that the lime material of sediments has two main sources of origin. One part (30 to 60%) is formed by the erosion of lime rocks of modern and ancient reef plateaus and of lime skeletons of animals – molluscan shells, corals, urchin's needles, etc. This erosion proceeds by the action of the boring, scraping and grazing reef animals, such as boring clams, sponges, worms, coral-eating and periphyton scraping fish and urchins. Thesefore, this part of sediments is formed mainly by "chips" of borers and by fecal material of coral feeders and scrapers (Emery 1956; Bakus 1968; Randall 1974; Gygi 1975; Ogden 1977).

The second part of lime reef sediments is formed by small skeletal elements of reef animals and sea weeds – calcite plates, globules or spicules. The same elements are contained in sponges, soft corals, holothurians, asteroids as well as in calcareous algae. The third part is formed by calcareous microshells of benthic foraminifera and of planktonic coccolithophorids (Mc Kee et al. 1959; Flood 1983). Reef sediments contain fractions of rubble particles (>10 mm), crude sand particles (2–10 mm), of medium sands (0.5–2 mm), and a fine sediment fraction (0.1–0.5 mm). The rubble fraction is formed from the dead coral colonies and from the crusts of coralline algae eroded by boring animals and then crushed by surf. The rubble and the sand material are sorted out by surf and currents. Its fractions are resedimented in definite reef zones. The rubble is accumulating mostly on the surface of flats where it forms the ramparts, banks and islands. Large masses of rubble material accumulate in the moat on the outer reef slope under the reef edge and also in shallows of the large lagoons exposed to surf action. The most widespread fractions of reef sediments are medium coral sands with particle sizes of 0.3–0.6 mm (Fig. 1.7). The medium coral sands are formed mainly from calcareous plates of Halimeda, from the spiecules of soft corals, from shells of forams, and from lime

Table 1.7. Shares of various sources of carbonate material in the formation of medium and fine fractions of reef bottom sediments; data by Hoskin (1966) Stoddart (1969) Maxwell (1968) Thomassin and Masse (1985)

Location	Corals	Coralline algae	Halimeda	Molluscan shells	Foraminifera	Other skeletal material
Johnston atoll	20	51	1	10	2	16
Rongerik atoll (lagoon)	19	–	25	6	11	39
Enivetok atoll (lagoon)	12	–	28	3	30	26
Rongelap atoll (lagoon)	7	–	43	8	16	26
Bikini atoll (lagoon)	1	–	43	8	8	35
Cocos atoll (lagoon)	45	18	11	15	3	8
Barrier reef Batabano (Cuba)	10	10	30	10	10	–
Ragged I. reef (Bahamas)	12	18	19	18	13	4
Andres reef, same	24	33	17	6	12	6
GBR, Australia (flat)	20–40	17–40	10–30	Up to 15	8–20	Up to 5
Same (lagoon)	Up to 15	Up to 15	Up to 65	20–30	15–40	15–30
Reefs off New Caledonia	21	11	12	35	9	10

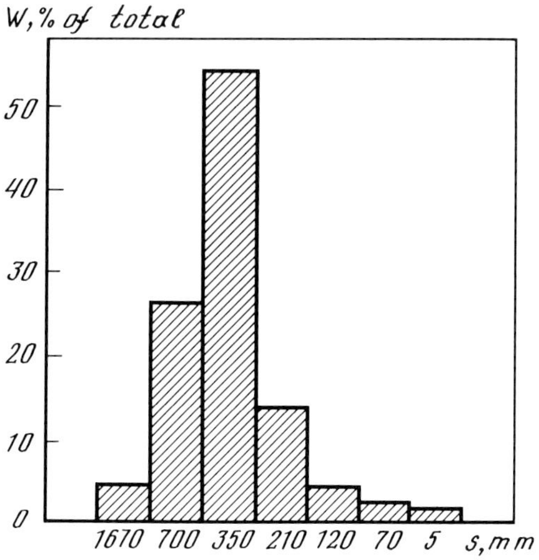

Fig. 1.7. Granulometric composition of coral sand from the foot of outer reef slope in the fore-reef zone of the Jamaican reef; *S* size of particles; *W* weight of particles in given size fraction. (After Goreau and Goreau 1973)

material sorted out by surt and produced by bioerodes, boring and scraping animals (Wallace and Schaefersman 1977). The fine fraction of reef sediments is formed during the further reworking of the medium fraction of sediments by the sediment-feeding and digging bottom animals, such as gastropods, holothurians, fish, shrimps, eunicid polychaetes. The fine particles thus formed are resuspended, redistributed by waves and currents and resedimented in less turbulent reef bottom zones, such as the bottom of deep lagoons, or around the leeward patch reefs.

The processes of reworking of coral sands proceeds on the reef on a large scale (Bakus 1973) and takes place also in lagoon bottom areas inhabited by dense populations of holothurians. The rate of this process can attain 10^4 tons km^{-2} $year^{-1}$ (Bonham and Held 1963). The benthos-feeding fish swallow and thus rework 300–600 tonnes of sands km^{-2} $year^{-1}$ (Cloud 1962). An important role in bioturbation and redistribution of the reef sediments is played by digging animals, such as shrimps, polychaetes and fish (Roberts et al. 1981). The most active in this field are sturgeon, the balistides and box fish (Bardach 1961). In passing through the guts of animals, under the action of acid excreted by their intestines and also owing to mechanical rubbing, the sand particles are ground to a fine fraction. The fine fraction of reef sediments is also produced by calcareous algae in the form of crystals of calcite (cf. Sect. 1.2). The alcyonacean octocorals and tunicates have also very fine microspicules or microglobules which participate in the formation of fine sediments. They are also formed by the aerial erosion of exposed remains of Pleistocene reefs and from the terrigenous clay brought by rivers (Neumann 1965; Morelock and Koenig 1967). In particular, the terrigenous components comprise the fine sediments which cover the bottom of the GBR lagoon (Maxwell 1968) and the bottom joining the outer slopes of the monsoon fringing reefs off the coast of central Vietnam. The fine downy sediments consisting mostly of the needle crystals of aragonite, glued with coral mucus, often cover bottom of the patch-reef areas of some Pacific atolls at the leeward edge of the reef flat (Thorp 1936). The soft aragonite and calcite oozes occupy large areas in the reef zones around the Great Bahama Bank (Black 1933). They are formed by the fine globules and crystals of skeletal lime material produced by calcareous algae (Lowenstam 1955), and probably by the carbonates sedimented from sea water under the influence of intensive photosynthesis of reef bottom plant associations, followed by an increase in pH in water columns up to pH 9 (Broecker and Takahashi 1966; Chave et al. 1972). The sedimentation of fine carbonaceous material produced during this chemogenic crystallization of calcium carbonate in water columns over the activety functioning reef proceeds largely under the action of the filtering and sedimentary feeding activity of animals. An important role is played by the corals themselves. During their sedimentary feeding, they accumulate the mineral particles on their surface, then glue them together with their mucus and shed them off to the ground, thus avoiding siltation. During this process, the soft fine carbonaceous sediments are formed. The same kind of sediments are formed by filtering bivalves or by sedentary polychaetes. Their dense populations produce a mass of pseudofeces enriched with particles of the chemogenic calcyte. This materials form often the thickness of the soft fine sediments, which are enriched with mucus and have a slimy texture (Roy and Smith 1971).

A significant part of the calcareous particulated material formed on flats during the processes of bioerosion, and which are accumulated there by biofiltration from water columns, is then driven out to the lagoon (Smith et al. 1971). The rate of sedimentation in the lagoons is therefore very high – 1–1.2 m 10^{-3} years (Cloud 1962) and could sometimes reach 3 m (Bardach 1961). The medium and the fine sediments occupy often also the foot of the outer reef slope. They are driven out by tidal currents from the flat and from the lagoon through the channels, which cut the reef slopes (Fairbridge 1950). A significant amount of suspended carbonate material is driven out from the lagoon by currents through the passes in the ring reefs of atolls (Smith et al. 1971). The flow of this material from the reef down to its outer slope could be evaluated as being up to 1–4 kg m^{-1} day^{-1}. During storms it could be much more (Hubbard et al. 1981a). The annual transport of calcareous sediments from the reef to the fore-reef foot zone amounts to 100–1000 kg m^{-1}. This material will reach even the bottom of the surrounding deep-sea floor, which is evidenced by the presence of the shallows-dwelling foram Homotrema in deep-sea sediments (Mackenzie et al. 1965).

The zonal distributions of size fractions of sediments can be observed in lagoons of atolls. Their shallows are covered with the crude sands mixed with rubble material, sorted by the surf. The bottom of lagoon slopes is occupied by medium coral sands which often consist of the abrased lime plates of the calcareous alga Halimeda – the so-called Halimeda sands. The bottom of the central deep part of the lagoon is covered by fine coral sands of carbonate silts. A significant part of these sediments is transported from flat to lagoon by fish (Emery 1956; Randall 1974). They feed during high tide on the flat with periphyton, coralline algae and corals. While migrating during the ebb-tide to the lagoon they excrete with feces large masses of the swallowed carbonate material. The size of its annual transport was evaluated to be about 2 tonnes ha^{-1}, it being commensurate with that carried by surf and currents (Bardach 1961).

The upper layer of coral sand, being well aerated and illuminated, is inhabited by dense populations of microalgae, bacteria and microzoobenthon (cf. Sects. 4.1, 5.6), this making the coral sands one of the most active elements of the coral reef ecosystem, despite the fact that the absolute content of organic carbon in them is relatively low: 0.15–1% per dry weight. The latter seems to be larger in reef areas with an abundant coral growth (Sorokin 1986a; cf. Table 4.1). The corals shedding mucus obviously enrich the nearby soft sediments with organic matter. Indeed, the content of

Table 1.8. Content of organic carbon in reef bottom sediments as percentage of dry weight; data by Thomassin et al. (1976), Sorokin (1978b, 1991), Hubbard et al. (1981b), Vaugelas (1981), Vaugelas and Naim (1981), Propp et al. (1983), Thomassin and Cauwet (1985)

Location	Biotope		
	Coral sands among thickets of macrophytes and seagrasses	Coral sands in lagoons	Silts or silted coral sands with admixture of terrigenous material
Aqaba, northern Red Sea.	0.14–0.34	0.09–0.26	0.11–0.16
Coast of central Vietnam	0.7–1.2	0.12–0.35	0.7–1.0
Moorea I., French Polynesia	0.17	0.17	0.45
Kaneohe Bay, Hawaii	0.27	0.37	0.29
Velangilala atoll, Lau, Fiji	0.17	0.47	–
Saint Croix, Virgin Islands	0.4–0.6	0.4–0.6	0.2–0.4
Barrier reef of Tulear, Madagascar	0.11–0.16	0.16–0.23	–
Phantom bank, Timor Sea	0.24–0.55	–	–
Heron Island, GBR	0.32	–	–

organic carbon in coral sands depends upon the degree of coral growth on the reef. In reef zones with a low turbulence this shedding activity of corals results in the formation of detritus sediments, enriched with organic matter and with microbial biomass (Di Salvo and Gundersen 1971; Sorokin 1973c). The content of organic matter in reef sediments remains relatively low even with a significant flow from the water column (Charpy and Charpy-Rouband 1988; Le Borgne et al. 1988). It is a consequence of intensive decomposition processes by microbial populations, which comprise in them an extremely high biomass, forming 5 to 40% of their total organic matter (cf. Sect. 4.1). The content of organic carbon usually varies in coral sands within 0.3–0.6%. Rarely does it exceed 1% and the same content of C_{org} was found in silty sediments, which often occupy the deep parts of the lagoon at barrier reefs and which therefore contain a significant portion of terrigenous material (Table 1.8).

2 Reef Environments

2.1 Hydrodynamics and Thermohaline Features of Reef Waters

Among the most important hydrophysical factors interacting with the reef biota and influencing reef growth and its morphometry are currents, wave stress, tides, internal waves, thermohaline structure of the water column. These factors themselves, and also their influence upon the reef biota, have been so far insufficiently studied because of their extremely complex and variable nature. Each individual reef is unique in its morphology, and thus has its own peculiar relationship with the physical environments, which are moreover very variable in different regions of the tropical zone. The currents interfering with reefs form a multitude of reciprocally directed eddies. The first attempts to clear up the complex nature of hydrodynamics in the reef areas and to investigate the interference of currents with their geomorphological structures had been made by American hydrologists on some atolls of the Marshall Islands during the nuclear tests in 1947–1950 (Von Arx 1948; Munk and Sargent 1948). Later, investigations of the same kind were made at the Caribbean reefs (Shinn 1963; Kühlmann 1974, 1975; Roberts 1974, 1977; Roberts et al. 1975). Studies on the tide currents and measurements of the residence time of water in lagoons were accomplished at the Fanning atoll (Gallagher et al. 1971), the Takapoto atoll (Sournia and Ricard 1976a), the Canton atoll (Roberts 1974), and the Tarawa atoll (Kimmerer and Walsh 1981). Studies on the hydrology of the GBR had been started in the late 1920s by Orr (1933). Later, extensive studies were carried out by French Research Insitution ORSTOM (Brandon 1973) and by the Australian Institute of Marine Science (AIMS). Hydrologists from AIMS gained fundamental insight into reef hydrodynamics (Pickard et al. 1977; Andrews 1983b; Andrews and Gentien 1982; Wolanski and Pickard 1985, 1986). Circulation in the lagoon of the barrier reef off New Caledonia was studied by Rougerie (1985).

These investigations showed that among the most important factors of the hydrodynamic stress in coral reef areas are the tides (Church et al. 1985). They induce regular and powerful tidal currents, which have a great impact upon the formation of reef constructions, as well as on zonal distribution of reef bottom communities and on their functioning. The amplitudes and the periodicity of tides are very variable in different regions. On the Caribbean reefs the tides are so small that they are called microtides (Kjerfve et al. 1982). Their amplitudes are only 15–30 cm (Roberts et al. 1975). But even such microtides may induce significant tide currents with a velocity of up to 30 cm/s^{-1} (Geer and Kjerfve 1982). At the Pacific atolls, the heights of tides are usually within 0.5–2.5 m (Gallagher et al. 1971; Sournia and Ricard 1976a; Atkinson et al. 1981). The average height of tides in the GBR area is 1–3 m, but in some places it might be up to 9.5 m (Ludington 1979; Frith 1983). A peculiarity of tides in lagoons of atolls and even of barrier reefs lies in their retardation as compared with their time schedule in the surrounding ocean (Gallagher et al. 1971; Ludington 1979). In lagoons the tides begin 1–3 h later than in the surrounding ocean and their heights are there much less, while the eb level in the lagoon is usually higher than in the ocean because of water trapped in it. Thus the amplitudes of tides within lagoons are 0.3–1.5 m moreover than outside in the ocean (Fig. 2.1). In the lagoons of semi-closed ring reefs the tides are retarded by one-half of the tide cycle – by 3 h, and start only when water begins to fill the lagoon, overflowing the reef flat (Fig. 2.1). Regions with small tides have 24-h cycles or mixed 24-h and 12-h cycles. In the last case most often only the 24-h maxima are well defined. In the zone of the GBR, with a significant height of tides, there are clear-cut 12-h cycles with alternating amplitudes (Fig. 2.1). On the backsurface of tidal waves, by measuring sea-level fluctuations the existence of varions other internal waves was established which resulted also in sea-level oscillations (Andrews 1983b; Wolanski and Bennett 1983; Wolanski and Pickard 1983). The periodicity of these waves, additionally to the basic tidal waves, could be several minutes (internal Kelvin's waves) to several hours (solar tidal waves), and up to several tens of days (barotrophic coastal internal waves). The latter cause the fluctuations of sea-level by 30–35 cm, having a periodicity of 60–90 days (Wolanski and Pickard 1983b). All these fluctuations of the sea-level together with the tides induce the fluctuations in depth of the boundary of the thermal discontinuity layer, and thus accelerate the upwelling of nutrient-rich deep waters to the reef biotopes (see below).

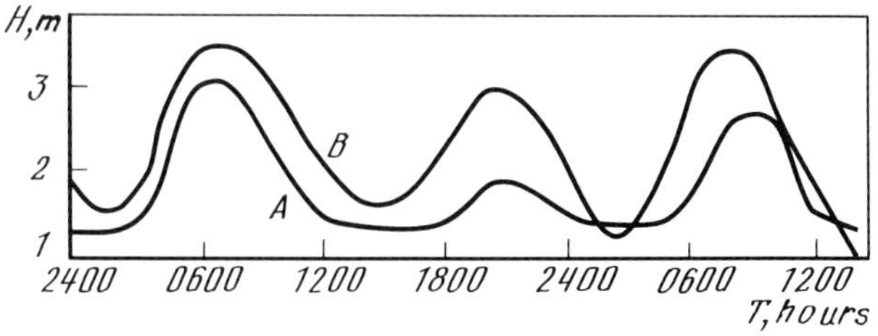

Fig. 2.1. Tidal fluctuations of sea level (*H*, *m*) in lagoon of the One Tree I. reef (*A*), and in the surrounding waters of the Coral Sea (*B*); *T* time of day, h. (After Lundington 1979)

Most reef systems are situated in zones of permanent and vigorous wave stress, caused mainly by the action of trade winds, which have a general western direction and speed of 2–10 m s^{-1}. The height of waves induced by such winds is 1–4 m. Trade winds of average strength, inducing waves 2–3 m high, act on a medium-sized atoll with the power of half a million hp (von Arx 1948). The average power of a medium-high wave acting upon a windward reef equals $1-4 \times 10^9$ erg s^{-1} (Roberts 1974). During storms the power of the wave stress increases many times. A storm of medium strength with a wind velocity of 10–20 m/s drives waves whose height near the reef edge will reach 5–8 m. Many reef regions suffer constantly from the periodical action of destructive tropical storms (hurricanes or typhoons) when the speed of wind exceeds 20 m/s and the height of waves reaches 10–15 m (Stoddart 1969, 1974; Woodley et al. 1981; Kjerfve and Dinnel 1983). Wave stress is among the most powerful physical factors acting upon the reef ecosystems, including its geomorphology, zonal distribution of its structural elements, distribution of carbonate material and topography of its biotopes (Roberts 1974; Heckel 1974; Adey 1978; Geister 1977; Bradbury and Young 1981; Dollar 1982). This influence is especially displayed in the frontal zone of windward flats and on the outer slopes.

Interacting with the elements of a complex relief of windward reefs the oncoming waves are transformed, changing their periodicity and height (Thornton et al. 1976). Largely under the action of waves on the outer windward slope a system of buttresses and channels is formed which actually compose a system of natural wave-brakers, called also the spoor and groove system (Munk and Sargent 1948). Interacting with waves, they dissipate their energy, thus decreasing their destructive stress upon the reef construction and the biota. The degree of development of spoor and groove systems on reefs directly depends upon the wave stress acting upon them (Roberts 1974). The dissipation of wave energy proceeds also by their interaction with the channels which cut through the outer frontal part of the reef flat. Often these channels are actually the prolongations of grooves of the buttress system (Fig. 2.2). The channels with langths up to 100–150 m actually serve as a trap for waves, where they finally loose their energy. The waves loose a part of their energy in supporting an elevated sea-level above the submerged flat where the pressure is maintained by water pumped up by waves through the channels and grooves. This difference between sea level above the flat and that of the ocean is about one-fifth of the height of oncoming waves. It means that at an average height of waves of 2–4 m this elevation of level above the reef flat could be 40–60 cm (Munk and Sargent 1948; Tait 1972). But during storms with wave heights of 8–10 m it could reach 1.5–2 m, and could exceed the height of the tide itself. The wave stress as well as the differences in sea-level caused by it induce powerful pulses of currents, having speeds up to 1.5–2 m s^{-1} (Roberts et al. 1975, 1983; Atkinson et al. 1981; Church et al. 1985; cf. Fig. 2.3). Such currents easily drag along not only gravel and rubble but also large coral heads,

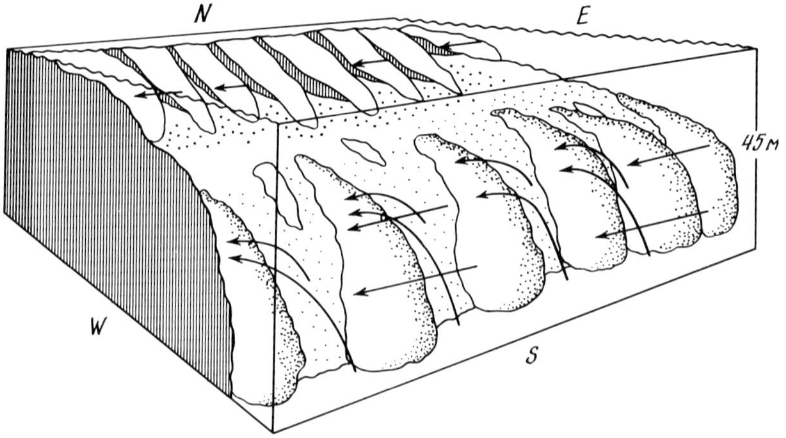

Fig. 2.2. Vectors of currents on outer reef slope in the buttress zone. (Modified after Roberts et al. 1975)

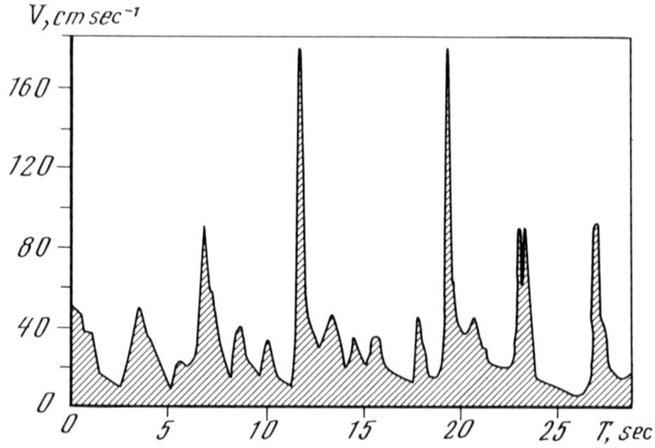

Fig. 2.3. Pulses of current over the flat to the lagoon; *V* current velocity; *T* time; (After Roberts et al. 1983)

causing the formation of ridges and keys on the reef crest. The flows, induced by waves on the flat, pass up it into the lagoeon, thus influencing water circulation also in the lagoon. The debit of water into the lagoon through the flat reachs 20–30% of total inflow at medium wave stress even in such a semi-open atoll as the Bikini (von Arx 1848). At the Enivetok atoll, which has two wide passes, nevertheless up to 80% of the water total comes into the lagoon over the flat, driven by the wave stress. The tides, being high enough there (1–1.5 m), provide anyway only 20% of the inflow (Atkinson 1981).

The impact of wave stress upon the geomorphological features and on the biological community structure of reefs is extremely significant in regions situated in the traditional paths of hurricanes (Stoddart 1962, 1974; Hernandez et al. 1977; Dollar 1982). These are the Caribbean reefs, those of the Philippines and of the southern GBR (Stoddard 1969; Woodley et al. 1981; Kjerfve and Dinnel 1983). In these regions the strong tropical cyclones with pressure in the centre of 945–955 mm are recurring every 3–10 years. After the passing of such a cyclone with waves more than 10 m high near the reef edge not only the ramose but also the massive corals are swept off the outer reef slope and flat. On the flat and in the lagoon huge ridges of rubble appear. Inside the lagoon during such a storm the height of waves will reach 3–5 m. These waves smash up the fine ramose corals. Destruction of corals proceeds not only through stress of water and waves but also by the moving masses of rubble and broken-out coral heads. The recovery of reefs after passing of a hurricane takes 5–10 years, but some reefs cannot repair the damage even in 10 years (Stoddart 1974).

Most coral reef biotopes exist in conditions of permanent and intensive hydrodynamical movement of water masses, caused by the interactions of various multidirectional and fluctuating currents with a complex relief of reef constructions. For the measurement of currents speed three main methods were used: winches (Novozhilov 1980; Wolanski and Jones 1980; Andrews and Gentien 1982), buoys, floats, empty bottles, drogues (Hamner and Hauri 1981; Atkinson et al. 1981), and dyes (vor Arx 1948; Roberts et al. 1975; Marsh et al. 1981; Atkinson et al. 1981; Oberdorfer and Buddemeyer 1985). The best results were obtained by their combinations. The turbulence exchange is quentified via analysis of the vertical thermohaline structure of the water column and also with the use of dyes and model analysis (Bode and Starck 1983).

The main driving forces in reef hydrodynamics are the oceanic currents, wind and wave stress, the short-period oscillations of sea level caused by tides, and by the long-period internal waves caused by the heterogeneity of basic fields in tropical shelf areas (Atkinson et al. 1981; Frith 1981). Their relative importance varies, being dependent upon the reefal morphometry, weather conditions in the region, and on the seasonal successions in the tide heights. Most reef regions are situated in areas of the wind-driven, reverse or compensative oceanic currents, such as the trade wind currents, equatorial counter-currents, and the East-Australian current. Their velocity varies in shelf areas within 10 to 50 $cm\,s^{-1}$. The interaction of an atoll situated in open ocean with a dominating current results in the upwelling of deep waters at its windward side. One of the causes of upwelling is the divergence of the oncoming current near the reef by two flows encircling the reef body (Hamner and Hauri 1981). The upwelling can be also induced by the interactions of the currents encircling the windward reef slope, with its systems of spoors at depth of 20–40 m (Roberts et al. 1975; cf. 1.1. Fig. 2.2). The oncoming current and the upward flows going through the grooves

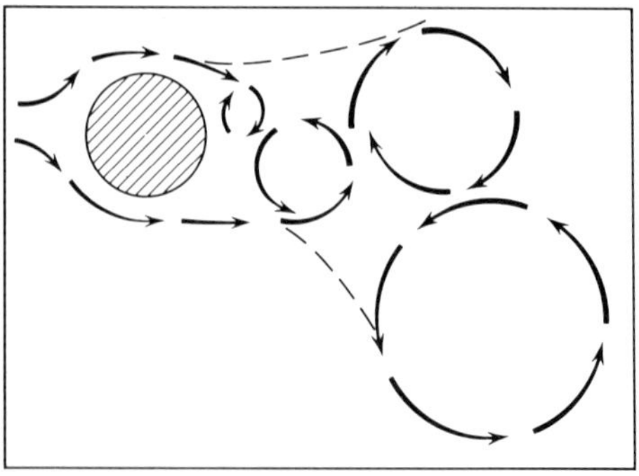

Fig. 2.4. Scheme of interaction of a ring reef with the coming oceanic current. (After Hamner and Hauri 1981)

of the windward slope induce also strong currents in the moat at 3–6 m depth with velocities up to 1–1.5 m s^{-1}. Where the moat is crossed by channels cutting the reef flat, the flows from it go up these channels and through them into the lagoon (Marsh et al. 1981). At the leeward side of the reef the encircling currents form a complex system of reciprocal eddies, which enhance the turbulence and thus also induce the vertical transport of deeper waters, bringing up the nutrients (Wolanski 1982; Fig. 2.4). These local upwellings, caused by the interactions of reef constructions with the oceanic current, result in the so-called island mass effect, the effect of enhanced productivity of plankton in waters near islands in zones of large quasi-permanent oceanic currents (Gilmartin and Relevante 1974; Hamner and Hauri 1981; cf. Sect. 3.2.2).

The action of wave stress as a driving force in hydrodynamic processes manifests itself mainly in the windward flat at the expense of dissipation of wave energy. As mentioned before, the sea-level in its frontal zone is supported 20–40 cm higher than in the surrounding ocean at wave heights of 1.5–4 m. This difference induces permanent currents across the flat to the lagoon with average velocities of 20–40 m s^{-1}, while that of the pulses along the channels could be 1–2 m (von Arx 1948; Roberts 1974; Maragos 1978; Marsh et al. 1981; Atkinson et al. 1981). The strength of these currents varies within the diurnal tide cycle (Fig. 2.3). It increases at high tide and decreasses at the ebb-tide. According to the last-mentioned authors the velocity of current crossing the windward flat of the Enivetok atoll varied within 10 to 150 cm s^{-1}. The rate of water transport during high tide was 1.5 and during the ebb – 0.05 (an average of 0.56 $m^3 s^{-1} m^{-1}$). During one tide phase via windward flat ca. 66 × $10^8 m^3$ of water or around 60% of its total daily debit arrived into the lagoon. For the Bikini atoll the corresponding share was 30% (von Arx 1948).

On reefs of atolls and on barrier reefs, situated in areas of a weak wave stress, the main driving force of flows over the flat are the tide currents.

Their velocity and direction change, these being dependent upon the height and on the phase of the tide. They move the water mass back and forth. During the incoming tide water goes into the lagoon first through the passages, then also up to the flat. During the ebb the current goes out from the lagoon. These inflowing currents are especially strong in the passages of the ring reefs and over their leeward reefs (von Arx 1948; Ludington 1979). The velocity of tidal currents could be high (up to 30–80 cm s^{-1}) even on reefs with relatively small tides of 20–40 cm (Maxwell 1968; Roberts et al. 1975; Thompson and Golding 1982). An average velocity of outflow currents over leeward reefs of an atoll could be 20–40 cm s^{-1} (Ludington 1979), but in the passages it could be much more – up to 2–5 m s^{-1} (von Arx 1948; Michel et al. 1972). The velocity of currents in passages of the elevated Fanning atoll was 40–80 cm s^{-1} (Gallagher et al. 1971). Over the flat and in channels of barrier reefs the average velocities of tidal currents varied within 30–50 cm s^{-1}, but in passages they could be up to 2 m s^{-1} (Maxwell 1968; Roberts et al. 1975; Cresswel and Greig 1978; Maragos 1978; Wolanski and Jones 1980; Geer and Kjerfve 1982; Thompson and Golding 1982; Marsden 1983). The periodicity of these currents sequential to the periodicity of tides (Fig. 2.3). On barrier reefs along the continental shelf the long-period barotrophic tidal waves, while causing the fluctuations of sea-level, also induce the long-period oscillations of currents with periods of 10 to 30 days (Cresswell and Greig 1978; Wolanski and Bennet 1983; Wolanski and Pickard 1983b). In lagoons of the ring reefs the main driving force of currents is the wind stress (Atkinson et al. 1981). Some influence on the lagoonal circulation, especially in shallow lagoons, is exerted also by wave-driven currents crossing the windward flat and tidal currents going through the channels (Frith 1981). In deep lagoons their influence is limited to zones near the windward flat and close to the passages (Fig. 2.5). In the surface layer the wind-driven drift prevails, while in deeper layers the compensation currents are induced. The speed of surface currents in lagoons varied within 5 to 20 cm s^{-1} (von Arx 1948; Ludington 1979; Atkinson et al. 1981). But in large atolls with deep and vast lagoons they could attain 40 cm s^{-1} and more (Gladfelter et al. 1978). In the lagoon of the Enivetok atoll the velocity of wind driven currents, as measured by the use of dye, was 50 cm s^{-1}, attaining sometimes even 1.4 cm s^{-1} (Odum and Odum 1955). In lagoons of

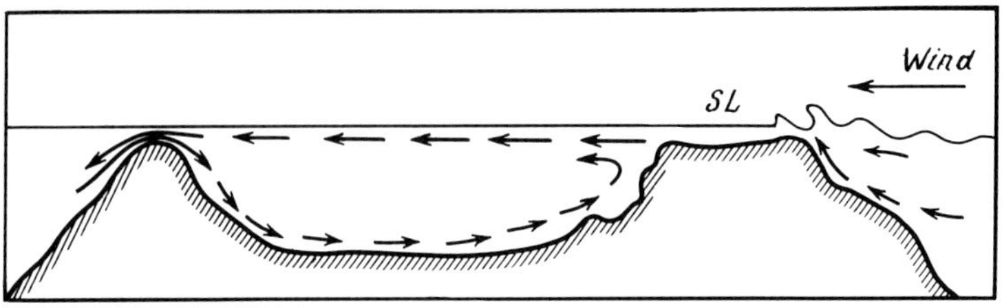

Fig. 2.5. Scheme of currents in the lagoon of atoll. (After von Arx 1948)

the barrier reefs the rate and especially the directions of currents, which are dependent upon the wind stress and the tide flows through the passages between the barriers, nevertheless are dominated by the influence of the main general oceanic current, which prevails in the shelf area. Thus in the lagoon of GBR the dominating direction of the current (SSE) coincides with the direction of the here dominating East-Australian current. In the lagoon of the barrier reef along New Guinea near Port Moresby the direction of current is determined by the here dominating oceanic SE current (Moore 1981). The rates of surface currents in lagoons of barrier reefs vary within 5–30 cm s^{-1} (Orr 1933; Roberts et al. 1975; Cresswell and Greig 1978; Andrews 1983a).

The vertical thermal stratification in reef waters is practically absent even in deep semi-closed lagoons (see below). This proves that there is a high level of vertical turbulence in their water columns. The zone of the most active turbulence is situated near the outer windward reef slopes. Measured by stroboscopic photography of a patch of dye, the coefficient of turbulence up the windward reef slope was more than $2 \times 10^2 cm^2 s^{-1}$ at depths of 20–30 m at a current velocity of 20 cm s^{-1} (Roberts 1977). But locally this coefficient in the vicinity of reefs can be even more – up to $3–5 \times 10^2 cm^2 s^{-1}$. In deeper layers the turbulence is usually more intensive, it being enhanced by the interaction of currents with elements of reef constructions. For example, near the reef off Koetivi Island (Seychelles) the turbulence coefficients near the surface were $0.5–1.7 \times 10^2$ and near the bottom $3.5 \times 10^2 cm^2 s^{-1}$ (Novozhilov 1980). The turbulence coefficients were maximal during the periods of incoming tide.

The residence time of water in the lagoons of the ring reef depends on the heights of tides, rates of currents, as well as upon their morphometry. Its estimations were based upon measurements of the rate of water transport across the flat and through the channels. In the semi-closed lagoon of the Bikini atoll during one tide cicle some 4% of the total water volume in it is exchanged. The residence time for this atoll was estimated to be within 40 days and the time for complete mixing of the water column by currents in its 60 m deep lagoon 6–12 days (von Arx 1948). The longest exchange rates were recorded in the shallow lagoon of the One Tree ring reef – less than 1 week (Ludington 1979; Frith 1981). At the Enivetok atoll the rate was 30 days, in the Canton atoll 50 days (Smith and Jokiel 1978), in the Tarawa atoll 10–15 days (Kimmerer and Walsh 1981) and in the elevated Fanning atoll, which has only one narrow pass, about 8 months (Gallagher et al. 1971; Atkinson et al. 1981). The exchange of water between lagoon and the surrounding ocean proceeds not only via surface hydrodynamics, but also through the porous walls of reef via the moving groundwaters (Lam 1974; Rougerie 1985). Their flows are induced by the fluctuations of sea-level during the tides. The flows of groundwaters inside the lime body of the reef dissolve the carbonates and thus perforate it, creating its permeability (Johannes 1980). The most permeable for the underground passage of oceanic waters into the lagoon and back appeared to be the lime rocks of ancient Pleistocene reefs (Oberdorfer and Buddemeyer 1985).

The salinity in reef waters is controlled by several factors, mainly by evaporation and by their mixing with the and with fresh water from rivers. The evaporation occurs on reefs very extensively, especially in overheating shallows and on flat areas which drain during the ebbs. The refreshing of reef waters happens most often on reefs near continents or high islands, such as Tahiti, New Guinea, and Hawaii. The rains are especially intensive during the winter, increasing the discharge of fresh water by rivers to the reef areas, and thus decreasing drastically the salinity of reef waters, as it often happens in the region of the GBR (Wolanski and van Senden 1983). Periodial drops in salinity are often the cause of the mortality of corals in the same regions. The seasonal fluctuations of salinity in lagoons of barrier reefs under the influence of rain and river waters could attain 3–8‰. In the lagoon of the GBR in summer salinity grows by up to 36‰, while in late autumn it drops to 28–32‰ (Fig. 2.6), when the surrounding coral sea has a salinity of 35.4‰ (Brandon 1973). During periods of river floods in winter it could also drop by 1–3‰ (Wolanski and van Senden 1983). In lagoons of atolls with an intensive water exchange, like the Bikini atoll, the salinity in the lagoon is equal or slightly more than in the surrounding ocean (Ford 1949). But in the closed or semi-closed lagoons under the influence of

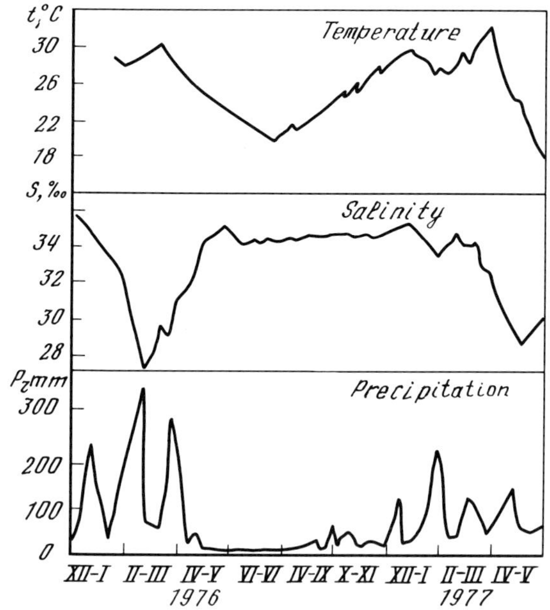

Fig. 2.6. Seasonal changes in water temperature ($t°$, C), salinity (S) and precipitation (P_r) in the lagoon of the GBR, near Townsville in 1976–1977. (After Relevante and Gilmartin 1982)

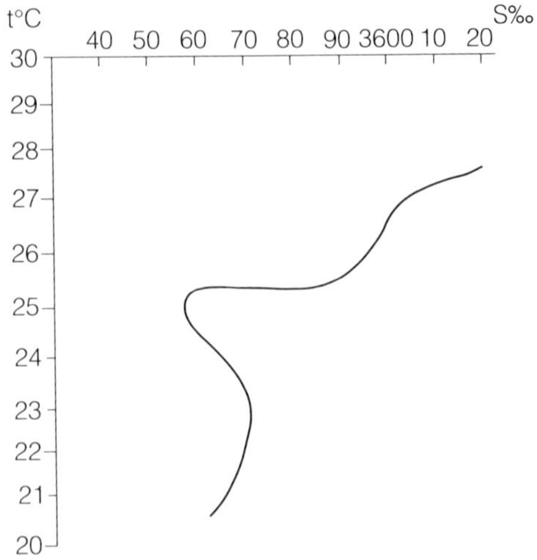

Fig. 2.7. T/S curve in surface waters of the southern GBR; average data in Sept.–Oct. 1967. (After Brandon 1973)

evaporation it could increase to as much as 38–40‰, as in the lagoon of the Canton atoll (Smith and Jokiel 1978). Further increase of salinity in lagoons of such atolls with a residence time of water of 1–2 years is prevented only by the water exchange through the porous lime body of the reef (Salvat 1981b; Rougerie 1985). The vertical gradients of salinity in lagoons are usually very low because of an extensive turbulence. They vary within 0.05–0.2‰ per 50 m (Ford 1949). The gradients may increase during the rainy seasons in their coastal areas, but they are quickly leveled by turbulence.

Degrees of temperature and salinity variations in reef waters are reflected well by the T/S curves. One of these for the GBR lagoon is given in Fig. 2.7. It denomstrates significant fluctuations of t° and S even within one region and during one season. The fluctuations are caused by large diurnal changes of water temperature in shelf areas as well as by extensive evaporation. The temperature of shelf water and the thermal stratification are also influenced in this area by fluctuations of the East Australian current, the barotrophic oscillations of water level, and also by the variations in the t° and S of waters coming from the ocean into the lagoon (Andrews and Gentien 1982; Wolanski and Bennett 1983). The scale of fluctuations of the temperature itself could be large even within the same region (Pickard 1983), and especially on the high-latitude reefs situated near the boundaries of coral reef growth (Potts and Swart 1984). For example, on reefs of the Persian Gulf seasonal temperature of the water varies from 16 to 40°C (Kinsman 1964), at the Abrolhos reefs (S.W. Australia, 28° S) 17 to 28 °C, and in the southern GBR 23° – 24° S from 18 to 30 °C (Pickard 1983). But in shallows of the reefs in low latitudes even the diurnal fluctuations of temperature reach 5–8 °C, while in deep lagoons and up the outer slopes they are negligible. The vertical stratification there is also slight or absent. The difference in t° even at depths of 40–60 m does not exceed 1–2° and even this change starts from depths of 20–30 m downwards. Some decrease of t° of the deep lagoonal waters is caused by the inflow of the deeper cold water during the tide (Andrews 1983b).

The lime reef constructions are subjected to powerful hydrodynamic stress, which partially destroys them and enhances the processes of erosion. But in its interactions with the reef this physical stress also stimulates many important biological processes, induces the development of defence reactions in reefal flora and fauna. These reactions of the biota finally result in compensation of the physical damage caused by hydrodynamic stress to the reef, thus providing and even enhancing its growth and morphogenesis (Heckel 1974). Among such compensative reactions are the acceleration of coral growth in reefal zones with active hydrodynamics – on buttresses, on the fore-reef, and at the outer windward reef slope (Wells 1957; Goreau and Goreau 1973; Roberts et al. 1975). Extensive hydrodynaimcs and high transparency of the waters over the outer slope provide better conditions for autotrophic and heterotrophic feeding of corals and a decrease in their siltation. Being adapted to the wave and currents stress corals form in these zones dense communities. A rich coral growth stimulates the growth of the reef body on the outer slope and especially the growth of buttresses – one of the key geomorphological structures of the reef. Wave stress and the elevation of sea level up the frontal reef edge are the main factors instrumental in the development of algal ridges constructed by the coralline algae (cf. Sect. 1.1). The wave stress up the ridge keep out fish and urchins, which feed on coralline algae, thus preventing their grazing. It is the elevated sea level up the reef edge which permits algae to construct the elevated ridge along it. Being cross-cut by channels, the ridge dissipates the wave energy and thus preserves the corals that grow on the submerged flat from destruction.

Under the influence of the wave stress and currents a sorting out and redistribution of the carbonate material occur (Maxwell 1968; Gallagher et al. 1971; Roberts et al. 1975; Smith 1973; P.J. Davies 1977). Most important in these processes are the pulses of currents with a velocity of $1–2\,m\,s^{-1}$, generated by waves on the reef flat with a periodicity of 3–5 s (cf. Fig. 2.3). These powerful pulsing currents transport the rubble, collect it in ridges had heaps. By flushing them the currents prevent their being covered by soft sediments. In these conditions the rubble is quickly consolidated by sponges and by calcareous algae which fix it onto the flat surface, thus providing for the growth of the reef body. In submerged parts of the reef flat on the consolidated rubble material starts the growth of corals. These processes compensate for the erosion and destruction of reef at sites of extensive hydrodynamic stress (Marshall 1968).

Active hydrodynamics on coral reefs is one of the most important factors ensuring their high productivity. The interaction of reefs with the oceanic currents, as well as the tidal and the barotrophic long-period oscillations of the sea-level in the vicinity of reefs bring about the elevation of deep oceanic waters rich in nutrients to the surface (Wolanski 1982). It proceeds not only via formation of eddies and divergences of current after these interactions, but also by the driving up of waters from deeper layers by strong tidal currents. The possibility of this increases in regions of barrier reefs subject to the influence of large vertical movements of the upper boundary of the thermocline caused by long-period barotrophic shelf-internal waves. For example, the amplitude of vertical shift of the thermocline boundary off the outer edge of the GBR up to the depth of 300 m could be 100 m (Wolanski and Pickard 1983b). The input of nutrients through the hydrodynamic processes of advection of deep waters in the vicinity of reefs is one of the important factors supporting high plankton productivity in reef waters (cf. Sect. 3.2.2). Interaction of the ring reef with incoming oceanic current results in the formation of a series of rings. When going downstream their diameter increases. They cease at a distance 10–15 times exceeding the diameter of the ring reef itself (Fig. 2.4). The formation of these rings supports the development of plankton and influences the distribution of pelagic fish in shelf areas off atolls and small high islands (Alldredge and Hamner 1980; Hamner and Hauri 1981).

The oceanic currents as well as the local tidal- and wind-driven surface currents support a permanent flow of plankton over and through the communities of benthic planktonovorous animals including corals and planktonovorous fish, being thus one of the most important factors of trophodynamics in the reef ecosystem (Williams et al. 1984). The back-and-forth tidal currents accelerate the exchange of the dissolved and particulated organic matter between the reef ecosystems and the pelagic ecosystems of the surrounding ocean (Le Borgne et al. 1989). Transporting and redistributing the organic matter and plankton inside the reef system, currents provide energy interconnections between different biotopes and different sites of the reef ecosystem, thus uniting it as a whole (Sorokin 1977a).

2.2 Nutrients

Ever since Charles Darwin the enigma of the existence of a rich life on coral reefs surrounded by surface oligotrophic oceanic waters depleted of nutrients became a point of heated discussions and investigations (Stoddard 1969; Andrews and Gentien 1982; Crossland 1983; Atkinson 1983; Wiebe 1985). And, in fact, the photosynthetic primary production of the powerful reef autotrophic associations of plants and symbiotic animals approaches the highest level known for marine and terrestrial ecosystems: 5–10 $g\,C\,m^{-2}\,day^{-1}$, and sometimes even more, while the ambient concentrations of inorganic nutrients (NO_3^-, NO_2^-, NH_4^+, PO_4^{2-}) in waters of the surrounding ocean usually happens to be very low. Most often it drops below a level that limits the growth of phytoplankton (<0.5 μ mol NO_3 and 0.2 μ mol PO_4). Moreover, the level of primary production of reef-bottom autotrophs actually does not depend much upon the level of standing stocks of inorganic nutrients in reef waters (Kinsey 1977, 1983; Sorokin 1977a, 1984b; Wiebe 1985; Atkinson 1988). It seems that rich reef benthic communities have to lose the nutrients as salts and also as organic forms, being readily flushed out by blue oligotrophic waters. The outflow of organic detritus and plankton driven out by currents from the reefs was also proved (cf. Sect. 2.3). But the very existence of the coral reef ecosystems a priori needs the positive balance of nutrients. While they actually exist and flourish, a positive balance of nutrients in their exchange between the reef and the ocean should be axiomatic. Thus we need not even prove it, but we do have to understand its mechanisms. Already a first approach to the problem proves that these mechanisms should be mostly biological rather than physical, e.g., such as are controlling the input of nutrients to the euphotic zone in the pelagic communities.

At present, we also know in general the biological mechanisms which govern the input and the turnover of nutrients and provide for their positive balance in the reef ecosystems (Wiebe 1985; Sorokin 1986b, 1990b). The level of their content in waters over the reef is also well known (D'Elia 1978; Crossland 1983), but data on the stock of their organic forms in organic matter and in the biomass of organisms are still scarce (Hatcher 1985; Crossland et al. 1991). Probably, in some reef areas, iron could also be a limiting element, besides N and P (Entsch et al. 1983).

2.2.1 Ambient Concentration in Waters

Basic data are available mostly for the reefs of the Pacific basin: the Great Barrier Reef (Relevante and Gilmartin 1982; Andrews 1983b; Crossland and Barnes 1983; Crossland 1983; Hatcher 1985; Sorokin 1989, 1990a; Furnas et al. 1990), French Polynesia (Sournia and Ricard 1976b; Ricard 1977c, 1981; Salvat 1981b), some atolls (Johannes et al. 1972; Pilson and Betzer 1973; Wiebe et al. 1975; Marshall et al. 1975; Smith and Jokiel 1975; Kimmerer and Walsh 1981; Charpy and Charpy-Ronband 1988), Hawaiian reefs (Gilmartin and Relevante 1974; Henderson 1981; Atkinson 1981b, 1983), and high latitude reefs of Abrolhos (Crossland et al. 1984). The ranges of corresponding data are summarized in Table 2.1, which shows that the level of the standing stock of nutrients is equally low up the reefs, situated in waters of a very diffent level of planktonic productivity, and even up polluted reefs (Hawaii). This leads to the conclusion that the ambient concentrations of inorganic nutrients in waters over the reef, being values in equilibrium, reflect only the direction of the turnover but say nothing about the flow rates or about the real size of its pool in the system. Thus, they

Table 2.1. Level of ambient concentrations of dissolved organic and inorganic nutrients in waters over coral reefs and in surface waters of the surrounding ocean ($\mu mol\, l^{-1}$)

Region	Inorganic nutrients			Organic nutrients	
	PO_4-P	NO_3-N	NH_4-N	P	N
Lagoons of atolls	0.03–0.8	0.05–2.4	0.1–1.3	0.10–0.20	2.0–6.0
Lagoons of barrier reefs	0.02–0.7	0.02–2.7	0.05–2.7	0.15–0.25	3.1–4.6
High latitude reefs off East and West	0.03–1.8	0.1–5.0	0.1–11.0	0.14–1.6	3.0–21.9
Australia	0.16–1.36	0.05–2.6	0.4–2.4	0.37	3.4–7.5
Lagoon of polluted reefs	6.1–0.4	0.2–0.8	0–0.4	0.10–0.15	1.0–4.0
Tropical oceanic surface waters					

cannot be actually used for solving the basic problem concerning nutrients – the problem of nutrients supply of the primary products. The content of PO_4-P in reef waters usually is the same as or little more than in surface waters of the surrounding open ocean: $0.1–0.4\,\mu mol\, l^{-1}$, which is close to the level limiting phytoplankton growth – $0.1–0.2\,\mu mol$. Therefore, the phosphorus alone should be the main limiting nutrient in reef planktonic communities. As to the bottom autotrophes, they seem not to suffer at all from nutrient limitation, as this is supplied from their deposits through processes of regeneration (Entsch et al. 1983; Smith 1984). Actually, the plant-macrophyte components of the autotrophic bottom communities intercept mostly the flow of nutrients from the bottom as a result of the processes of decomposition of their organic fractions caused by bacteria and benthic fauna (see below). Thus, they are much less dependent upon the horizontal flow of nutrients in oceanic waters passing over the reef (Atkinson and Smith 1987). The periphytonic communities take organic nutrients and also recycle them internally inside the system as corals do as well. Finally, the sea-grasses with their rhizoids and the microphytobenthos take in the nutrients directly straight from the bottom sediments.

The standing stock of inorganic nitrogen in the form of NO_3, NO_2 and NH_4 in reef waters varies between $0.6–1.5\,\mu mol\, l^{-1}$. It is usually more than in the surrounding ocean (Webb et al. 1975; Andrews and Muller 1983; Szamant-Froelich 1983; Hatcher and Frith 1986; Furnas et al. 1990; cf. Table 2.1). Thus it exceeds the level of phytoplankton limitation ($0.5\,\mu mol\, l^{-1}$). The dominant form of nitrogen in reef waters is NH_4^+, while in oceanic surface waters nitrate. The highest content of ammonia in water was recorded up the high-latitude Abrolhos reefs and up the One Tree reef: up to $9–11\,\mu mol$ (Johannes et al. 1983b; Hatcher and Hatcher 1981). In shallow tide pools it might reach over $16\,\mu mol$. The reef and the pool biotopes, overgrown by macrophytes and by periphyton and enriched also with plant detritus, are the sites of intensive microbial nitrogen fixation, which is the main source of this ammonia enrichment. The NH_4^+ thus produced is then oxidized to NO_3 by nitrifying microflora which is localized in the upper layer of bottom sediments and in the periphytic overgrowth (Webb and Wiebe 1975; Webb et al. 1975; Atkinson 1981b, 1983).

Ambient nutrients concentration in water is relatively stable up the different reef biotopes (Odum and Odum 1955; Pilson and Betzer 1973). A significant increase – up to 1 to 3 μmol of PO_4-P – was observed only in the lagoons of barrier reefs due to the influence of the drainage of rivers during the rainy season, and in lagoons of atolls situated in the equatorial upwelling area (Ricard 1981; Kimmerer and Walsh 1981). Nutrients, and especially the limiting one – phosphorus, are present in reef waters also in dissolved and suspended organic matter in their organic forms. Oceanic surface waters contain 0.1–0.2 μmol of dissolved organic phosphorus and 1–4 mol of organic N (Table 2.1). In reef waters the content of P-org is about the same, while the N-org content could be even more: 3–6 μmol. Making use of organic nutrients proceeds in coral-reef ecosystems by the participation of microbial communities and through the action of the coral-reef biofilter (Sorokin 1978b, 1986; cf. Sect. 3.1.2).

2.2.2 Mechanisms of External Input of Nutrients

The input of nutrients into the coral reef ecosystem, or the debit part of their balance, is provided by several interacting physical and biological mechanisms. The nutrients are brought to the reef with oceanic waters driven up the reef by the wind and by the geostrophic currents (Johannes et al. 1972; Kimmerer and Walsh 1981; Hatcher 1985). The importance of this source of input increases in areas where the surface waters are enriched with nutrients through their elevation from deeper layers in areas of quasi-permanent upwellings and divergences in the equatorial zone. The same happens in the vicinity of an island's systems interacting with the geostrophic oceanic currents – so-called "island mass effect" (Gilmartin and Relevante 1974; Andrews and Gentien 1982). The nutrients of the deep oceanic water may penetrate into the closed or semi-closed lagoon of the ring reef traveling with the water that is filtering into it through its porous base during high tide, when the level of water in the lagoon is lower than outside in the ocean (Rougerie 1982, 1985). The waters of the barrier and fringing reefs off the high islands and continental coasts get a significant input of nutrients with the river and ground drainages (Marsh 1977; Johannes 1980; D'Elia et al. 1981; Johannes and Hearth 1985). At atolls that are situated in productive waters a significant nutrients input is provided by the sea birds, which form there big colonies. The PO_4-P content in such reef areas could be up to 1 $\mu mol\,l^{-1}$. The ancient reefs off some islands contain large deposits of phosphorite ores, which are 200–800 thousands of years old (Veeh 1985). They were discovered at the Nauru and the Christmas Islands. On a smaller scale phosphorites can be found in reef rocks of atolls. On being eroded they release phosphates, which are driven out with groundwaters and thus enrich the water column over the reef with phosphorus (Gordon et al. 1971). An important role in the mobilization of the nutrients deposited in bottom sediments is their permanent resuspension by means of wave action (Fanning et al. 1982; Entsch et al. 1983). In some reef areas a significant input of nutrients is produced by accumulation of masses of the floating wind-driven blue-green algae Trichodesmium (Sorokin et al. 1982).

The processes listed above provide a permanent flow of nutrients, both inorganic and organic in their forms, from the external sources and from the deposits within a coral reef ecosystem. The consumption of nutrients from this flow, their retention and regeneration are accomplished by biological mechanisms. One of them is the consumption of inorganic N and P by phytoplankton and bacterioplankton, which are present in coral-reef waters. They consume PO_4-P in about the same proportion of 1–2% (by weight) to their production in carbon units per day (Sorokin 1985, 1990a). Direct estimations of its consumption by microplankton of coral-reef waters by the use of labeled phosphate showed (Table 2.2) that it is quite significant: $0.03–0.15\,\mu mol\,l^{-1}\,day^{-1}$. The microplankton, which consumes phosphorus from sea water passing over the reef, is itself then grazed by reef filter feeders. Thus, the nutrients of oceanic waters become included via the food web into the nutrients pool of the reef ecosystem mainly in the form of biomass and feces.

The nutrients in oceanic waters coming over the reef are also utilized by benthic plant communities, such as periphyton, microphytobenthos and seaweeds (Muscatine et al. 1979). The periphytonic associations include numerous nitrogen-fixing bacteria and cyanobacteria which assimilate the atmospheric nitrogen. They harbor also a rich heterotrophic microflora, which uses for its growth mainly the dissolved organic matter taken straight from the oceanic waters, and thus utilizes also the organic forms of N and P which it contains (cf. Table 2.1). The macrophytes consume inorganic nutrients from oceanic waters by the surfaces of their thalloms (Pilson and Betzer 1973; Nelson 1985) and may accumulate them as stock (Robishaux 1985). The seagrasses get nutrients mainly from the sediments through their roots. But their leaves usually bring with them a lot of epiphytic macroalgae, which consume nutrients also from water columns (Short 1981).

The symbiotic animals, such as corals, gorgonaceans, tridacna clams, may consume inorganic nutrients in their ambient concentration in waters passing over the reef: $0.3–1\,\mu mol\,PO_4$, $0.4–2\,\mu mol\,NO_3$ and $0.8–5\,\mu mol\,NH_4$ per liter. The consumption of inorganic nutrients by these animals was established by observations of the decrease in concentration in their presence (Kawaguti 1953; Webb and Wiebe 1978; Muscatine and D'Elia 1978; Propp 1981; cf. Fig. 2.8), or by the measurements of consumption of radiolabeled N and P compounds (Pomeroy et al. 1974; Sorokin 1973c; D'Elia 1977; Propp 1981; Burris 1983; Wafar et al. 1985; Atkinson and Smith 1987). The ahermatypic corals, which do not have zooxanthellae, only

Table 2.2. Rate of consumption of inorganic phosphate by total microplankton (P_{at}) and its consumption only by bacterioplankton (P_{ab}) in waters of Heron I. and Wistari ring reefs, GBR, during the mid ebb-tidal phase; C_p – primary phytoplanton production; K_p – PO_4-P contents in water; T – its turnover time, days (original data by author)

Place, time, temperature of water °C	Biotope	Depth, m	K_p µ mol P l^{-1}	C_p mg C m^{-3} day^{-1}	P_{at} µg P l^{-1} day^{-1}	P_{ab} µg P l^{-1} day^{-1}	P_{ab}/P_{at} %	T, days
Heron I[1], August; t° 21–22 °C.	Deep part of lagoon	6.5	0.90	14.5	0.33	0.27	83	9.3
	Edge of lagoon, patch reef	1.5	0.34	11.2	1.22			9.0
	N-W corner of lagoon	4.5	0.40	6.8	1.18	0.80	68	11.0
Heron I., October; t° 24–25 °C	Open sea 200 m off reef edge	18.0	0.18	13.7	3.44	2.33	68	1.7
	Over reef crest	0.2	0.13	14.2	1.57	1.17	74	2.5
	Shallows near the island's shore opposite tourist resort	0.5	0.30	7.7	2.0	1.92	48	4.5
Wistari atoll, lagoon; August t° 21–22 °C	Deep part of lagoon	6.0	0.38	9.5	2.91	2.50	86	4.1
	Windward slope, patch reefs	2.0	0.22	4.2	1.02	0.88	87	7.0
	Leeward shallows	0.7	0.18	12.4	1.06	0.84	79	5.7

Fig. 2.8. Consumption of ammonia from water by coral *Pocillopora* in the light (*1*) and in the dark (*2*), as well as its release by ahermatypic coral *Tubastrea* in the light (*3*); *N* NH_4 content in water; *T* exposure time. (After Muscatine and D'Elia 1978)

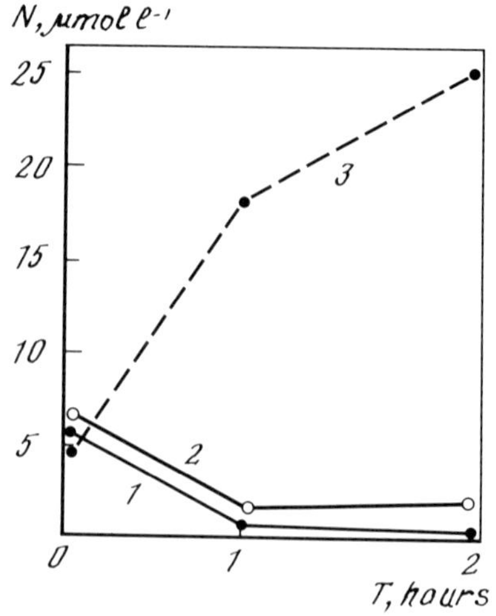

excrete N and P but cannot consume them (Muscatine et al. 1979; Muscatine 1980; cf. Fig. 2.8). The saturation curves of consumption of PO_4, NO_3 and NH_4 by corals are similar to the Michaelis-Menten kinetic curve (D'Elia 1978; Webb and Wiebe 1978; Muscatine and D'Elia 1978). This indicates that the consumption proceeds mainly via the enzymatic active transport mechanisms. The threshold concentrations at which the consumption starts are within 0.1–0.2 $\mu mol\,l^{-1}$, and that of V_{max} 1–2 $\mu mol\,l^{-1}$, which is the same range as in the oceanic phytoplankton. The rate of consumption of ammonia by corals was twice as much as that of nitrates in their ambient concentrations (D'Elia and Webb 1977). The ions of NO_3, being consumed by the epithelial cells of polyps, are then transferred into the cells of zooxanthellae, where they are reduced to ammonium and involved in the biosynthesis via glutamine as an intermediary. A large portion of the amino acids thus produced is then translocated back to the polyp's cell (Muscatine 1980; Crossland and Barnes 1977a). The incubation of coral with $N^{15}O_3^-$ resulted in the accumulation of N^{15} in its tissues in organic form (Marian 1979).

The inorganic phosphate consumed by corals is accumulated in the cells of zooxanthellae, where it is incorporated into ATP and nucleotides, which are then largely translocated into the hosts's cells. In them this phosphorus is mineralized to PO_4-P, which is again consumed by zooxanthellae, thus closing the cycle. The above mechanism provides the semi-closed cycle of nutrients in symbiotic animals, which drastically decreases their loss by such animals to the surrounding waters. Nevertheless, even the symbiotic animals (corals) excrete some amounts of PO_4-P and NH_4-N. Therefore, the above-mentioned estimations of their uptake by corals by measurements of the decrease in their content present in water are conventional. The values thus received have reflected not a real uptake, but an indefinite equilibrium value of uptake and release. Moreover, corals excrete not only inorganic but also a sighificant amount of organic nutrients (D'Elia 1977). The real uptake could be measured only with the use of labeled N and P compounds (Pomeroy et al. 1974; D'Elia 1977; Burris 1983; Atkinson 1987a). We have measured flow rates of PO_4-P between corals and surrounding water using radiolabeled phosphate $^{32}PO_4$. The experiments showed that at low concentrations of PO_4-P, close to the ambient ones in coral-reef waters ($0.1-0.3\,mol\,l^{-1}$), its change in water during the exposure of corals does not reflect real rates of PO_4-P uptake at all, because the consumption is masked by its simultaneously proceeding excretion. Moreover, at low concentrations of PO_4-P the loss often exceded the uptake and thus the content of PO_4-P in water in the presence of coral was not decreasing, but even increasing (Table 2.3, Fig. 2.9). Even with concentrations of PO_4-P higher than the ambient ones the values of uptake obtained by the simple recording of the decrease in its content in water would be in any case underestimated. The absolute rates of the uptake flow of PO_4-P (Ac) from water to coral in its concentration close to natural (0.16–0.30 mol) were significant: 10–$75\,\mu g\,P\,kg^{-1}\,h^{-1}$, while the net result of uptake-release was oscillating around zero: +14, $-28\,\mu g\,P$. The uptake rates as measured with the radiophosphorus method in dark were a little lower than in the light, but the release increased (Fig. 2.9) and both reciprocal processes were at equilibtium. In this case the content of PO_4-P in water in the presence of coral was stable, while its flow rates between coral and water remained about the same as in the light (Fig. 2.9).

Besides the release of organic nutrients corals excrete also the organic N–P compounds (Johannes and Webb 1970; Muscatine and D'Elia 1978).

Table 2.3. Flows of PO_4-P between corals and surrounding water in experiments with radiolabeled $^{33}PO_4$; t° of water 22 °C; K_P – initial PO_4-P content in water, μmol; A_c – uptake; A_e – release; $\pm A_t$ – net changes of PO_4-P content in water (after Sorokin 1989)

Species of coral	K_p	Elements of PO_4-P balance (flow rates) $\mu g\,kg^{-1}\,h^{-1}$		
		A_c	A_e	A_t
Pocillopora damicornis	2.0	170.3	82.5	−87.8
	0.3	75.1	46.5	−28.6
	0.06	3.6	6.4	+2.8
Stylophora pistillata	3.0	53.0	36.6	−16.4
	0.3	28.9	13.9	−14.5
Porites andrewsi	0.16	29.5	44.2	+14.7
Aeropora squamosa	0.16	12.6	11.4	−1.2
Cladiella sp. (symbiotic alcyonacean)	0.16	92.8	97.5	+4.7
Pacifigorgia (ahermatypic gorgonacean)	0.26	78.6	2019.6	+1941.0
Leptogorgia sp. (ahermatypic gorgonacean)	0.26	34.7	2166.6	+2132.0

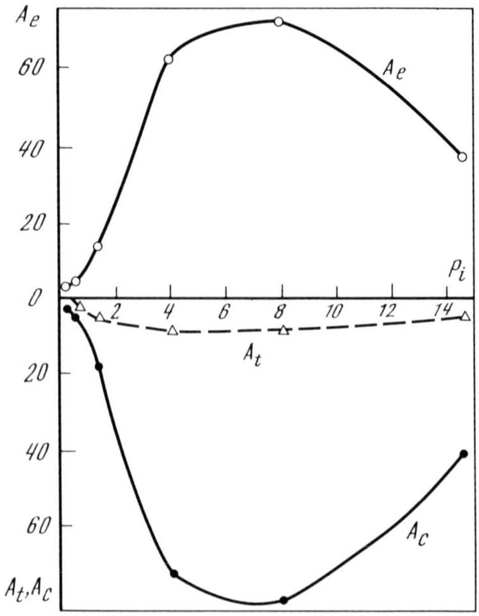

Fig. 2.9. Dependence of the consumption (A_c) and release (A_e) of PO_4-P ($\mu g\,P\,kg^{-1}\,h^{-1}$) by coral *Pocillopora damicarnis* upon its content in water (P_i, $\mu g\,l^{-1}$); A_t net consumption of PO_4-P by coral. (After Sorokin 1990c)

But the net release of N and P by symbiotic animals is much less than by asymbiotic ones (Lewis 1973; Muscatine and Porter 1977). Thus, the semi-closed nutrients cycle in corals and other symbiotic animals improves their total balance in the coral-reef ecosystem. It provides input of nutrients from oceanic waters via their uptake and the preservation of their stock, decreasing their losses from the system as regards the corals themselves. Their uptake of the inorganic nutrients from water is only an additional way of supply, which only compensates their loss during metabolism (D'Elia 1977; Propp 1981). The main source of nutrients supply for corals remains their heterotrophic feeding.

Finally, one of the most powerful mechanisms of nutrient input from external sources into the reef ecosystem is operating by food-web trophodynamics. It was also pointed out above that inorganic nutrients of oceanic waters, after being assimilated by phytoplankton and bacteria, enter the reef food-web through the activity of filter-feeders. The same mechanisms provide a permanent flow of external nutrients in their organic forms. The dissolved organic nutrients, such as nucleic acids. amino acids and amines, enter the reef food-webs via their assimilation by bacteria and by their subsequent uptake by filter-feeder. The second important mechanism is the direct assimilation of DOM by reef filter-feeders which have ciliated epithelium, such as trochophores, echinocardiums, corals, clams, serpulid and sabellid worms (Stephens 1962; Sorokin 1977b, 1984b; Sorokin and Wyshkwartzev 1973). The suspended organic nutrients are consumed by sedimentary and filter-feeders. Nutrients contained in the biomass of zooplankton, in passing over the reef oceanic waters, enter the nutrients pool of the reef ecosystem through their consumption by planktonic and benthic predators – mainly by corals and planktonovorous fish (Hamner et al. 1988). These fish and also larvae of many bottom fish and various bottom fauna while grazing zooplankton exrete feces, containing a bulk of nutrients consumed, which then are mineralized via the food-web by animals, which feed of feces, including the corals themselves (Meyer et al. 1983).

2.2.3 Nitrogen Fixation

The assimilation of atmospheric nitrogen by nitrogen-fixing microbial and cyanobacterial periphytonic and endolithic associations wide-spread in bottom reef biotopes is the basic process in building up the positive balance of nitrogen in reef ecosystems. The first observations on the dynamics of nitrogen in waters passing over the actively functioning reefs, even with a predomination of macrophytes that had to consume inorganic nitrogen from them, nevertheless invariably recorded a permanent and intensive release of nitrogen by them, mainly in the form of ammonia and in organic compounds (Johannes et al. 1972; Webb et al. 1975; Wiebe et al. 1975; Wiebe 1976). The conclusion that this enrichment of oceanic waters with nitrogen passing over the reef is the sequel of an intensive N_2-fixation in its bottom biotopes seemed to be quite natural, accounting for the abundance of nitrogen fixing cyanobacteria in them (Hatcher and Hatcher 1981; Paerl et al. 1981). The reef-bottom biotopes enriched with organic matter are also optimal sites for the growth of azotobacter. Direct estimations of rates of N^2 fixation were made with the aid of N^{15} (Burris 1976) and with the acetylene method (Hanson and Gundersen 1976; Capone 1977; Pehnhale and Capone 1981; Wilkinson et al. 1984). They showed that the most active N_2 fixation takes place in periphyton, which covers the rubble, the dead corals and the flat rocks, as well as in the upper layer of coarse Halimeda sands. The periphyton overgrowing these substrates contains a large quantity of various N_2-fixing blue-green algae such as: Stychonema, Calotrix, Nostoc, Hermothamnion, Hyella, Oscillatoria, Symploca, Lyngbia. The rates of N_2-fixation in these sites varied within 1–9 (N^{15}-method) and 10–30 $\mu g\,N\,g^{-1}\,h^{-1}$ (acetylene method). Very high rates of N_2-fixation were recorded in zones where the macrophytes had been grazed out by fish and were then replaced by periphyton (Wilkinson et al. 1984). Some sponges harboring the cyanobacterial symbionts appeared to be able to fix atmospheric N_2 (Wilkinson and Fay 1979). In various reef biotopes the

Table 2.4. Calculated values of annual N_2-fixation (kg N hectare^{-1}) according to results of estimations in samples of bottom sediments and in periphyton

Substrate, biotope	N_2-fixation		Reference
	Ranges	Average	
Periphyton on rubble	18–368	–	Mague and Holm Hansen (1975)
Same	41–506	493	Wiebe et al. (1975)
Same	3–14	–	Burris (1976)
Periphyton on rocks (windward flat)	41–2500	684	Hanson and Gundersen (1976)
Same (leeward reef)	0–20	4.4	Wilkinson et al. (1984)
Same (outer slope)	2–153	20	Wilkinson et al. (1989)
Coral sand	3–384	91	Potts and Whitton (1977)
Same	4–40	15	Odintsov and Propp (1985)

2.2 Dynamics of Nutrients

production of bound initrogen via N_2-fixation was evaluated within 500–900 kg hectare^{-1} year^{-1} (Table 2.4). These data, obtained employing the acetylene techniques could be 1.5–2 times overestimated (Burris 1976). But even at its more real rate of 100 kg ha^{-1} year^{-1} the production of bound N_2 will be 30 mg N m^{-2} or 0.2–0.3 μ mol l^{-1} day^{-1}, that is 30–40% of its stock in a water column 2–3 m thick. The share of N_2-fixation in the nitrogen supply of the reef bottom plant associations could be evaluated within 10–30%.

2.2.4 Dynamics of Nutrients

The basic turnover of nutrients within the coral reef ecosystem proceeds via their transformations in the food-webs. The processes of nutrients turnover going on in the water column starts with the consumption of their inorganic salts by the phytoplankton and bacterioplankton. Thus the dispersed dissolved nutrients are transformed into the concentrated particulated matter, in the biomass of microorganisms, and in part into the excreted, by microplankton dissolved, organic nutriens. Its larger portion is mineralized via respiratory decomposition, by planktonic and benthic filterers and by bacteria of bottom sediments (Atkinson 1987a). The efforts of researchers were directed mostly toward studies of nutrients cycling only in bottom communities, while the processes which take place in water columns over the reef were, in fact, ignored (Pilson and Betzer 1973). But the measurements of the uptake of PO_4-P in waters over the reef and the calculation of its turnover times showed a fast cycling of phosphorus precisely in the water column. The estimations have been made in waters over the reefs of Heron Island and the Wistari ring reef at the south of the GBR with the aid of radiolabeled phosphorus ^{33}P (Table 2.2). The rates of total PO_4-P consumption by microplankton appears to be very high in reef waters even at very low ambient content. It varied within 0.3 to $3.4\,\mu g\,P\,l^{-1}\,day^{-1}$; which resulted in a turnover time of 2 to 11 days (Sorokin 1990a). The rates of total PO_4-P consumption by microplankton (A_t) in general were higher in spring (October), when the water was warmer, than in winter. Seasonal changes in nutrients cycling were also detected by Hatcher (1985). The A_t-values were not significantly dependent upon the level of primary production of phytoplankton. It was often high even with very low primary production. This happened because a larger portion of PO_4-P was consumed by bacteria, but not by the phytoplankton. The share of bacterioplankton in A_t in coral reef waters exceeds usually 50% and varies within 50 to 80% (Table 2.2). Thus the bacterioplankton appears to be one of the most important and active agents of nutrients cycling in water. Its function could be seen in the consumption of inorganic P and N from oceanic water with the use of energy of the organic matter depleted by nutrients, excreted by bottom animals (corals) and by macrophytes. The misrobial biomass thus produced is used by filtering the fauna, which includes the nutrients contained in it, into the nutrient pool of the reef ecosystem (cf. scheme; Fig 2.12).

In conditions of relative stability of ambient PO_4-P concentrations in waters over the reef, and by their relatively low net exchange rates (see below), the A_t values should be in an equilibrium with the reciprocal values of the rates of regeneration of PO_4-P from organic forms during the processes of heterotrophic decomposition. Calculations of the latter, using the data on the rate of organic matter destruction (total plankton respiration measured as BOD at in situ t°) and on the ratio C:P = 60 (by weight), gave values of possible rates of PO_4-P regeneration in water columns as 1–$3\,\mu g\,P\,l^{-1}\,day^{-1}$ at rates of destruction of 50–$150\,\mu g\,C\,l^{-1}\,day^{-1}$ (cf. Sect. 2.3). So the calculated rates of regeneration are within the same range as rates of PO_4-P consumption. This proves a tight equilibration of a main limiting nutrient in reef systems – phosphorus. As for the cycling nitrogen, the processes of N_2-fixation provide a permanent excess of its organic and inorganic compounds which are released into and brought out by passing oceanic waters.

The processes of exchange of nutrients between water and bottom sediments were studied by a team working at the Enivetok atoll (Johannes et al. 1972; Pilson and Betzer 1973; Webb et al. 1975). The team intended to solve this problem by measuring the changes in their ambient concentrations in water along the standard cross-sections through windward reef flats in areas where corals and macrophytes are dominant. It was supposed that in the oceanic waters, passing over the reef along the transsect, the content of inorganic nutrients should decrease as a consequence of their consumption by autotrophic communities of benthic plants and coral. But these observations revealed a picture of indefinite fluctuations in them, which were difficult to interpret. The same indefinite character of change in the concentration of inorganic nutrients in oceanic water passing over benthic autotrophic communities of reef flats was recorded in analogous studies accomplished later on other reefs (Hanson and Gundersen 1976; Kimmerer and Walsh 1981; Hatcher and Hatcher 1981; Crossland and Barnes 1983; Johannes et al. 1983b). In most cases a weak release of inorganic N and of organic N and P, as well as a slight uptake of inorganic Poccurred (Atkinson 1983, 1987a; Atkinson and Smith 1987;). But in the diurnal balance even the changes of PO_4-P content were found to be near zero. Moreover, in the biotopes of lagoonal patch reefs dominated by living corals the net release of phosphorus also most often was recorded with a diurnal balance (Andrews and Muller 1983). Thus all attempts not only to quantify, but even establish the direction of the flows of nutrients between the water masses passing over the reef and its bottom communities appeared to be in general unsuccessful (Atkinson 1983; Hatcher 1985). But at least the results thus obtained confirmed statements about the importance of N_2-fixation as one of the main sources of nitrogen supply in coral reef communities (Webb et al. 1975).

Additional attempts to gain insight into the problem-of nutrient flows were made in experiments with plastic transparent enclosures secured to the bottom and covered with sand or rubble. In these the changes of nutrients in enclosed volumes of water were also measured in diurnal cycles (Henderson 1981; Propp 1981; Propp et al. 1983; Harrison 1983). The results were again indefinite. In most cases during the day and also at night the concentration of inorganic N and P in water inside an enclosure did not change at all or it slightly increased, and not only in the concentration of N but even of inorganic P, instead of their expected consumption by autotrophic communities of periphyton and microphytobenthos.

The faults in the above approaches seem to be quite obvious. The bottom communities of reef flat as well as communities of rubble and sand comprise together the autotrophic and the heterotrophic components. So they should simultaneously consume nutrients and release them. Therefore, the above-listed measurements of the change in their ambient concentration in oceanic waters passing over the reef or in the in situ enclosures reflect only the resulting equilibrium values of these two reciprocal flows, i.e., between the uptake of nutrients by the autotrophic part of the community or the cellular systems of symbiotic animals, and their release by the heterotrophic ones. But these measurements cannot reflect the size of these flows per se. These flows, the same as is the case with corals, could be measured only by employing the isotopic traces techniques. For example, with the aid of labeled PO_4-P it is possible to measure the real value of the uptake flow between water and bottom, because this uptake will not be masked by the release of nonlabeled phosphate from the bottom. If we will know the uptake flow rate A_c and the net change of PO_4-P concentration ($\pm A_t$), we may calculate the release flow rate as: $A_e = A_c + A_t$. The corresponding studies we have carried out at the Heron Island reefs with the use of radioisotope ^{32}P as the label. Experiments with the periphytonic

Table 2.5. Phosphate flows between components of bottom biotopes and water column during daylight; K_p – initial PO_4-P content in water, μmol l^{-1}; A_c – uptake flow; A_e – release flow; $\pm A_t$ – net changes of PO_4-P content in water (After Sorokin 1990c)

Component	K_p	Flows of PO_4-P				
		μg P kg^{-1} h^{-1}			mg P $m^{-2}day^{-1}$	
		A_c	A_e	A_t	A_c	A_t
Piece of reef-flat rock, covered with periphyton	1.00	84.3	27.3	−57.0	10.1	−6.8
	0.35	18.2	15.1	−3.1	2.2	−0.4
Coral rubble from reef flat covered with periphyton	1.0	59.1	20.1	−38.4	11.3	−7.4
	0.35	31.0	18.0	−13.0	3.7	−1.6
	0.16	11.5	21.5	−10.0	1.4	+1.2
Dead branch of coral Stylophora covered with periphyton	1.0	85.6	44.4	−41.2	18.5	−8.9
	0.35	56.0	53.9	−2.1	12.1	−0.5
	0.16	14.6	17.4	+2.8	3.2	−0.6
Living branch of the same coral colony	1.0	110.5	31.9	−78.6	15.9	−11.3
	0.35	51.7	28.8	−22.9	11.2	−4.9
	0.16	12.6	11.4	−1.2	2.3	−0.2
Coral sand from leeward reef-flat	1.0	25.0	19.3	−5.7	4.8	−1.1
	0.35	15.2	23.2	+8.0	2.9	+1.5
	0.16	6.1	11.2	+5.1	1.2	+0.9

overgrowths over the flat rocks, dead coral colonies and rubble were accomplished in aquaria. Studies of the PO_4-P exchange flows between the coral sands and the water column were performed in situ in dark and light plastic enclosures.

These experiments proved the existence of the above-mentioned reciprocal flows of PO_4-P from and into the periphytonic communities (Table 2.5, Fig. 2.10). The same was recorded for coral sands. Both these

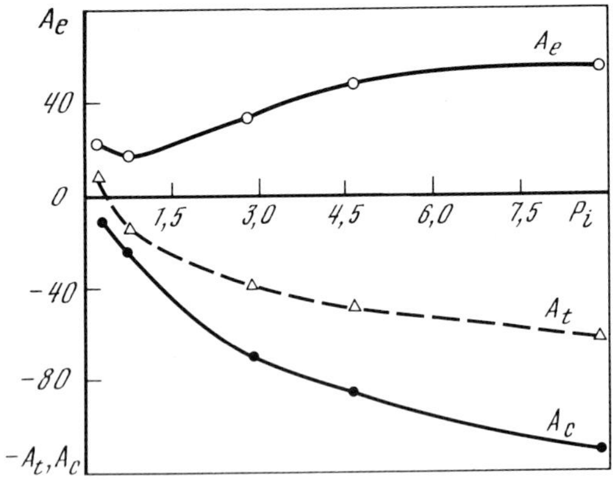

Fig. 2.10. Dependence of the consumption (A_c) and release (A_e) of PO_4-P (μg P kg^{-1} h^{-1}) by coral *Pocillopora damicarnis* between rubble overgrown by periphyton and the water column; for designations, cf. Fig. 2.9. (After Sorokin 1990c)

compounds of bottom biotopes have in their microstructures the autotrophic micro- and macroalgae, which consume PO_4-P, and the heterotrophies, which release it (cf. Sect. 4.1). These compounds function within these structures in a narrow coincidence, providing a semi-closed cycling of nutrients, the same as between zooxanthellae and polyp tissues in coral. Moreover, not only the absolute rates of A_c and A_e flows of PO_4-P, but even their ratios and the 24-h balances of PO_4-P coincided well in periphyton and in the living corals, being calculated per 1 g of dry weight of hard substrate (Table 2.5). The best coincidence was found in experiments with the living and the dead parts of the same ramose colony, if the latter is overgrown with the periphyton. This coincidence looks more logical if we take into the account that the rates of photosynthesis and respiration in living corals and in dead corals overgrown with periphyton also coincide (cf. Sect. 4.2).

An analysis of the table shows that the flow rates of PO_4-P between periphyton and water and the ratios between its consumption and release depend upon its concentration in water. The same was shown by Atkinson (1987). At low PO_4-P content, similar to the ambient ones in waters over reefs of the Heron I. (0.1–0.3 $\mu mol\,l^{-1}$), the rate of consumption of PO_4-P (A_p) by the periphyton over the piece of flat rocks and over the coral rubble was within 10–60 $\mu g\,P\,kg^{-1}$ of dry matter h^{-1}. That is commensurable with the same values in living corals (Tables 2.3, 2.5), as measured at the same low ambient concentrations. At these low concentrations of PO_4-P in water the rate of the reciprocal excretion flow (A_e) was equal to the rates of the consumption (A_t) and sometimes it was even higher. In these cases, which are usual at a low PO_4-P content, the net release of PO_4-P proceeded, and thus the balance value A_t was positive (Table 2.5, Fig. 2.11). In the enriched water, where the PO_4-P was 0.5–3.5 μmol, the consumption flow in periphytic microcosms increased up to 80–120 $\mu g\,P\,kg^{-1}\,h^{-1}$, while the excretion flow was not so dependent upon the external PO_4-P content and increased much less. Thus, in the enriched water consumption was 1.5–3 times more than excretion, and the net uptake of PO_4-P readily took place. Thus at high PO_4-P concentrations the balance values A_t were always negative – the periphytonic communities did not release any PO_4-P into the water, but only consumed it. One could deduce that this feature of its dynamics in the enriched experimental environement has nothing to do with the functioning of ambient phytobenthic communities, because such high PO_4-P concentrations are exclusive in waters over the reef. But actually it is not so. The reef benthos icludes rich populations of big slowly moving or sedentary invertebrates as well as a lot of territorial fish, which live permanently in a definite place of their bottom biotope. All these stay-at-home animals excrete locally significant portions of nutrients – and of PO_4-P as well. Thus its concentration in some local, especially in semi-closed zones in which these animals like to hide, could be rather high. Therefore, the ability of the phytobenthic communities and also of corals to consume

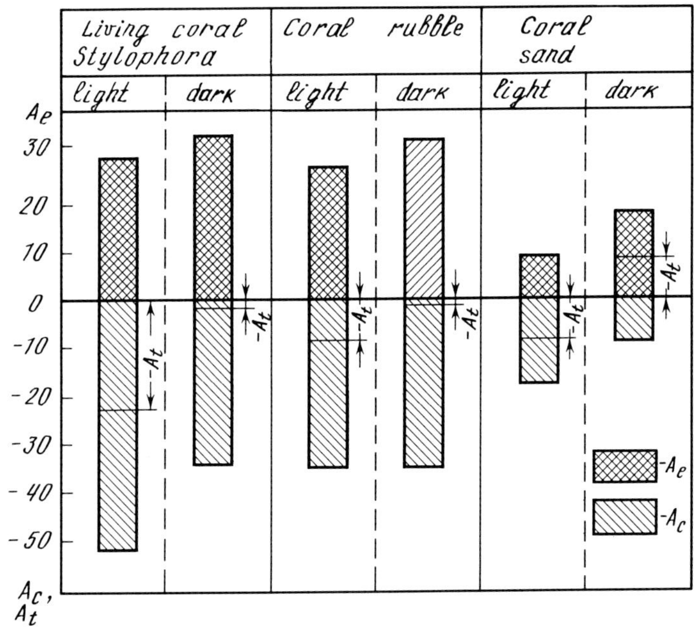

Fig. 2.11. Flows of PO_4-P between some components of reef bottom biotopes and the water column. For designations, cf. Fig. 2.10. (After Sorokin 1990c)

readily the inorganic nutrients in their high concentrations is an important feature of them which actually increases the possibility of the use of nutrients sources of reefs by their benthic autotrophic communities. The increase of PO_4-P concentration in experiments up to 2–3 μmol resulted in a linear growth of its uptake rate (A_c) by periphyton (Fig. 2.10), while the release flow (A_e) increased slightly. Further elevation of the concentration of PO_4P caused the bending of the curve, which had the tendency to come to level out at 9–10 μmol. The rate of release stabilized at the level of 30–70% of that of the uptake. The balance values A_t at high PO_4-P content were negative (net consumption continued). It comprised correspondingly 30–70% of actual PO_4-P consumption flow (A_c). The same as in living corals, the PO_4-P uptake in the periphytonic communities does not depend much on light conditions. This was observed also in microphytobenthos of sands (Fig. 2.11). In dark the consumption flow slightly decreased, while the release rates increased so that the net release increased and the A_t value became more positive, especially in experiments with coral sands, when at low PO_4-P concentrations the A_t at night became positive, changing from slightly negative in daylight (Table 2.6).

Calculations of PO_4-P flows between the bottom biotopes of rubble covered with periphyton and a water column m^{-2}, day^{-1} gave values of A_c and A_e within 3 to 10 mg P, while the balance values A_t were negligible –

Table 2.6. Elements of phosphate exchange balance between water and bottom sediments in various bottom biotopes at reefs of the Heron atoll obtained in the in situ experiments; for designations, see Table 2.5 (After Sorokin 1990c)

Biotope	Depth (m)	PO_4-P contents in water ($\mu mol\ l^{-1}$)	Condition of illumination	Elements of PO_4-P balance				
				in $\mu g\ m^{-2}\ h^{-1}$			in $mg\ m^{-2}\ day^{-1}$	
				A_c	A_e	A_t	A_c	A_t
Coral sand covered by microalgae (diatoms) on the flat near Island	0.5	0.22	Light	326.6	190.8	1358	7.68	−1.77
			Dark	313.3	302.9	−10.4	–	–
Coral sand on flat opposite the Resort	0.7	0.22	Light	195.2	168.8	−26.4	3.58	+0.68
			Dark	103.5	186.0	+82.5	–	–
Coarse sand in the Shark Bay near the shore of Isl.	0.8	0.16	Light	165.8	145.4	−20.4	4.7	+0.92
			Dark	117.9	192.3	+79.4	–	–
Coral sand covered by microalgae among patch reefs	1.2	0.21	Light	201.9	282.5	80.8	4.6	+3.81
			Dark	136.1	373.9	+237.8	–	–
Fine coral sand in central part of lagoon	6.0	0.17	Light	161.0	87.0	−74.7	3.79	−1.07
			Dark	144.9	135.0	+9.3	–	–
Coral sand on outer leeward slope	8.0	0.19	Light	248.0	197.3	+50.7	4.70	+0.92
			Dark	144.2	270.8	+126.7	–	–

within 0–0.5 $mg\,m^{-2}$ (Table 2.6). The same flow rates were measured in lagoonal coral sands containing a rich microphytobenthos. The in situ enclosure experiments in this biotope gave the values of flows A_c and A_e as 2 to 7 $mg\,P\,m^{-2}\,day^{-1}$. A significant net release of PO_4-P in 24 h balance ($+A_t$) was recorded in most sites of this biotope: 1–3 $mg\,P\,m^{-2}$. In the same kind of experiments by Harrison (1983) the A_t values were ±0.1–1.5, changing sign in different parts of the lagoon. A low level of the balance values A_t in coral reef bottom biotopes actually reflects the above-mentioned semi-closed character of nutrients cycles in their main compartments – in corals, in periphytonic communities, and in microphytobenthos of coral sands. In bottom biotopes occupied by corals and periphytonic overgrowths (Tables 2.3, 2.5; Fig. 2.10) at low ambient PO_4-P concentrations (0.1–0.5 mol) the balance A_t value is approaching zero. Thus, it seems that the communities do not release any inorganic PO_4-P, while the reciprocal A_e and A_c flows could be significant. But we leave unaccounted for in this balance an intensive consumption flow of the organic P, which is consumed by periphyton as its dissolved, and by corals as both dissolved and planktonic fractions. It is just this unaccounted for part which covers a slight release of PO_4-P by benthic communities and participates as well in the building up of the stock of organic phosphate stocks and mineral phosphorite (apatite) deposits in bottom sediments and in reef rocks (Entsch et al. 1983). The bottom reef community then as a whole consumes phosphorus as well in its inorganic as its the organic forms, thus providing a significant portion of the necessary nutrients supply to its autotrophs.

The calculated ratios between the flows of PO_4-P and rates of metabolism in bottom communities occupied by corals and rubble look like the following. At average values of these lows (A_c, A_e) 5–20 $mg\,P\,m^{-2}$, and by rates of photosynthesis and respiration of 2–4 $g\,C\,m^{-2}$ per day the C:P ratio will be 100–400 by weight. A high degree of equilibrium in the processes of nutrients dynamics between bottom and water seems to explain the fact of the coral reef ecosystem's ability to control it via its biological mechanisms. In fact, a low ambient level of the concentrations of inorganic nutrients is preserved even in reef areas subjected to permanent pollution (Johannes 1972b; Kimmerer and Walsh 1981; Crossland 1983) or in the artificially fertilized reef biotopes (Kinsey and Domm 1974), while drastic changes in the structure of their benthic communities could happen (Wiebe 1985). This ability to keep a low level of nutrients in water when building up their stock in bottom biotopes and in rocks is probably related to the very existence of coral reefs, because an elevated level of nutrients in reef waters increases the competition from the side of algae, and also inhibits the growth and undermines the health of corals. Elevated nutrients levels will cause a decrease in their calcification and enhance their infestation with boring animals (Kinsey and Davies 1979b; Highsmith 1979b).

Main pools of mobile nutrients exist mainly in the living biomass of organisms, and (much less) in organic matter of detritus and sediments

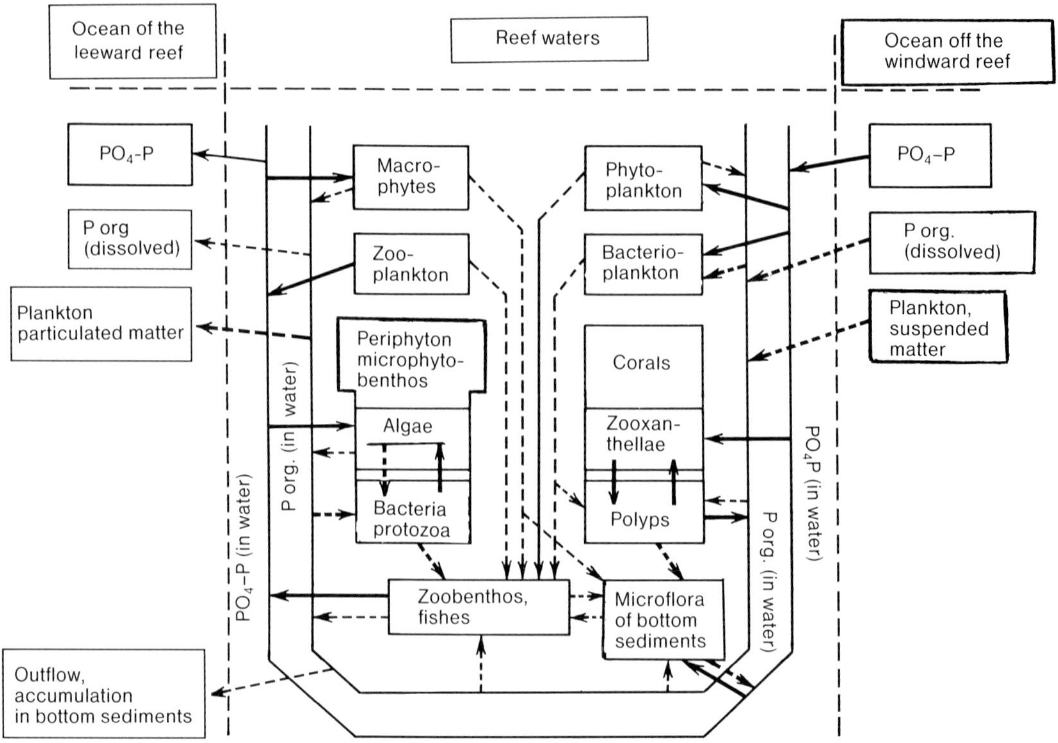

Fig. 2.12. Scheme of phosphorus dynamics in coral reef ecosystem and the interconnection of pools of organic phosphorus and inorganic phosphate; *dotted lines* flows of organic phosphorus; *solid lines* flows of PO_4-P. (By the author)

(Hatcher 1985), as well as in waters that fill the internal microcavities and channels in the lime body of the reef (Szamant-Froelich 1983). Direct estimations of the possible size of these pools have not been done yet (Crossland et al. 1991), but there are some approximations for the biomass pool, based on data on total amounts of living biomass in reef biotopes and on the C:N:P average ratios in it (Bunt et al. 1975; Lee et al. 1975). These approximations give its value as follows: 3–10 g N and 0.5–2.5 g P m^{-2} of bottom. If the estimated rates of phosphate flows between bottom and water column are at level 3–10, and the calculated ones for inorganic nitrogen 40–150 mg m^{-2} day^{-1}, then their approximate turnover times should be rather short: 2–4 months. Some of the main patterns of phosphorus dynamics in a reef ecosystem are demonstrated in Fig. 2.12.

2.2.5 Conclusion

As we have seen above, the processes of input of the main nutrients, nitrogen and phosphorus, into the reef ecosystem as well as their recycling proceed mainly through the functioning of biological mechanisms based of trophodymanics within the reef biota (Crossland 1980, 1983). The same mechanisms drive the flows of nutrients inside the planktonic and the benthic communities as well as between them via consumption an excretion of their mainly organic forms within the food-webs (cf. scheme, Fig. 2.12). Therefore, the measurements of standing stocks and flows of only their inorganic forms cannot give correct information even on their direction, let alone their size. The domination of flows of organic nutrients inside reef-bottom communities over the flows of their inorganic forms is a consequence of the abundance of compartments in them with semi-closed nutrients cycles, as in corals and periphyton (Yonge and Nicholls 1931; Lewis 1973; Pomeroy et al. 1974). The processes of nutrients regeneration in the reef ecosystem are accomplished mostly by planktonic and benthic animals (Ikeda et al. 1982) as well as by fish, with a minor participation of bacteria, which largely rather consume than regenerate them. Their role in inorganic nutrients regeneration from organic forms can be seen mainly in the consumption of dissolved organic phosphorus from the oceanic waters passing over the reef. They take it up into their biomass and thus into the food-web. Being grazed by filter-, detritus- and periphyton-eaters, these nutrients consumed by bacteria are finally mineralizing. Bacteria participate also in the mobilization of the stock of organic nutrients in the so-called regenerative sediments which consist of detritus (Di Salvo and Gundersen 1971). But the final regeneration of nutrients is accomplished mainly by animals and among them by the benthic filterers and detritovores as well as by reef fish (Webb et al. 1977; Meyer et al. 1983). In accordance with the data of the latter authors, the rate of N-regeneration by reef zoobenthos attains around $75\,mg\,m^{-2}\,day^{-1}$. Fish, having an average biomass of about $70\,g\,m^{-2}$, could excrete some $30\,mg\,m^{-2}$ of $NH_4\,day^{-1}$, and also the zooplankton according to Ikeda et al. (1982) should excrete at least 15 mg $NH_4\,m^{-2}\,day^{-1}$. Thus the total regeneration rate could be evaluated to be within $115\,mg\,N\,m^{-2}$. At an average depth of waters of 5 m over the reef it will be around $1.5\,\mu mol\,l^{-1}$. This rate of regeneration might then provide the turnover time of N in reef waters as around 2 days. As for phosphorus, regeneration could be estimated as $20\,mg\,m^{-2}\,day^{-1}$, which is around $0.15\,\mu$mol, which means the same 2 days of turnover time.

With reference to the scheme (Fig. 2.12) it can be seen that while the nutrients stocks are permanently and many times recycled inside the reef communities and are also kept in pools of living biomass, detritus and sediments, reef autotrophic communities should not suffer nutrients limitations even when flushed by the depleted oligotrophic oceanic waters (Pilson and Betzer 1973; Wiebe 1985; Atkinson 1988). A large storage capacity of the coral reef ecosystems in relation to nutrients enables them to build up their stock during the periods of their extensive input with drainage (Marsh 1977) or deep oceanic waters, and then use this stock in periods of when their supply is poor (Tsuda 1974; Kimmerer and Walsh 1981;

Crossland 1983). In any case, the fact of a permanently high primary antotrophic production on reefs existing in the flow of oligotrophic oceanic waters proves that the annual balance of nutrients in the system "reef–ocean" should be positive. That is to say, the input of the main limiting nutrient, phosphorus in its organic and inorganic forms from oceanic waters, exceeds its loss from reef to ocean. Its key role as limiting nutrient is clearly seen in its regulatory function of community structure in reef benthic macrophytes (Littler et al. 1991). The problems of the dynamics of nutrients in reef ecosystems need further attention. It should be important to accumulate more data on their flows in the food-chains and in the regenerative systems with the use of isotopic techniques. This kind of information is extremely important for the monitoring of coral reef environments in areas of pollution and eutrophication.

2.3 Organic Matter in Reef Waters

A large part of the organic matter produced in coastal marine communities by benthic plants, and on reefs also by symbiotic animals, is introduced into the food-web via the "detritus chain", i.e., dead organic matter (Darnell 1967). In coral reef ecosystems, even at their high level of direct grazing of plants and corals, which has been evaluated as within 50% (Hatcher 1983b), more than 80% of organic matter would pass via the "detritus chain", if the assimilability of plant materials by animals is 30–40%. This part of primary plant production enters into the food chain as suspended, and also as dissolved, organic matter. The latter is produced largely through its direct excretion by plants and by corals, which comprise up to 20–30% of their primary production. The suspended organic matter is formed from the disintegrated faeces of phytophages, from the crushed talloms of dead macrophytes, as well as from the microbial and algal biomass washed out from bottom sediments and periphyton by surf. The importance of this stock of organic matter in reef trophodynamics was stressed also by the first investigators of this problem (Odum and Odum 1955; Bakus 1967). They pointed out the abundance of filter feeders in reef benthic communities and the domination of the detrital mass in their guts. These conclusions drew attention to the study of suspended organic matter and organic aggregates in reef waters (Johannes 1967; Klim 1969; Marshall 1968; Coles and Strathmann 1973; Quasim and Sankaranarayanan 1970; Gordon 1971; Gerber and Marshall 1974; Le Borgne et al. 1989).

As regards the stock of dissolved organic matter, data are still scarce (Westrum and Meyers 1978) perhaps mostly because of methodological obstacles. For example, the most widespread wet combustion persulfate method after Menzel-Vaccaro underestimates it 1.5–2 times. Moreover, the informative value of data on its total stock is limited because 70–80% of it is represented by its inert humic fraction. Thus the data on its total stock could be anyway hardly used directly for evaluations of the production of dissolved organic matter and of involving it in the biological processes. This situation induced attempts to measure just the labile part of its stock using the BOD (biological oxygen demand) methodology. But even these data appear to be difficult to interpret because the content of labile organic matter in reef water is a result of the dynamic equilibrium of its production and of its consumption – decomposition mostly by the bacterioplankton. Therefore, it was important to estimate not only the stock (S), but also the rate of at least one of these flows of labile organic matter. Measuring the flow of its consumption (D), for example, and accepting a relatively stable equilibrium of the above reciprocal processes, we can calculate its turnover time T (T = S/D), and evaluate the rate of release flow (Sorokin et al. 1983). In accordance with this method, the labile organic matter (LOM) fraction is part of the total organic matter accessible to immediate microbial action, calculated from the experimentally estimated decrease in oxygen in bottled BOD samples after 20–30 days of exposure at t° 20–25 °C. During this time, the BOD curve approaches plateau, thus showing the complete exhaustion of the stock of labile organic matter. The calculations of stock of LOM-carbon (S) account for the part consumed in the sample by bacteria and protozoa for their growth (Sorokin et al. 1983). Special attention was paid to the study of the production and the chemical composition of coral mucus (Richman et al. 1975; Ducklow and Mitchell 1979b).

Data on the content of organic matter are given in Table 2.7. They were obtained by wet combustion and thus could underestimate real values 1.5–2 times. The uncorrected estimations gave a value of 1.3–1.9 mg C l^{-1}, while in the surrounding ocean it was 1 mg. The corresponding corrected numbers would be 3–4 mg C and 1.5 mg C l^{-1}. Thus, waters over the reef are enriched with the dissolved organic matter (DOM) 1.5–2 times, and sometimes even

Table 2.7. Content of dissolved organic matter in reef waters (Corg, mg C l^{-1})

Place	Reef zone	C_{org}	Reference
Enivetok atoll	Surrounding ocean	1.03	Johannes and Gerber (1974)
	Windward slope	1.08	Webb et al. (1975)
	Flat	1.31	Webb et al. (1975)
	Flat	2.5–3.5	Johannes and Gerber (1974)
	Lagoon	1.2	Webb et al. (1975)
Fanning atoll	Lagoon	1.6–1.9	Gordon (1971) Gordon et al. (1971)
Andros atoll	Lagoon	0.60	Marshall et al. (1975)
Reef Grand Cayman (Caribbesn)	Flat	0.82	Westrum and Meyers (1978)
Abrolhos reefs (West Australia)	Surrounding ocean.	0.15–0.39	Crossland et al. (1984)
	Flat	2.10–2.88	Crossland et al. (1984)
	Leeward flat	3.61–4.18	Crossland et al. (1984)
	Lagoon	1.56–3.66	Crossland et al. (1984)
Kaneohe Bay (Oahy, Hawaii)	Flat	1.64	Marshall et al. (1975)
	Lagoon	1.77	Marshall et al. (1975)

Table 2.8. Suspended organic matter (C_p, $\mu g\,C\,l^{-1}$) in reef waters

Place	Reef zone	C_p	Author
Enivetok atoll	Surrounding ocean	10–30	Marshall et al. (1975)
	Flat	20–240	Gerber and Marshall. (1974)
	Lagoon	20–30	Gerber and Marshall (1974)
Fanning atoll	Flat	150	Gordon (1971)
	Lagoon	80	Gordon (1971)
Kawaratti atoll	Surrounding ocean	75	Quasim and Sankaranarayanan (1970)
	Flat	1035	Quasim and Sankaranarayanan (1970)
	Lagoon	355	
Abrolhos reefs (Western Australia)	Surrounding ocean	14–42	Crossland et al. (1984)
	Reef crest	870–940	Crossland et al. (1984)
	Lagoon	330–870	Crossland et al. (1984)
Lizard I. (GBR)	Windward slope	150–280	Crassland and Barnes (1983)
	Leeward reef	180–310	Crassland and Barnes (1983)
Abako reef (Bahamas)	Flat	130–240	Marshall et al. (1975)
South Caicos reef (Caribbean)	Flat	200	Marshall et al. (1975)
	Lagoon	130	Marshall et al. (1975)
Kaneohe Bay (Hawaii)	Flat	500	Marshall et al. (1975)
	Lagoon	420	Marshall et al. (1975)
Grand Lameshur (Virgin Islands)	Flat	710	Hickel (1972)

more (Gerber and Marshall 1974; Westrum and Meyers 1978). Up the reefs of Abrolhos it increased more than ten times – from 0.3 to 4.2 $mg\,C\,l^{-1}$ (Crossland et al. 1984). On the polluted reefs of the Kaneohe bay it was also high. The largest content of organic matter in the reef waters on the cross-sections were recorded up the reef flat. In waters of the lagoon it decreases (Fig. 2.13).

The content of suspended organic matter (SOM) is also much higher in reef waters compared with the surrounding oceanic waters (Table 2.8), and attains its maximum over the flat, where it is 5–10 times higher than in the open ocean: 100–150 $\mu g\,C\,l^{-1}$. Up the flat of some barrier reefs it reached even 200–500 $\mu g\,C\,l^{-1}$ (Glynn 1973a), and over the flat of the Kawaratti atoll and up the reefs of Abrolhos it attained 1–1.5 $mg\,C\,l^{-1}$. The same high content of suspended organic matter was recorded in waters of Hawaiian reefs in the Kaneohe bay and of the reefs of the Virgin Islands which are subjected to anthropogenic pollution (400–700 $\mu g\,C\,l^{-1}$). In lagoonal waters its content decreases because of settling of the organic particles. Still it remains several times more than in the open ocean. The ratio "DOM/SOM", in oceanic waters 50–100, decreases in water over the reefs to 10–30, and in polluting reefs even to 3–7. The content of carbon in suspended material 10–15% per dry weight in oceanic waters, decreases in reef waters to 4–8% (Hickel 1972). Thus the dry weight of total suspended matter in reef waters is large, 1–20 $mg\,l^{-1}$, which is 10–20 times more than in oceanic waters.

The stock of LOM after the BOD data could be evaluated in reef waters as being within 0.6–2.5 $mg\,C\,l^{-1}$ (Motoda 1940; Sorokin et al. 1983; see Table 2.9), while in waters of the open ocean it is 2–5 times less: 0.3–0.5 $mg\,C\,l^{-1}$ (Sorokin and Mamaeva 1980). In oceanic waters this relatively

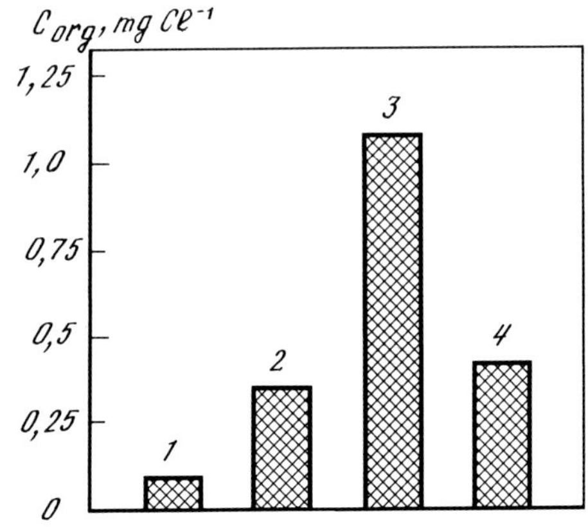

Fig. 2.13. Content of suspended organic matter in waters of the open ocean 6 km (*1*) and 1 km (*2*) off the windward flat of the Kawaratti atoll, as well as in waters over the reef flat (*3*) and in the lagoon (*4*). (After Quasim and Sankaranrayanan 1970)

high stock of LOM is maintained chiefly by its excretion by phytoplankton. In reef areas the oceanic waters, which pass over the reef-flat, get a new input from plants and symbiotic animals. It is clearly reflected by a drastic increase in DOM and LOM contents (Fig. 2.13). Larger contents of LOM were recorded up the coastal or leeward reefs and reef flats. The percentage of LOM in total DOM in reef waters is much higher than in oceanic waters, comprising 30–40%, as against 15–25% in the latter. This also reflects the immediate enrichment with LOM of waters in contact with the reef benthic communities.

The LOM excreted by benthic autotrophes and thus introduced into the water column is then consumed by planktonic, periphytonic and benthic heterotrophic microflora. The rates of its consumption by benthic microflora are difficult to measure because of the reciprocal process of its excretion. But estimations of the use of LOM by planktonic microbial communities could be done by measuring their respiration rates. They are usually very high in reef waters: 0.2–0.6 $\mu g\, O_2\, l^{-1}\, day^{-1}$ (Table 2.9). The turnover time of LOM, which was estimated by taking into account only its consumption in the water column, was at most of the reefs investigated rather short: 2 to 8 days. Its real value was more than 10 days. Moreover, a real LOM turnover time should be even shorter if it could be possible to take into account in its estimation also the LOM consumption by benthic and periphytonic microflora. An extremely quick turnover time of LOM stock in reef waters reflects high metabolic rates of their planktonic communities.

Data on the composition of DOM in reef waters are scarce (Meyer et al. 1978; Fontaine et al. 1981). But if we agree that it might not be very different from the composition of DOM in oceanic waters, it should contain 20–30% of peptides and amino acids, 15–30% of carbohydrates and 10–

Table 2.9. Stock of labile organic matter (S) rates of its decomposition (D) and its turnover time (T), original data by author

Place	Reef zone, depth, m	D, mgO_2l^{-1} day^{-1}	S, $mg\,C\,l^{-1}$	T, days
Praslin I. (Seychelles)	Coastal reef, 2 m	0.21	0.61	7.8
	Reef edge; 16 m	0.23	0.67	7.9
Kotivi I. (Seychelles)	Flat; 2 m	0.32	1.31	10.9
	Leeward reef edge; 7 m	0.66	1.23	5.0
	Lagoon; 15 m	0.43	0.98	6.1
Kargados-Karahos super atoll (Mauritius)	Flat; 1 m	0.35	1.04	8.0
	Lagoon; 28 m	0.54	0.69	3.4
Thu I. (Vietnam)	Flat; 2 m	0.67	0.64	2.5
	Lagoon; 8 m	0.78	1.28	4.4
Siton atoll (same)	Leeward reef edge; 10 m	0.51	0.97	5.1
	Lagoon; 20 m	0.12	0.68	1.5
Wungro Bay (same)	Coastal reef; 2 m	0.39	2.54	17.4
	Reef edge; 12 m	0.56	1.28	6.1
	Lagoon; 10 m	0.39	0.55	3.8
Tortu I. (same)	Lagoon, 18 m	0.50	0.79	4.2
	Coastal reef; 1 m	0.68	1.21	4.7
Kulaohon I. (same)	Coastal reef; 2 m	0.38	1.79	12.6

20% of lipids (Romankewich 1979). Reef waters contain also the vitamins and the antibiotics.

The phytoplankton, the benthic algae and corals excrete a significant part of their photosynthetic production as DOM. It varies within 5 to 40%. The excreted products contain glycine and other amino acids, sugars, and phenols. The percentage of its release increases in stressed environments. Accepting that primary production in actively functioning reef biotopes, such as flats covered with macrophytes, corals, and coralline algae, is around 4–6 $g\,C\,m^{-2}\,day^{-1}$ and an average percentage of excretion of photosynthates is 10–15%, the release flow of DOM from such a biotope should be 400–600 $mg\,C\,m^{-2}\,day^{-1}$. At an average depth of waters of 1–2 m over the reef flat this flow could provide an increase in the content of DOM in the water column amounting to 200–600 $mg\,C\,l^{-1}$. This is 2–4 times more than the rate of its decomposition in a water column, which is usually 50–150 $mg\,C\,m^{-3}\,day^{-1}$ (Table 2.9). Direct observations on distribution of DOM across the reef showed (Table 2.10) that over the flat covered with living corals its concentration increased 1.5–4 times compared with oceanic waters coming to the reef during the tide. For 4–5 h of its contact with reef the content of DOM increased per 0.5–0.8 $g\,C\,m^{-3}$ as an average. Such an increase was recorded up the reefs of the Sinton atoll, Wungro Bay, and Tortu Island in the South China Sea (Sorokin et al. 1983). During the complete tidal cycle, the release of DOM, when intensively functioning, could reach 1–1.5 $g\,C\,m^{-2}$, which comprises 20–30% of its primary

Table 2.10. Excretion of the labile organic matter by corals, calculated per wet weight of colonies, after Zorokin et al. (1983)

Corals	Photosynthesis $\mu g\,C\,g^{-1}h^{-1}$ (P)	Excretion of LOM $\mu g\,C\,g^{-1}h^{-1}$		E/P,%
		Limits in experiment	Medium value (E)	
Lobophytum sp.	28	3.4–7.5	5.2	19
Alcyonium sp.	48	10.2–14.9	12.8	27
Sarcophyton sp.	29	4.7–8.5	6.2	21
Lythophyton sp.	48	12.3–23.7	17.1	36
Acropora squamosa	26	3.6–5.8	4.8	18
Pocillopora damicornis	28	6.1–8.2	6.9	25
Porites sp.	12	1.8–3.3	2.5	21
Dead coral branches overgrown by periphyton	20	1.3–3.6	1.8	9

production. A significant portion of DOM thus released was comprised by the corals themselves via their mucus production. The excretion of mucus serves two important functions: the shedding of sediments and the sedimentary filter-feeding (cf. Sect. 8.5). The coral mucus contains glicanes, sulfosugars, lipids, cetyl palmitates and proteins (Daumas and Thomassin 1977; Benson et al. 1978; Richards et al. 1983). These products are the components of LOM, utilizable by bacteria. The release of DOM by corals was established in experiments (Johannes and Webb 1970; Richman et al. 1975; Crossland et al. 1980; Sorokin et al. 1983). In our experiments the rate of release in some coral species was up to 150–200 $\mu g\,C\,g^{-1}\,day^{-1}$, which is close to 20–30% of the rate of their photosynthesis. Soft corals also were found to release DOM (Table 2.10). The same ability to a lesser extent was found in periphytonic communities overgrowing coral rubble. An approximate evaluation of the rate of release of LOM by a bottom reef community dominated by corals with their mass of 5–10 $kg\,m^{-2}$, based upon the above data, ranges within 0.4–0.8 $g\,C\,m^{-3}\,day^{-1}$ with a depth of 1 m of the water column. These numbers correspond to those obtained by the BOD method (see above).

The composition of suspended organic matter depends upon the character of the dominating source of its formation. In reefs situated in the open ocean its autochthonous sources dominate (Gerber and Marshall 1974; Lewis and Price 1975; Sournia 1977), such as plant detritus, faecal material, pseudoplankton, microflora, microalgae, washed out from sediments, mucus excreted by corals and other reef animals, and the DOM. It contains also colloidal and mucoid components, which subjected to flocculation in reef environments form the flakes of the "marine snow" (Johannes 1967; Richman et al. 1975). The coral mucus comprises a significant amount of the total input of organic matter into the water column. The flakes of mucus absorbing the microplankton, the skins of crustaceans, the loricas of photozoans and appendicularians and fine inorganic particles grow to sizes of 0.3–1 cm forming thus the "marine snow" (Marshall 1968). The mucus contains proteins, lipids and a significant amount of zooxanthellae (Marshall 1972). It provides so food of high quality from these organic aggregates for zooplankton and fish which feed on it (Johannes 1967; Hatcher 1983a). The autochthonous suspended matter, enriched with the proteinaceous compounds of mucus, pseudoplankton, microplankton and the microbial populations on the detrital particles, have low C/N ratios: 4–8 compared with the open ocean (15–30) (Coles and Strathmann 1973; Gerber and Marshall 1974; Caperon et al. 1976; Crossland and Barnes 1983).

Among the allochthonous sources of particulated matter there should be mentioned the suspended matter of oceanic waters passing over the reef, as well as the wind-driven drifting masses of blue-green algae mentioned above. In waters of barrier and fringing reefs near the continents or near high islands the particulated material is also supplied by the rivers, which drive to the reefs the terrigenous material (Wolanski and Pickard 1983a; Sorokin and Tyapkin 1984). It arrives also from mangroves (Benner and Hodson 1985). Attempts to quantify the rate of release of suspended matter from the reef bottom biotopes up to the water column have given very high values. Over the flat of the atoll Enivetok dominated by living corals this release was estimated to be 300–400 mg m^{-2} of dry matter per day (Johannes 1967). On flats with domination of macrophytes it was 270 mg C m^{-2} day^{-1} (Johannes and Gerber 1974), and on flats of the atoll Kavaratti 2.3 g C m^{-2} day^{-1}, or about 20% of the primary photosynthetic production (Quasim and Sankaranarayanan 1970). The release of suspended matter is most intense in the zone of windward flats which are subjected to wave action and have a high primary organic production. The suspended matter elevated up to the water column over the flat is then driven to its lagoonal back-part (zone of patch reefs) and to the lagoon. At the Enivetok atoll this outflow of suspended matter from 1 m of the length of the flat has been evaluated by Johannes and Gerber (1974) and by Johannes et al. (1972) as follows: 800 g C, 100 g N and 2 g P per day. At the Kawaratti atoll it was found to be 600 g m^{-1}, on reefs of Puerto Rico 2.5 kg m^{-1} of dry material and on the high latitude reefs of Abrolhos, with a rich growth of macrophytes, it was up to 9 kg m^{-1} day^{-1} (Hatcher 1983a).

The above data as well as the energy balance scheme (cf. Fig. 11.7) clearly show that the larger part of autotrophic production created in the coral reef ecosystem passes into the trophic pool via the stage of DOM and SOM. The suspended matter (SOM) coming to the reef and also produced within the reef system is readily used by filtering reef-bottom fauna and by planktonic heterotrophs, including microzooplankton and bacteria. The energy flows into the food-webs involving these sources dominate in the balance scheme. This is just the reason why through the intensive use of the sources of dead organic carbon its contents in reef bottom sediments is usually extremely low (0.15–0.5%). Most of the organic matter produced by autotrophs is consumed and oxidized in the water column where it is many times resuspended. Its settled remains are consumed and decomposed by an abundant benthic fauna and microflora. The biomass of bacteria alone in coral sands might comprise up to 3–4% of the total organic matter which they contain, and the turnover time of this living microbial biomass is only 4–5 days.

The organic matter suspended in reef waters has a high nutritional value because of the abundance of microbial biomass and mucus in it. The suspended matter is the source of feeding for corals (Johannes and Gerber 1974; Lewis and Price 1975), and zooplankton (Johannes 1967; Richman et al. 1975; Sumner et al. 1975; Gerber and Gerber 1979; Gottfried and Roman 1983; Le Borgne et al. 1989). The flakes of "marine snow" and of aggregates are ingested by soft corals (Coffroth 1984) and even by fish (Hiatt and Strasburg 1960; Benson and Muscatine 1974). The LOM and DOM of reef waters are also readily used and not only by planktonic periphytonic and benthic bacteria, but also directly by many planktonic and benthic animals, which have the ciliated epithelium, like some larvae (trochophores,echinocardium, veligers), and like benthic filterers: bivalves, sponges, sedentary polychaetes, tunicates and corals. These animals, by feeding on LOM in concentration close to that of the ambient sphere, could compensate by its direct assimilation for 10–30% of their energy expenditures (Sorokin 1968, 1978b, 1984a, 1986a; see Tables 9.17–9.19).

The stocks of LOM and SOM in the water column over the reef being mobile and moving to and fro, driven by tidal currents, act as an energy interconnection link between the reef biotopes and, before all, between the biotopes of the windward flat where the autotrophic production predominates, and of the lagoon where the heterotrophic processes are more active (Johannes et al. 1972; Smith and Marsh 1973; Sorokin 1971b, 1978b, 1986a,b). The flows of detrital suspended matter and LOM connect the biotopes of flats with planktonic and benthic communities of the outer reef slope (Hubbard et al. 1981) and of the surrounding ocean (Marshall 1965; Coles and Strathmann 1973; Zieman et al. 1979; Darnell and Soniat 1979; Sorokin et al. 1982, 1983; Barnes et al. 1976).

3 Plankton in Coral-Reef Waters

For a long time an absolute predomination of bottom communities in functioning coral-reef ecosystems has been accepted axiomatically (Odum and Odum 1955; Johannes et al. 1970). But this opinion contradicted observations of the obvious abundance of planktonovore animals in reef communities. In fact, about half of the reef benthic animals are the filterers, or the sedimentary feeders, which feed on plankton. Among them are bivalves, some gastropods, sedentary pelychaetes, brittle stars and crinoids, many crustaceans, bryozoans, tunicates, and also corals, hydroids, and zoantharians. The zooplankton is a main source of food for the pelagic larvae of most bottom animals, of fish larvae, and of a dense population of such common reef planktonovore fish like Alanetta, Chromis, Caesio. Nevertheless, the plankton of coral-reef waters has been studied less than that of oceanic waters. Its real quantitative studies have started comparatively recently in 1968–72, and those of the microzooplankton only in 1981. The data, which we now have, are still scarce and patchy, and not only on metereotrophic microplankton but also on the traditional objects of plankton studies, phytoplankton and zoo-plankton. Still, even so we can also recognize that plankton in reef-waters is abundant, has high specific productivity and plays a sighificant role in reef trophodynamics and metabolism.

3.1 Bacterioplankton

3.1.1 Biomass and Production

The organic primary production of reef is extremely large. Being as high as 30–80 $g\,m^{-2}$ of wet biomass per day, it cannot be grazed directly by phytovores. The nongrazed organic matter enters the water column and bottom sediments as detritus. The enrichment of reef waters with dissolved organic matter, organic aggregates and detritus is well known (Gerber and Marshall 1974). It can enter the reef food-web mainly via the microbial link, and so bacteria should be a priori a key component in reef trophodymanics. This opionion was expressed by Odum and Odum (1955) long ago, but at that time corresponding quantitative evaluations were lacking (Wood 1960).

First attempts to quantify bacteria in reef bottom sediments were made by Di Salvo in 1969 on Hawaiian reefs. We have measured the microbial biomass and its production in waters and sediments of some Pacific reef systems in 1968–1969 (Sorokin 1971a, 1972). These studies demonstrated an exclusively rich microflora in reef biotopes, which as for its abundance in bottom sediments could be compared to fertilized ponds and the water columns up the reefs – with eutrophic oceanic waters. At the same time it was shown that filtering reef animals as well as the corals themselves feed on suspended bacteria. These data proved the important role of bacteria in reef trophodynamics and metabolism and thus stimulated further investigations in this field. The main directions of these studies constituted estimations of the standing stock and production rates of microbial biomass of the stocks and sources of labile organic matter for microbial production, and of the availability of bacteria as food for the reef fauna, including fine and crude filters and sedimentary feeders, detritovores, and sediment-eaters. For estimations of the microbial biomass, the total number of bacteria has been counted and the size of their cells determined microscopically, first on the membrane filters stained with erythrosine and cleared in immersion oil after Rasumov (Sorokin 1971b; Sorokin and Kadota 1972), and later on the black Nucleopore filters stained by a fluorochrome and analyzed under the epifluorescence microscope, after Hobbie-Wright (Moriarty 1979; Ducklow and Mitchell 1979b; Moriarty et al. 1985b). The production of microbial biomass was first estimated by the use of the ^{14}C-dark assimilation method by Romanenko (Sorokin 1971a,b, 1978b, 1986b). Later the thymidine method by Fuhrman and Asam was used (Moriarty and Pollard 1982). The first method was criticized by the former authors, nevertheless the production rate values obtained by both methods in general coincided (Table 3.1). Moreover, after my own experience with intercalibration methods I have arrived at the following conclusion. The nature of the basic coefficients, which are used in the formula for recalculation of the radioactivity of taken up ^{3}H-thymidine remains still indefinite and is selected by each investigator more or less arbitrarily: such as the ratio between the thymidine uptake and number of cell divisions (shift between 1 and $16 \cdot 10^{18}$ div.n mol^{-1}), content of carbon per cell (shift 3–4 times). Also, the specific activity of ^{3}H-thymidine preparations in each batch is largely indefinite and cannot be checked. Thus the thymidine method had to be calibrated each time it was used after some other method which gave more direct absolute values of microbial production in carbon units. In my experience the best method for this purpose appears to be just the dark uptake ^{14}C method in its modern modified form. Parallel estimations of microbial production by both methods properly intercalibrated showed no essential difference (Table 3.2).

The rates of respiration of bacterioplankton (M) were calculated as follows: $M = [P \cdot (1 - K_2)]/K_2$, where P is microbial production, and K_2 efficiency coefficient of microbial biosynthesis, which in planktonic bacteria

Table 3.1. Parameters of biomass and activity of bacterioplankton in waters over Pacific reefs; designations: N – total number of bacteria, $10^6\,ml^{-1}$; B_b – microbial biomass $mg\,C\,m^{-3}$; P – production ($mg\,C\,m^{-3}\,day^{-1}$), and M – respiration ($\mu g\,O_2\,l^{-1}\,day^{-1}$) of bacterioplankton; R_b – its food ration, and P_b – primary phytoplankton production, both in $mg\,C\,m^{-3}\,day^{-1}$. Data on Lizard I. by Moriarty et al. (1985b), on One Tree I. by Linley and Kopp (1986) rest by Sorokin (1971a, 1973, 1978b)

Reef	Abundance of living corals	N	B	P_b	P_b/B	M	R_b/P_p
Fringing reef off Lizard I. (GBR)	High	1–2.5	15–40	28–58	1.5–2.8	160–330	–
Tarawa atoll, Gilberts	High	0.89–1.36	49–110	14–42	0.2–0.4	80–240	1.8–2.5
Majuro atoll, Marshalls	High	0.69–2.60	14–78	10–36	0.2–0.7	60–200	1.5–19.0
Barrier reef off Bili-Bili I., New Guinea	High	1.39–3.32	53–123	14–30	0.2–0.3	80–180	5.6–24.5
Butaritari atoll, Gilberts	High	1.97–2.83	79–170	24–41	0.2–0.4	130–230	3.4–4.4
Ring reef, Heron I. GBR	High	1.26–1.78	40–72	25–38	0.5–0.6	140–215	2.1–11.0
Ring reef, One Tree GBR	High	0.12–1.30	2–32	1–147 (av. 14.6)	1.4–4	6–800	–
Saint Andrew reef, Admiralty I.	Average	0.37–0.74	11–21	7–14	0.6–0.7	40–80	6.2–7.6
Ninigo atoll, Admiralty I.	Average	0.29–0.62	13–32	7–21	0.5–0.7	40–120	3.0–7.0
Funafuti stoll, Ellis I.	Average[a]	0.22–0.45	8–18	2–6	0.3–0.6	14–37	3.1–5.0
Ngelelevu atoll, Lau I.	Weak, many dead corals	0.16–0.41	6–22	3–9	0.4–0.6	18–50	3.3–17.0
Fringing reef off Paama I., New Hebrides	Weak algal reefs	0.08–0.32	2–7	1–5	0.3–0.6	5–25	0.9–4.3
Acting volcano Lopevi I., New Hebrides	Weak[b]	0.11–0.14	2–3	1–5	0.5–0.8	10–12	0.5–0.9
Open ocean, trade winds area	–	0.05–0.1	1–2	1–3	1.0–1.5	5–20	0.5–2.0

[a] Reef was destroyed by cyclone.
[b] Rear corals directly colonizing the lava slopes, reef is absent.

Table 3.2. Abundance and production of bacterioplankton in waters of some reefs of GBR; numbers in brackets – microbial production after ^{3}H-thymidine method estimations. In – inshore zone of sandy shallows. For other designations see Tables 3.1, 3.4. (Data by the author)

Reef	Zone	Depth m	N	B_b	P_b	P_b/B_b	R_b/P_p
Heron I. ring reef	C_s	22	0.22	12.9	17.6 (8.8)	1.4	1.4
	R_s	17	0.35	20.7	20.2 (10.1)	1.0	1.8
	F	0.4	0.91	77.5	38.2 (29.0)	0.5	9.5
	R_e	0.6	0.41	34.8	20.0 (21.8)	0.6	12.0
	L	5.5	0.81	41.3	43.3 (24.6)	1.0	12.8
	L_e	0.7	0.46	31.6	24.6 (21.2)	0.8	18.0
One Tree I. ring reef	C_s	20	0.31	10.3	6.3	0.6	0.3
	R_e	3	0.46	27.5	31.9	1.2	9.3
	L	6	0.57	19.5	10.8	0.6	1.7
	F	0.5	0.31	18.4	27.1	1.5	5.9
	I_n	2	0.23	15.3	21.2	1.4	4.4
Lady Musgrave ring reef	R_e	2	0.82	19.5	10.5	0.5	2.5
	L	5	0.62	21.1	13.8	0.7	4.3
	I_n	0.8	0.65	15.0	11.6	0.8	3.3
Wistary ring reef	R_e	3	1.04	26.5	14.0	0.5	0.8
	L	5	0.61	15.6	14.7	0.9	0.5
	I_n	1	0.94	39.3	26.6	0.7	1.5
Lizard I. fringing reefs	C_s	20	0.58	11.9	8.5	0.7	0.5
	F	1	1.28	26.2	12.2	0.5	0.6
	L	8	0.81	22.0	7.4	0.3	0.4
	L_c	4	1.28	26.4	9.1	0.3	0.5
	R_s	3	0.75	25.5	6.9	0.3	0.4
	I_n	2	0.87	19.7	5.6	0.3	0.8

was 0.32 (Sorokin 1971b). The food ration of bacterioplankton (R_b) was calculated using the formula $R_b = P/K_2$. The rate of microbial decomposition of organic matter was measured with the BOD bottle technique by incubating the dark bottles in situ for 24 h.

The main results of quantifications of number, biomass and production rates of bacterioplankton in waters of some Pacific reefs are given in Tables 3.1–3.4. They demonstrate a high level in its development, which is comparable with its abundance in the mesotrophic – eutrophic waters, the most productive ones in upwelling areas. The total number of bacteria in reef waters varied within $0.5–3 \times 10^6\,ml^{-1}$, which is usually 4–6 times more than in the surrounding oceanic waters if reefs are situated in flows of oligotrophic waters, or 1.5–3 times more if the reefs are situated in more productive mesotrophic waters of large lagoons or in the vicinity of continents or large islands. The number and biomass of bacteria in waters over the reefs seems to be dependent upon the level of abundance of living corals on them. The biomass of bacterioplankton of flourishing healthy reefs with an abundant coral growth was usually high – 15 to $60\,mg\,C\,m^{-3}$, while up reefs that are barren and poor in living corals, or up the algal reefs, the microbial biomass in water columns was low – $1–10\,mg\,C\,m^{-3}$. It was also poor in waters of the formerly rich but cyclone-damaged reefs of the Funafuti atoll. In the vicinity of the Lopevi I., where real reef was absent, up the rocky slopes with rear corals the microbial biomass was as low as in the surrounding oceanic waters: $1–2\,mg\,C\,m^{-3}\,day^{-1}$. High values of microbial number and biomass were recorded on such reefs rich with living corals as the atolls Tarawa, Majuro, Butaritari, Heron, One Tree, Scott reefs and the barrier and fringing reefs off Lizard I. and New Guinea.

To evaluate the significance of the above levels of biomass standing stock of bacterioplankton as a food source, it is necessary to compare them with values of levels optimal for feeding or filtering reef fauna. For the crude filterers this is $20–40\,mg\,C\,m^{-3}$ and for the fine filterers $6–10\,mg\,C\,m^{-3}$ (cf. Sect. 3.1.2). Thus even at an average level of its standing stock the bacterioplankton of reef water could meet the food demand of most filter-feeders. Comparison of the standing stock of bacterioplankton and of phytoplankton in reef waters proves that bacteria comprise a dominating part of the total plankton biomass. Its biomass often exceeds the biomass of planktonic algae up to several order of values (Table 3.3). The cause of this seems to be quite obvious. The basic energy source for the bacterioplankton in reef environments is supplied by the prduction of bottom autotrophic communities which exceeds the production of phytoplankton by 1–2 orders (cf. Sect. 4.3.3).

The above-mentioned difference in the number of bacterioplankton in waters over the reef as compared with surrounding oceanic waters exists with a very intensive water exchange between reef and ocean. High gradients in the abundance of bacterioplankton under such conditions of active hydrodynamics could be supported by high rates of microbial production in warm reef waters enriched with labile organic matter. These gradients are also supported by an active washout of bacteria from the sediments, permamently resuspended by waves, and also through the rich

Table 3.3. Abundance of bacterioplankton and ratio of its biomass to the biomass of phytoplankton (B_b/B_p) in waters up the different zones of reefs. For designations, cf. Tables 3.1, 3.4. Reef zones: R_s – over the outer reef slope; C_s – 50–100 m off the outer reef edge. Data by Sorokin (1986a)

Reef	Zone	Depth, m	N	B_b	B_b/B_p
Atoll Velangilala,	L	5	0.56	18	36
Lau	F	1	0.66	21	9
	R_s	15	0.17	4	1
Barrier reef, S.E.	L	2	0.64	25	3
coast off New	R_s	15	0.86	64	12
Guinea	C_s	30	0.22	6	2
Fringing reef	F	0.5	0.09	2	5
Pentacost,	R_s	3	0.12	3	1
New Hebrides	C_s	30	0.06	1	1

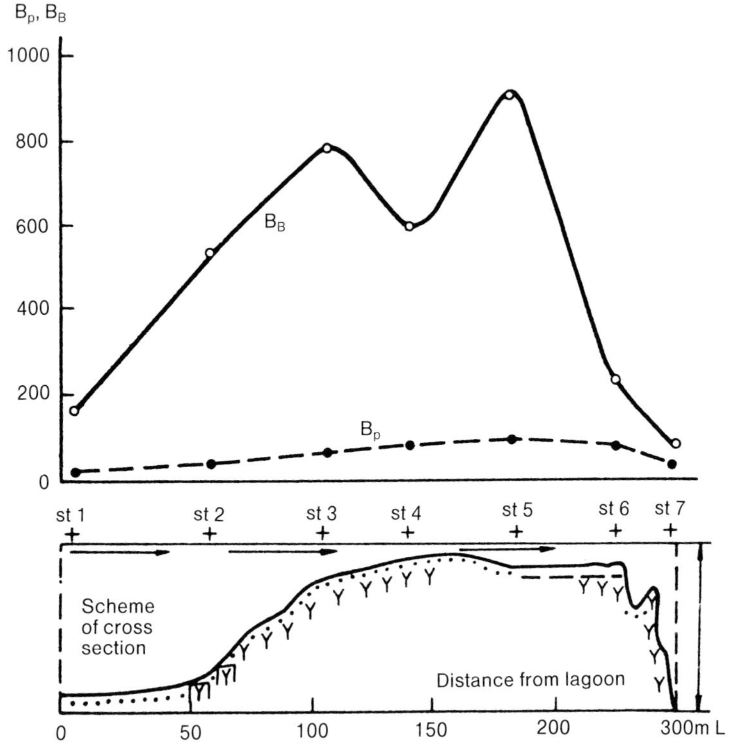

Fig. 3.1. Biomass of bacteria (B_b) and of phytoplankton (B_p) mg m^{-2} of wet weight at the section across the windward reef of the Conflict atoll (Louisiada I.); *L* distance; *arrows* show direction of current. (After Sorokin 1986a)

periphytonic communities. The most intensive multiplication of bacteria proceeds in waters near the outer reed edge, which use the fresh organic matter driven here from the flat by the ebb currents. When the tide comes these waters return again to the reef where the bacterioplankton thus produced is consumed by the reef filtering fauna (cf. Fig. 3.5). A high biomass of bacterioplankton could most often be found in turbulent waters up the outer reef slope, in waters up the lagoonal patch reefs, as well as in shallow lagoons (Fig. 3.1). In the water column of a deep lagoon, as in the Majuro atoll or the Conflict atoll, the microbial biomass was found to be usually low. Examples of the horizontal distribution of microbial biomass in water in different zones of reef at the cross sections are given in Figs. 3.1, 3.2 and in Tables 3.2, 3.3.

The microbial production in reef waters is usually also high (Sorokin 1978b, 1986b; Burns et al. 1984; Moriarty et al. 1985; Linley and Kopp 1986). Being at a level of 10–50 mg C m^{-3} day^{-1} it is commensurable with or, most often, exceeds the primary production of phytoplankton which in

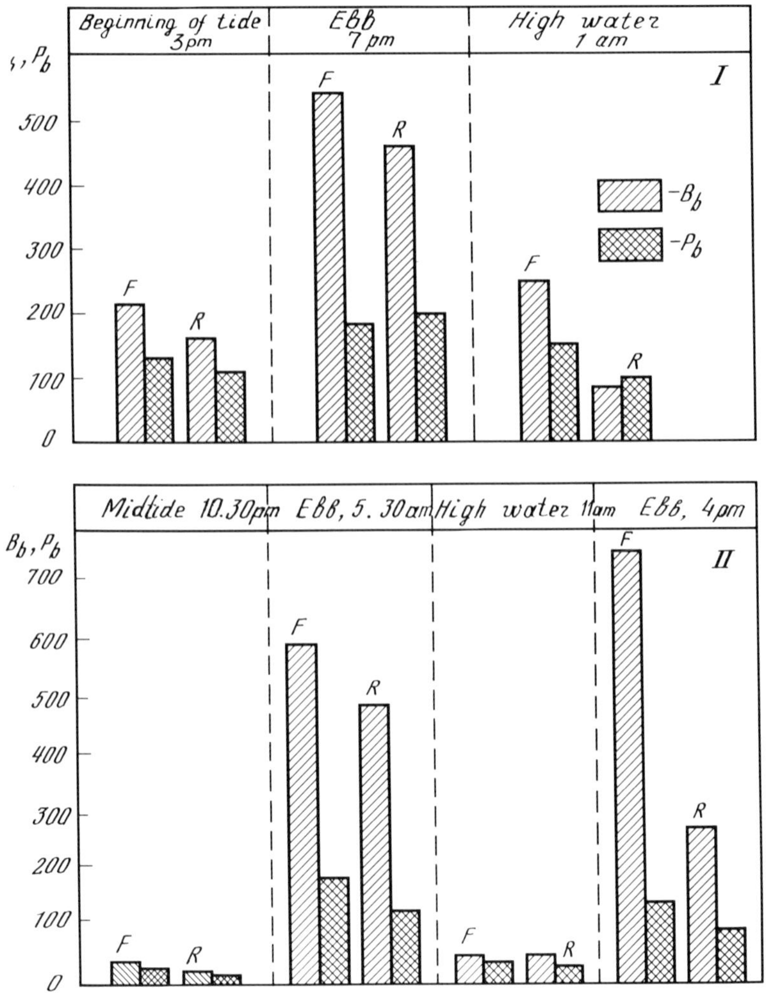

Fig. 3.2. Diurnal changes of the bacterioplankton biomass (B_b) and its production per day (P_b), both in mg m^{-3}, in waters over different reef zones depending upon phases of the tide; *F* reef flat; *R* patch reefs; *I* Scott reef, Timor Sea; *II* Koetivi I., Seychelles. (Data by the author)

reef waters usually comprises 10–30 mg C m^{-2} day^{-1} (cf. Sect. 3.2.2). Higher microbial production was recorded most often in waters with higher biomass, e.g., up the reefs abundant in living corals, such as the atolls and ring reefs Majuro, Butaritari, Heron, or the barrier reefs off New Guinea or off the Great Barrier Reef (Tables 3.1, 3.3). On reefs with weak coral growth the microbial production in the water column was low (Sorokin 1986). At the cross sections the maxima of microbial production were usually observed in waters of shallow lagoons, up the lagoonal patch reefs, and up the outer reef edges (Tables 3.1–3.3; Fig. 3.1). The coefficients of specific production (P/B) per day in the reef waters were on the average relatively low: 0.3–1.0 (Tables 3.3, 3.4; Fig. 3.3). Thus, they were most often lower than for bacterioplankton in the open ocean (Table 3.1). We explain this decrease in P/B by an admixture to the reef bacterioplankton of significant portions of less active benthic bacteria washed out from the bottom sediments. More recent estimations of microbial production and of its growth rates employing modern methods of ^{3}H-thymidine uptake (Landry et al. 1984; Moriarty et al. 1985; Linley and Kopp 1986) or FDC gave even higher P/B values of 0.06–0.17 per hour, or 1.6–4.0 per day. Thus the doubts about overestimations of microbial production rates by using the ^{14}C dark uptake method, critically referenced by Moriarty et al. (1985), seem to be premature.

The calculated rates of microbial respiration (Table 3.2) in waters over flourishing reefs were high: 0.1–0.3 mg $O_2 l^{-1} day^{-1}$. It was equal there to 60–80% of total respiration of the microplankton, as measured by the BOD bottle method. The respiration rates of bacterioplankton and its food rations calculated in carbon units per day usually exceed several times the primary phytoplankton production (Table 3.4). Only in the waters of the Lizard I. reefs, situated in productive waters of the GBR lagoon, were their ratios in winter less then 1 (Table 3.2). Thus only a minor part of the local primary production could be a source of energy for the bacterioplankton in reef waters. Basic energy it obtains from bottom biotopes, excreting labile dissolved and particulated organic matter (cf. Sect. 2.3).

Table 3.4. Bacterioplankton biomass and activities in waters over some Pacific reefs. Designations of reef zones: L_c – coastal lagoon; R_e – patch reef zone ("rear"); F – reef flat; B_r – brakers zone; L_c – central lagoon; A – percentage of bacteria in aggregates of size 3–5 μ. For other designations see Table 3.1. (After Sorokin 1971a, 1978b, 1986a)

Reef	Reef zone	B_b	P_b	P_b/B_b	M	R_b/P_p	A %
Coconut I., Hawaii,	F	43	28	0.7	160	2.5	17
inside the Bay	L	21	31	1.5	175	1.5	22
Same, entrance of	F	63	46	0.7	260	6.2	10
the Bay	R_e	45	56	1.2	320	2.6	23
Same, external	F	11	16	1.5	90	1.9	18
barrier reef	B_r	5	6.5	1.4	37	4.1	11
Majuro atoll,	F	2.3	1.6	0.7	9	1.9	21
external reef	L_c	1.8	2.7	1.5	15	2.5	24
Same, lagocnal reef	L_c	29.0	5.4	0.2	30	6.9	11
	F	19.0	7.4	0.4	41	5.7	15
Scott reefs	L_c	116	14	0.1	80	1.5	25
(northern)	F	51	12	0.2	68	4.0	15
	R_s	146	25	0.2	140	17.0	19
	B_r	14.6	57	4.0	322	–	–
One Tree I., ring	F	14.3	60	4.2	339	–	–
reef, GBR	R_e	11.3	22	2.0	124	–	–
	L	12	19	1.6	107	–	–

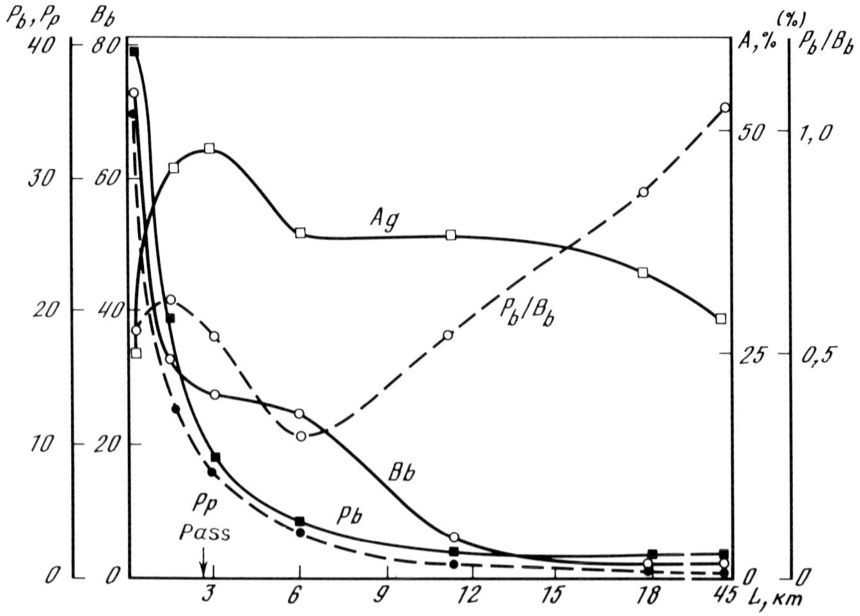

Fig. 3.3. Changes in the indices of biomass and production of bacterioplankton at the N.E. cross-section from the lagoon of the Heron I. ring reef out to the open Coral Sea; B_b and P_b biomass of bacterioplankton, its production per day, and P_p primary phytoplankton production per day, all in mg C m^{-3}; *Ag* bacterioplankton in aggregates, % of total; *L* distance from the lagoons. (After Sorokin 1973c)

Data on the biomass and production of bacterioplankton in waters over some fringing reefs and atolls of the Indian Ocean are given in Table 3.5. The highest biomass level – up to 120–170 mg m^{-3} – was recorded in waters over reefs with a rich coral growth situated either in mesotrophic waters of the equatorial countercurrent (Koetivi I.) or in oligotrophic waters of the trade wind current (Kargados-Karahos Islands). The same was observed in waters over the reefs in the South China Sea off Vietnam surrounded by mesotrophic – eutrophic waters and at the Sinton atoll surrounded by oligotrophic waters. The rates of microbial production varied within 10 to 30 mg C m^{-3} day^{-1}. The specific production coefficients P/B were usually relatively low there (0.1–0.7 per day). An abundance of bacterioplankton in waters over reefs with an abundant coral cover becomes more understandable in the light of recent findings by Paul et al. (1986) of an abundant microflora in surface microlayers over the colonies of living corals; up to sevenfold higher for microbial numbers and production.

The biomass of bacterioplankton in reef waters fluctuates during the day depending upon the tidal periods (Fig. 3.2). A lower level was observed during high tide and a higher at the ebb, and this difference sometimes was up to one order of values. The production rates of bacterioplankton also

Table 3.5. Limits of microbial biomass and production in waters over reefs in the Indian Ocean. For designations see Table 3.1. Original data

Reefs	B	P	P/B
Fringing reef off Socotra I.	7–26	3–17	0.2–0.7
Fringing reef of Praslin I., Seychelles	24–62	4–15	0.2–0.6
Fringing reef of Koetivi I., Seychelles	110–170	4–18	0.05–0.1
Kargados-Karahos super atoll	12–120	6–19	0.1–0.8
Fringing reefs, Kuavan bay, Vietnam	20–60	10–15	0.2–0.7
Fringing reefs, Van Fong bay, Vietnam	6–21	11–21	0.3–0.7
Fringing reefs, Vung-Ro bay, Vietnam	115–260	20–45	0.1–0.4
Fringing reefs off The Thu I. Vietnam	17–110	6–26	0.2–0.4
Sinton atoll, South China Sea	16–70	4–7	0.1–0.4
Fringing reefs off Spratley I., South China Sea	14–35	6–12	0.2–0.4
Fringing reefs off Namsu I., Gulf of Siam	12–30	10–18	0.6–1.0

increase during the ebb up to 3–5-fold. Thus, the increase in its biomass during the ebbs is brought about not only by its washout from the sediments and from the above-mentioned coral surface microlayers, but also at the expense of its accelerated production. An increase in the microbial activity in reef waters during the ebbs was established also by measurements of diurnal changes of rates of ^{14}C-glucose assimilation (Meyer Reil et al. 1981). Significant diurnal changes in specific rates of microbial production were detected by Moriarty et al. (1985) in waters over the reef flax of the Lizard I. reefs. At 6 a.m. the number of cells dividing per hour was $5 \times 10^7 l^{-1}$, and at 4 p.m. $38 \times 10^7 l^{-1}$. These changes were most probably connected with the photosynthetic production and excretion of organic matter by bottom autotrophes because in the lagoonal waters these fluctuations were minor.

A significant part of the reef bacterioplankton is associated with aggregates, with particles composing "marine snow", and with the reef pseudoplankton, so abundant in the vicinity of coral reefs (Marshall 1968; Lewis and Price 1975). The same aggregates participate in the formation of "marine snow" in oceanic surface waters. This is formed there by planktonic bacteria, which produce films and flakes during their growth independently of the presence or absence of other particles (Sorokin 1970). In reef waters the participation of the coral mucus in this process of aggregates formation can be supposed (Johannes 1967). Experimentation with labeled protein hydrolizates (Sorokin 1978b), as well as by direct microscopy (Moriarty et al. 1985), showed that some 20 to 50% of the total bakterioplankton present in aggregates are of 3–10 μ sizes. Experiments with reef water samples filtered through the 1-μ filter to keep out particles and aggregates showed that in the presence of an inhibitor of microbial growth the aggregates did not form even in the presence of coral mucus, added as homogeneous prefiltered solution. But without inhibitor these aggregates formed readily, especially in samples enriched with glucose, which enhances microbial growth (Sorokin 1978b). The effect of stimulating their formation by coral mucus was detected only after 5 days of exposure. Mucus flakes gradually became inhabited by bacteria, which used them also as a source of nutrition (Ducklow and Mitchell 1979b; Rublee et al. 1980; Paul et al. 1986). Nevertheless, the formation of aggregates is an immanent feature of microbial growth and proceeds also without suspended particles or mucus, which were supposed to be the centers of microbial growth. The aggregates grew well even in prefiltered water. The aggregation of a significant part of bacterioplankton makes it accessible for grazing by crude filterers and even by seizers.

It was shown above that the microbial biomass and production in waters over the reefs are several times greater than in the surrounding oceanic waters, especially in oligotrophic regions. The influence of large reefs upon the abundance of bacterioplankton in oceanic waters was detectable at a distance of 10–12 km, as it could be seem from Fig. 3.3. The microbial biomass and also its production will gradually decrease within this interval to a very low level, but the specific production rates (P/B) will increase. The latter is connected with the enhanced trophodynamics in pelagic ecosystems of tropical ocean, as well as with the absence of an additional input of inert microbial biomass from the sediments, which happens in waters over the reefs. The enrichments of oceanic waters, in contact with reefs, by the dissolved organic matter could be seen from the elevated values of the potential microbial production, which is an indirect indicator of the relative abundance of labile organic matter in sea water. For example, at the cross section from the Tarawa atoll to the open ocean elevated values of potential microbial production – up to 83 mg C m^{-3} – were found at a distance of over 6 km from the atoll. Further, it decreases to 9 mg C m^{-3} at a distance of 17 km (Sorokin 1973c). We may conclude that the highly productive reef ecosystems enrich waters of the surrounding oceanic and shelf waters with bacterioplankton and with labile organic matter, thus increasing their food sources, which are used by fish.

3.1.2 Bacteria as Food for Coral Reef Animals

Coral reefs are inhabited by an abundant filtering fauna. Being an important component of the plankton in reef waters, the bacterioplankton was supposed to be an important source for its feeding. This was proved experimentally by Di Salvo (1971b, 1973). He placed dead coral heads

Table 3.6. Filtration rates (F) and the food rations per day (R) in benthic filterers feeding on bacterioplankton; initial concentration of bacteria in experiments 2–3 × 10^6 cells ml^{-1}. (After Sorokin 1978b)

Animals	F: $l\,g^{-1}$ of wet weight/h	F: $l\,g^{-1}$ carbon per day	R, % carbon in consumer's body
Polychaetes			
Cladopterum sp.	0.70	130	1.70
Megalomma sp.	3.00	650	8.40
Serpulobris sp.	4.00	670	8.70
Sponges			
Toxadocea violacea	0.45	380	5.70
Myxiella rosacea	0.10	85	1.10
Halichordia melanodovia	0.14	103	3.50
Oysters			
Crassostrea gigas	0.73	147	2.00
Ostrea sardwichiensis	0.18	33	0.43
Tunicates			
Herdmania sp.	0.05	68	0.90
Ascidia nigra	0.05	32	0.40
Didemnium sp.	0.30	410	5.60

inhabited by various sessile infaunal filterers into water containing ^{35}S-labeled suspended bacteria and recorded a rapid decrease in its contents. The same phenomenon was observed when he placed into this water the heads of living corals. Thus he established the ability of corals for sedimentary filter-feeding with planktonic bacteria. Di Salvo also demonstrated the same ability in the tridacna clams. These studies were prolonged with the use of ^{14}C-labeled natural bacterioplankton (Sorokin 1972, 1973c, 1974, 1978b, 1984a). Experiments have been accomplished with common reef filterers: with oysters (*Crassostrea gigas*, *Ostrea sandwichiensis*), with ascidians (*Ascidia nigra*, *Hedmania* sp.) with sponges (*Toxadicea violacea*, *Myxiella rosacea*, *Halichordia melanodovia*) and with the sedentary polychaete worms Megalomma, Hydrodies and Cladopterum (Table 3.6; Figs. 3.5, 3.7). All these filterers, besides their ability to feed on phytoplankton, feed also on bacterioplankton in its ambient concentrations in reef waters (Fig. 3.7). The fine filterers, such as sponges and polychaete worms, feed on bacterioplankton more efficiently than the crude filterers, being able to filter out from the reef waters both the aggregates and the single cells of bacteria. In fact, the efficiency of their feeding did not decrease when the aggregates were taken out from the water by prefiltration. But in experiments with crude filterers, as the bivalves, the *Cladopterum polychaeltes* and the tunicates, the same prefiltration decreased the rate of their feeding 2–5 times (Fig. 3.5). Thus, they made use mostly of the unaggregated part of the natural bacterioplankton.

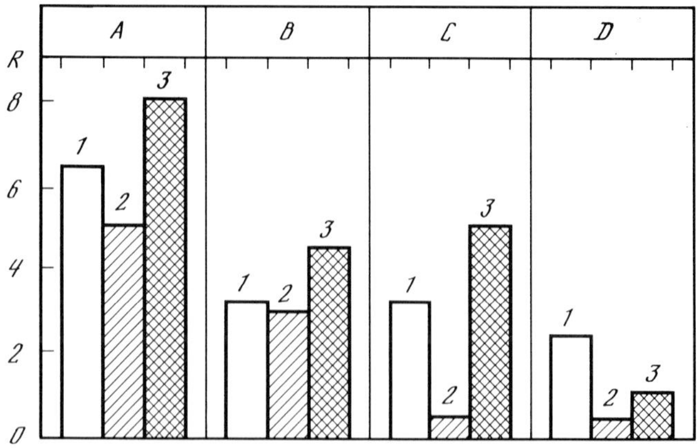

Fig. 3.4. Spectrum of feeding in some reef filter-feeding benthic animals; kinds of food offered: *1* natural bacterioplankton; *2* the same devoid of the aggregates by prefiltration via filter with 3 μ pores; *3* phytoplankton. Animals in experiments: *A* sedentary polychaete worm *Megalomma* sp.; *B* sponge *Toxadocea*; *C* ascidian *A. nigra*; *D* oyster clam *Crassostrea*; *R* food ration per day, % of carbon content in the body of animal. (After Sorokin 1978b)

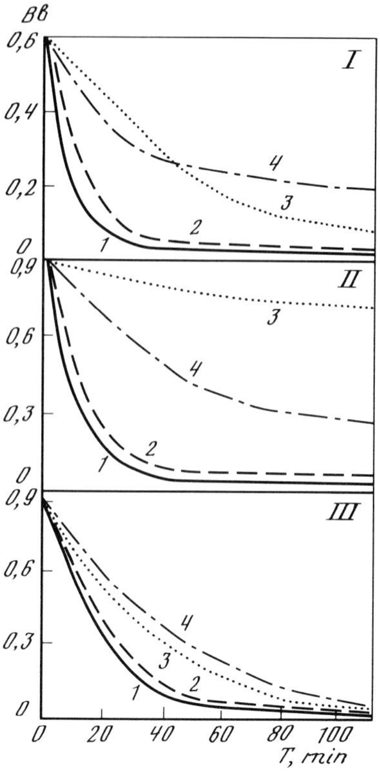

Fig. 3.5. Decrease of concentration of microplankton in water in presence of reef filter-feeders: *1* sponge *Toxadocea*; *2* sedentary polychaete *Megalomma*; *3* ascidian *A. nigra*; *4* oyster *Crassostrea*. Kinds of microplankton: *I* natural bacterioplankton; *II* the same devoid of aggregates by prefiltration through filter with pores of 3 μ; *III* phytoplankton (dinoflagellates *Prorocentrum*); *T* time of exposure; B_b biomass of microplankton in water, mg m^{-3} of wet weight. (Data by Sorokin 1978b)

The fine filterers decreased the bacterioplankton concentration in experiments from 2×10^6 to $2–4 \times 10^4$ cells ml^{-1}, i.e., by 1–2 orders (Fig. 3.6). The latter level of bacterioplankton density is severalfold lower than in the oligotrophic oceanic waters surrounding the Hawaiian reefs, where the experiments had been carried out ($20–40 \times 10^4$). These data prove that the fine reef benthic filterers may use the bacterioplankton from the oceanic waters passing over the reefs, and also that near the islands their trophic level and the microbial biomass increase. The crude filterers in accordance with these data may utilize 30 to 70% of the total bacterioplankton standing stock in reef waters, including its aggregated part plus some portion of single microbial cells (Figs. 3.4–3.6).

The same experiments proved that the bacterioplankton as source of feeding for reef filterers is comparable to phytoplankton (Fig. 3.4, Table 3.6). Their food rations from feeding with the bacterioplankton at its ambient concentrations appeared to be commensurable with those from their feeding with algae (dinoflagellates). The stock of bacterioplankton biomass is used also by pelagic filterers, such as the veligers, which feed on it even more efficiently than on phytoplankton (Sorokin 1973c). Among the fine planktonic filterers the most important are the appendicularians Oicopleura, which belong to the most efficient fine filter feeders of marine zooplankton. The filtering copepods Paracalanus are intermediate between fine and crude filterers, and the calanids of the genus *Calanus* are crude filterers and can use only the aggregated part of bacterioplankton. Thus, because of the abundance of bacterial aggregates, the bacterioplankton can

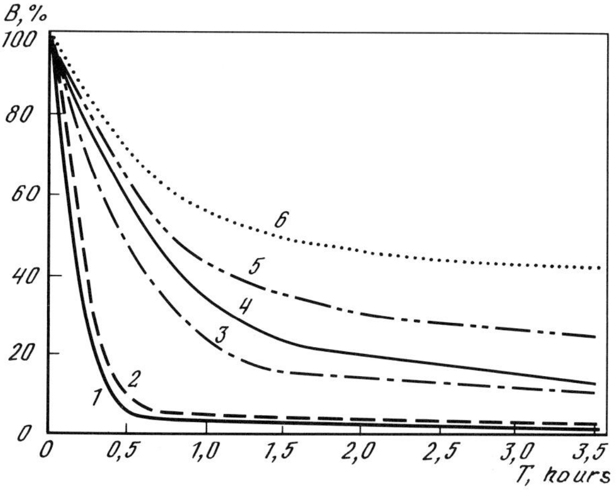

Fig. 3.6. Decrease of concentration in labeled bacterioplankton in water in the presence of reef filter-feeders: *1* sponge *Toxadocea*; *2* sponge *Halichordia*; *3* colonial ascidian *Didemnium*; *4* ascidian *A. nigra*; *5* ascidian *Herdmania*; *6* oyster *Crassostrea*. Initial biomass of bacterioplankton (*B*), accepted as 100% on ordinate, was equal to 0.84 mg l^{-1} of wet weight; *T* exposure time. (Data by Sorokin 1978)

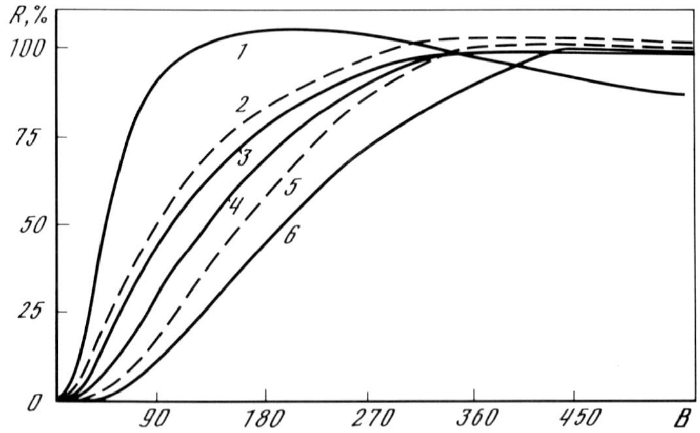

Fig. 3.7. Dependence of the feeding rate (relative values, *R*) of some filter-feeders upon the bacterioplankton biomass in water (*B*, $\mu g\,l^{-1}$, wet weight): *1* sponge *Toxadocea*; *2* veligers of *Gastropoda*; *3* sedentary polychaete *Megalomma*; *4* coral *Pocillopora*; *5* *Crassoatrea*; *6* copepod *Eucalanus*. (After Sorokin 1978b)

be an important food source even for the crude-filtering zooplankters, which dominate in the reef zooplankton (Rieper 1978). A large amount of microbial aggregates may be ingested even by the catching mixed-feeders like Acartia and also by fish larvae.

The efficiency of filter-feeding of reef animals could be evaluated by measuring the rates of filtration. Estimations were made in experiments with natural bacterioplankton labeled with ^{14}C. The most efficient among them appeared to be the fine sessile filterers: sponges, sabellid and serpulid polychaetes. Per 1 g of their body carbon they were able to take out 0.3–0.8 μ sized microbial cells from 400–600 l of seawater per day (Table 3.6). It means that with their density of $200–300\,g\,m^{-2}$ in a bottom biotope they can strain off all the suspended food including bacteria from a water column 2–3 m thick. The rate of filtration of natural bacterioplankton in crude filterers like clams and the Cladopteran polychaetes was found to be also significant: more than $100\,l\,g\,C^{-1}\,day^{-1}$. The avoidance of aggregates by prefiltration decreased them in crude filterers 5–10 times (Table 3.7). The rates of filtration of the natural bacterioplankton by fine and even by crude filterers appeared to be comparable with those for phytoplankton, which has been recognized not so far ago as the main or even as the single source of food for reef filter-feeders (Table 3.7).

The food rations of benthic filter-feeders feeding on bacterioplankton have been estimated by us in experiments either by calculations based upon the values of the rates of filtration, or by their direct estimation as quantities of ingested food during the short-term exposures in water with labeled bacterioplankton (Tables 3.6–3.8). In oysters, sponges and ascidians values per day, as related to the carbon in their bodies, varied within 0.5 to 5%. In

Table 3.7. Rate of filtration (F), food ration per day (R) in benthic filter feeders from the Hawaiian reef feeding with: a) natural bacterioplankton; b) bacterioplankton prefiltered, to avoid the microbial aggregates; and c) phytoplankton (dinoflagellates). Data by Sorokin (1978b)

Animals	Kind of suspended food	F $1\,g^{-1}\,C\,day^{-1}$	R % of carbon content in body of consumer.
Sponge			
Toxadocea violacea	a	276	3.6
	b	335	3.7
	c	426	5.8
Polychate worm			
Megalomma sp.	a	780	10.2
	b	755	8.3
	c	915	13.3
Tunicate			
Ascidia nigra	a	182	2.4
	b	19	0.3
	c	120	5.9
Oyster			
Crassotrea sp.	a	152	2.0
	b	15	0.2
	c	231	3.2

Table 3.8. Food rations (R) and assimilability of food (I) in filtering reef animals feeding on ^{14}C-labeled natural bacterioplankton; duration of experiment 30 min. Data by Sorokin (1978b)

Animals	R		I, %
	$\mu g\,C\,h^{-1}$ specimen^{-1}	Body carbon per day, %	
Sponge Halichordia	86.00	3.40	82
Tunicate			
Ascidia nigra	41.00	1.60	83
Polychate *Megalomma*	7.10	7.80	73
Oycter *Crassostrea*	22.00	2.20	68
Coral *Pocillopora*	0.95	5.50	76
Veligers of gastropods	0.013	51.00	74

the fine-filtering polychaetes it reached 8.7%. In the crude-filterer polychaete Cladopterum it was several times less: 1.7%. But even in crude-filterers, such as oysters, the relative food rations with their feeding on bacterioplankton were not much less than those with their feeding on planktonic algae (Table 3.7). The above values of relative food rations in some filtering bottom animals feeding on bacterioplankton were on the ordinary level for macrobenthic animals of corresponding size. Relatively low rations in sponges and in ascidians were mostly the consequence of a high content of inert skeletal organic carbon in their bodies, which was taken into account during the calculations of relative rations.

As regards optimum concentrations of bacterioplankton for the feeding of filtering benthic and planktonic animals, corresponding values estimated with the use of labeled bacterioplankton are given in (Fig. 3.7). For most fine filterers, like sponges and Oicopleura, this was attained with a bacterioplankton biomass of 18–30 $mg\,C\,m^{-3}$ which represented an average level in most reef waters. For crude reef filterers, which used only the aggregated part, the optimum was 3–4 times more: 60–80 $mg\,C\,m^{-3}$, while ordinary biomass in water over the reefs reached 30 to 100 $mg\,C\,m^{-3}$. Thus not only fine- but crude-filterers too can acquire their food rations or a significant part of it by feeding only on the bacterioplankton.

The corals themselves can actively feed on bacterioplankton (Table 3.8; Fig. 3.7). Their relative rations with feeding on bacterioplankton attain up to 5%, i.e., the same values as of powerful filtering animals, such as the sponges or oysters. More detailed data are given in the chapter on feeding of corals (cf. Sect. 9.2.1).

The assimilability of ingested bacterioplankton by filter-feeders appeared to be very high: 60–80% (Table 3.8).

A scheme generalizing the results of studies on the trophic importance of bacterioplankton in reef ecosystems is given in Fig. 10.7. It shows that bacteria accomplish an important function, being an intermediary energy link between the bottom autotrophs and benthic secondary producers, the filtering animals. Quantitative evaluations of this function of bacterioplankton are given elsewhere (cf. Sect. 10.3).

3.1.3 Luminescent Bacteria

Luminescent bacteria, which inhabit oceanic waters, emit light without external stimulation at the expense of energy of their metabolism resulting in the accumulation of ATP in them. The latter participates in light emittance via the luciferin-luciferase system of luminescent bacteria (Tchumakova and Gitelson 1975; Primakova 1985). They are represented in the sea by the rod-shaped cells and are related to the group of enterobacteria – inhabitans of animal intestines. Their habitat main is indeed the intestines of marine animals and their special luminescent organs. Being excreted by animals with feces, they enter the bacterioplankton, and being ingested again by filter-feeders, can get into the intestines again, thus distributing themselves among the populations of marine animals. The luminescent bacteria appeared to be very abundant in coral-reef habitats. In waters above the reef their numbers are more than by an order over those in oceanic waters, comprising 10^3–10^4 cells ml^{-1} according to counts of their colonies on nutrient broth agar medium. Sometimes they attained there more than half the total of colony counts of heterotrophic bacteria (Primakova 1985). Most of them inhabit detritus particles or the surface of seaweeds. Dense populations of them were found in the intestines of different reef animals, such as fish and holothurians, as well as on the surface of living corals. Thus, the luminescent bacteria in reef habitats are mostly connected with animals and plants.

3.2 Phytoplankton

Phytoplankton is second in its importance as component of microplankton in coral-reef waters. Correct estimations of its number and biomass, of its production and even of its species composition meet with methodological caveats, which we have only recently started to overcome. But then most of the existing data had been collected with the use of methods which were far from perfection. Thus, when discussing the results given here we must consider these realistically. The phytoplankton samples were collected in reef waters with plankton nets of over 50 mesh (Shirota 1966; Glynn 1973b; Ricard 1977b), while through the nucleopore filter of 5 μ up to 50% of the total phytoplankton biomass may pass (Sorokin et al. 1977; Legendre et al. 1988). Most phytoplankton data were collected by the Utremöl sedimentation method. But the fixative, which during several months separates the collection and the processing of samples ruins 70–90% of planktonic algae, especially the small pennats, naked dinoflagellates, and coccoliths. This method also drastically underestimates the blue-green algae, which are floating instead of settling and therefore evade the counts (Sorokin 1986b). Definite and well-known problems arose around the interpretation of C^{14}-primary production data (Sorokin 1971b; Eppley 1981). Early estimations in reef waters made after the original Steeman-Nielsen version may have resulted in underestimates by a factor of 2 to 4.

No wonder then that the first estimates of phytoplankton primary production have led to the opinion that phytoplankton in reef waters was negligibly little (Doty and Capurro 1956; Jones 1963). But the following investigations gave quite a different picture (Angot 1968; Sournia 1968; Ricard 1977c). They showed that at least sometimes during the season phytoplankton in reef waters could be abundant. But the controversy in evaluating its role as a functional component of reef ecosystems is still alive and needs impartial discussion and further investigation.

3.2.1 Taxonomy

The best-studied taxonomic group of phytoplankton in reef waters is the diatoms. Nevertheless they are not always dominating. But they are the best survivals after the Utremöl procedure. For example, in a lagoon of the barrier reef off Tahiti I, they comprised only 10–30% of the total primary production. The larger part of it was composed by nanoplanktonic algae-flagellates and coccoliths (Ricard 1977b).

Fundamental studies of phytoplankton in reef waters off some islands of French Polynesia were accomplished by Ricard (1976, 1977b). He described 580 species of algae in them, which belonged to 70 genera. The most common of them, which were found in samples all around the year, were *Actinocyclus ehrenbergii*, *A. subtilis*, *Ethmodiscus appendiculatus*, *Gyrosigma balticum*, *Melosira sulcata*, *Navicula longilyra*, *Nitzschia seriata*, *N. longissima*, and *Surirella fastuosa*. During the seasons the phytoplankton communities were periodically dominated by some common *Chaetoceros* species: *Ch. curvisetus*, *Ch. atlanticum*, *Ch. compressum*, and by some other diatoms, like *Synedra acus*, *Amphora lineolata*, *A. angulata*, *Coscinodiscus moelleri*, *Surirella gemma*, and *Asterionella notata*. In lagoonal phytoplankton during some seasonal periods the blue green algae Lyngbia, Nostoc and Oscillatoria were numerous.

In the phytoplankton samples collected by French researchers in lagoons of some atolls of the Society and the Tuamotu Islands often the dinoflagellates Goniaulax, Ceratium, Peridinium were more numerous than the diatoms (Sournia and Ricard 1975a,b, 1976a). Dense populations of them were recorded in the lagoon of the Takapoto atoll. The cocolithophorids and blue-green algae were also numerous. Total numbers of planktonic algae, recorded in the lagoonal waters were high even with the use of Utremöl sedimentation: 10^4–10^6 cells l^{-1}, to say nothing of the possible loss of a larger part of them during 8 months storage of the Utremöl samples before their processing in Paris. The same authors pointed out a high percentage of bentho-planktonic diatoms in the samples. But this statement, which could be found also in earlier publications (Marshall 1933; Jeffrey 1968), may be as well an artifact, caused by better resistance of these algae to fixatives and to storage. If the delicate cells of other algae had been ruined during this procedure, the ratio was artificially moved to the firm bentho-planktonic diatoms.

The seasonal observations revealed a real "bloom" of phytoplankton in lagoons of atolls, such as Takapoto, Rangiroa, Moruroa and Rongelap (Johnson 1954; Michel et al. 1972). So even by employing insufficient sedimentation methodology, leading to a large loss of phytoplankton cells, the above research demonstrated that in waters of the lagoons of some atolls a varied and abundant phytoplankton exists. Large seasonal variations in its number and composition finally disproved former ideas about the homeostasis in plankton communities of tropical seas.

Observations on phytoplankton in the lagoon of the Great Barrier Reef revealed its seasonal maxima to be in Spring and in Autumn (Relevante and Gilmartin 1982). The first was dominated by the diatoms *Chaetoceros compressus*, *Bacteriastrum comosum* and *Nitzschia pungens*. During the second maximum in autumn besides the above mentioned also *Thallasiotrix frauenfeldii*, *Chaetoceros tetrastycon*, *Rhyzosolenia setigera* and *Dactylosolen mediterranea* were numerous, and together with nanoplanktonic algae numbered up to 3×10^7 cells l^{-1}. The total numbers of diatoms and dinoflagallates were 10^4–10^5 cells l^{-1}. The spring blooms of the blue-green alga *Trichodesmium rubrum* caused there sometimes real "red tides". In December 1975 – and also in July and August – I have observed from the plane the brownish stripes of masses of floating *Trichodesmium* stretching in the lagoon of the GBR between Bundaberg and Townsville for several tenths of kilometers.

Our observations accomplished in different reef regions, taking counts of algae on the membrane filters, showed that especially this alga and some other bentho-planktonic filamentous blue-greens are an important and sometimes even dominating feature in phytoplankton. This alga often attains mass blooms in oligotrophic tropical oceanic and coastal waters (Bowman and Lancaster 1965; Quasim 1970; Sorokin and Tsvetkova 1972). It was dominating in waters of New Guinea, New Hebrides, Lau, and New Guinea reefs (Table 3.9). This alga actively grows in open ocean up the thermocline at depths of 40–60 m. At a late phase of its growth the cells form coenobia of 0.5 – 3–5 mm size, which float up to the surface. At this stage they are poorly photosynthesizing (Relevante and Gilmartin 1982) and transfer to heterotrophic metabolism (Karnaukhov and Yashin 1980). The wind-driven floating algae accumulate in large stripes tenths and hundreds of kilometers long (Quasim 1970). Their biomass in these stripes could attain tenths of grams per m^{-3}. Coral reefs are natural barriers in the way of their drift. They are accumulating over them in large masses. Thrown out by waves up to the supralittoral zone, they comprise an additional source of detritus for reef detritovores. This phenomenon we observed on a large scale at reefs of the GBR and of Vietnam (Sorokin et al. 1982, 1991).

Table 3.9. Phytoplankton number (N) and range of domination of main groups: Bg – blue-greens; Dt – diatoms; Di – dinoflagellates

Reefs	N, $10^3 l^{-1}$	Domination of groups by biomass
New Guinea	12–44	Bg, Di, Dt,
Katafang atoll, Lau	6–12	Bg, Di, Dt,
Velangilala atoll, same	10–14	Bg, Di, Dt,
Ngelelevu atoll, same	3–8	Bg, Di, Dt,
Tuvuta Island, same	4–8	Dt, Di, Bg,
Titia Island, same	6–14	Bg, Dt,
Panarora Island, New Hebrides	5–15	Bg, Dt, Di,
Paame Island, same	13–58	Dt, Di,
Calvados atoll, Louisiada	3–26	Dt, Di, Bg,
Conflict atoll, same	5–25	Dt, Di, Bg

The phytoplankton in waters over the reefs of the Indian Ocean was investigated by French researchers. According to Sournia (1968) the diatoms comprised about 30% of the total species, omitting the nanoplanktonic algae. In lagoonal waters of the barrier reef of Tulear, Madagascar in total 152 species of diatoms and 158 species of dinoflagellates were listed. Most of the mass species were usually in the surrounding oceanic waters, such as the diatoms *Chaetoceros danicum*, *Ch. decipiens*, *Ch. curvisetus*, *Coscinodiscus lineatus*, *C. nodulifer*, *C. radiatus*, *Asterionella japonica*, *Navicula lyra*, *Nitschia closterium*, *N. seriata*, *Rhizosolenia habirara*, *Thalassiosira nitzschioides*, *Thallassiotrix frauenfeldii*, and the dinoflagellates *Ceratium furca*, *C. fusus*, *C. horridum*, *C. pentagonum*, *Peridinium conicum*, *P. oceanicum*, *P. depressum*, *Goniaulax polygramma*, *G. kofoedii*, *Prorocentrum micans*, *Pyrocistis lunnula*. A significant part of phytoplankton communities in waters of the Madagascar and Mauritian reefs was comprised by small silicoflagellates, coccolithophorides, naked flagellates and picoplankton cyanobacteria (Travers and Travers 1965). The blue-green alga *Trichodesmium* dominated in reef waters around the Nosy Be Island Its dense blooms had been recorded in March, when its biomass attained 1.6 g m^{-3} (Fig. 3.8). We have observed the same there at the end of March 1983. Masses of wind-driven *Trichodesmium* were recorded accumulating up the reefs of the Kavaratti atoll (Quasim 1970).

More recent studies of phytoplankton in reef areas of the Indian Ocean and of the South China Sea were made with the use of more modern and

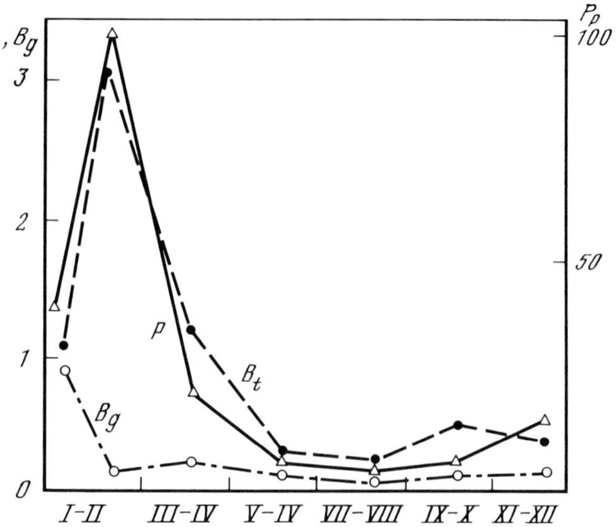

Fig. 3.8. Seasonal changes in phytoplankton in waters over reefs in the D. Ambaro bay, Nosy Be I., Madagascar; B_t and B_g total biomass of phytoplankton and biomass of blue-green algae, g m^{-3} of wet weight; P_p, primary production, mg C m^{-3} day^{-1}. (Modified from Angot 1968)

adequate methodology (Orlova 1985). The phytoplankton cells were concentrated by gentle reverse filtration at large (90 mm diameter) 1 μ Nucleopore membranes (Sorokin 1986a). The concentrates fixed with a weak Utremöl solution were instantly counted in the shipboard laboratory. In addition, epifluorescence microscopy was used to count picoplanktonic cyanobacteria. These studies revealed a real blooms of diatoms, in waters up the reefs off Socotra and Praslin Islands. The blooms were dominated by the pennates *Nitzschia closterium* and *N. delicatissima* and by the centric diatoms *Chaetoceros affinis* and *C. compressum*. They comprised over 40% of the total cells numbers, which attained up to $1.8–4.2 \times 10^6\,l^{-1}$ (biomass $3–9\,g\,m^{-3}$). The phytocone included there 74 species of diatoms, 20 of dinoflagellates, 3 of cyanobacteria, and 2 of chrysomonads. Significant portions of biomass were composed by picoplanktonic Synechococci of 1–2 μ size (Table 3.10). Their biomass attained in some samples of reef waters $0.5\,g\,m^{-3}$, or up to 70% of the total phytoplankton biomass. The same was recorded by Moriarty et al. (1985) in waters off Lizard I. where they numbered $1–20 \times 10^7\,l^{-1}$. According to Linley and Kopp (1986) in waters over the One Tree I. ring reef the number of cyanobacteria was 23×10^3 cells ml^{-1} and production about $1–4\,mg\,C\,m^{-3}\,day^{-1}$. Same was established on reefs of Polynesia (Legendre et al. 1988).

The composition of the diatoms populations there, in general, resembled that described in waters off Madagascar. Over 50% of diatoms were made up of pennats, in part, benthic washed out from the sands. Cosmopolitic and tropical species comprised 81% of the total diatom species while tropical-boreal 16%, arctic-boreal 3%. The most numerous species were represented here by the genera *Chaetoceros* (9 species) and *Nitzschia* (8 species).

In the lagoon of Koetivi I. of the Seychelles phytoplankton was also abundant: $0.2–1.2 \times 10^5$ cells l^{-1} (Table 3.11), while in surrounding oceanic waters it was less numerous, which proves the incident character of its

Table 3.10. Number and biomass of cells of picophytoplankton (containing chlorophyll, as estimated by epifluorescence) in reef waters of some islands in the Indian Ocean. (After Kopylov and Orlova 1985)

Place	Depth (m)	Numbers ($10^6\,l^{-1}$)	Biomass	
			$mg\,m^{-3}$	Total phytoplankton, %
Praslin I. (Seychelles)	2	3.70	280	38
	7	0.20	30	0.3
Koetivi I. (Seychelles)	2	0.80	116	53
	6	3.68	52	42
	14	2.80	159	61
Gubbet-Kallansia Bay (Socotra I.)	1	4.80	405	74
	2	1.30	93	9
	28	2.56	578	12

Table 3.11. Number and composition of phytoplankton populations in reef areas of some islands in the Indian Ocean. (After Orlova 1985)

Place	Depth (m)	Total number of species of algae in population	Numbers of main taxonomic groups of phytoplankton $10^3 l^{-1}$		
			Diatoms	Dinoflagellates	Cyanobacteria
St. Anne's Bay,	2	25	55	8	590
Praslin I.	5	26	1672	15	2475
(Seychelles)	16	20	16	0.1	0
Northern reef,	4	16	19	0	0.2
Koetivi I.	5	20	19	0.7	0
(Seychelles)	24	18	99	28	0
Leeward reef,	0.5	18	22	2	0
Raphael I.,	13	19	98	12	0
Kargados-Karachos I.	30	30	1007	19	50
Gubbet-	1	18	28	0.4	0
Kallansia Bay (Socotra I.)	28	29	803	0.5	0

developing just in lagoonal waters. The phytoplankton was dominated there by diatoms (70% of the total number). The blue-greens *Trichodesmium* and *Nostoc* were also present. The species spectrum of dinoflagellates was about the same as described by Sournia (1968). The diatoms were also dominating in waters of a large reefal area of the superatoll Kargados-Karachos (Mauritius). They were represented mostly by the genera Phyzosolenia (nine species), Chaetoceros (nine species) and Nitzschia (seven species).

Studies of phytoplankton in coral-reef waters off the coast of central Vietnam in the South China Sea were accomplished in the French Institute of Marine Sciences in Nhatrang (Dawydoff 1936; Nguen Thuong and Le Thi Ngoc 1972; Shirota 1966). The last-named author published a record of the phytoplankton species of this region collected in the plankton nets. More recent investigations were accomplished there later with the use of more modern methodology (Sorokin et al. 1982; Orlova 1985). They showed that the phytoplankton in these waters in March during its seasonal minimum was nevertheless abundant, dominated by small dinoflagellates Peridinium, Gymnodinium, Prorocentrum and the diatom *Nitzschia delicatissima*. In April the bloom of *Trichodesmium* started all along the coast from south of Nhatrang to the Mekong river mouth. Their numbers attained in the surface layer $100–300 \times 10^3 l^{-1}$ (biomass up to $40 g m^{-3}$). The photosynthesis of these algae was low, proving that they were here in a late phase of their successional growth (Quasim 1970), being driven there by trade winds from vast areas of the South China Sea. We have observed the same phenomenon in the vicinity of the Scott reefs in the Timor Sea (Sorokin 1986). Thrown out by waves they formed green stripes on shores of sandy cays.

Data on the phytoplankton of Atlantic reefs are still scarce (Glynn 1973; Hargraves 1982). The latter author colected phytoplankton samples with the aid of a plankton net 10 and 60 μ mesh and recorded in waters up the barrier reef Belize 73 species of algae. Among them the most numerous were the diatoms Chaetoceros (16 species) Rhyzosolenia (10 species), Coscinodiscus (7 species), and the dinoflagellate *Triceratium* sp.

Thus, an analysis of the taxonomic composition of plankton in reef waters shows that it is formed by three main components. One is comprised of the mostly common oceanic or neritic diatom species. Being opportunistic, they adapt to new reefal environments and start to grow there. This could be proved by data showing an increase of phytoplankton density in the vicinity of reefs or in the lagoons (Angot 1968; Sournia and Ricard 1975a). Sometimes the growth of diatoms in these waters attains a level of "blooms", as it was found in waters off the reefs of Socotra, the Seychelles and Kargados-Karahos I. (cf. Table 3.11), and in lagoons of some Pacific atolls (Michel et al. 1972). The second important contribution to the reefal phytoplankton is comprised by the periphytonic and benthic microalgae washed out by waves and tidal currents from bottom biotopes. From this source it is enriched with large benthic diatoms, filamentous blue-green algae, and even with fragments of red algae (Glynn 1973a). This washout of benthic algoflora is a significant phenomenon from the point of view of reef trophodynamics because it makes available a part of the production of bottom microalgae for being directly consumed by the benthic filter-feeders and by zooplankton. Together with the algae there are washings out also of fregments of filamentous algae and detritus – forming thus the "pseudoplankton". Its biomass in turbulent zones of reef may exceed the biomass of the phytoplankton itself (Quasim and Sankaranarayanan 1970; Johannes and Gerber 1974; Sournia 1977). The pseudoplankton itself often prevails in the guts of filtering zooplankters (Gerber and Marshall 1974), being ingested even by the polyps of corals (Lewis 1977a). Finally, the third component of the phytoplankton is the wind-driven blue-green algae mentioned above. It is especially important for such areas as the Coral Sea, the Timor Sea and other Indonesian seas, the South China Sea, and many reefs of the western Indian Ocean, where the blooms of *Trichodesmium* are an ordinary phenomenon (Chindambaram and Mukundan 1944; Quasim 1970).

3.2.2 Biomass, Production, Chlorophyll

Data on the biomass of phytoplankton in water over the reefs are still scarce. Most researchers gave their quantitative data in number of cells, but rarely estimated the biomass of the phytoplankton. Moreover, as mentioned above, the use of ordinary Utromöl sedimentation method leads to the underestimation of up to 50 or more. As regards the separate groups of the biomass (Sukhanova 1983) an adequate values could be obtained perhaps only by a combination of the above-mentioned reverse Nuclepore filtration with the direct epifluorescence counts of 0.4-μ black Nuclepore filters. But such estimations are still rare (Orlova 1985; Kopylov and Orlova 1985). The chlorophyll quantifications are more informative than the number of cells. But using the clorophyll content in water for calculations of phytoplankton biomass or production may given only some indefinite approximations because its content in cells as well as the assimilation numbers of photosynthesis may vary by an order of values, especially in reefal phytoplankton, composed largely of benthic and pseudoplanktonic elements, which might be rich in chlorophyll but functionally inactive.

The first measurements of primary photosynthesis production in lagoons of some Pacific atolls resulted in very low numbers: 0.5–2 $mg\,C\,m^{-3}\,day^{-1}$ (Sargent and Austin 1949). In fact, its rates in lagoons of many atolls may be very low, especially during the seasonal phytoplankton minima (Table 3.12). And in general, primary phytoplankton production in waters of atolls whether in lagoons or over the reefs is less than in waters of the barrier and fringing reefs near big islands or continents (Table 3.13). The data

Table 3.12. Ranges of primary phytoplankton production in lagoons of atolls and of semicircle and ring reefs

Place	Primary production per day	
	$mg\,C\,m^{-3}$	$g\,C\,m^{-2}$
Majuro atoll (Marshalls)	4–36[d]	0.01–0.40
Rongelap atoll (Marshalls)	5[a]	–
Enivetok atoll (Marshalls)	2[a]	0.03
Fanning atoll (Tuvalu)	70[b]	0.3–0.7
Takapoto atoll (Tuamotu)	4–31[c]	0.02–0.4
Tarawa atoll (Kiribati)	28–52[d]	0.01–0.03
Butaritari atoll (Kiribati)	20–35[d]	0.02–0.3
Ngellelevu atoll (Lau, Fiji)	1–3[d]	0.01–0.03
Ninigo atoll (Admiralty)	6–12[d]	0.16–0.22
Heron I., ring reef (GBR)	3–37[d]	0.02–0.25
One Tree I., ring reef (GBR)	7–41[d]	0.02–0.20
Wistari ring reef (GBR)	35–68[d]	0.2–0.35
Lady Musgrave ring reef (GBR)	7–20[d]	0.02–0.1
Sinton atoll (South China Sea)	10–40[d]	0.03–0.5
Scott semicircle reef (Timor Sea)	5–29[d]	0.02–0.1
Koetivi semicircle reef (Seychelles)	9–16[d]	0.03–0.24
Kargados-Karahos super atoll (Mauritius)	6–250[d]	0.02–1.0
Curtaun and Albuquerque semicircle reefs (Caribbean Sea)	0.5–8[b]	0.02–0.1

[a] Data by a) Sargent and Austin (1949).
[b] Gordon et al. (1971).
[c] Sournia and Ricard (1976a).
[d] Data by the author.

Table 3.13. Ranges of primary phytoplankton production in barrier and fringing reef areas

Place	Primary production per day	
	$mg\,C\,m^{-3}$	$g\,C\,m^{-2}$
Vairao and Faa lagoons (Tahiti)	8–58[a]	0.01–0.68
Tiahura (Moorea I)	7–58[a]	0.03–0.23
Lakeba I. (Lau, Fiji)	9–28[b]	0.03–0.19
Paama I. (Law, Fiji)	3–4[i]	0.01–0.05
Kaneohe bay, Oahu I. (Hawaii)	12–60[c,i]	0.01–0.3
St. Andrew reef, Paliali I. (Admiralty)	5–14[i]	0.04–0.2
Lizard I. (GBR)	5–40[d,i]	0.01–0.2
Lagoon opposite Townsville (GBR)	10–130[i]	0.15–1.0
Langemack bay, New Guinea	27–49[i]	0.2–0.9
Praslin I. (Seychelles)	213–315[i]	0.5–2.8
Tulear (Madagascar)	46–57[e]	0.2–0.7
Nosy Be (Madagascar)	4–103[f]	0.1–2.4
Gubbet-Kallansia Bay (Socotra I.)	14–357[i]	0.1–3.6
Ben Goy bay (Central Vietnam)	28–125[i]	0.28–0.65
La Parguera (Caribbean Sea)	100–700[g]	–
Florida keys	–	0.04–0.06[h]
Thu I. (Vietnam)	23–57[i]	0.11–0.48

References: a) Sournia and Ricard (1976a); b) Ricard (1977a); c) Gilmartin and Relevante (1974); d) Griffiths (1976); e) Maestrini and Pizarro (1966); f) Angot (1968); g) Burkholder et al. (1967); Gonzales (1965); h) Jones (1963); i) data by the author.

generalized in those tables show that its level varied drastically even in waters of the same reef in its different zones. In shallow areas of the reef its values are usually lower then in deeper waters of lagoons or in the vicinity of the outer reef slope. Other causes of variations are the seasonal and diurnal changes in phytoplankton biomass usual in reef waters.

In lagoons of some Pacific atolls of the Lau, Marshalls and Tuamotu Islands the primary production was usually low (2–4 $mg\,C\,m^{-3}$), however even there seasonal blooms of phytoplankton occurred (Sournia and Ricard 1976a). But most often in lagoons of atolls production exceeds by up to an order that in the surrounding oceanic waters, being 10 to 50 $mg\,C\,m^{-3}\,day^{-1}$, which is its level in mesotrophic oceanic waters. For example, in the lagoon of the superatoll Karados-Karahos, situated in the stream of blue oligotrophic waters of the south trade wind current, the primary production was up to 250 $mg\,C\,m^{-3}\,day^{-1}$, and the phytoplankton biomass up to 5.2 $g\,m^{-2}$ (Table 3.12), corresponding to its level in upwelling areas. High primary production (70–100 $mg\,C\,m^{-3}\,day^{-1}$) was recorded in the semiclosed lagoon of the Fanning atoll during the development of the dinoflagellates and the Nostoc-type blue-greens in its waters (Gordon et al. 1971). Thus, in lagoons of some ring reefs abundant communities of phytoplankton are developing periodically. The chlorophyll content in waters of ring reefs is also usually 5–10 times more than in surrounding oceanic waters, being 0.2–1 $mg\,m^{-3}$ (Kollentz-Mishke and Fortus 1980). Seasonal observation in the lagoon of the Takapoto atoll showed its fluctuations to be within 0.22 to 0.77 $mg\,m^{-3}$ (Sournia and Ricard 1976a).

In waters of barrier and coastal fringing reefs primary production values were very variable in their different zones. The lowest primary production and phytoplankton biomass were recorded in shallow areas of reefs – in zones of flats, patch-reefs and in littoral shallows. The maxima corresponded to waters of lagoons and outer reef slopes (Table 3.13; Figs. 3.9, 3.10), where waters are out of direct contact with the bottom communities of the reef. High primary production close to those in eutrophic oceanic waters was observed in waters of some barrier and coastal reefs in the Indian Ocean, such as the reefs off Madagascar, the Seychelles, and Socotra I., as well as in the lagoon of the Australian Great Barrier Reef (Figs. 3.3, 3.8). It attained there 50–300 $mg\,C\,m^{-3}\,day^{-1}$, or 0.7–2.8 $g\,C\,m^{-2}$, calculated per total water columns. The latter numbers are not very far from the primary production in the reef bottom biotope. Therefore, an axiomatic opinion about the negligible input of phytoplankton into the primary organic production in reef ecosystems does not seem to be correct, especially as far as the barrier and coastal reef systems are concerned. The share of phytoplankton could be there up to 20–30%. Even in lagoons of the barrier

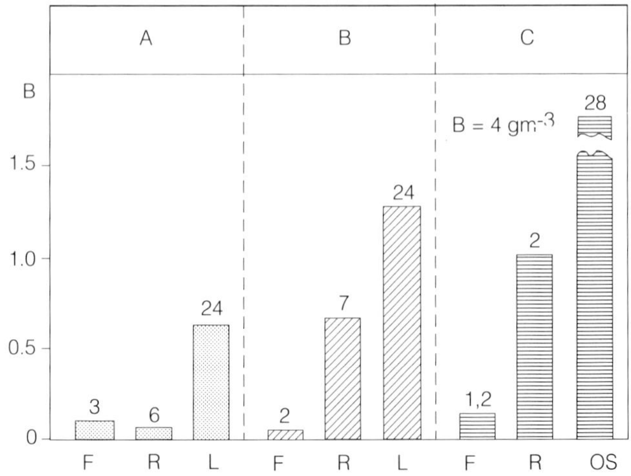

Fig. 3.9. Changes in phytoplankton biomass (B, $mg\,m^{-3}$) at the corss-section from the reef flat to the lagoon (L) or to the open sea (OS): **A** Koetivi I., Seychelles, **B** Kargados-Karahos I., Mauritius; **C** Socotra I.; *F* reef flat; *R* reef edge. (Data by Orlova 1985)

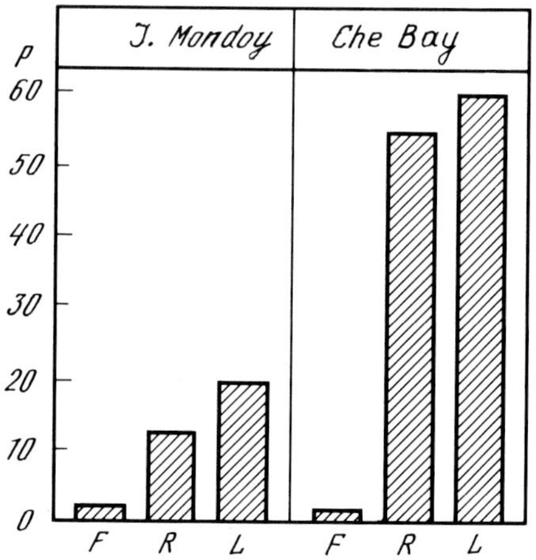

Fig. 3.10. Changes in primary phytoplankton production (P, mg C m^{-3} day^{-1}) at the cross-sections from the reef flat to the lagoons of the reefs to the lagoon of some Vietnam reefs; *F* reef flat; *R* outer reef edge; *L* lagoon. (After Sorokin et al. 1982)

reefs off islands surrounded by oligotrophic waters primary production often attains values of 50–60 mg C m^{-3} day^{-1}, which is an order more than in the nearby open ocean.

In the reef areas off the central Vietnam coast and around the Thu Islands in the South China Sea the primary production of phytoplankton was high even when measured during its seasonal maximum in the spring (Table 3.13). In zones off the reef edges up the reef slope it was often 30–40 mg C m^{-3} attaining sometimes 100 mg C m^{-3} day^{-1}. In shallow areas, where waters directly and intensively contacted with bottom biotopes with a dense coral cover, the primary production decreased by an order to 2–5 mg C. Waters up Namsy Island (Gulf of Siam) were rich in phytoplankton. Its primary production was high – up to 160 mg C m^{-3} day^{-1}, even just up the reef. A comparison of the primary phytoplankton production in lagoons and up the reef edge (0.5–1.1 g C m^{-2} day^{-1}) with the photosynthesis of microphytobenthos of coral sands at depths of 6–10 m showed that it amounts to 20 to 80% of the latter. In waters off the Sinton atoll in the South China Sea phytoplankton production attained about 10% of primary production of microalgae in coral sands.

As regards values of the phytoplankton biomass in reef waters, the available data are still very fragmentary (Table 3.14). A high level for it was recorded by Angot (1968) in waters off the Nosy-Be Islands – 3–8 g m^{-3} (Fig. 3.8). Much lower values were recorded in waters of barrier and coastal reefs with scarce coral growth dominated by coralline algae, such as reefs of the Panarora and Paama Islands of Vanuatu, Fiji (Sorokin 1986a). The chlorophyll content in reef waters around Tahiti and Moorea varied during the seasons between 0.03–0.17 mg m^{-3} (Table 3.14). In waters of the

Table 3.14. Phytoplankton biomass in waters of coral reefs (B, $mg\,m^{-3}$, wet weight) and coefficients of its specific production per day (P/B)

Types of reefs	Place	B	P/B[a]
Atolls and ring reefs	Velangilala (Vanuatu)	5–39	0.2–0.9[b]
	Funafuti (Tuvalu)	5–12	0.5–0.8[b]
	Ngellelevu (Lau, Fiji)	10–40	1.0–1.9[b]
	One Tree I. (Capricornia, GBR)	35–670	1.1[e]
	Sinton (South China Sea)	40–440	1.2–1.3[b]
Solitary semicircle and ribbon reefs in open ocean	Scott reef (Timor sea)	28–100	0.4–3.9[b]
	Kargados-Karahos (Mauritius)	30–5200	0.3–2.6[b]
	Koetivi I. (Seychelles)	80–640	0.3–1.0[b]
Coastal and fringing reefs	Panarora I. (Vanuatu)	13–23	0.8–1.2[b]
	Paama I. (Vanuatu)	25–90	0.7–1.4[b]
	Praslin I. (Seychelles)	450–8700	0.5–1.5[c]
	Socotra I. (Yemen)	600–8300	0.3–0.7[c]
	D'Ambaro Bay (Madagascar)	190–3300	0.02–0.5[d]
	Kuavan Bay (Ben Goy Gulf, Vietnam)	1250–3120	0.5–1.1[d]
	Vanfong (same Gulf)	590–1070	0.7–1.3[d]
	Tortu I. (Nhatrang Bay, Vietnam)	150–230	0.7–1.4[d]
Barrier reefs	Tuvuta (Lau, Fiji)	6–20	0.4–1.1[b]
	Lizard I. (GBR)	65–380	1.1–2.4[b]

[a] Value of B expressed in carbon units with 6% of wet biomass.
[b] After Sorokin (1986a).
[c] After Orlova (1985) and by the author.
[d] After Angot (1968).
[e] After Linley and Kopp (1986).

Hawaiian reefs and of the Great Barrier Reef it was 0.1–0.3 $mg\,m^{-3}$. The largest values were recorded in waters of the Nosy Be Islands during the seasonal phytoplankton maximum. The resuspension of coral sands in reef shallows may also cause an increase in the chlorophyll content in reef waters (Walker 1981).

The physiological state of phytoplankton could be characterized by parameters of specific production (P/B) and by the coefficients of the relative content of phaeopigments (Fo/Fa). The first is calculated as a ratio of its primary production per day to its biomass (both being expressed as carbon units). The values of P/B usually vary in tropical oceanic phytoplankton within 0.5 to 3. In the reef waters its lower limits are as a rule less – down to 0.01 – when the accumulation of the mass of inactive blue-greens proceeds in them. The washout of benthic algoflora, which ceases its physiological activity in a strange milieu, also results in a decrease in P/B (Table 3.14). But up the outer reef slopes and in the lagoon, where this influence is minimal, the P/B coefficients of phytoplankton could be as high as in the open ocean – up to 1.5–4.0.

The coefficients Fo/Fa are measured as the ratios of the fluorescence of plant pigment before (Fo) and after acidification of a water sample (Fa). The acid transforms chlorophyll into the less fluorescing phaeophytin. These coefficients vary in natural populations between 1.4 and more in a healthy, readily multiplying phytoplankton community where the dead chlorophyll-phaeophytin is practically absent, and 1 which reflects the abundance of phaeophytin in degrading its community which, overgrazed and nutrient depleted, is enriched with the dead detrital pigments. The results of estimations of Fo/Fa coefficients in waters off barrier reefs around Tahiti and Moorea Islands showed that they were the same as in the surrounding ocean. Thus they proved that the phytoplankton grows normally in lagoonal water rather than being only brought there with the incoming tides from the open ocean (Table 3.14).

The above data prove that the phytoplankton in reef waters is more abundant than in waters of the surrounding open ocean (Michel 1969; Sournia and Ricard 1976a; Ricard 1976; Kollentz-Mishke and Fortus 1980; Sorokin 1986b). In Fig. 3.3 a typical picture is presented of the change in primary phytoplankton production measured on the cross sections from the Heron Island ring reef to the open Coral Sea. At distances of 6–9 km from the reef it decreased about 10 times, while on same kind of cross from the lagoon of the Tarawa atoll to the open ocean it decreased by about 2 orders at a distance of 16 km (Fig. 3.3). Within the water column of even deep lagoons the vertical distribution of phytoplankton is usually uniform. Sometimes phytoplankton is scarcer in the surface layer of deep lagoons filled with waters arriving from the flat during the ebb, depleted of planktonic algae, or with waters coming from the ocean with the tide. The white disk transparency in reef waters in lagoons or at outer reef slopes usually is within 10–15 m. Thus the phytoplankton photosynthesis in them is not limited by light deficiency down to the depths of 15–25 m. A typical curve of light attenuation in the water column of some Indooceanic reefs is given in Fig. 3.11.

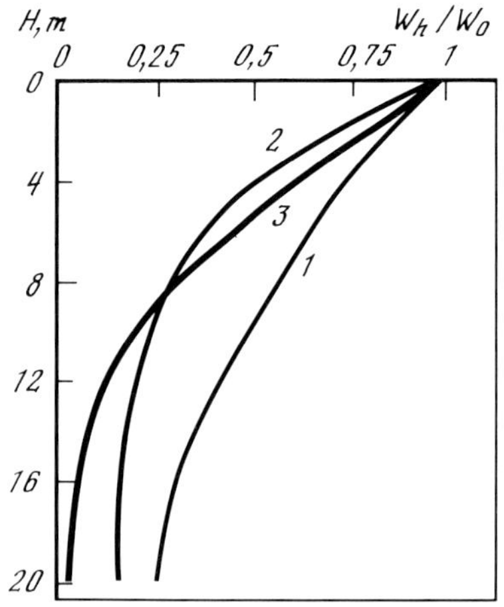

Fig. 3.11. Decrease in light intensity (Wh/W_o) in water column off the reefs of *1* Praslen I.; *2* Koetivi I.; and *3* Kargados-Karahos I.; *H* depth

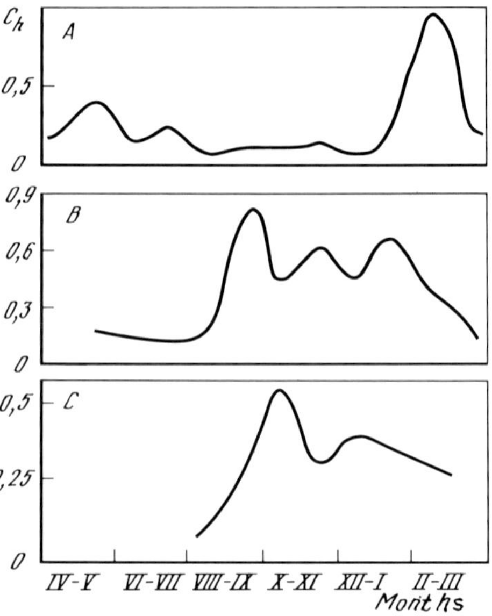

Fig. 3.12. Seasonal changes in chlorophyll content (C_h, mg m^{-3}) in coral-reef waters; *A* off Nosy Be I., Madagascar; *B* off the Laurel reef; Puerto Rico; *C* in the lagoon of the Takapoto atoll (Tuamotu I.). (Data by Angot 1968; Glynn 1973b; Sournia and Ricard 1976b)

Observations by Angot (1968), Sournia (1969), Glynn (1973b), Ricard et al. (1979) and Sammarco and Crenshaw (1984) demonstrated a wide range of seasonal fluctuations in the abundance and taxonomic composition of phytoplankton in reef waters (Figs. 3.8, 3.12). They could be caused by the changes in general oceanic circulation that take place usually in spring and autumn. These changes are most clearly manifest in the Indian Ocean. The second cause may be the periodic character of successions of planktonic communities in tropical, even in homeostatic environmental conditions, which stands in connection with the relationships between phytoplankton grazing control and the processes of nutrients regeneration (Sorokin 1985). Significant seasonal changes of chlorophyll from 0.13 to 2.5 mg m^{-3} in waters of the lagoon of the Great Barrier Reef near Townsville were caused mainly by fluctuations in river discharge. Its maximum was recorded in the summer-autumn period coinciding with the rain period and with the increase in rivers discharge. During this maximum the phytoplankton was dominated by the blue-green alga *Trichodesmium*. The same phenomenon has been observed also in waters off reefs of New Caledonia (Binet 1984). The reef phytoplankton seems to be subjected also to diurnal fluctuations (Milliman and Mahnken 1972; Sournia and Ricard 1976a; Sorokin 1986b; Fig. 3.13). The diurnal fluctuations are caused by nocturnal grazing by demersal zooplankton, by tidal changes of water masses and, possibly, by diurnal

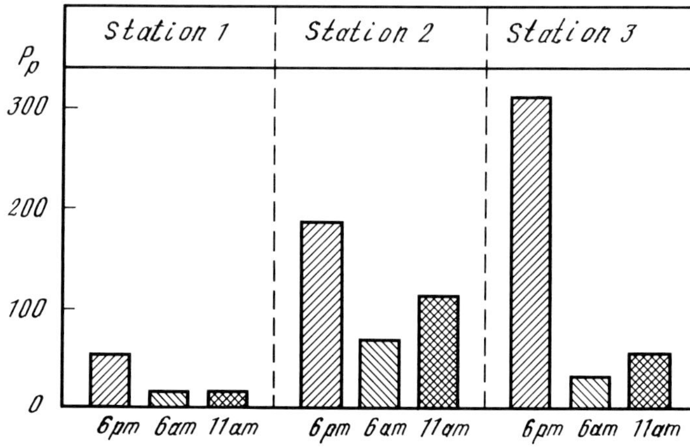

Fig. 3.13. Diurnal changes in primary phytoplankton production (P_p, mg C m^{-3} day^{-1}) at the cross-section over the fringing reef off Chuong I. (Namsu I., Vietnam): *Station 1* reef flat, 1 m depth; *Station 2* reef edge, 4 m depth; *Station 3* 100 m off the reef edge to the open sea, 18 m depth. (Data by the author)

cycles of multiplication of planktonic algae. Durnal observations in reef waters of the Namsu Islands showed that at night the demersal zooplankters which emerge from the bottom biopopes grazed out a significant portion of phytoplankton (Fig. 3.13, station 2). But till the next midday its concentration was restored again, especially near the reef edge where the algae multiplied faster.

3.3 Zooplankton

The pioneering study of zooplankton in reef waters had been accomplished by Kramer (1897). This study, now almost forgotten, nevertheless contained basic conclusions which are still valid. He has established that the reef zooplankton includes the oceanic as well as the endemic species. He has found also that the zooplankton of reef waters is more abundant quantitatively than in the surrounding oceanic waters. These conclusions were proved later by Johnson (1949, 1954) and Bakus (1964). Significant insights into the composition of reef zooplankton, into its distribution, abundance, and seasonal dynamics were gained by studies carried out in waters of the Great Barrier Reef (Russel 1934; Farran 1949), of the Hawaiian reefs (Edmondson 1937), in lagoons of the Palau Islands (Motoda 1940), of some Pacific atolls (Michel 1969; Michel et al. 1972; Le Borgne et al. 1989) and of atolls of the Indian Ocean (Tranter and George 1972). The above researchers used for the collection of zooplankton samples vertical or horizontal daytime tow-nets. Some of them have mentioned the limitations and insufficiency of this method because a significant part of the reef zooplankton is composed of demersal and meroplanktonic forms, which hide in the bottom biotopes during daytime and move to the water column only at night (Renon 1977). Therefore, the daytime tow-nets employed in those studies were not suitable for providing a real picture. Another method was used by Johannes et al. (1970). They fixed plankton nets up the reef flat surface in a direction meeting the incoming current. But the zooplankton biomasses they thus obtained were negligible – 5 mg m^{-3} in daytime and only 8 mg m^{-3} at night. This method was obviously invalidated by disregarding the ability of demersal animals to move on the solid surface and so to crawl out from the net.

New developments in methodology were the use of nocturnal tows up the bottom by scuba divers (Emery 1968; von Hentig 1971; Hickel 1972; Porter 1974b) and airlift to get samples of demersal zooplankton (Emery 1968). Later various traps were used, similar to those which are employed by entomologists to collect flying insects emerging from water or from the soil (Porter and Porter 1977; Alldredge and King 1977; Mc William et al. 1981; Sale et al. 1976; Rützler et al. 1980). But, actually, trap data cannot give adequate quantifications of zooplankton number and biomass (Hobson and Chess 1979). The traps catch and accumulate an excess of holoplanktonic zooplankters or mysids, which use the trap as shelter. Thus the use of traps could lead to overstimating holoplanktonic zooplankters or underestimating a portion of demersal animals, like some of the gammarids, which do not rise far from the bottom even at night, or like animals with negative phototaxis. The author used a simple but from his viewpoint representative method of collecting quantitative samples of zooplankton by passing 100–250 l of water, taken at night from the boat with a bucket, through the plankton net. This method was especially efficient in shallows over the reef where it was difficult to catch zooplankton with the aid of a net, especially at night.

As regards the microzooplankton, including planktonic animals less than 0.2 mm in size, the early researchers have used for this purpose methods which were not actually quantitative meaning the sedimentation method for the counting of planktonic ciliates, or the method of catching them and also the larvation with the plankton nets. Recently, new methods were offered and used, which made a real quantifications of the main groups of microzooplankton possible. The fraction of larger eucariotic microzooplankton – the larvaton (40–200 size) – has been counted quantitatively by the author in 3–1 samples concentrated by reverse filtration through 2–4 μ Nuclepore filters into a volume of 30 ml, and then – by sedimentation after a weak Utremöl fixation – into a volume of 5 ml. Another method used by the author was the direct epifluorescence microscopic counting of these microzooplankters stained with acridine orange on the black Nuclepore filters (4 μ pore size). The ciliates could be quantitatively counted in fresh intact water samples (within 1–2 h after sampling) in glass box chambers of 20–30 ml capacity and 4 mm high (Sorokin 1980). Another really quantitative method is their epifluorescence microscopy on 0.4 μ black Nuclepore filters after Sherr and Sherr (1983). Both latter methods were used also to count in reef waters other small microzooplankters of 2 to 10 μ size, like the zooflagellates, the planktonic amoebae, and the small radiolarians (Sorokin and Tyapkin 1984; Kopylov 1985; Moiseev 1985).

3.3.1 Mesozooplankton

The composition of reef zooplankton varies in the estimates of different authors, depending upon the sampling techniques used and the time and place of sampling, even on the same reef. The daytime tows, especially those made in reef zones influenced by incoming oceanic waters, were enriched with the holoplanktonic species brought from the ocean by currents (Johnson 1954; Gerber and Marshall 1974; Moore and Sander 1976; Ferraris 1982; Goswami 1983). The total number of zooplankton species in such samples could be 100 to 120. In the nocturnal samples, because of an additional catch of demersal and meroplanktonic species, it increased about twice, and in the trap samples it might be as much as 260 (Sale et al. 1976a; Mc William et al. 1981; Tables 3.16–3.18). The night tows gave samples enriched with the demersal and the meroplanktonic forms, which were quite different by species composition from zooplankton samples collected in surrounding oceanic waters (Tables 3.16, 3.17). The species composition of zooplankton samples collected in light traps and with the aid of airlift differed from the samples collected by night tows or by the use of dark traps (Emery 1968; Sale et al. 1976a). In the light traps, for example, the zoea

Table 3.15. Taxonomic composition of zooplankton in the lagoon of the atoll Moruroa and surrounding oceanic waters: average data of daytime net tows. (After Michel 1969)

Type of zooplankton	Taxonomic groups	Number of zooplankton (m^{-3}) in lagoon (N_1)	Number of zooplankton (m^{-3}) in ocean (N_2)	Ratio N_1/N_2
Holoplankton	Copepods	52.5	17.9	2.9
	Chaetognaths	23.0	0.7	32.8
	Nauplii, copepodids	7.0	2.0	3.5
	Appendicularians	4.8	0.1	48.0
	Medusae, ctenophores	2.1	0.05	42.0
	Euphausiids, amphipods	0.1	0.7	0.7
	Siphonophores	0	0.7	0
Meroplankton (larvae)	Annelid larvae	0.36	0.26	1.4
	Zoea of crabs	0.7	0.1	7.0
	Veligers of Mollusca	1.7	0.2	8.5

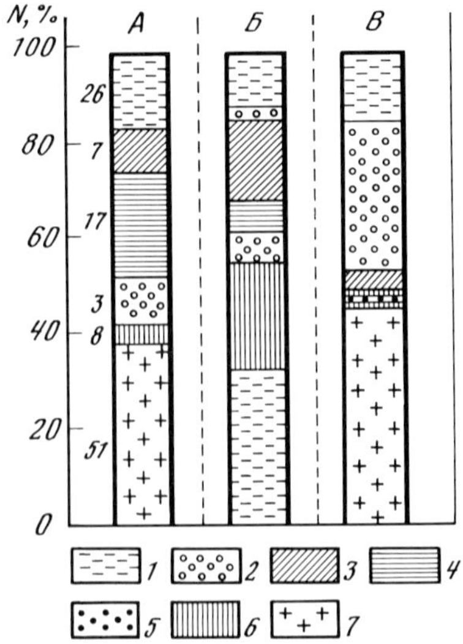

Fig. 3.14. Taxonomic composition of zooplankton caught by light traps from different bottom biotopes of the Heron I. reef: *A* lagoonal bottom; *B* outer reef slope; *C* 300 m off the reef edge in the strait, at depth of 20 m. Zooplankton groups; *1* mixture of rare taxa; *2* larvaton; *3* decapod larvae; *4* amphipods; *5* isopods; *6* young of cumaceans; *7* copepods; *N* share of a group in total number of zooplankton in the catch. *Numbers* in the first column correspond to number of zooplankton species of a given group in the catch. (After Sale et al. 1976)

and stomatopods species instead of the copepods were most numerous (Fig. 3.14). The same dominance of the demersal species has been observed by us in nocturnal samples, obtained by passing volumes of water taken over the reefs off Koetivi Island through the plankton net (Tables 3.17, 3.18). In samples collected by light traps over the Heron Island reef, of 106 species of zooplankters 66 were typical demersal forms, which were absent in surrounding waters, and 40 others were species common to reef and to the nearby open sea. The taxonomic composition of zooplankton depends upon the type of reef biotope up which it was collected (von Hentig 1971; Sale et al. 1976a; Porter et al. 1977; Mc Williams et al. 1981; Birkeland and Smalley 1981; Sammarco and Crenshaw 1984; see Tables 3.18, 3.19; Fig. 3.14). These observations have proved that different species of zooplankton prefer their own sheltering places at the bottom (Emery 1968; von Hentig 1971).

The holoplanktonic species of reef zooplankton are dominated by the cyclopoid and calanoid copepods (Tables 3.15, 3.17, 3.18). Among other crustaceans the cladocerans, amphipods, mysids, harpacticids and the hyponeustonic copepods pontellids are also present (Renon 1977; Le Borgne et al. 1989). The holoplankton includes the chaetognaths, the appendicularians, the pteropods and also same gelatinous forms (Table 3.20). Many of the holoplanktonic species, present in the samples of reef

Table 3.16. Number of specimens of different taxonomic groups in zooplankton samples collected in the lagoon of the atoll Takapoto (vertical net tows by day and by night; after Renon 1977)

Type of plankton	Taxonomic groups of zooplankton	Number of zooplankton (l^{-1})	
		Day	Night
Holoplankton	Copepods	91	1460
	Appendicularians	4	54
	Chaetognats	0.6	32
	Amphipods	0	26
	Ostracods	0.3	23
Total holoplankton		96	1600
Meroplankton and demersal zooplankton	Decapods (larvae,	0.7	43
	juveniles)		24
	Veligers	10	7
	Foraminifera	4	10
	Fish larvae	0.3	
Total meroplankton and demersal zooplankton		14	84

zooplankton, are common to the reef as well as to the oceanic waters. But it does not mean that the reef zooplankton is a part of the oceanic zooplankton, passively transported above the reefs by current, as it had been supposed before (Johannes et al. 1970). Actually, it forms a separate specific community of reef zooplankton, which represents a part of the coral reef ecosystem, included in its food-webs. Indeed, even the holoplanktonic species form in reef waters much denser populations than in the surrounding oceanic waters, thus using the resources of the reef itself. Moreover, some typical oceanic zooplankton crustaceans, such as Oithona or Acartia in reef environement acquire features of behavior, which are typical for components specifically of reef zooplankters. They migrate at daytime to the bottom to escape the grazing. Migrating there in this way are not only the mibile holoplanktonic crustaceans but also the slow-moving species like Oicopleura (Tables 3.15, 3.16). Many of the typical oceanic holoplankters, when in reef environments, start to hide at daytime in shelters or to form swarms near the caves or up the corals or on sea urchins (Emery 1968; von Hentig 1971; Hamner and Carleton 1979).

The demersal species of zooplankton appear in water columns mostly at night (Tables 3.16–3.19; Figs. 3.15–3.17). Among them the most numerous are crustaceans copepods, amphipod mysids, and sometimes polychaetes (Birkeland and Smalley 1981; Ohlehorst 1982; Goswami 1983). The emergence of the demersal zooplankters from their bottom shelters results in a 10- to 100-fold increase in number and biomass of nocturnal

Table 3.17. Day and night zooplankton numbers over the North reef of Koetivi Island (Seychelles) (data by the author)[a]

Time	Station[b]	Numbers of main zooplankton groups (m^{-3})						Total biomass, $mg\,m^{-3}$ (wet weight)
		Copepoda	Mysidae and euphausiids	Zoea of crabs	Polychaetes (young)	Veligers	Fish eggs, others	
1100 h a.m.	1	24	0	0	0	8	16	50
	2	24	0	0	0	16	8	20
	3	16	0	8	8	24	16	40
1000 h p.m.	1	190	700	300	22	130	60	1820
	2	220	1290	840	65	1150	40	3650
	3	120	980	380	15	240	110	2800

[a] Water (150 l) was passed by a bucket through the plankton net of 60 μ mesh; biomass was estimated by wet combustion.
[b] *Station 1*: 30 m offshore up sandy bottom at 2 m depth; *Station 2*: 60 m offshore, patch reef, at 4 m depth; *Station 3*: 250 m offshore, patch reef at 6 m depth.

Table 3.18. Numerical abundance of main taxonomic groups of zooplankton collected by day and by night on a cross section through the fringing reef of Chuong Island (Gulf of Siam) (Data by the author)

Time	Station[a]	Number of zooplankton (m^{-3})						
		Copepoda	Ostracoda	Mysidae	Zoea	Polychaetes	Chaetognatha	Total zooplankton
By day 1600 h	1	190	0	20	0	0	0	210
	2	80	0	10	0	0	0	90
	3	120	0	10	0	0	0	130
	4	850	0	0	0	20	10	880
At night 2200 h	1	1700	510	750	20	30	20	3030
	2	1360	150	960	50	5	120	5675
	3	3100	0	160	20	50	60	3390
	4	4200	0	70	50	80	120	4520

[a] Position of stations: 1 – reef flat; 2 – reef edge; 3 – reef slope; 4 – 200 m off reef to the sea.

Table 3.19. Zooplankton composition in waters over and in the vicinity of the Heron I. reefs (data by the author)

Time of sampling; tide phase	No. of stations	Depth, m	Fractions[a]	Biotopes[b]	Taxa[c]	Biomass, $mg\,m^{-3}$
	8	0.6	D 70%, H 30%	r.c.	Decl. Mysl. Cop.	72
22–24 h, 25 Sept;	9	1.0	D 80, H 20%	fl.p.r.	Decl. Gam. Mys. Pol. Cop.	253
Middle ebb	10	1.5	D 100%	l.p.r.	Decl. Gam. Mys. Pol.	358
	11	1.5	D 100%	fl.p.r.	Decl. Gam. Mys. Pol.	312
	12	0.7	D 100%	r.c.	Decl. Harp.	1150
	7	26	H 90%	o.s.	Cop. Ch.	490
19–21 h, 5 Oct;	8	2.0	H 90%	r.c.	Cop. Si.	240
High tide	9	3.2	H 70%, D 70%	fl.p.r.	Cop. Decl.	116
	11	3.5	H 50%, D 50%	fl.p.r.	Cop. Ch. Decl. Mys.	266
	1	1.2	H 100%	l.f.	Cop. Ch. Ap.	400
	2	1.0	H > 90%	l.f.	Ap. Si. Cop.	223
20–22 h, 9 Oct;	3	1.0	H > 90%	l.f.	Cop. Ap.	500
Coming tide	4	1.5	H 80%, D 20%	l.f.	Ap. Cop.	138
	5	2.0	H > 90	H	Cop. Ap.	124
	6	25.0	H 100%	o.s.	Ch. Cop. Ap.	835

[a] Fractions: H – holoplankton; D – demersal zooplankton.
[b] Biotopes: o.s. – surrounding water outside off the reef; r.c. – reef crest; fl.p.r. – patch reefs on the reef flat; l.p.r. – patch reefs on outskirts of lagoon; L. – lagoon, l.f. – leeward flat.
[c] Cop – copepods; Harp – harpacticoids; Ap – appendicularians; Ch – chaetonaths; Si – siphonophores; Decl – decapod larvae; Gam – gammarids; Mysl – mysid larvae; Mys – mysides; Pol – polychaetes.

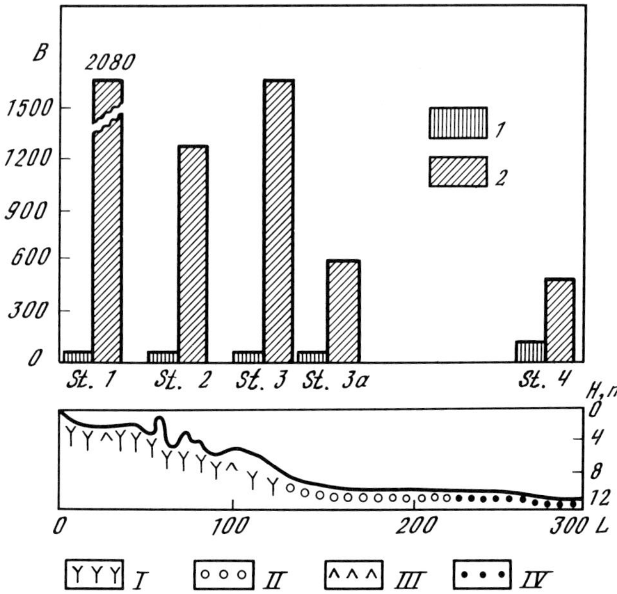

Fig. 3.15. Biomass of zooplankton (B, $mg\,m^{-3}$) in the daytime (*1*) and in the nocturnal (*2*) samples collected up the north reef off Chuoong I. (Namsu I. Vietnam). Positions of stations, cf. *profile below the graph*. Biotopes: *I* living corals; *II* rubble; *III* reef-flat rocks; *IV* coral sand; *L* distance from the shore, m; *H* depth

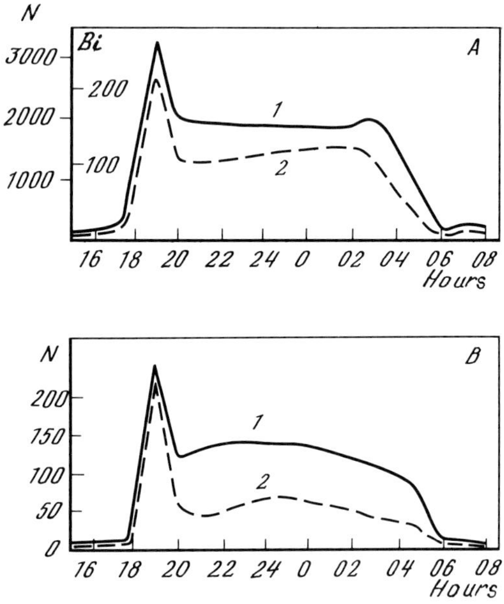

Fig. 3.16A–B. Diurnal changes in the number (N, l^{-1}) and biomass (Bi, $mg\,m^{-3}$) of zooplankton, caught by net tows in the lagoon of the atoll Takapoto (Tuakapoto) **A** Total zooplankton; **B** most common holoplankters *Acartia fossae* (*1*) and *Sagitta oceania* (*2*). (Modified from Renon 1977)

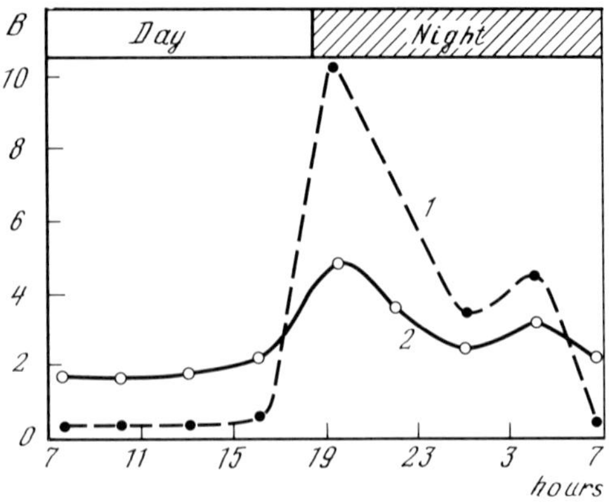

Fig. 3.17. Diurnal changes in zooplankton biomass, *B*, $mg\,m^{-3}$ estimated by measuring the total volume of catch at the San Blas reef (Puerto Rico); *1* net towed above the bottom by scuba diver; *2* vertical net tows out from the reef up to depth 100 m, in the upper 0–13 m. (Based on Porter 1978)

zooplankton (Tables 3.17–3.19; Figs. 3.15, 3.17). The range of dominance of different groups in samples of demersal zooplankton varies depending on the place and on the method of their collection (Alldredge and King 1977; Hobson and Chess 1979). As shelter in daytime most of them use the porous limestone surface at the reef flat, the rubble and the sand, the thalloms of macrophytes, the periphytic turfs, and the colonies of living or dead corals. The better part of the night they spend in the water column, feeding as filterers or as predators on other zooplankters. The demersal zooplankton includes also the bentho-planktonic animals, such as some gammarids or some polychaetes. They live permanently on the bottom and emerge into the water, periodically at night for propagation, or for distribution to new reef areas. For a list of common taxa of demersal and meroplanktonic zooplankters in reef waters cf. Table 3.2.

As for the conclusion on the leading role of copepods in demersal reef zooplankton (Porter and Porter 1977; Porter et al. 1977; Alldredge and King 1977) it could not be accepted as quite correct because of the above-mentioned insufficiency of the trap methodology used by these authors (Hobson and Chess 1979). They have caught demersal zooplankton using traps fixed at some distance upwards from the bottom or with gaps between the frame and the ground, and counted the zooplankton thus collected as it emerged from the bottom under the trap. But these traps collected also copepods which penetrated the trap from the sides and used it as a shelter for hiding in daytime. The above authors have counted as a demersal form even such a holoplanktonic copepod as Acartia. In their trap samples copepods comprised 80–90% of the total zooplankton species, while in samples collected in traps more or less isolated from the surrounding water it decreased to 50–60%. In samples collected with the aid of completely isolated traps the copepods contributed only about 20% of the total zooplankton (Hobson and Chess 1979).

In the trap samples the copepod populations are often dominated by cyclopids and mostrillids rather than by calanids as in the open ocean. In such samples taken over the outer reef slope off Heron Island the reef copepod species *Calanopia* and *Thaumalens* were the most numerous, but not the calanoids *Eucalanus* or *Undinula*, most common in the nearby oceanic waters (Sale et al. 1976a). The trap samples collected over the Philippine reefs were dominated by the *Cyclopoida*. Over the fringing reef of one the Namsu Islands the nocturnal zooplankton was dominated by the copepods which numbered over $4000/m^{-3}$. This reef was short and therefore the plankton community was here strongly influenced by the rich zooplankton of the surrounding waters of the Gulf of Siam. Nevertheless, even there a larger portion of the zooplankton was comprised by demersal mysids and amphipods (Table 3.18). In the vast shallow lagoon of the Koetivi Island, where the planktonic community was separated from the direct influence of the oceanic waters, in the night samples the demersal reef species of mysids amphipod and euphausiids absolutely predominated (Table 3.17). In the nocturnal zooplankton samples collected over Heron Island during high tide, the holoplankton comprised up to 90% of its total number in waters up the reef-flat and closer to the reef edge where the plankton community was strongly influenced by inflowing oceanic waters, while up the patch reefs it comprised only about 50%. During ebb tide, when the external influence was minimal, the zooplankton samples collected over the patch reefs and reef flat contained almost exclusively the demersal and meroplanktonic zooplankters – gammarids, mysids, polychaetes and the decapod larvae (Table 3.19). A list of the commonest taxa of the demersal and meroplanktonic reef zooplankton is given in Table 3.21.

Table 3.20. Common holoplanktonic genera of reef zooplankton

Taxonomic groups	Genera
Copepoda, (Calanoida)	*Acartia*, *Eucalanus*, *Calocalanus*, *Clausicalanus*, *Paracalanus*, *Calanopia*, *Canthocalanus*, *Nanocalanus*, *Candacia*, *Euchata*, *Undinula*, *Temora*, *Centropages*, *Labidocera*, *Bestiola*
Copepoda (Cyclopoida, Monstrilloida)	*Pseudocyclops*, *Coricaens*, *Farranula*, *Oithona*, *Oncaea*, *Sapphirina*, *Scolecetrichella*, *Monstrilla*
Cladocera	*Penilia*, *Evadne*
Mysidae	*Mysidium*, *Halichoeres*, *Thalassosoma*
Amphipoda	*Brachicelsus*, *Eupronoe*, *Hyperietta*, *Hyphenia*, *Lestrigonus*, *Synopia*
Chaetognata	*Sagitta*, *Spadella*
Tunicata	*Doliolium*, *Salpa*, *Thalia*
Appendicularia	*Oicopleura*, *Fritillaria*
Syphonophora	*Abylopris*, *Chelophyes*, *Diphyes*

Table 3.21. Main taxa of demersal zooplankton and meroplankton of reef waters (Johannes and Gerber 1974; Emery 1968; Alldredge and King 1977; Sale et al. 1976a; Mc William et al. 1981; and others)

Demersal zooplankters	Meroplanktonic zooplankters
Copepods-calanoids and cyclopids Stephos sp., *Pseudodiaptomus cornutus*, *Lichmolgidus* sp., *Paramisophria* sp., *Cyclopina schneideri*	*Eaphausiids Pseudoeuphausia latifrons* (juveniles)
Copepods-harpacticids Peltiidium sp., *Tisbe* sp., *Thalistridus* sp., *Eudactilopus stratus*, some of Tegastridae	*Decapods Natanita* sp., *Raptania* sp., (larvae)
Amphipods-gammarids (mostly juveniles) *Synopia variabilis*, *Aoroides* sp., some of Eusiridae, Oedocerotidae, Caprellidae, Ampeliscidae, Phoxocephalidae, Dexaminidae, Melphidippidae	*Hydromedusae Lirope* sp., Aglaura Eirene
Isopods. Euridice littoralis, *Munna* sp., some of Anthuridae, *Corolana* sp., juveniles *Gnatia* sp.	*Nemeretine and polychaete worms* Syllidae, Hesionidae, *Exogene* sp., *Armandis* sp., (juveniles), Serpulidae, Sabellidae, Nereidae (larvae)
Mysids Siriella chierchiae, *Mysidium gracile*	*Larvae and juveniles* planulae of corals; veligers; juveniles and larvae of benthic crustaceans (carideans, cumaceans, stomatopods, tanaids, Thalassinids, crabs, barnacles); larvae of tunicates, sipunculids, asteroids, urchins; larvae and juveniles of nematods
Ostracods Conchoecia spinorostris, some of Cylindroleberdinae	
Euphausiids E. tenera	
Decapods Lucifer chacei	
Phoraminiferans Globigerina sp., *Tretomphlus* sp.	

An obvious reason for the above-mentioned diurnal migrations of reef zooplankton, which results in a drastic decrease in its density in water columns in daytime, could be seen in its effort to escape grazing by pelagic predators, such as fish, fish larvae and predatory chaetognath Sagitta, which are numerous in reef waters (Vivien 1973). Similar diurnal migrations are known also in many oceanic holoplankters, which descend to the dark water layers during the day. But the reef bottom is well illuminated. And we observe that even typical oceanic holoplankters, such as copepods, either the quickly moving Sagitta or the slow-moving Oicopleura, on entering reef environments all start to migrate early in the morning down to the bottom and form there near the possible shelters dense monospecific schools or swarms, which could contain 1000 and more specimens. They use as shelters caves or crevices in lime rocks, colonies of ramose corals, thickets of macrophytes and sea-grasses (Emery 1968). The formation of schools decreases in itself the possibility of grazing. Observing the behavior of zooplankton in daytime at reefs of the Namsu Islands, the author has seen the swarms of mysids at the base of ramose corals, in the caves and in crevices of the flat. When approached with the hand, such a swarm immediately retreated into its refuge, acting as a whole body.

The swarms form not only endemics of reefs such as mysids, but also the typical pelagic crustaceans living also in the open sea, such as the above-mentioned Oithona and Acartia (Emery 1968; von Hentig 1971; Hamner and Carleton 1979). For example, the typical pelagic copepods *Acartia spinosa*, *A. tonsa* and *Oithona nana*, randomly distributed when living in open sea, in reef environment form during the day dense swarms near the edge of the thickets of sea-grass or in the holes amid them. Their swarms could contain several thousands of specimens (von Hentig 1971). The behavior of these swarms resembles that of the typical demersal reef endemics, such as mysids or pelagic planktonovorous fish. The swarm of copepods behave as a single body, opposing the currents. When dispersed by a diver, the swarm soon collects again in one place. The swarms were observed to collect each day near the same shelters at the bottom. The copepod swarms were observed in caves, perforating the patch reefs and the outer reef slopes. *Acartia spinata* formed sometimes a large (2 m in diameter) dense swarms between the patch reefs over piles of rubble and also on outer slopes in grooves and caves of the buttress zone. Up the outer slope there were observed also schools of some other copepod species like *Oithona oculata* and the harpacticoid *Metis* sp. These pelagic copepods behave in schools like typical shoaling animals. At night, their swarms dispersed but in the morning they re-formed the schools at the same place. Thus, these typical pelagic copepods, when living in reef waters, adopt the behavior features peculiar to the endemic reef residents like mysids or the other demersal zooplankters.

The reef mysids usually form small schools close to their shelters – holes or crevices. The mysid *Mysidium gracile* is shoaling at daytime near the aggregations of the urchin Diadema, which has long poisonous needles. On being endangered, the school will hide instantly inside the thickets of needles. Some mysid shoals were observed to keep also near the holes, in which the pomacentrid fish live, and hiding in them when endangered. The host fish did not hunt the mysids but attacked the intruder (Emery 1968).

The reasons of migrations of the demersal zooplankton might be different in different taxa. Most of them migrate at night from the bottom to the water column for feeding since there they are less endangered by the pelagic predators. Another reason for their emergence from the bottom at night, which may be not less important, is to avoid the grazing of bottom predators – molluscs, crustaceans, polychaetes, most of which feed at night, as well as many benthic fish. For demersal predatory zooplankters, such as the demersal mysids, cumaceans, cyclopoids and harpacticoids, the reason for migrating at night up to the water column could be the hunting for holoplankters and other demersal plankters, which are abundant at night coming out from their shelters. Some demersal species go up to the water column only for propagation, like some amphipods and a variety of polychaetes (Allen 1957). The benthoplanktonic young gammarids periodically go to the water column from the bottom and swim to occupy new bottom areas (Alldredge and King 1980). Some bottom-dwelling cyclopoids go to the water column in the daytime to hunt in the schools of holoplankters, crowding near their shelters. The time schedule of night migrations appears to be different for different species of demersal zooplankters. After sunset the first to ascend to the water column are the cumaceans and polychaetes. They return at 2–3 a.m. Amphipods and isopods ascend in the water several times during the night (Alldredge and King 1980; Ohlhorst 1982).

A significant influence is exerted on the distribution of reef zooplankton by the tendency of many zooplankters to move to the center of eddies (Alldredge and Hamner 1980). Thus, it concentrates in the centers of the numerous rings forming during the interaction of currents with reef constructions at their windward sides (cf. Fig. 2.4). In the centers of such rings the concentration of zooplankton might be by an order greater than only a few tens of meters away from it (Hamner and Hauri 1981).

First attempts at quantification of reef zooplankton by means of daytime net tows gave relatively low values (50–100 mg m^{-3}), which nevertheless were higher than those in the surrounding ocean (Johnson 1949). Later, when the role of the night migration of demersal and meroplanktonic zooplankters was recognized, the nocturnal tows and traps gave total numbers and biomass values 10–50 times greater than the daytime tows (Emery 1968; Tranter and George 1972; Alldredge and King 1977; Renon 1977; Porter et al. 1977; Porter 1977; Ohlehorst 1985). The number of mesozooplankton collected by night tows was $1–4 \times 10^3$ sp. m^{-3} (Tables 3.16, 3.17). Because the undivided weights of demersal species were 10–100 greater than those of copepods, and dominated in the daytime samples, the total biomass thus established was by 2–3 orders more than that found

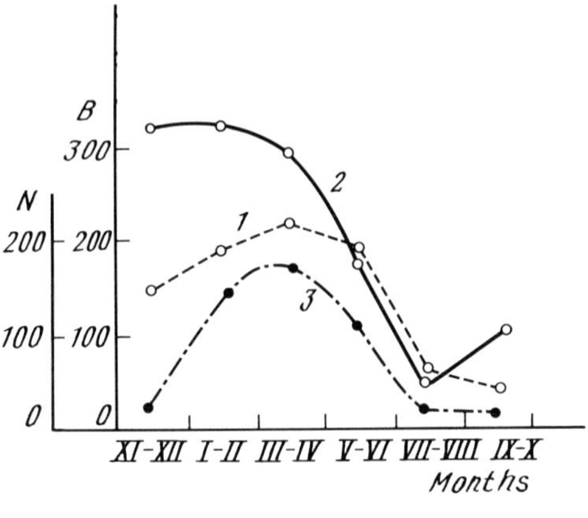

Fig. 3.18. Seasonal succession of the zooplankton biomass (*B*, mg m^{-3}) in the GBR lagoon near Townsville: *1* in the eastern part of the lagoon; *2* in the central part of the lagoon; *3* also in the central part – number of chaetognaths (N, l^{-}). (Modified from Sammarco and Crenshaw 1984)

Table 3.22. Biomass of net zooplankton in coral-reef waters and in waters of surrounding open ocean. Data by the author and Johnson (1949, 1954), Renon (1977), Michel (1969), Michel et al. (1972), Glynn (1973b), Porter (1974b), Porter and Porter (1977), Moore and Sander (1976), Geptner (1979)

Place	Zone, where reef zooplankton was collected	Technique[a]	Time of collection	Biomass of zooplankton ($mg\,m^{-3}$)		
				In reef waters		In nearby open ocean
				maximum	average	
Bikini atoll (Marshall I.)	Lagoon	n.b.	Day	–	147	71
Ninigo atoll (Bismarck Archipelago)	Lagoon	n.b.	Day	340	230	40
Moruroa atoll (Tuamotu)	Lagoon	n.b.	Night	783	425	24
Rangiroa atoll (Tuamotu)	Lagoon	n.b.	Night	2750	–	13
Takapoto atoll (Tuamotu)	Lagoon	n.b.	Night	170	54	12
Reef (atoll), Heron I., GBR	Leeward reef edge	n.v.	Night	800	720	55
Reef (atoll), Heron I., GBR	Lagoon	n.v.	Night	1150	850	55
Koetivi Island, (Seychelles)	Leeward reef	n.v.	Night	3650	2750	–
		n.v.	Day	50	35	–
Chuong I. (Namsu I., Gulf of Siam)	Fringing reef	n.v.	Night	1380	970	170
		n.v.	Day	20	15	35
Ulaksucul Reef (Panama)	Patch reef	n.b.	Night	1500	–	–
San Blas Reef (Puerto Rico)	Flat	n.a.	Night	10200	6000	2600
		n.a.	Day	2200	1800	30
Laurel Reef (Puerto Rico)	Flat	n.a.	Night	10000	1000	115
Western Reef (Barbados)	Flat	n.a.	Day	600	250	40

[a] n.a.: net tows by divers; n.b.: net tows from the boats; n.v.: passing of a definite volume of water taken by a bucket through the net.

Table 3.23. Biomass of zooplankton emerged from different bottom biotopes of barrier reef Bohol (Philippines) by day and night, caught in traps. (After Porter et al. 1977)

Type of bottom	Rate of emergence, biomass $g\,m^{-2}\,day^{-1}$		Total emergence
	by day 10.00–16.00	by night 16.00–22.00	
Branching corals	0.14	0.68	13.1
Flat corals	0.09	0.41	7.9
Coral rubble	0.08	0.24	6.7
Sand	0.05	0.21	4.1

through daytime net collections. It varies within 0.3 to $10\,g\,m^{-3}$ (Tables 3.22, 3.23). Our estimates of zooplankton biomass arrived at by passing 150 l at night through the plankton net carried out at reefs off Vietnam, the Seyshelles and at the Great Barrier Reef gave values of $0.5–4\,g\,m^{-3}$. This level of the zooplankton biomass is similar to that in waters of the most productive oceanic and neritic areas. The abundance of reef zooplankton could vary with the seasons (Lewis and Fish 1969; Le Borgne et al. 1989). For example in the lagoon of the Great Barrier reef the biomass of zooplankton collected by daytime net tows varied from minimal in summer ($50–60\,mg\,m^{-3}$) up to $200–300\,mg\,m^{-3}$ in winter. This winter maximum was directly connected with an increase in the discharge from rivers bringing the excess nutrients to the lagoon (Farran 1949; Sammarco and Crenshaw 1984; Fig. 3.18). The same was observed in waters up the reefs off New Caldeonia (Binet 1984). In the ring reefs, which were not subjected to the influence of rivers, zooplankton maxima in the same kind of samples were found in summer (Sale et al. 1976a; Ricard et al. 1979; Mc William et al. 1981).

The above data show that the zooplankton is extremely abundant in reef waters and so is an important component of the coral reef ecosystem. But really quantitative data are still scarce. Most quantifications are given in the numbers of specimens and say nothing about their biomass (Sale et al. 1976a; Alldredge and King 1977; Mc William et al. 1981; Walter et al. 1981). Also, traps and even plankton nets do not provide really quantitative zooplankton samples.

Such information as is available shows that the largest zooplankton biomass could be found at night up the shallow reef areas with patch reefs covered with living corals or with rubble (Tables 3.18, 3.22, 3.23; Fig. 3.15), i.e., in places where in accordance with earlier data (Johannes et al. 1970; Glynn 1973) it ought to be lowest, being depleted by the grazing of bottom-sessile predators, and especially by corals. This grazing could indeed be observed (Porter 1974b; Johannes and Gerber 1974). But the period of nocturnal grazing cannot be long, because it proceeds mostly at the initial and final phases of zooplankton migration: during its emergence and during its descent (Porter 1974b). Actually, as it could be seen from data displayed in the tables, grazing does not seriously decrease the zooplankton content in water. The calculated rate of grazing give a value of several tenths of $mg\,m^{-3}$ or 100–150 $sp.\,m^{-3}\,night^{-1}$ (Glynn 1973a; Johannes and Gerber 1974), while the size of migration is at least by one order greater: 0.5–2 $g\,m^{-3}$ or $1–3 \times 10^3$ $sp.\,m^{-3}$ during the same night (Tables 3.18, 3.23).

3.3.2 Microzooplankton

The microzooplankton of reef waters includes planktonic organisms less than 200 μ in size: the zooflagellates, the ciliates and the multicellular small larvae of planktonic and benthic animals. In minor quantities it contains also small radiolarian sarcodines, the planktonic amoebae and the colorless dinoflagellates, which together with the zooflagellates form the group of heterotrophic nanoplankton (sizes 1–20 μ). Among the zooflagellates the representatives of their naked forms are most numerous. They belong to their genera common from the orders Protomonadida and Kinetoplastida (Bodo, Parabodo, Monas, Oicomonas). The zooflagellates feed mostly on bacteriaplankton and on picoplanktonic cyanobacteria. They are consumed themselves in turn also by planktonic and benthic filter-feeders and by the ciliates, forming so an important intermedial trophic link between the bacterioplankton and the reef animals. The planktonic ciliates of reef waters include the nake holoplanktonic and benthoplanktonic ciliates and the loricated tintinnids. For a long time the latter were erroneously considered a main group of ciliates in reef waters, because their large specimens (100–150 μ) were present even in the net zooplankton samples. They also survived better the Utremöl fixation procedure, used as a main method of concentration of samples for ciliate counts (Kimor et al. 1981). But even the use of the sedimentation method resulted in a loss of 90% or more even of the tintinnids. Their biomass thus counted in reef waters was so small that this group of reef zooplankton was not deemed to deserve attention. The use of new methods resulted in proof that the nake ciliates alone comprised 70–90% of the total ciliate populations. Their size is 15–150 μ, and their dominating taxa are represented by the spirotrichs Strombidium and Strobilidium, and by the holotrichs Tiarina and Tontonia. The most numerous among them are usually the different strombidia, among which the small ones feed on bacteria and the larger ones are phytophages, feeding on small planktonic algae. The bentho-planktonic ciliates are represented mostly by a large – up to 150 μ sized – hypotrichs, like Trachelostyla or Urostyla, and happen to be numerous in turbulent areas of the reef, where they are washed out from their native environments: near bottom water layers, detritus, periphyton and surface layers of soft bottom sediments.

The multicellular microzooplankton includes mostly the small larvae of reef animals, like early nauplii of copepods, early veligers, rotifers, and juvenile Oicopleura, which pass through the ordinary plankton nets and the quantification of which needs special methodology.

The first attempt at quantification of the reef microzooplankton was made in the net samples (Marshall 1933). Later, the same was done by Sammarco and Cranshaw (1984). They counted large tintinnids and nauplii and found their biomass negligible, which was far from reality. In fact, it could be supposed to be high, when one took into account the abundance of bacteria, detritus and microalgae as potential food for organisms in the microzooplankton. Counting them with the use of special methodology finally confirmed this statement (Sorokin and Tyapkin 1984; Landry et al. 1984; Kopylov 1985; Moiseev 1985; Le Borgne et al. 1989; Blanchot et al. 1989). Their group dominant by biomass most often is formed by the naked ciliates (Tables 3.24–3.26), the same as in oceanic waters (Sorokin 1981c). By number, the most numerous are the zooflagellates which comprise a larger part of the above-mentioned joint group of heterotrophic nanoplankton. The number of organisms in this group in reef waters could attain $0.5 \times 10^6 l^{-1}$ and their biomass $2-10 mg m^{-3}$ (Sorokin and Tyapkin 1984; Landry et al. 1984; Moiseev 1985; Tables 3.25, 3.26). At some continental reefs it could be even more – up to $20-60 mg m^{-3}$. The larger of the last values was recorded in reef waters enriched with detritus, formed after the decay of wind-driven blue-green algae near the reefs (Table 3.25).

The number of ciliates in reef waters was usually high: $0.3-2 \times 10^3 l^{-1}$ (Tables 3.24–3.26). Their biomass varied within 20 to $60 mg m^{-3}$, but sometimes it attained $100-300 mg m^{-3}$, which is high even for the productive

Table 3.24. Number (N, l^{-1}) and biomass ($mg m^{-3}$, wet weight) of components of microzooplankton as estimated by epifluorescence microscopy counts in reef waters off Lizard I., GBR; positions of stations and depths: 1 50 m out from the windward reef, 25 m; 2 over the windward reef flat, 1.5 m; 3 center of lagoon 8 m; 4 patch reefs, 2.5 m; 5 slope of leeward reef, 4 m. Original data

Time, tide phase	No. of station	Zooflagellates		Ciliates		Nauplii		Total biomass
		$N, 10^5$	B	$N, 10^3$	B	N	B	
Midday; ebb	1	6.8	22	16.0	135	9	3	160
	2	0.1	0.3	0.5	5	5	2	7.3
	3	8.8	28	7.0	25	40	12	65
	4	1.0	0.3	0.5	0.5	13	4	4.8
	5	3.1	11	2.5	8	16	5	24
Night; incoming tide	1	6.1	20	18	320	8	4	344
	2	4.0	13	1.5	22	31	16	51
	3	10.0	33	11.5	218	20	10	261
	4	2.6	9	3.0	50	4	2	61
	5	8.5	28	9.7	155	7	3	186

Table 3.25. Number and biomass of planktonic protozoa in reef waters of Vietnamese coastal areas and in the South China Sea; samples were taken near the outer reef edge from the surface. (After Kopylov 1985; Moiseev 1985)

Place	Zooflagellates		Ciliates		Total biomass of protozoa ($mg\ m^{-3}$)	Total biomass of microheterotrophs including bacteria ($g\ m^{-3}$)
	Number $10^3\ l^{-1}$	Biomass $mg\ m^{-3}$	Number $10^3\ l^{-1}$	Biomass $mg\ m^{-3}$		
Kua Van Bay	160	10	1.7	130	140	2.8
Ben Goy I.	290	6	5.8	120	126	0.42
Hon Gom Cape	80	0.2	1.0	1.5	15	0.23
Vung Ro Bay (corner)	120	4	2.5	64	68	0.65
Vung Ro Bay (entrance)	460	26	10.6	230	256	0.40
Grand Bank, Nhatrang Bay	400	25	0.7	90	115	1.35
Che Bay	140	6	3.4	180	186	0.39
Torthu Island, Nhatrang Bay	120	6	1.4	60	66	1.03
Sinton atoll	120	4	1.9	80	84	0.13
Spratly I.	900	32	0.9	307	373	0.12
Thu I.	2000	66	10.6	110	170	0.81

Table 3.26. Ranges of number and biomass of main groups of microplankton in reef areas of some islands in the Indian Ocean and in the Gulf of Siam. (After Kopylov 1985; Moiseev 1985)

Place	Reef zone	Range of depths at stations (m)	Nanoheterotrophs		Ciliates		Multicellular micrzooplankton		Total microzooplankton biomass ($mg\ m^{-3}$)
			Number $10^3\ l^{-1}$	Biomass ($mg\ m^{-3}$)	Number $10^3\ l^{-1}$	Biomass ($mg\ m^{-3}$)	Number sp l^{-1}	Biomass ($mg\ m^{-3}$)	
Gubbet-Kallansia Bay (Socotra)	Coastal patch reef	1–2	130–350	2–10	0.1–1.0	3–35	10–34	2–7	7–53
	Seaward, outside the reef	24–26	220–370	3–12	1.2–1.8	16–52	17–42	3–9	31–65
St. Anne's Bay, Praslin I. (Seychelles)	Patch reefs	2–7	200–400	6–8	1.0–2.0	75–100	14–96	7–17	90–135
North reef, Koetivi I. (Seychelles)	Leeward reef	3–14	120–180	1–13	0.8–1.0	30–40	1–30	0.2–6	37–44
Coco I. reef, Kargados-Karahos I.	Leeward reef	1–6	2–180	1–9	0.8–1.0	30–40	1–2	0.4	30–42
South reef, Chuong I. (Namsu I.)	Flat and reef edge	1–8	260–550	20–38	0.3–0.5	11–170	–	–	47–208
Western reef, Mongtai I. (Namsu I.)	Reef edge and slope	3–10	136–280	7–22	0.3–0.4	10–21	–	–	27–45

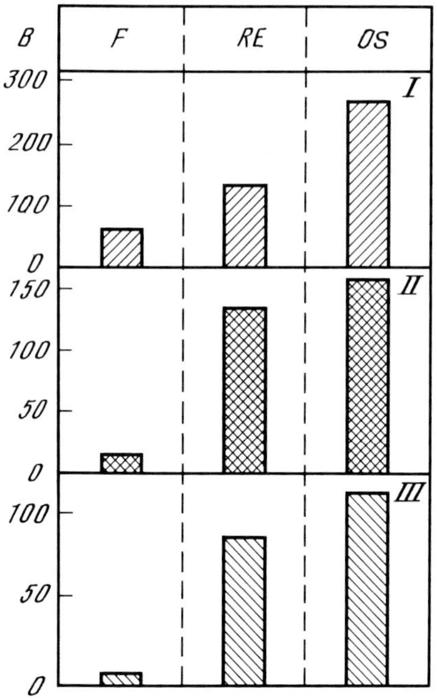

Fig. 3.19. Changes in biomass of planktonic protozoa (B, mg m^{-3}) at the section across some reefs of Vietnam in the South China Sea; *I* Van Fong bay; *II* Kua Van bay; *III* Sinton atoll. Reef zone depths: *F* reef flat, 0.5–1.5 m; *RE* reef edge, 8–12 m; *OS* open sea, 100–200 m off the reef edge. (After Sorokin and Tyapkin 1984)

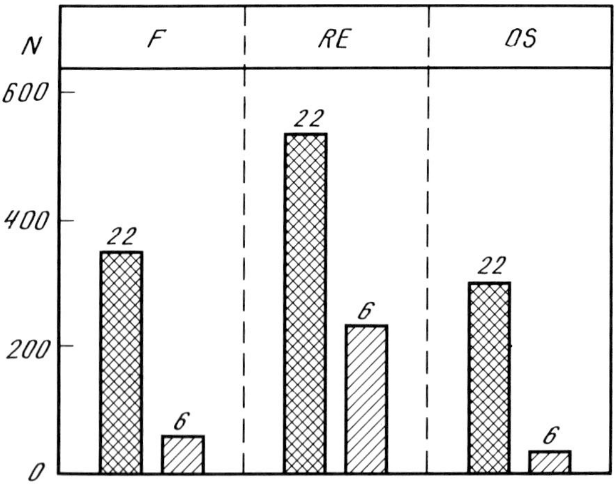

Fig. 3.20. Diurnal changes in the number of planktonic ciliates (N l^{-1}) in waters from different reef zones; off Chuing I. (Namsu I., Vietnam) for designation of zones, cf. Fig. 3.19. *Numbers* above the columns give the hour of sampling. (Data by Kopylov 1985)

upwelling areas (Table 3.25). Thus, their biomass in reef waters is 3–10 times greater than in surrounding oceanic waters. The number of ciliates is very variable in different reef zones and during the day (Fig. 3.19). In waters up reef shallows they were less abundant than off the reef edge or in the lagoon. In the late evening their number was at its maximum, and early in the morning at its minimum (Fig. 3.20). Among the factors controlling their number are grazing by migrating zooplankton and by sessile benthic animals, as well as the variations in their growth rates. Multicellular microzooplankton is represented mostly by the micro-larvaton (Le Borgne et al. 1989). It comprises 10–20% of the total microzooplankton biomass (Tables 3.24, 3.26). Its numbers range within 10–40 sp. l^{-1}, with a biomass of 3–15 $mg\,m^{-3}$. Total biomass of microzooplankton in reef waters varied within 20 to 200 $mg\,l^{-1}$ (Tables 3.24, 3.26). Thus, it is several times less than the nocturnal biomass of mesozooplankton. But their functional activities could be commensurable, because the specific production rate of microzooplankton (P/B) is about an by order more than that of mesozooplankton. The generation time of planktonic ciliates is 10–40 h, and so the coefficients of P/B per day were 0.4–1.5, and in zooflagellates 1.5–3.2 (Landry et al. 1984; Kopylov 1985; Le Borgne et al. 1989). It means that the protozoan populations reproduce themselves approximately in 1 day. Thus the production of both these components of coral reef zooplankton appears to be of the same magnitude, 30–50 $mg\,m^{-3}$ of the biomass. The calculated respiration rates of microzooplankton in reef waters off Vietnam gave values of 30–100 $ml\,O_2\,l^{-1}\,day^{-1}$. So its share in total plankton respiration could be 20 to 40% (Sorokin and Tyapkin 1984). A point of special importance of microzooplankton in reef trophodynamics consists in its exclusive role as a main source of feeding for early fish larvae, ensuring their survival at this critical period of their ontogenesis.

3.3.3 Conclusion

The results of studies on coral reef zooplankton prove that it forms its own communities within the reef ecosystems, which differ from the communities in the surrounding oceanic waters. The fauna of reef zooplankton includes endemic reef holoplanktonic, demersal and meroplanktonic species, and contains also typical oceanic species. Many of the open-sea zooplankters behave within the reef planktonic communities as typical reefal forms. They employ the reef refuges to escape daytime grazing, and use the food resources of the reef as well. A bulk of reefal zooplankton is reproduced at the expense of these recources, and thus the zooplankton brought to the reef with passing or incoming oceanic waters comprise a minor part of its nocturnal biomass.

The data on a high density of the nocturnal zooplankton, which could be expressed by values of 0.5–1.5 $g\,m^{-3}$, lead to a conclusion about possible high rates of heterotrophic respiration of the zooplankton alone in water columns up the reef at night, which at the above biomass lewel should be 0.2–1 g O_2 m^{-2} day^{-1}. Counting the respiration of bacterioplankton, it should be 1.5–2.5 g O_2 m^{-2} day^{-1}. Thus the traditional approach to measure reef heterotrophic metabolism, counting only the respiration of benthos and neglecting possible respiration of plankton, should be carefully reconsidered, because with uncounted phytoplankton photosynthetic production of oxygen (0.3–0.6 g O_2 m^{-2} day^{-1}), and uncounted respiration of plankton (1.5–2.5 g O_2 m^{-2}) alone the range of the daily fluctuations of oxygen content in waters over the reef could attain 1.5–2 g O_2 m^{-2}. But the diurnal fluctuations of oxygen are still accepted as being induced by metabolic processes, which go on in bottom communities.

The same concerns the roles of zooplankton in secondary production. Its calculated approximate production, taking an average biomass of 0.5–1.5 g m^{-3}, a specific production (P/B) of 0.15 per day, and water depth over the reef of 3 m, should be 0.3–0.7 g m^{-3}. Thus it could not be very far from the production of zoobenthos even if its biomass is around 50–100 g m^{-2}. So the roles of zooplankton and of zoobenthos in the reproduction of reef trophic resources seem to be quite comparable. The reason for this conclusion could be deduced a priori from the fact of the high density and variability of planktonovorous fish and fish larvae of most pelagic and bottom reef fishes.

4 Benthic Microflora, Periphyton and Plant Associations

The plant benthic communities and the bottom microflora are responsible for more than half the total autotrophic production and of beterotrophic metabolism in coral-reef ecosystems. The plant communities include the microphytobenthos of soft sediments, the periphytonic overgrowths of the solid surfaces and rubble, the thallophyte seaweeds and the seagrasses. The bottom microbial associations are represented by bacteria in soft sediments (soft sands), by bacteria in the detrital sediments, by periphytonic bacteria and by bacteria symbiotic to benthic animals (sponges). The relative importance of each of the components listed varies in different reef zones.

4.1 Bacteria and Microalgae in Reef Bottom Sediments

The first proof of an important role played by benthic microflora in reef metabolism was demonstrated by Di Salvo (1969) and Di Salvo and Gundersen (1971). Studying the detrital sediments of Hawaiian recefs, they obtained extremely high plate-count numbers of heterotrophic bacteria – up to 10^6 cells g^{-1}. The author made in 1968–1971 estimations of the total number, biomass and production of bacteria in coral sands and periphytonic overgrowths in a variety of Pacific reefs (Sorokin 1971a, 1973c, 1978b).

The soft reef sediments are represented mostly by coral sands. The sand particles, being formed from the skeleton material of reef animals and calcareous algae, have a porous structure (Fig. 4.1). This heightens the possibility of their overgrowth by bacteria. Being imbedded into the mucous cover, which they form around the particle, they comprise a kind of microbiocoenosis (Weise and Rheinheimer 1978; Sorokin 1981c). Among bacteria, which inhabit the sand particles, the most numerous are rods which attach to the particle surfaces with the aid of polysaccharide fibrils (Weise and Rheinheimer 1978; Fletcher and Marshall 1982). Disk-shaped bacteria are also attached to the surfaces of sand grains. The coccoid bacteria form on the surface of the sand particles the mucous microcolonies. Many single cells in these microbiocoenoses are also within the mucus capsules (Moriarty and Haywood 1982). Prostecobacteria and Caulobacteria attach to the surface by means of their stalks. Among the physiological groups recorded

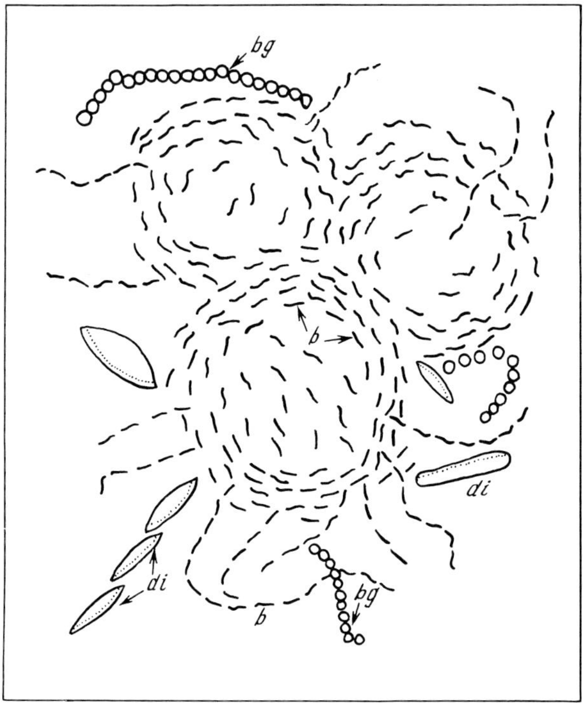

Fig. 4.1. Disposition of the cells of bacteria (*b*), diatoms (*di*), and blue-green algae (*bg*) around a particle of coral sand

in the reef sediments there were the saprophytic bacteria decomposing proteins, chitin or agar (Di Salvo and Gundersen 1971; Gundersen and Stroupe 1976), sulfate-reducing and nitrifying bacteria (Sorokin 1973c; Webb and Wiebe 1975; Barnes et al. 1976; Skyring and Chambers 1976).

The total number of bacteria in coral sands by microscopic direct counts was usually around 10^9 cells g^{-1}, attaining sometimes $2–3 \times 10^9$ (Tables 4.1–4.3), while the content of organic carbon in them was very low – less than 1% (Table 4.2).

In the detrital (or "regenerative") sediments, which accumulate in crevices of dead coral heads or in trenches of lime rocks of flats, the total number of bacteria approaches that in the slime of a sewage plant – up to $10^{10} g^{-1}$, and their biomass up to $1 mg C g^{-1}$, that is 5–8% total organic carbon of a sediment (Di Salvo 1973; Sorokin 1973c). That was the highest level of microbial density ever recorded in a natural nonpolluted habitats. The wet microbial biomass in reef soft sediments varied within $0.3–2 mg g^{-1}$ or $0.05–0.4 mg C g^{-1}$. That comprises some 2–5% of the total stock of organic matter in the sediments (Sorokin 1973c, 1986a; Moriarty 1982; Moriarty et al. 1985b, 1990; Hansen et al. 1987). The microbial biomass in

Table 4.1. Total number of bacteria (N) in reef bottom sediments, their biomass and production, calculated per 1 g of dry sediment

Location	Reefal zones	Sediments, sites	N, 10^9 g^{-1}	Biomass		Production	
				mg C (B)	C_{org}, %	μg C g^{-1} day^{-1} (P)	P/B
Coconut I.[a] Hawaii	Reef flat	Coral sand among living corals	3.6	0.44	5.1	40	0.1
		Coarse coral sand near reef edge	2.8	0.30	3.3	13	0.04
		Detrital "regenerative" sediment	5.6	0.48	2.5	16	0.03
	Lagoonal shallows	Silty-detrital coral sand among dead corals	9.7	1.82	8.0	14	0.05
		Coral sand	0.3	0.008	0.3	0.8	0.1
		Fine coral sand among seaweeds	3.1	0.20	1.6	14	0.07
Majuro atoll,[a] Marshalls	Windward flat	Coral sand	1.7	0.066	1.0	12	0.18
	Lagoonal coastal reef	Coral sand among corals	1.6	0.040	1.5	7	0.18
	Lagoonal patch reefs	Fine mucous coral sand under living corals	3.5	0.18	3.0	35	0.20
		Fine detrital sand among seaweed	3.8	0.36	6.0	71	0.20
		Detrital "regenerative" sediment	7.2	0.36	1.2	98	0.27
Davis reef,[b] GBR	Reef flat	Coral sand	1.5	0.10	2.6	14	0.14
	Patch reefs	Coral sand	1.2	0.084	1.4	28	0.33
	Lagoon	Silted coral sand	1.8	0.133	3.4	24	0.18
Lizard I.[c] GBR	Reef flat	Coral sand among rear-living corals	–	0.07	3.5	40	0.57
		Coral sand within the seagrass bed	–	0.12	2.8	40	0.33

[a] After Sorokin (1978b).
[b] After Y.A. Hansen et al. (1987).
[c] After Moriarty et al. (1985b, 1990).

Table 4.2. Contents of organic carbon (Corg), total number of bacteria (N, $10^9\,g^{-1}$), their biomass and production per day (P) in coral sands of reefs, depending on their cover of living corals

Features of reefs	Location	C_{org}	N	B		P	
				$mg\,C\,g^{-1}$	% of Corg	$\mu g\,C\,g^{-1}$	P/B
Dense cover of living corals	Scott reefs (Indian Ocean)	0.80	4.20	0.63	7.9	43	0.07
	Bili-Bili I., New Guinea	0.65	2.20	0.54	8.3	39	0.07
	Barrier reef off Panarora Louisiada (rich coral growth)	0.45	2.80	0.39	8.6	45	0.12
	Lizard I., GBR[a]	0.48	–	0.34	7.1	92	0.27
Average coral cover	Conflict atoll, Louisiada	0.27	0.81	0.074	2.7	12	0.16
	Funafuti atoll, Tuvalu	0.81	0.31	0.080	1.0	12	0.15
Poor coral growth	Lizard I., GBR, bare reef flat[a]	0.43	–	0.12	2.8	30	0.25
	Vekai atoll, Lau archipelago	0.37	0.37	0.056	1.5	7	0.12
	Fringing reef off the Pentacost I., New Hebrides	0.16	0.37	0.018	1.1	0.9	0.05
Degrading algal reefs, living corals absent	Katafang atoll, Lau	0.66	0.41	0.026	0.4	3	0.12
Reef absent, rare corals on lava rocks	Lopevi I., active volcano, New Hebrides	0.06	0.04	0.0015	0.25	0.2	0.13

[a] Data by Moriarty et al. (1985b); Other data by Sorokin (1986b).

Table 4.3. Total number (N), biomass (B) and production (P) of bacteria in reef soft-bottom sediments

Location	Method[a]	N, 10^9 g^{-1}	B, mg C g^{-1}	P, g C m^{-2} day^{-1}	Reference
Kaneohe bay, Hawaii	mic	0.9–4.9	0.13–0.44	12–40	Sorokin (1978b)
	atp	–	0.66–1.90	325–1670	Burns et al. (1984)
Lizard I., GBR	mur	–	0.072–0.145	40–92	Moriarty et al. (1985)
Fringing reefs, central Vietnam	mic	0.9–11.3	0.14–1.46	20–120	Original data
Majuro atoll, Marshall	mic	0.8–3.5	0.04–0.36	5–35	Sorokin (1978b)
Enivetok atoll, Marshalls	atp	0.5–5.0	0.1–1.0	–	Wiebe (1978)
Fanning atoll, Lain	mic	1.84	0.84	23	Original data
	mur	–	0.03–0.04	–	Moriarty (1982)
Atoll Sinton, South China Sea (coral sand)	mic	0.12–0.67	0.011–0.07	–	Original data
Same (detrital sediment)	mic	10.4	1.60	–	Original data

[a] Designations of methods employed for biomas estimations: mic – direct microscopy on stained slides or membrane filters; efm – epifluorescence microscopy; mur – via measurements of muramic acid; atp – ATP method.

reef sediments is by about an order more than in silts of temperate coasts (Sorokin 1978b). These extremely abundant populations of bacteria in reef sediments could be supported only by their high production and metabolism based energetically on a high flow-rate of suspened organic matter, which settles a from the water column (cf. Sect. 2.3). In fact, the production of bacteria in them is very high. As estimated by different methods, it was 10 to 70 μg C g^{-1} day^{-1}. That is, within an active layer 2–3 m thick it would be 0.2–2 g C g^{-2} day^{-1} or 1–10 g m^{-2} of wet microbial biomass. According to Moriarty et al. (1985b), production of bacteria, estimated by the thymidine method in coral sands of Lizard Island ranged in summer within 0.12–0.37 g C g^{-2} day^{-1}, comprising 30–40% of the gross photosynthesis production of their microphytobenthos. Taking the total stock of microbial biomass as 10–50 g $m-^{2}$, the coefficients of specific production of benthic bacteria (P/B) could be estimated within 0.1–0.2 per day. That means that the turnover time of the stock of microbial biomass in coral sands should be rather short: 5–10 days.

The abundance of microbial populations in coral sands appears to be dependent upon the density of coral growth (Table 4.2). The largest numbers of microbial biomass were found in sands of reefs with a rich or medium coral cover, and the smaller in sands of "algal" reefs, where the living corals were rear ones, such as the reefs of New Hebrides and some reefs of the Lau archipelago. These data prove that the mucus shed by corals is a most important source of energy, which supports a rich microbial population in reef sediments (Rublee et al. 1980; Pascal and Vacelet 1981; Paul et al. 1986; Sorokin 1986a; Wahbeh and Mahasnen 1988). The mucus enriched with settled particulated matter, when shed by corals, accumulates at the bottom under thickets of living corals and forms there downy mucous sediments, which are especially peculiar to the patch-free areas of some Pacific atolls (Sorokin 1973a). The density of microflora in them was about as high as in the detrital sediments (Tables 4.1–4.3). The rate of microbial respiration in these sediments was so high that also at depths of 1–2 cm on vertical profiles the Eh dropped to 100 mv, and at depths 10–15 cm dropped below 200 mv (Fig. 4.2), but what was very curious was that it happened without any sign of sulfate reduction in them (see below). A high microbial biomass (0.3–1 mg C g^{-1}) was also found in silted coral sands of the fringing reefs off the coast of central Vietnam (Table 4.3), and in the reef areas of the GBR covered with seagrasses (Moriarty and Pollard 1982). In coral sands of Vietnamese reefs in the open South China Sea it was 5–10 times less: 0.02–0.15 mg C g^{-1}. Extremely high values of microbial biomass, as measured with the use of the ATP-method, reported by Burns et al. (1984), were surely overestimates because this method counted together with bacteria also all other living components of the present in the sediment samples. In total, the density of microbial populations in coral-reef sediments is usually by 1–2 orders more than in the upper layer of silts of temperate coasts (Sorokin 1978a).

Fig. 4.2. Redox potential (E_h, *mv*) in columns of soft sediments of the Majuro atoll, Marshalls; *1* mucous sediment under the thicket of living coral; *2* coral sand of the lagoonal slope; *H* depth in column of sediment, cm. (Data by Sorokin 1973c)

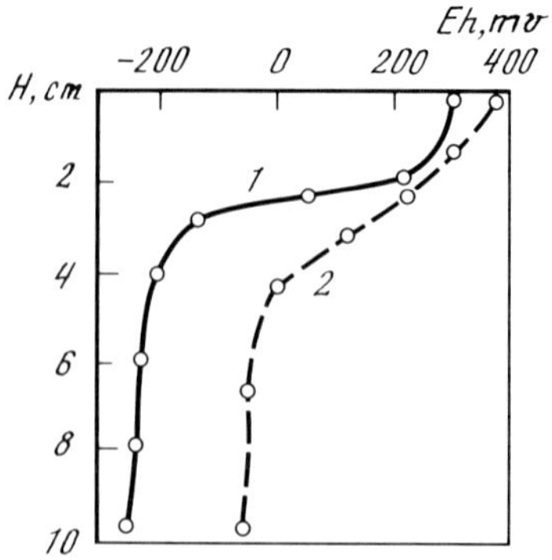

The rate of respiration of microbial populations in coral sands is correspondingly very high (Table 4.4). Measured by different methods, it ranges in coral sands within 0.1–0.6 mg $O_2\,g^{-1}\,day^{-1}$. It means that only 10–20 g of sand could consume all oxygen from 1 l of the oxygen-saturated sea water during one day. Per 1 m^2 the respiration of bottom communities (including other microbenthic plant and animal elements) in the coral sand areas varied within 0.8–8 g $O_2\,m^{-2}\,day^{-1}$. The microbial populations take up 70–80% of these values. Thus average values of microbial respiration in coral sand after measurements in situ or in vitro are either at the level of 0.1–0.3 mg $O_2\,g{-}^{1}\,day^{-1}$ or of 0.5–5 g $O_2\,m^{-2}$ (Table 4.5).

Intensive respiration in coral sands results in the developing of anaerobic conditions within a few centimeters of sediment cores (Figs. 4.2, 4.3). A

Table 4.4. Rates of respiration (M) of microheterotrophic organisms in reef bottom sediments, as measured by different methods

Sites	Types of sediments	Method[a]	M, day^{-1}		Reference
			mg $O_2\,g^{-1}$	g $O_2\,m^{-2}$	
Kaneohe bay, Hawaii	Detrital sediment	rwr	2.4–12.0	4.8–24.0	Di Salvo (1971b)
	Detrital sediment	bod	1.2	2.4	Sorokin (1978b)
	Coral sand	bod	0.05–0.1	0.7–15	Sorokin (1978b)
Majuro atoll, Marshalls	Coral sand	km	0.17–0.65	2.5–9.5	Sorokin (1978b)
Fanning atoll, Lain	Coral sand	km	0.13	2.0	Sorokin (1970)
Heron I., GBR	Coral sand	bod	0.05–0.1	0.7–1.5	Sorokin (1984a)
	Coral sand	enc	0.05	0.8	Propp et al. (1983)
	Coral sand with rubble	bod	0.07–0.12	1.0–1.8	Sorokin (1984a)
Moorea I., Society I.	Coral sand	enc	0.20	3.0	Sournia (1976b)
Coastal fringing reefs, central Vietnam	Carbonaceous silt	bod	0.65	3.2	Original data
	Silted coral sand	enc	0.26–0.55	2.6–5.5	Original data
	Coral sand	bod	0.13–0.53	2.0–8.0	Original data
Sinton atoll and Spratley, South China Sea	Coral sand	enc	0.13–0.26	2.0–3.5	Original data
	Coral sand with rubble	enc	0.21–0.23	4.2–4.5	Original data

[a] km – Indirect calculation using data on production rate of bacteria and K_2-coefficient of efficiency of use of assimilated food for growth; bod – bottle oxygen method; rwr – use of flow water respirometer; enc – use of in situ enclosures.

Table 4.5. Generalized data on levels of gross photosynthesis (P) and respiration rates (M) of microphytobenthic and microbial communities of soft reef sediments by measurements employing different methods

Sites	Method[a]	P	M	P/M (average)	Author
The Phantom bank, Timor Sea	rwr	0.055	0.26	0.2	Propp et al. (1983)
Heron I., GBR	rwr	0.26	0.32	0.8	Propp et al. (1983)
Heron I., GBR	bod	0.57	0.64	0.9	Sorokin (1984b)
One Tree I., GBR	enc	0.90	1.40	0.6	Kinsey (1978)
Takapoto atoll, Tuamotu	bod	1.0–3.2	–	–	Sournia (1976b)
Sinton atoll, South China Sea	enc	0.45–0.92	0.42–0.59	1.4	Original data
Scott reefs, Indian Ocean	bod	1.58–2.16	1.28–3.12	0.8	Sorokin (1986)
Majuro atoll, Marshalls	C^{14}+bod	0.40–2.22	0.28–3.96	0.6	Sorokin (1978b)
Nosy Be, Madagascar	C^{14}	0.23–0.39	–	–	Plante-Cuny (1973)
Kaneohe Bay, Hawaii	C^{14}+bod	0.08–0.21	0.68–1.90	0.1	Sorokin (1973c)
Central Vietnam	enc	0.06–0.36	0.46–0.68	0.4	Original data
Same site	bod	0.50–1.70	0.52–1.70	1.0	Original data

[a] C^{14} – radiocarbon method; for other methods see Table 4.3; P and M values are given as $g\,C\,m^{-2}\,day^{-1}$.

quick decrease in Eh occurs in them some 2–5 cm below within the sediment even in turbulent zones, thus creating conditions for the anaerobic microbial sulfate reduction. The author first observed this phenomenon in the sands of the Kaneohe bay, Hawaii (Sorokin 1973c). Up the layer of sulfate reduction was found the layer of active Thiobacilli, which oxidizes the hydrogen

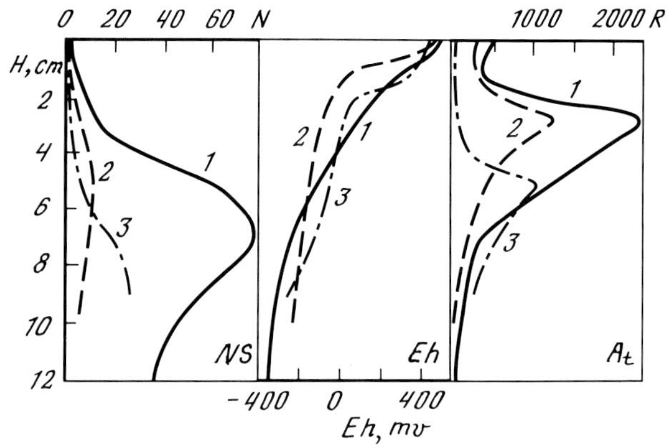

Fig. 4.3. Redox conditions and the distribution of bacteria of the sulfur cycle in columns of soft sediments – coral sand in Kaneohe bay, Hawaii; *Eh* redox potential; *NS* number of sulfate-reducing bacteria (colony counts), N, ml^{-1}; A_t relative activity of Thiobacili; R radioactivity of chemosynthetic $^{14}C_2$ assimilation in the presence of thiosulfate, cpm g^{-1}

sulfide thus formed (Fig. 4.3). Studies of sulfate reduction in sediments of the Kaneohe bay, the Majuro atoll, and off the coast of Vietnam showed that this process is most intensive in sediments of reefs subjected to pollution or to terrigenous siltation. It was quite odd that sulfate reduction was absent in the "mucous" coral sands mentioned above, which accumulate under the thickets of living corals; nevertheless, the Eh dropped below 100 mv also at 2–4 cm under the bottom surface. But the sulfate reduction in the same sediments occurred in polluted areas close to the village. In silty coral sands off the Vietnam coast, influenced by the terrigenous input of organic matter, sulfate reduction is widespread. This process appears to be an important ecological factor there. It causes the intoxication of benthic fauna and of fish feeding on benthos, and lead to the development of oxygen deficiency in near-bottom layers (Sorokin et al. 1982). In the lagoon of Lizard Island influenced by terrigenous outflow of detritus, sulfate reduction proceeds actively in coral sands at depths of only 1–2 cm below the surface. The maxima of sulfide content within the sediments were found at depths 3–4 cm, and maximum rates of SO_4^{-2} reduction by bacteria at 2–3 cm below the bottom surface (Skyring and Chambers 1976; Skyring 1985). This process was also discovered within the column of lime rock at depths of 2–3 m under the bottom surface of fringing reefs in the Kaneohe bay, Hawaii (Tribble et al. 1988).

White coral sands harbor dense populations of micro-algae which exhibit high phothosynthesis rates (Sorokin 1971a; Gribb 1973). The chlorophyll content in them reaches hundreds of mg m^{-2} (Sournia 1976b). Thus, besides their function in the heterotrophic production and in nutrients regeneration, soft bottom sediments are the sites of an intensive primary production, which attains there 1–3 g C m^{-2} day^{-1}. The micro-algae are living in the mucous micro-habitats around and within the porous sand graines (Fig. 4.1). In calm places they form a kind of greenish-brown algal mats over the sand bottom surface. These communities of microphytobenthos contain various benthic diatoms, cyanobacteria and zooxanthellae living within their symbiotic foraminiferans (Plante-Guny 1973; Gribb 1973; Sournia 1976b, 1977; Miller et al. 1977). The diatoms are represented by their genera *Navicula*, *Nitzschia*, *Surirella*, *Acanthes*, *Cocconeis*, *Amphora*, *Gyrosigma*, *Diploneis*, Bacillaria, *Donkinya*. Some of them are attached to the surface of sand graines (*Acanthes*, *Cocconeis*). Others are moving between them (*Nitzschia*, *Navicula*, *Gyrosigma*). Their number attains 0.5×10^6 cells g^{-1}.

Among the cyanobacteria the genera *Schizotrix*, *Calotrix*, *Rivularia*, *Oscillatoria*, *Chroococcus*, and *Lyngbia* are the most numerous. In coarse coral sands of turbulent zones, on reef shallows or in channels the symbiotic foraminifera harboring zooxanthellae are responsible for the larger part of photosynthesis production, not just the free living micro-algae (Plante-Cuny 1973; Sournia 1976b). The content of chlorophyll *a* in the upper layer of coral sands is usually 1.5–5 mg g^{-1}. The biomass of algae in sands comprises about 1–2% of their total organic matter which is at a level of 0.4–1.3 mg wet weight g^{-1} (Sorokin 1978b). In the detrital ("regenerative") sediments it could be ten times more (Sorokin 1973c). The gross primary production of microphytobenthos varied from 0.1–0.3 $g\,C\,m^{-2}\,day^{-1}$ in the coral sands of the polluted or deteriorating reefs with a poor coral growth, up to 1–3 $g\,C\,m^{-2}$ in sands of the flourishing reefs. Its top value (11 $g\,C\,m^{-2}$) was recorded in the greenish sand from the beach of the atoll Majuro (Sorokin 1978b). At its rates of 2–3 $g\,C\,g^{-2}$ the gross photosynthesis in sand per 1 g will be 100–150 $\mu g\,C\,g^{-1}\,day^{-1}$, which is not very far from the same values in living corals calculated per weight of their colonies (150–300 $\mu g\,C\,g^{-1}\,day^{-1}$; cf. Sect. 4.2). The P/B ratios in most coral sands investigated was most often below 1, reflecting so the domination of heterotrophic processes in them. Additional sources of organic matter are obviously needed to support them. The latter could be the coral mucus and the detritus transported from the autotrophic reef-flat biotopes. Thus the soft reef sediments, which cover 60–80% of the total reef-bottom area, are the sites of intensive promary and heterotrophic production, which supports in them an abundant zoobenthos. Microphytes and microheterotrophs of the soft bottom are responsible for approximately 20–30% of the total primary production and for 40–% of the total respiration of coral reef ecosystems as a whole.

The biomass of bacterial and algal microbenthic communities is a basic food source for numerous sediment-eating benthic animals, like holothurians, various worms and clams, as well as for many representatives of microbenthos and meiobenthos (Sorokin 1973c; Moriarty et al. 1985b; Hansen et al. 1985; Hansen and Skilleter 1988). Experimental evaluations of the grazing rates of bacteria demonstrated visible effects of decrease in the density of microbial populations by holothurians, clams and microbenthic copepods, and especially by the harpacticoids. Benthic bacteria are consumed also by nematods, protozoans, and turbellarians. Benthic grazers consume the most part of their daily production. Their assimilation by holothurians was proved in experiments with the use of radiolabelled bacteria (Sorokin 1978b).

4.2 Periphytonic Overgrowth

Vast surfaces of the porous lime reef rocks, as well as the rubble and the heads of dead corals, are densely overgrown with periphytonic turfs. The areas of these overgrown surfaces on the reef exceed 2–3 times the projective ones (Dahl 1973; Odum and Odum 1955). The periphytonic turfs are 2–5 mm thick. They look like brushes or like greenish or reddish mucous covers depending on the kinds of dominating algae. The turfs are structured like a microhabitats or microbioeoenoses, which unite the autotrophic and the heterotrophic components in one mucoid structure. Among these components different bacteria and algae, ciliates and forams as well as other microbenthic animals are present. The microflora in periphytonic communities contains the attached filamentous or chain-forming bacteria like *Cladotrix*, *Leptotrix*, *Crenotrix* and the flexibacteria. Among single-cell bacteria the prostecobacteria and caulobacteria are numerous, being attached to the surfaces of algae with the aid of their stalks. The communities of periphytonic algae include small thalloms of macrophytes 0.5–1.5 cm long, crustose coralline algae (Fig. 4.4), filamentous and attached microalgae. Among the macroalgae the green algae *Enteromorpha*, *Caulerpa*, *Chaetomorpha*, *Acetabularia* and *Bryopsis* could most often be met with. The brown algae are represented most often by *Ectocarpus*, *Giffordis*, and *Sphacelaria*, and the redalgae by *Gelidiella*, *Polisiphonia*, *Chondria*, *Champa*, *Laurencia*, Hypnea, and *Wurdemannia*. The filamentous algae, which dençely inhabit the periphytonic overgrowth, include the green algae *Cladophora*, *Cladopnoropsis*, *Trichosolen*, and the cyanobacteria *Calotrix*, *Schizotrix*, *Kyritrix*, *Oscillatoria*, and *Mastigocoleus*. Among other cyano-

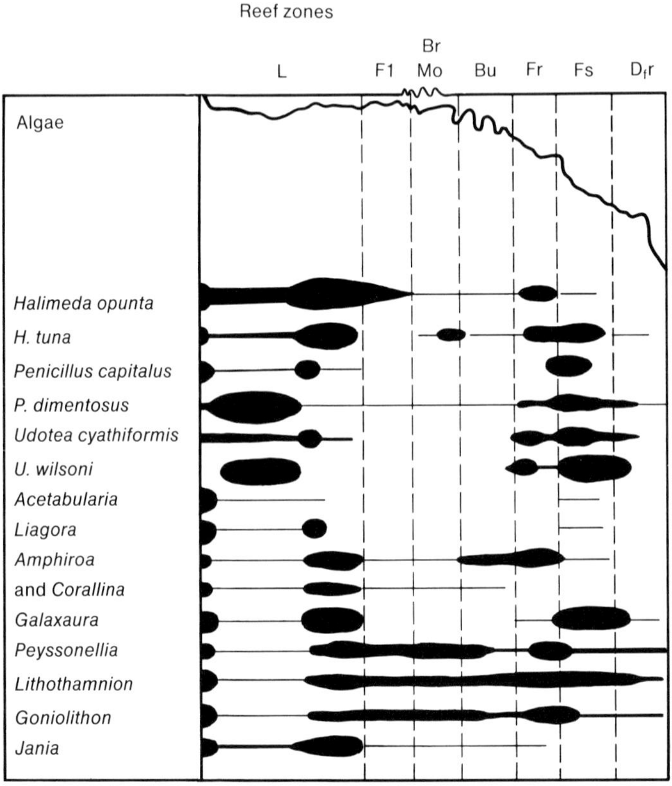

Fig. 4.4. Percentage of bottom cover by the key species of coralline algae at the section across the Waikiki fringing reef, off Oahu I., Hawaii; in the upper scheme, the bottom profile is shown. (After Littler 1973b)

Table 4.6. Comparative abundance and metabolic rates of algae and bacteteria in the periphytonic turfs over dead coral heads in the Kaneohe bay, Hawaii. N – total number of cells; B – biomass; P – production, and M – respiration rate, all as $\mu g\,C\,cm^{-2}\,day^{-1}$. (After Sorokin 1978b)

Biotope	Algae			Bacteria			B_1/B_2	M	P_1/M
	N, $10^4\,cm^{-2}$	B_1	P_1	N, $10^7\,cm^{-2}$	B_2	P_2			
Flat of coastal reef	60	54	24	32	34	11	1.6	38	0.6
Reef edge	40	35	19	24	12	4	2.9	15	1.2
Patch reef	90	93	28	17	14	5	6.6	16	1.7
Outer reef edge	420	450	88	12	4	1	112.5	5	17.6

bacteria *Spirulina*, *Lyngbia*, *Nostoc*, and *Rivularia* are usual and form often dense turfs covering pieces of rubble. Some of the cyanobacteria, such as *Nostoc*, *Rulfsia*, and *Entophysalia*, form the gum – black-or dark-green – covers on surfaces of dead coral heads and on rubble. Some filamentous algae bore into lime rock as well as into lime skeletons of living corals (Bertold et al. 1982). Among them are the green alga *Ostreobium reineckei* and the cyanobacteria *Entophyzalis deusta*, *Heyella* sp. and *Mastigocoleus testarum* (Odum and Odum 1955; Gribb 1973; Kobluk and Risk 1977; Le Campion-Alsumard 1989). The coralline algae from crusts a top the surfaces of rubble, dead coral heads and the solid surfaces of the flat rocks. (Larkum 1983) Most common among them are Porolithon, Jania, Litothamnion, and Neogonioliton. Periphytonic communities include also numerous unicellular algae, such as the diatoms (Camylodiscus, Podocystic, Navicula, Nitzschia) and the small cyanobacteria from the order Nostocales. Some of these algae attach to the surfaces either of living or dead corals (Montgomery et al. 1977). The microbenthic heterotrophic part of periphytonic communities includes ciliates represented mostly by Vorticella and by benthic Hypotricha, small nematods, and forams.

The joint biomass of bacteria and of microalgae in periphyton varies within 0.04–0.4 $mg\,C\,cm^{-2}$, or 0.4–5 $mg\,cm^{-2}$ of wet biomass. It comprises 5 to 15% of total organic matter in the rubble or in branches of dead corals (Sorokin 1978b). Estimated per 1 m^{-2}, the biomass of periphytic turnfis varies within 10–40 g C or 80–300 g of wet biomass (Klumpp and Mc Kinnon 1988; Sorokin 1986a). The ratio of biomass of algae and bacteria varied from 112 in clean parts of the reef (Kaneohe bay, Hawaii) down to 1.6 in its polluted part (Table 4.6). It means that in polluted reef the development of periphytonic algae is inhibited and they are replaced by bacteria. The total number of bacteria in periphytonic covers attains $10^9\,cm^{-2}$ of their surface or $2–6 \times 10^9\,g^{-1}$ of scraped material (Tables 4.7–4.9), being independent of the level of coral growth, unlike in the microbial populations in coral sands (Table 4.2). These data show, that the communities of reef periphyton are

Table 4.7. Total number (N) of periphytonic bacteria, their biomass (B) and production (P_b), and metabolic rates of periphytonic turfs ever the dead branches of ramose coral colonies, collected at reefs of the Majuro atoll, Marshalls. Data are given per 1 g of scraped material. (After Sorokin 1978b)

Site	Biotope	Density and production of bacteria				Metabolic rates $\mu g\,C\,g^{-1}\,day^{-1}$		P_P/M
		N, $10^9\,g^{-1}$	B, $\mu g\,C\,g^{-1}$	P_b $\mu g\,C\,g^{-1}\,day^{-1}$	P_b/B	Photosynthesis (P_p)	Respiration (M)	
Ajurotake I., coastal reef	Bare flat	3.10	180	120	0.7	124	330	0.4
	Flat with rare dead corals	2.80	170	110	0.6	302	311	1.0
	Flat covered with dead and living corals	1.92	120	35	0.3	87	101	0.9
Arurakku I., lagoonal reef	Dead coastal reef	7.20	360	100	0.3	280	281	1.0
	Flat covered with living corals	6.20	500	215	0.4	258	600	0.4
	Reef edge, thickets of living corals	2.3	140	43	0.3	139	109	1.3
Eroj I., lagoonal reef	Reef edge, thickets of living corals	4.3	260	106	0.4	93	300	0.3

Table 4.8. Number (N) and biomass (B) of bacteria (per gram of scraped material) in periphytonic turfs over dead corals or flat rocks at reefs with different levels of coral growth. (After Sorokin 1986a)

Level of coral growth	Place	N, $10^9\,g^{-1}$	B, $mg\,C\,g^{-1}$
Abundant	Scott reefs (north), Indian Ocean	2.5	0.36
	Scott reefs (south)	4.6	0.84
	Bili-Bili I., New Guinea	4.2	0.63
Average	Conflict atoll, Louisiada	3.9	0.31
	Panarora I., New Hebrides	4.7	0.38
Weak	Ngelelevu atoll, Lau	2.7	0.67
	Funafuti atoll, Tuvalu	3.7	0.92
	Paama, New Hebrides	3.6	0.32

using an external source of organic matter and nutrients for their growth, and this could be only their stock in the oceanic waters passing over the reef. These waters contain 0.1–0.2 μmol of dissolved organic phosphorus and 0.2–0.4 $mg\,C\,l^{-1}$ of labile organic matter, accessible for the periphytonic bacteria. The nitrogen does not limit their growth, bacause they contain abundant communities of nitrogen-fixing bacteria and cyanobacteria (cf. Sect. 2.2.3). These features of periphyton are extremely important for the functioning of coral ecosystems, providing for the positive balance of nutrients and organic matter in them during their interaction with the surrounding ocean.

The measurements of photosynthesis and respiration (Tables 4.6–4.9) showed that the periphytonic communities are among the most active metabolizing compartments of the reef system. The specific production of bacteria in them is 2–3 times more than that of the microflora in coral sands. The photosynthetic production in periphitonic overgrowth on dead corals comprised 0.4–1.1 $mg\,C\,m^{-2}$ (Tables 4.6–4.9). Calculated per 1 m^2 and taking into account bottom covered by rubble and dead corals, the photosynthesis of periphyton could attain up to 5–10 $g\,C\,m^{-2}\,day^{-1}$. The ratio of photosynthesis to respiration (P/M) was decreasing in the polluted areas, being below 1, while in healthy reefs the photosynthesis much exceeded respiration (Table 4.7): P/M ratios were there 2 to 6 (Majuro atoll, Sinton atoll; Table 4.9). The production of algae was here very high: 2–4 mg of wet algae biomass per gram of scraped material per day (or 120–250 $\mu g\,C\,g^{-1}\,day^{-1}$). The corresponding data available are summarized in Table 4.10. At an average, the production and metabolism of periphytonic communites in places of their domination were 1.5–2 times more than those in microphytobenthos of soft sediments. As in most cases they balance each other, they are equal most often to 2–5 $g\,C\,m^{-2}\,day^{-1}$. Only the net photosynthesis production of periphyton was estimated by Klumpp and Mc Kinnon (1988) to be within 0.8–1.9 $g\,C\,m^{-2}\,day^{-1}$. Taking into account then their extremely wide distribution in reef bottom biotopes

Table 4.9. Number (N) and biomass (B) of bacteria and the metabolic rates in periphytonic turfs over dead coral heads and rubble at reefs of Vietnam; data are given per 1 g of scraped material. (After Sorokin 1978b)

Region	Site	Bacteria		Photosynthesis		Respiration		P/M
		N, 10^4 g^{-1}	B, mg C g^{-1}	μg O_2 g^{-1} h^{-1}	g C m^{-2} day^{-1} (P)	μg O_2 g^{-1} h^{-1}	g C m^{-2} day^{-1} (M)	
Continental Vietnam	Hon-Ke I., fringing reef	4.5	0.27	75.7	2.06	12.0	0.86	2.4
	Van-Fong bay, fringing reef	2.8	0.45	37.2	1.26	23.0	2.07	0.6
	Vung-Ro bay, fringing reef	1.1	0.26	105.9	1.80	20.2	0.90	2.0
South China Sea	Sinton atoll, reef flat	0.6	0.08	148.5	3.03	38.8	2.02	1.5
	Spratley I., outer slope	0.5	0.06	131.2	2.23	7.3	0.33	6.8
	Thu I., fringing reef	0.6	0.09	36.1	0.57	4.6	0.21	2.7
	Sud I., fringing reef	0.8	0.05	149.0	3.58	31.5	1.96	1.8

and also their high production and metabolic rates, we may conclude that they are among the key coral reef communities besides the corals themselves. It was spectacularly proved by our experiments showing the equality of the photosynthesis and respiration rates in living ramose corals and in dead branches of the same coral colonies covered with the periphyton (Table 4.11).

The periphytonic turfs are one of the main sources of feeding for herbivorous invertebrates and fish (Harmelin-Vivien et al. 1989). They are grazed by numerous surgeon-fish, scarides, chaetodontids, as well as by crabs, shrimps, gastropods, and urchins (Bakus 1967). It is just the high biomass and production of epilithic algal and bacterial turfs which provide quick restoration of food sources for these important groups of reef animals, thus supporting their extremely dense populations, unknown in other coastal ecosystems. Another important function of periphytonic microflora could be seen in its role in the decomposition of the stable fraction of DOM in passing over the reef oceanic waters. Its dense cover over vast surfaces of reef rocks and rubble acts like the microbial communities in aerotanks of sewage plants, functioning under optimal conditions of temperature and oxygen supply. It could be supposed that possibly they could decompose even a stable fraction of the DOM of oceanic waters, which is not used by the planktonic microflora. This hypothesis had been forwarded on the basis of data showing the level of development of periphytonic microflora being independent of the level of the coral's cover (Table 4.8). Then we have chequed it by direct experiment (Table 4.12). So it could be supposed that this function of the periphytonic microflora in coral reefs could have a global importance for the turnover of organic matter in the world ocean, coral reefs being a site of microbial biodegradation of the stable fraction of the DOM in sea water. In this regard it could be supposed that the modern level of DOM content in it – about $2\,\text{mg C l}^{-1}$ – may be definitely controlled by

Table 4.10. Photosynthesis and respiration (P, M, $\text{g C m}^{-2}\text{day}^{-1}$) in periphytonic turfs overgrowing different types of solid surfaces

Place	Kinds of substrate[a]	P	M	P/M	Reference
Phantom bank, Timor Sea	prm	1.08	1.13	0.9	Propp et al. (1983)
Hon Ke I., Vietnam	pdc	1.95	1.48	1.3	Sorokin et al. (1983)
Heron I., GBR	pfs	0.75	0.30	2.5	Sorokin (1977b)
Curacao I., Caribbean	pfs	2.42	1.42	1.7	Wanders (1976b)
Enivatok atoll, Marshalls	pdc	1.23	–	–	Bakus (1967)
Velangilala atoll, Lau	pfs	2.12	2.65	0.8	Sorokin (1986b)
Conflict atoll, Louisiada	pdc	1.34	1.80	0.7	Sorokin (1986b)
Funafuti atoll, Tuvalu	pdc	2.90	1.41	2.1	Sorolin (1986b)
Sinton atoll, South China Sea	pfs	3.03	2.09	1.5	Original data
One Tree I., GBR	pfs	1.38	–	–	Klumpp and Mc Kinnon (1988)

[a] prm – rubble; pfs – surface of flat rocks; pdc – dead corals.

Table 4.11. Rates of photosynthesis (P) and respiration (M), both in $\mu g C g^{-1} day^{-1}$, in branches of ramose corals from the living part of the colony (lpc) and from the dead part of the same colony (dpc); K_c – contents of organic carbon in branches. (After Sorokin 1984b, 1986a)

Location	Example	K_c % of dry weight	P	M	P/M	P/K_c, %
Conflict atoll	lpc	3.13	190	170	1.1	0.6
	dpc	2.70	270	360	0.8	0.9
Funafuti atoll	lpc	0.55	296	204	1.5	5.1
	drc	1.09	182	98	1.8	1.5
Ngelelevu atoll	lpc	2.09	641	506	1.3	2.9
	dpc	0.86	580	282	2.2	2.6
Heron I.	lpc	1.40	408	295	1.4	2.9
	dpc	1.17	295	155	1.4	1.3

Table 4.12. Influence of periphytonic microflora, overgrowing the coral rubble, upon the decomposition of labeled dissolved humic acids in samples, taken in the open sea 3 km off the Heron I. reef. Volume of bottles: 300 ml; weight of rubble in them: 50 g; exposure time 12 days. (Original data)

Type of experiments	No. of experiments	Radioactivity of CO_3^{2-} in the samples, 10^3 cpm/10 ml
Seawater only	1	1.72
	2	1.54
Seawater + rubble	3	4.17
	4	4.03
Seawater + formalin	5	0.23
(control)	6	0.35

the time the total volume of oceanic waters has passed over the coral reefs. On a geological scale, this time should not, in accordance with my calculations, be so large – some 40×10^3 years (Sorokin 1977b).

4.3 Bottom Plant Associations

Benthic plant communities of coral reefs at first sight seem to be impoverished compared with the rich thickets of macrophytes and seagrasses of temperate coasts. In fact, the flora of the latter is several times more various than that of coral reefs. Most reef macrophytes have a small, sometimes filamentous, thalloms only 3–6 cm long. A significant part of their biomass is represented by unspectacular crustose Corallinanceae. The above-listed superficial observations led early investigators of coral reefs to a conclusion about the secondary role of benthic plants in coral-reef ecosystems. But this was soon changed (Doty 1974), as it was proved that even on reefs with a rich coral growth such as the Enivetok atoll or the One Tree reef, the primary production of bottom plant associations comprised 40–80% of the total primary production (Odum and Odum 1955; Borowitzka et al. 1983). Their roles in primary production and in calcification are moreover dominant in the so-called "algal" reefs, where corals themselves are rare (Stoddart 1969; Womersly and Bailey 1969). Peculiar features of reef plant associations are the high percentage of calcareous red and green algae in them (e.g., of corallines and Halimeda), a high taxonomic variability of their mini-thalloms or filamentous brown and red algae, which often form kinds of periphytonic turfs, and a significant abundance of epiphytic forms in them. These features of them are directly connected with a high ratio between the actual areas occupied by reef plant associations and the protective ones, which on reefs could be as high as 3 to 5 (Odum and Odum 1955; Dahl 1974). Additional areas are provided by the dead ramose corals, by walls of caverns and trenches, rubble, the surface of thalloms of large fleshy macrophytes, and by leaves of seagrasses. For example, on a Caribbean reef the surface area of leaves of seagrass was evaluated to be within 3.7–4.7 m^{-2} 1 m^{-2} of bottom area, and about half of the latter was occupied by epiphytic algae. The actual areas of solid substrates per 1 m^{-2} of bottom surface partly covered with rubble was evaluated to be within 2–3 m^{-2}, in reef crest zones 1.4–1.6 m^{-2}, and in thickets of ramose corals over 5. These excess surfaces are usually inhabited by periphytonic turfs, mini-thalloms of brown and red algae, and coralline algae. Inside numerous caverns and trenches poor light conditions limit growth of other algae, exept the red ones. Thus they attain at the reef great diversity and abundance.

The main factor controlling growth and composition of reef plant associations is intensive grazing of them by herbivorous fish and invertebrates (Earle 1972; Ogden 1976; Choat 1989). The grazing keep them permanently in a juvenile stage of growth and development and induces in them the domination of mini-thallomic fast-growing species, thus keeping their specific production high. This makes for a high level of their primary production in bottom reef biotopes, which, superficially viewed, does not seem to be so abundant (Gribb 1973; Wanders 1976a). Plant associations of coral reefs are a main source of organic detritus and dissolved organic matter, as well as an important source of various biochemical metabolites, like vitamins or antibiotics (Burkholder 1973). The plant detritus on many reefs is produced also by mangroves (Colley et al. 1962).

4.3.1 Composition of Reef Plant Associations

The reef plant associations include five basic taxa of algae: diatoms, blue-green algae, green, red and brown algae, and also the marine angiosperms –

the seagrasses. The first two groups are the components of periphyton and microphytobenthos, and have been reviewed above (cf. Sect. 2.4.2). The total number of thallomic macrophytes, represented by green, brown and red algae, in phytocoenes of the open sea reefs attains 100–250, while in reef areas near the continents, especially at the boundaries of tropics, it could be twice as much (Table 4.13). As an example of the latter there are the phytocenes of the Red sea reefs and of some atolls in the Indian ocean (Fig. 4.5; Table 4.14). The benthic flora of coral reefs is taxonomically much less abundant than that of the temperate coastal regions, where the total number of species attains 700–1000. Among these groups of reef macrophytes the poorest is the group of brown algae: 20–30 species compared with 300 at temperate coasts. The total numbers of macrophyte species in plant associations of separate reef systems of atolls varies within 5 to 70, while in those off the continental shelf 100 to 140 (Table 4.13). This increase is connected mostly with stimulation of development in the latter of the green algae, many of which prefer biotopes enriched with nutrients and organic matter, which arrive from land and from mangroves (Rützler and Macintyre 1982).

The green algae (Chlorophyta) are represented on reefs largely by specialized calcareous algae from the genera *Halimeda*, *Penicillus*, and *Udotea*, as well as by the filamentous algae *Cladophora* and *Chaetomorpha*. *Halimeda* is the most widespread reef calcareous alga with hard thalloms heavily incrusted with lime. It may attach to solid rocks and to anchor to the

Table 4.13. Number of species of basic taxa in phytocoenes of different reef regions compared with the temperate coasts of the British Islands and of southern Australia. (After Bakus 1969; Rützler and Macintyre 1982; Gribb 1973; Vinogradova and Kalugina-Gutnik 1985)

Region	Chlorophyta	Phaephyta	Rhodophyta	Total number of species
Bermuda	59	23	63	145
Florida	143	55	216	414
Belize	73	32	124	239
Islands of the Indian Ocean	50	26	46	122
South Vietnam	97	60	154	311
Maldives	28	8	22	58
Gilberts	21	5	29	55
Ifaluk atoll (Carolines)	46	3	37	86
Marshalls	108	24	117	249
Hawaii	21	8	35	64
Philippines	74	29	109	219
Solomons	–	–	–	243
Great Barrier Reef of Australia (southern part)	–	–	–	230
British Islands	198	198	353	749
Southern Australia (temperate zone)	94	191	725	1010

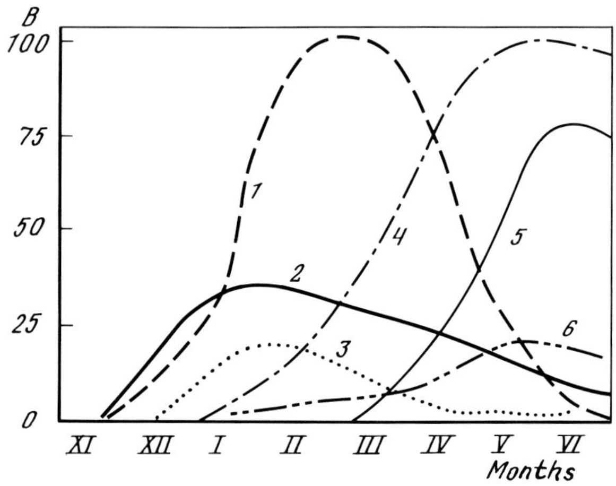

Fig. 4.5. Seasonal succession of phytocoene of macrophytes at fringing reef off Eilat, Red Sea: *1 Sargassum dentifolium*; *2 Cystoseira myrica*; *3 Colpomenia sinuosa*; *4 Styropodium zonale*; *5 Turbinaria eilatensis*; *6 Liagora turneri*; *B* biomass, relative values. Modified from Benayahu Loya 1977)

soft bottom as well with the aid of its rhyzoids (Colinvaux 1980; Drew and Abel 1983; Abel and Drew 1985). Often the species of *Halimeda* are the key taxa in the plant associations either of the hard or of soft bottom biotopes. On a separate reef 10 to 20 species of this genus could be found, each of which displays definite ecological specialization (Hillis-Colinvaux 1980). The cover of the bottom area with *Halimeda* could attain sometimes 60–80%, and its biomass up to 15–20 kg m.$^{-2}$ *Halimeda* appears to be one of the main producers of calcareous material (cf. Sect. 1.3) and of organic detritus (Maxwell 1968; Goreau and Goreau 1973; Drew and Abel 1983; Abel and Drew 1985). Among the most important representatives of reef green algae

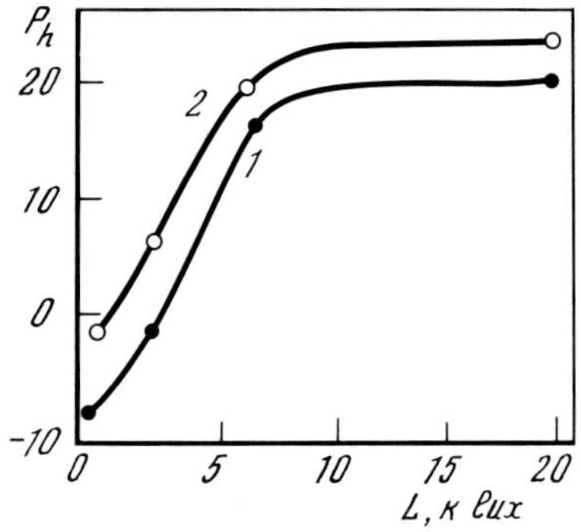

Fig. 4.6. Light dependence curve of net photosynthesis (P_h, $\mu g\,O_2\,cm^{-2}\,h^{-1}$) in coralline algae *Porolithon pachydermum* (*1*) and *Lithophyllum intermedium* (*2*). L illumination, klx. (After Wanders 1976)

Table 4.14. Average biomass (B) of mass species of algae in reef bottom biotopes of some islands of the Indian Ocean. (After Vinogradova and Kalugina-Gutnik 1985)

Reef	Mass species of algae	Reef zones, biomass					
		Coastal channel, sand		Reef flat		Reef edge, patch reefs, sand	
		kg m^{-2}	% of total	kg m^{-2}	% of total	kg m^{-2}	% of total
N-E reefs off Koetivi I., Seychelles	*Thalassiodendron ciliatum*	1.06	98.6	0.02	2.8	1.79	79.9
	Udotea flabellum	0.015	1.4	0.16	22.7	0.015	0.7
	Dictiosphaeria cavernosa	0	0	0.49	69.4	0.40	17.8
	Laurencia sp.	0	0	0.002	0.3	0.032	1.4
	Jania rubens	0	0	0.02	3.0	0	0
Leeward reef off Kargados Karahos I.	*Thalassiodendron ciliatum*	0.71	55.7	0	0	0	0
	Caulerpa cupressoides	0.31	24.5	0.19	1.0	0	0
	Dictiosphaeria cavernosa	0.12	9.4	0	0	0	0
	Halimeda opunta + *Halimeda* sp.	0	0	21.45	95.8	0.22	28.9
	Udotea sp.	0	0	0	0	0.55	71.0

from the order Siphonales is *Caulerpa*. The green algae *Ulva*, *Enteromorpha* and *Cladophora* are among the pioneer macrophyte species which appear after the destruction of reefs by cyclons (Dollar 1982), or are colonizing the artificial reefs (Tsuda and Kami 1973) Being a preferred plant food for many reef herbivores, they are actively grazed and this significantly limits their growth. But they can partly compensate this damage by their high growth rate (Borowitzka et al. 1983).

The brown algae (Phaephyta) list on reefs only 10–30 species. But their role in plant associations could be quite significant. Some of the brown algae, such as *Sargassum*, *Turbinaria*, *Padina*, *Dyctiota*, and *Pocockiella*, could be key species in some of them. Some brown algae have a definite affinity to reef biotopes with a weak or medium wave-stress, like leeward zones or patch reefs or littoral zones (Tsuda, 1977). Among these are *Ectocarpus*, *Sphacelaria*, *Ralfsia*, *Chnoospora*, and some species of *Sargassum* (*S. polycystum*). Other brown algae may live in turbulent zones of reef flats and windward slopes (*Padina*, *Hydroclathrus*, *Dictyota*, *Turbinaria*, *Sargassum navaehollandiae*). On outer slopes of Caribbean reefs *S. platycarpum* often forms dense thickets (Van den Hoeck 1969).

Taxonomically, the most numerous group in reef algal associations is the group of red algae (Rhodophyta). Most of the species of this group are represented by algae with tiny filamentous thalloms, which form together with other algae the periphytonic turfs. Among them are *Gelidiella*, *Gelidiopsis*, *Briopsis*, *Galaxaura*, *Laurencia*, *Ceramium*, *Polisiphonia* (Wanders 1976b; Borowitzka et al. 1983). The red algae *Polisiphonia* and *Campia* could often be met with among the epiphytes on other macrophytes and on seagrasses (Rhyne 1971; Fortes 1981; Penhale and Capone 1981). Red algae are abumdant in turbulent zones of the outer reef edge and the windward flat, as well as in trenches, and in poorly illuminated trenches and caves. Red algae include also important reef builders – the red coralline algae from the family Corallinaceae, which comprises hundreds of species (Littler 1972; Johaneen 1981). The reef plant communities might include 10–20 of their crustose and branching articulated species. Among the latter are *Amphiroa*, *Jania*, *Peyssonnelia*. The crustose forms are represented most often by the following genera: *Porolithon*, *Lithothamnion*, *Lithophyllum*, *Hydrolithon*, *Goniolithon*, *Neogoniolithon*, *Sporolithon*. Many crustose coralline algae form various knobs, lobes or protuberances on the surface of their colonies.

On the Pacific atolls, and especially on their reef flats and on the windward reef edges there are among the most widespread crustose coralline algae several species of the genus *Porolithon*: *P. onkoides*, *P. craspedium*, *P. gardineri* and also *Sporolithon erythraeum* and some species of *Lithothamnion* (Taylor 1950; Doty and Morrison 1954; Littler 1972, 1973b, 1976; Littler and Doty 1975 Fig. 4.4). On reefs of the western Atlantic the most important crustose Corallinaceae are *Hydrolithon boergesenii*, *Lithphyllum intermedium*, *Neogoniolithon solubile*, *Porolithon pashidermum*, *Archaeo-*

lithothamnion dinotus, *Lithothamnion ruptile*, and *Tanarea prototypa* (Van den Hoeck et al. 1978). These opportunistic crustose calcareous algae not only form basic carbonate constructions in turbulent reef zones at the windward reef edge and on their windward slopes, but also compose a main portion of the total algal biomass in these zones, especially on reefs with algal ridges along their windward edge (Womersly and Bailey 1969; Littler 1972; Adey and Macintyre 1973). Thus they accomplish an important function of reef builders, which mount and cement the reef constructions in the physically most stressed zones. The branching corallines are also widespread on reefs, inhabiting the trenches, the caverns, and places between the coral colonies. They participate also in the formation of periphytonic communities and in the consolidation of reef lime constructions, producing a bulk of crumbly calcareous material (Adey and Vassar 1975; Wanders 1976b).

The seagrasses inhabit mostly the soft bottom biotopes in lagoons or in reef-flat zones. They form often dense thickets, thus dominating there by biomass (Table 4.14). The parimary production in bottom biotopes with the domination of seagrasses largely exceeds that of the thickets of seaweeds or corals. Their leaves and litter, formed after their decay are a significant source of feeding for reef fauna (Zieman et al. 1979). The reef seagrasses include the genera *Thalassia* (*T. testudinium*, *T. hemprichii*), *Thalassiodendron* (*T. ciliatum*), *Zostera* (*Z. marina*, *Z. capricorni*), *Halophila* (*H. ballonis*, *H. ovalis*, *H. minor*, *H. stipulacea*), *Halodule* (*H. wrightii*, *H. universus*), *Cymodocea* (*C. serrulata C. rotundata*), *Syringodium* (*S. filiformis*, *S. isoetifolium*), and *Enhalus* (*E. ciliatum*), *Posydonia* (*P. australis*). On the Pacific reefs the dominating seagrasses are *Thalassiodendron ciliatum*, *T. hemprichii* and *Halophila stipulacea*, while at the reef of the Western Atlantic the turtle grass *Thalassia testudinum* is most widespread (Quasim and Bhattathiri 1971; Zieman 1975; Tsuda 1977b, 1979; Norris and Bucher 1982). The seagrasses appear usually at a late stage of succession of benthic plant communities in the soft bottom biotopes (Williams 1981).

At many reefs the plant associations include also the mangroves (Clough 1982; Lugo and Snedeaker 1974; Teas 1983). Their leaves and roots are an important source of organic detritus (Colley et al. 1962; Lugo et al. 1975). Mangroves are trees and shrubs 3–15 m high. Their base parts are formed by branching supporting roots and by air roots. They form dense forests and thickets at reef shallows and in the intertidal stripes and keys, especially in coastal lagoons of barrier reefs (Forsberg 1976) and on low intertidal plateaus of the Indo-Pacific reefs (Stoddart 1980). Different mangrove species taxonomically belong to different familes of angiosperms. Within a separate reef system 10–20 species of mangroves could be found (Bunt et al. 1979; Stoddart 1980; Buckley 1982; Boto et al. 1984). But their local forests are formed, as a rule, by some single species of one of the following genera: *Rhizophora*, *Avicennia*, *Bruguiera*, *Ceriops*, *Laguncularia*, *Sonneratia*, *Luminitzera*, and *Conycarpus*. Their different species are adapted to specific environemental conditions and to definite types of bottom sediments.

Among the most common mangroves are several species of the genus *Rhyzophora*, which prefer the muddy bottom permanetly covered with sea water, while *Bruguiera* and *Luminitzera* prefer habitats covered by brackish waters and isolated from the direct in inflow of sea water (Clough 1982). The most common mangroves in the Indo-Pacific reefs are: *Rhizophora stylosa*, *R. mucronata*, *R. apiculata*, *Bruguiera gymnorhiza*, *B. parviflora*, *Cariops tagal*, *C. decandra*, *Luminitzera littorea*, *Xylocarpus granatum*, *Avicennia marina*, and *Sonneratia caseolaris*. On Caribbean reefs are found: *Rhizophora mangre*, *R. harrisoni*, *R. racemosa*, *Avicennia tormentosa*, *A. germinaus*, *A. schaueriana*, *Conocarpus erecta*, and *Laguncularia racemosa*.

4.3.2 Zonal Distribution, Phytocoenes Composition

The distribution of reef bottom plant associations at normal cross sections clearly reflect the zonal structure of the reef benthic communities (Van den Hoeck et al. 1978; Norris and Bucher 1982). They include one or several key plant species. In the zone of rocky littorals these are the macrophyte species, which are able to fix firmly their thalloms to the rock surfaces, like *Ralfsia expansa*, *Enteromorpha iluthrata*, *Caulerpa* sp., *Laurencia microcladata*, *Galaxaura squalida*, *Syringodium* sp., *Valonia* sp., and *Botriocladia* sp., On the sandy bottom of lagoons the key species are usually the seagrasses, like *Thalassia*, *Thalassionema*, *Zostera*, and the green calcareous alga *Halimeda*. The places of lagoonal bottom, covered with rubble or with dead coral heads, often are dominated by seaweeds: *Sargassum*, *Penicillus*, *Dictyosphaeria*, *Valonia*, *Udotea*, *Dictyota*. In the patch reef zone, on patch-reefs and the surrounding sands and rubble the algae *Sargassum*, *Halimeda*, *Dictyota*, *Caulerpa*, and *Penicullus* are most abundant. The dead coral heads in this zone are covered with branching coralline algae, such as *Jania* and *Amphiroa*. On the sandy submerged bottom of reef flats *Halimeda*, *Turbinaria*, *Cladophora*, *Pocockiella*, *Gelidiella*, *Boodleea* and the filamentous blue-green alga *Schizothrix* are often dominating. At the stripes of solid flats exposed during the ebbs, the most common are, as a rule, the crustose Corallinaceas: *Porolithon*, *Neogoniolithon*, and *Sporolithon*. On upper windward reef slopes the coralline algae *Porolithon*, *Paragoniolithon*, *Amphiroa*, *Jania*, as well as the thallomic macrophytes *Laurencia*, *Acetabularia*, *Microdiction*, *Homotrema*, Galaxaura, *Boodlea*, *Halimeda*, and *Udotea* are among the most common. The biomass of 1–2 of the key species might comprise up to 95% of the total (Table 4.17). The zonal composition and abundance of reef macrophyte communities are definitely controlled by the conditions of nutrients (especially of phosphorus supply) (Littler et al. 1991).

Van den Hoeck et al. (1978) described the distribution of common taxa of seaweeds at the section across the barrier reef off Jamaica. The high degree of their specialization and their affinity to definite reef biotopes was shown.

The most opportunistic among them appeared to be the crustose red alga *Peyssonnelia*. The reef flat of frin-ging reefs near Lizard I., being subjected to wave stress, is covered mostly by periphytonic turfs, composed of mini-thallomic macrophytes, and with coralline algae (Table 4.15). Among the dominating macrophyte taxa here were *Halimeda opunta*, *Caulerpa*, *Laurencia*, *Amphiroa*. At the leeward side of the reef flat algae were dominating which prefer calm waters, such as *Halimeda cavernosa*, *Boodlea*, *Dictyosphaeria*. The domination of seagrasses and thallomic macrophytes in calm lagoonal biotopes and the key role of red corallines in turbulent zones of reef crest and outer slope could also be seen from data on the relative projective cover of the bottom surface by main groups of benthic plants (Tables 4.15, 4.16).

Another example of zonal distribution of benthic algae could be seen at the cross section through the reef zones of the Diego Garcia stoll in the Indian Ocean (Rhyne 1971). On the sand bottom of the littoral zone of the lagoon, with an admixture of rubble and dead corals, the phytocoene included the macrophytes *Enteromorpha*, *Giffordia*, *Dictyota*, *Pocockiella*, *Hypnea*, and *Halimeda*. The rubble stripes and ridges were occupied by *Microcoleus* and *Padina*. At the coastal patch reefs *Cladophora*, *Cryptonemia*, *Valonia*, *Botriocladia*, *Caulerpa*, *Bryopsis*, *Codium* and *Galaxaura* were the most numerous, while at the lagoonal patch reefs *Valonia*, *Dictiosphaeria*, *Caulerpa*, *Acetabularia*, and *Cymodocea*. On thalloms of *Cymodocea* the epiphytes *Sphacelaria furcigera*, *Amphiroa fragillissima*, *Jania tenella* and *Campia parvula* were abundant. The sandy bottom in the central part of the lagoon was dominated by *Halimeda*. The coral heads, the tridacna shells and large pieces of rubble were colonized by *Sargassum*, *Turbinaria*, *Padina*, *Jania*, *Galaxaura*, *Dictyota*. The reef flat, powdered with sand, was covered with dense algal turf, composed of filamentous blue-green and green algae, the mini-thallomic macrophytes *Dictiosphaeria*, *Boodlea*, *Gelidiella*, *Laurencia*, and *Heterosiphonia*, and of the coralline alga *Jania*. On the rocky surfaces of the reef flat blackish-green hillocks were numerous. They were formed by the blue-green algae *Rivularia*, *Microcoleus* and *Shizotrix*. The trenches in the surface of the flat were occupied by *Halimeda*. On the windward edge of the reef flat, stressed by surf, crustose corallime algae were dominating. Among them were *Porolithon*, *Neogoniolithon*, *Sporolithon*, *Lithophyllum*, and *Hydrolithon*.

The above-listed Corallinaceae are a basic element of phytocoenes in the surf zones of reefs. Their crustose colonies sustain a severe wave-stress. But even among them there are species with definite ecological adaptations which occupy their own niches, where they are dominating. The most opportunistic species of crustose corallines is *Porolithon oncoides*, which endures direct sunlight, long exposure and over-heating (Fig. 4.4). This species forms firm thick hill-shaped colonies which are subjected to a most powerful surf. Inhabiting the surf zone this alga escapes grazing by fish (Adey 1975). This species usually dominates among the crustose corallines

Table 4.15. Percentage of projective cover of the bottom surface by dominating species of macrophytes and their species diversity (H′ – Shannon's index) at the transsect across the reef flat of the S.E. fringing reef off Lizard I., GBR. (After Pichon and Morrissey 1981)

Stations index	Position of stations on reef	Dominating laxa of macrophytes	Diversity (H′)	Percentage of cover			
				Thallomic macrophytes		Coralline algae	Periphytonic turfs
				soft	calcareous		
A	Windward slope	*Halimeda opunta*	2.2	3.2	8.8	2.0	86.0
C	Windward reef edge	*H. opunta*, *Laurencia*, *Caulerpa*, *Amphiroa*	2.1	4.9	3.6	1.2	90.3
E	Reef flat	*H. opunta*, *Laurencia*, *Caulerpa*	1.9	4.7	3.8	0.9	90.6
G	Leeward edge of flat	*Laurencia*, *H. opunta*, *Boodlea*, *Dictiosphaeria*, *Turbinaria*	2.5	17.4	2.4	0.7	79.5
H	Lagoonal patch reefs at leeward reef edge	*Boodlea*, *Dictiosphaeria*, *Halimeda cavernosa*	2.8	19.3	1.4	0.5	78.8

Table 4.16. The percentage of projective cover of bottom surface in different reef zones at cross section via the barrier reef Belize by different groups of benthic plants. (After Rützler and Macintyre 1982)

Groups of plants	Lagoon	Patch reefs	Reef crest	Outer reef slope
Seagrasses	22.2	0.5	0	0
Thallomic macrophytes	13.6	8.2	3.7	3.8
Thallomic calcareous algae	0.6	6.8	2.6	1.7
Corallines	0	5.2	35.6	2.1

Table 4.17. Biomass of algae and seagrasses in reef bottom biotopes. (Data by Vinogradova and Kalugina-Gutnik 1985; Bakus 1967; Zieman et al. 1979; Wanders 1976b; Odum and Odum 1955; Wahbeh 1981; Burkholder et al. 1967; Williams 1981)

Location	Key species	Biomass, kg m^{-2} (wet weight)
Socotra I.	*Sargassum*, *Padina*	0.4–1.6
Praslin I., Seychelles	*Sargassum*, *Turbinaria*, *Padina*	5.4–9.0
Koetivi I., Seychelles	*Thalassiodendron*, *Syringodium*	1.0–16.0
Kargados-Karahos I., Mauritius	*Halimeda*, *Thalassiodendron*	3.0–21.0
Ben Goy Bay, Vietnam	*Galaxaura*, *Padina*	1.2
Thu I., Vietnam	*Sargassum*	2.5
Thu I., Vietnam	Turfs of mini-thallomic macrophytes	0.8
Namsu I., Gulf of Siam	*Turbinaria*, *Ralphasia*	0.2–4.5
Caribbean reefs	Blue-green algae	0.5
Caribbean reefs	*Thalassia*	0.4
Curacao, Antilles	*Sargassum*, *Dictyosphaeria*	1.8
Enivetak atoll	*Caulerpa*, *Dictiosphaeria*, *Dictyota*	4.0–7.0
Gulf of Aqaba, Red Sea	Seagrasses	0.7–1.5
Puerto Rico	Seagrasses	12.0–20.0
Florida	Seagrasses	1.6–7.6

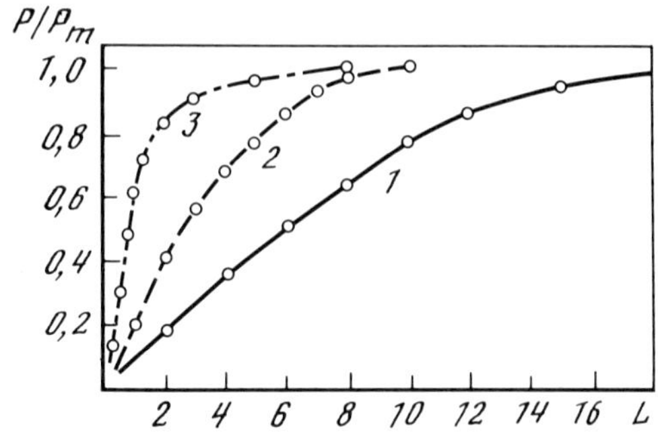

Fig. 4.7. Light dependence curves in thalli of macrophyte *Sargassum ilicifolium*, taken from different parts of the plant: *1* from upper part; *2* from middle part; *3* from basal, shadowed part; P/P_m gross photosynthesis – relative values; *L* light intensity, wt m^{-2}. (Data by Titlyanov et al. 1985)

at the windward reef edges and over the algal ridges, being a main reef builder in this zone of Indo-Pacific reefs. At the Atlantic reefs (Saint Croix, Virgin Islands) the same turbulent zones are occupied by another crustose coralline alga, *Lithophyllum congestum* (Adey 1975). This species is very similar to *P. oncoides* in its spectrum of ecological adaptations. Growing on reef edge, this alga constructs solid knolls, "boylers", exposed during the ebbs, which in their structure are similar to the algal ridges of Indo-Pacific atolls. On the algal ridges of numerous algal reefs in the Indian Ocean the same function is accomplished had by *Neogoniolithon frutescens*.

Another mass species of *Porolithon*, *P. gardineri*, is also largely similar to *P. oncoides* in its ecological characteristics, though it is less resistant to wave-stress. Its colonies are not so monolithic as in *P. oncoides*, having protuberances on their surface. The *Sporolithon* and the *Hydrolithon* are adapted to a medium level of illumination and do not endure the strong surf. They inhabit the shadowed calmer places under corals, under macrophytes, or in trenches of flatts. These species also could be met with in biotopes of outer reef slopees at depth of more than 10 m, where the surf and the light conditions are optimal for them (Littler 1973b; Doty 1974; cf. Fig. 4.4). Coralline algae sustain better also the strong currents which often happen to flow over the reef flat, and are less damaged by rubble which they are moving up from the bottom. Therefore, they often replace the fleshy macrophytes in parts of the flat subjected to strong wave-driven or tidal currents (Doty 1974). The antropogenic pollution also promotes the distribution of coralline algae over the reef because they endure better pollution compared with the thallomic macrophytes (Littler and Doty 1975). The distribution of different taxa of Corallinaceae along the outer reef slopes is controlled by illumination. Their species, which inhabit the deep zone of the outer slope, are adapted to very low light. In transparent waters they have been found down to depths of 350–400 m. They have a special red pigment, phycoerythrin, which is assumed to further their light adaptation (Littler 1972).

Reef bottom plant communities are subjected to the seasonal successional changes (Benayahu and Loya 1977; Borowitzka et al. 1977). These changes were especially well detectable at high-latitude reefs, such as those of Eilat (Red Sea), where the winter association "*Sargassum-Cystoseira-Colpomenia*" in summer was replaced by the association "*Stypopodium-Turbinaria-Liagora*" (Fig. 4.5). The succession was observed taking place also during the formation of plant communities at the artificial reefs. Their surface first was captured by the fast-growing filamentous or fleshy green algae *Calothrix*, *Cladophora*, *Sphacelaria*, *Microcoleus*, *Ulva*. They were then grazed out by fish and replaced by firmer slow-growing algae, such as *Dictyota*, *Laurencia*, *Turbinaria*, and by Corallinaceae (Tsuda and Kami 1973; Dollar 1982).

The phytocoene of bottom plant associations of coral reefs includes abundant populations of epiphytic algae, living on thalloms of macrophytes and on leaves of seagrasses. They are represented by all the main taxa of marine algae. Most common epiphytes among blue-greens are *Oscillatoria*, *Calotrix*, *Dichotrix*, and *Plectonema*. The thallomic epiphytes have mostly tiny filamentous thalloms 1–1.5 cm long. Among them the most numerous are the green algae *Cladophora*, *Enteromorpha*, *Chlorodesmus*, Ulvella, *Derbesia*, the brown algae *Ectocarpus*, *Sphacelaria*, *Myrionema*, *Giffordia*, and the red algae *Ceramium*, *Polysiphonia*, *Hypnea*, *Chondria*, *Gelidiella*, *Griffithsia*, *Jania*, and *Amphiroa*. The red algae are usually dominating also in the epiphytic plant communities (Fortes 1981; Norris and Bucher 1982). These communities have a high degree of species variability. For example on thalloms of only two species of *Sargassum* there were found over 70 species of epiphytic algae.

The reef seagrasses prefer shallow plain bottom areas of lagoons, covered with fine coral sands or with silty sands, with a depth of water during the ebb of 0.2–1 m. They form in such biotopes dense meadows or so called "seagrass beds" (Zieman 1975; Tsuda 1979). The thickets of seagrass beds are usually formed by one or two species of seagrass with of one of them dominating in the last case (Zieman et al. 1979). On the bottom of deeper parts of lagoons and on stripes of sandy bottom of reef flats they usually form mixed communities together with the macrophytes. Amid the above-mentioned seagrass beds are situated patches of bare sandy bottom – so-called "halos". The halos are situated usually around large corals, patch reefs or near the edges of lagoonal reefs. They are formed as a result of grazing activities of herbivorous fish and urchins, which have refuges in the above-mentioned corals or in patch reefs. In daytime the grazers are hidden in their refuges, and at night they go to their pasture areas, plucking off first the plants near their hiding places (Ogden et al. 1973; Anzari et al. 1991).

4.3.3 Biomass, Metabolism, Features of Ecology

The projective cover of the reef bottom surface by plant communities varies in its different zones from 10 to 100%. The largest coverage could be observed usually in areas of lagoons and reef flats, and the smaller at outer reef slopes (Tables 4.15, 4.16). On an average, the projective cover of bottom by plant associations varies within 20–30% in the lagoons, 40–80% on reef flats, and 5–20% on outer reef slopes. Their biomass varies correspondingly within the same scale (Table 4.17). On the reef flat, while 20–40% of its surface is covered by coralline algae, their wet biomass varies within 1–5 $kg\,m^{-2}$. But at flats covered more than 50% by the fleshy macrophytes, the plant biomass *is* more: 3 to 8 $kg\,m^{-2}$. On the bottom of lagoons and on submerged parts of reef flats covered with thickets of seagrasses or *Halimeda* the biomass could attain 15–22 $kg\,m^{-1}$. In thickets of *Sargassum* biomass is usually 1–4 $kg\,m^{-2}$. On seagrass beds it is on an average within 2 to 8 $kg\,m^{-2}$ (Tables 4.14, 4.17).

The primary production of benthic plant associations was measured in different individuals of their common species (Marsh 1970; Quasim and Bhattathiri 1971; Wanders 1976b; West and Larcum 1983; Titlyanov et al. 1983b, 1985; Morrissey 1985) as well as in their associations inhabiting different reef zones (Odum and Odum 1955; Marsh 1970, 1974; Wanders 1976b; Odum et al. 1959; Murphey and Kremer 1983; Littler 1979). The ^{14}C and oxygen release methods were employed (cf. Sect. 10.1). In crustose coralline algae the gross photosynthesis varies within 10–20 $\mu g\, C\, cm^{-2} h^{-1}$ (0.03–0.06 $\mu g\ O_2\, cm^{-2} h^{-1}$), which is equal to 100–230 $\mu g\, C\, cm^{-2} day^{-1}$ or to 1–2.3 $g\, C\, m^{-2}$ at 100% cover (Marsh 1970; Littler 1973a; Littler and Doty 1975; Wanders 1976b). The respiration rates in them varied within 40–100 $\mu g\, C\, cm^{-2} day^{-1}$, and the P:M ratio within 2 to 4. So their net primary production was relatively high: 40–150 $\mu g\, C\, m^{-2} day^{-1}$, thus providing their linear growth of 10–20 mm $year^{-1}$, which was comparable with the growth rates of massive corals (Agegian 1981). The diurnal course of photosynthesis in red Corallinaceae appeared to be resembling that in reef corals (Wanders 1976b; cf. Sects. 8.4, 9.1). The ratio of photosynthesis production per day to that per 1 h in coralline algae was 9–11 (Wanders 1976b), while our estimates for the periphytonic communities were within 9–10.5 (Sorokin 1986a). Having these an average value of 9.6, we use it for the recalculation of daily photosynthesis rates for generalizations given in Table 4.18.

The compensation point of gross photosynthesis and respiration, when its net production is zero in diurnal balance was attained in the coralline algae at a very low illumination: 1–2.5 × 10^3 lx, that is 1–3% of midday PARS (photosynthetically active radiation penetrated under the water surface). This level of illumination happens to reign in transparent oceanic waters (20–40 m Sechi disk visibility) at depths of 60–80 m. On outer reef slopes off the Marshall I. the compensation point in coralline algae was found at a depth of ~80 m (Marsh 1970). The light saturation point of photosynthesis in coralline algae was usually attained at 7–9 klx (Marsh 1970; Wanders 1976b; cf. Fig. 4.6). About the same level of saturation light intensity was recorded in other reef macrophytes: 9–11 klx (cf. Fig. 4.7). Thus, coralline algae are adapted also to a high level of illimination.

Associations of reef thallomic macrophytes have an extremely high level of gross primary production – up to 10–12 $g\, C\, m^{-2} day^{-1}$ (Tables 4.14, 4.18). This high level of communitary photosynthesis is provided by the high photosynthesis rates of individual plants, which in most macrophytes is 2–4 $\mu g\, O_2\, g^{-1} h^{-1}$, reaching in some of them even 10 $\mu g\, O_2\, g^{-1}$ (Table 14.19). The lower values of gross photosynthesis, calculated per gram dry weight, are peculiar to the calcareous thallomic algae, like Halimeda or Penicillus, because their thalloms are loaded with the inert calcareous skeletal material. But calculated per surface of thalloms, the rate of photosynthesis in *Halimada*, for example, is twice as high as in coralline algae. The coefficient of specific production (P/B) in *Halimeda* is around 0.07 (Drew and Abel 1983). The green calcareous alga *Penicillus* produces

Table 4.18. Gross photosynthesis and respiration (P, M, $g\ C\,m^{-2}\,day^{-1}$) in different reef bottom plant associations. Data by Marsh (1970), Wanders (1976a,b), Sournia (1976b), Littler (1973b), Hillis-Colinvaux (1974), Rogers and Salesky (1981), Smith and Marsh (1973), Murphey and Kremer (1983), Quasim et al. (1972), Quasim and Bhattathiri (1971), Odum (1957), Odum et al. (1959), Boto et al. (1984), Bunt et al. (1979), Lugo and Snedeaker (1974), Colley et al. (1962), Cintron et al. (1980), Clough (1982)

Place	Types of plant associations	P	M	P/M
Enivetok atoll	Crustose coralline algae	1.50	0.84	1.8
Enivetok atoll	Macrophytes on flat	11.60	6.00	1.9
Curacao I., Antilles	Coralline algae on reef flat	2.40	1.40	1.7
Curacao I., Antilles	Turf of mini-thallomic macrophytes and corallines	3.40	1.00	3.4
Moorea I., French Polynesia	Macrophytes and coralline algae on reef flat	3.8	1.20	3.1
Oahu I., Hawaii	Macrophytes and coralline algae on reef flat	3.4	0.80	4.4
Jamaica I.	Coralline algae	4.00	1.70	2.4
GBR, Australia	Halimeda (lagoon)	0.78	0.32	2.4
St. Croix., Virgin I.	Turf of mini-thallomic macrophytes	10.10	3.20	3.2
Curacao I., Antilles	*Sargassum* thickets	10.50	3.50	3.0
Clipperton atoll	Seagrass	8.60	8.00	1.1
Kavaratti atoll	Seagrass	12.9	9.50	1.4
Baja Fluorescente, Puerto Rico	Seagrass	5.6	6.70	0.8
Florida	Seagrass	10.6	7.70	1.4
Jamaica	Seagrass	13.6	9.60	1.4
GBR, Australia	Mangrove forest	2.20	–	–
Florida	Mangrove forest	8.20	6.80	1.2
Florida	Pigmy mangrove thicket	1.40	2.00	0.7

Table 4.19. Gross photosynthesis (P) and respiration (M) rates in different species of marine benthic plants; P and M values are given as $mg\,O_2\,h^{-2}\,g^{-1}$ dry weight. Data by Wanders (1976a), Quasim and Bhattathiri (1971), Titlyanov et al. (1983b)

Algae	P	M	P/M
Sargassum platycarpum	7.27	1.39	5.2
Dictyopteris justii	5.21	1.13	4.6
Dictyota dentata	10.89	1.07	10.2
Hypnea sp.	2.2	1.5	1.5
Cymodocea sp.	1.7	1.8	0.9
Cladophora sp.	6.5	2.4	2.7
Ulva sp.	7.5	2.3	3.2
Sargassum polycystum	5.5	1.5	3.6
Sargassum ilicifolium	10.5	3.5	3.1
Sargassum kjellermanianum	4.5	1.0	4.5
Thalassia testudinum	2.2	1.2	1.8

4–8 generations per year by an average lifetime of only about 2 months for individual thalloms (Vadas et al. 1980).

The form of light-dependence curves of photosynthesis in macrophytes varied even within one species, depending of the level of ambient illumination (Titlyanov et al. 1983b). Light saturation level of illumination was usually attained in them at 5 to 10% of the incident surface PARS, e.g., at 8–30 $\mathrm{wt\,m^{-2}}$ with penetrated PARS of 250–350 $\mathrm{wt\,m^{-2}}$. The level of light saturation may also change several times depending even on the place where sprouts for the experiments are obtained: from the top illuminated part of the thallom or from its basal permanently shadowed part (Fig. 4.7). The data prove the ability of the photosynthesic apparatus of macrophytes for quick light adaptation to ambient conditions of illumination, which could change during the individual life of an alga. This is one of the most important ecophysiological features of reef macrophytes which provides their high primary productivity. One of the ways of this adaptation is a drastic increase in the chlorophyll content in thalloms, which could be 1.5–3 times more in their permanently shadowed parts compared with those well illuminated (Titlyanov et al. 1983b, 1985). Another mechanism of light adaptation is displayed in the changes of the vital form of thalloms. For example, the green alga *Caulerpa*, when living on outer reef slopes at depths of more than 10 m has filamentous thalloms, while the specimens living in reef shallows have massive thalloms with short, thick branches (Jaubert 1981).

The available data on photosynthesis production and metabolic rates of the benthic plant associations of various reefs are summarized in Table 4.18. In bottom biotopes harboring macrophytes together with the coralline algae the gross primary production values were at their average level: 3–5 $\mathrm{g\,C\,m^{-2}\,day^{-1}}$ (Pauri 1988). Thus taking an average biomass of algae of 200 to 800 $\mathrm{g\,C\,m^{-2}}$ in this type of plant associations, the coefficients of specific production P/B are close to 0.02–0.04 $\mathrm{day^{-1}}$. In associations dominated by the green calcareous algae *Halimeda* and *Penicillum* primary production varie within 1 to 4 $\mathrm{g\,C\,m^{-2}\,day^{-1}}$. In bushes of macrophytes with large massive fleshy thalloms, like *Sargassum*, *Ulva*, *Caulerpa*, *Cystoseira*, as well as in the dense turfs of filamentous or mini-thallomic algae (excluding corallines), primary production happened to be very high, attaining the upper limit to its rate known in natural plant communities: 10–12 $\mathrm{g\,C\,m^{-2}\,day^{-1}}$ with P/M ratios 2–4 and net photosynthesis rates of 4–7 $\mathrm{g\,C\,m^{-2}\,day^{-1}}$.

Photosynthesis production of reef benthic plant communities in biotopes dominated by corals being 5–8 $\mathrm{g\,C\,m^{-2}\,day^{-1}}$, it is twice as much as in those dominated by coralline algae. But in plant associations dominated by fleshy macrophytes or by seagrasses it is commensurable to or even more than in communities dominated by corals. The average content of chlorophyll *a* in benthic plant communities varie within 100 to 250 $\mathrm{\mu mg\,m^{-2}}$ with plants biomasses 1.5–4 $\mathrm{kg\,m^{-2}}$ (wet weight). Thus 1 mg chlorophyll is contained in approximately 15 g of wet biomass. The content of chlorophyll per m^2 of bottom is about equal in plant associations dominated either by corallines or by thallomic macrophytes (Wanders 1976b).

The highest gross production of photosynthesis in reef communities was recorded in the seagrass communities – sea grass beds (Odum 1957; Westlake 1963; Quasim and Bhattathiri 1971; Zieman et al. 1979; Murphey and Kremer 1983). The gross photosynthesis in these communities could attain 12–20 $g\,C\,m^{-2}\,day^{-1}$. Their average values varied within 5–10 $g\,C\,m^{-2}\,day^{-1}$ and the P/M ratios 1–1.5 per day. The net photosynthesis production in reef seagrass communities is relatively low: 1–4 $g\,C\,m^{-2}$. Often it is negative because of high respiration rates of the plants themselves and of the abundant benthic heterotrophs which inhabit the seagrass biotopes (Table 14.8). The net production of reef seagrass associations was also estimated by measuring their crops (Zieman 1975; Greenway 1976; West and Larkum 1983). Its values within 1.2–2.4 $g\,C\,m^{-2}\,day^{-1}$ have been obtained in this way. The primary production of mangroves was measured experimentally (Colley et al. 1962) by the indirect method of calculating from data on the light energy absorption by leaves and the chlorophyll content in them (Bunt et al. 1979), and by analyzing the weight of fallen leaves and branches (Boto et al. 1984). The generalized estimates by Lugo and Snedeaker (1974) resulted in values of a gross production of 13–18 $g\,m^{-2}\,day^{-1}$ of dry organic matter (4–6 g C). The data obtained by this method of calculation in several sites of the Great Barrier Reef of Austrialia gave ranges of 2–3 $g\,C\,m^{-2}\,day^{-1}$. Per 1 cm^2 of leaves surface it was about 130 $\mu g\,C\,day^{-1}$, that is several times less than that of the macrophytes. The direct experimental estimates showed its average level as 8 $g\,C\,m^{-2}\,day^{-1}$, that is about twice as much as the indirect once (Colley et al. 1962). The fallen-leaves estimates gave range of net primary production of mangrove forests within 2 to 4 $g\,m^{-2}\,day^{-1}$ of dry weight or around 1–1.5 $g\,C\,m^{-2}$ (Colley et al. 1962; Pool et al. 1975; Cintron et al. 1980; West and Larkum 1983).

Significant roles in the ecology of benthic plant associations are played by epiphytes which are harbored by macrophytes, seagrasses, and by air roots of mangroves as well. The relationships between the host and the epiphytic plants are often more similar to symbiosis than to those of commensalism or parasitism. Thus, between them an exchange with the metabolites and photosynthates (Penhale and Smith 1977) as well as with the nutrients (Penhale and Thayer 1980) occurs. The epiphytic blue-greens supplying their hosts with fixed nitrogen, get from them the photosynthetically produced organic matter (Hanson 1977; Capone et al. 1979; Penhale and Capone 1981).

The above data demonstrate the important structural and functional roles of the benthic plant associations in coral reef ecosystems. They provide the bulk of total autotrophic organic production and of calcareous material. On an average, their production could be evaluated to be within 30–50% of the total, but on the "algal reefs" with a weak coral growth or at high-latitude reefs it could be up to 70–80%. Their most important ecological feature is a high net primary production as compared with other primary producers, like corals. The P/M ratios in plant associations are most often 2 to 4, while in associations dominated by corals it is close to 1.5–2 (see Tables 4.18, 4.19). From this point of view benthic plant communites are among the main suppliers of reef ecosystems with primary organic matter. This role is especially important in biotopes of lagoons and reef-flats. The function of calcareous algae as reef builders was discussed above (cf. Sect. 1.2).

The ecological role of reef plant associations is greatly enhanced by the domination of mini-thallomic or filamentous macrophytes and seagrasses in them, which are readily consumed and digested by herbivores (Ott and Mauer 1977). The same high trophic quality is possessed by numerous epiphytic and epilithic algae, as well as by coralline algae (cf. Sects. 5.6, 6.3.1). The direct grazing of a significant part (20–40% of total primary production created within the reef ecosystems) shortens and simplifies the food-chain, incrasing thus the efficiency of secondary production processes (Randall 1967; Bakus 1969). The apparent abundance of herbivorous fish and invertebrates in reef benthic communities is one of the main factors responsible for their unusually high density in bottom biotopes. On the other hand, a high grazing pressure, resulting in the cleaning of solid reef substrates from plants occupation, makes to a large extent their colonization possible by sessile reef-building organisms, including coralline algae (Bakus 1969; Littler and Doty 1975; Birkeland 1977; Wanders 1977; Hatcher 1983b).

Reef benthic plants are a main source of food for about a quarter of the species of reef fish (Earle 1972). Out of 212 species of reef fish off the Antilles, in 59 plants comprised a significant part of their food rations. Many fish feed on plant detritus. Among the herbivorous invertebrates there are many species of gastropods, urchins, crabs, and shrimps (Klumpp and Pulfrich 1989). A number of fish, as well as urchins and green turtles, feed on seagrasses, or on detritus produced by their dead leaves (Zieman 1975; Ogden and Zieman 1977; Ogden et al. 1980). Through the detritus food web, the production by mangroves is also incorporated into reefal trophodynamic processes (Lugo and Snedeaker 1974; Teas 1983).

5 Reef Zoobenthos

The bottom fauna of coral reefs is extremely rich and varied. It finds there good trophic conditions, plenty of hiding places to escape grasing for vagile animals and a great abundance of solid substratum for sessile benthic invertebrates (Paine 1974; Kohn and Levitien 1976; Reichelt 1982). A sample of 2–3 kg of the rock material may contain over 100 species of animals from several tens of families (Hutchings 1974). Main groups of reef zoobenthos could be represented on a separate reef by hundreds of species: 50–150 species of forams, 150–500 species of molluscs, 100–250 species of crustaceans, 50–100 species of echinoderms, 100–200 species of polychaetes, and 50 and more species of sponges. This high species diversity of benthic animals has its origin in their fine trophic and ecological partitioning and specializations, and in a wide distribution among them of commensalism, symbiosis, different morphological adaptations, chemical and color information, and chemical defence. The macrobenthos is probably one of the best-studied components of reef fauna from the chorological aspect. The meio- and the microbenthos have been much less studied, which is explained by the methodological obstacles of sampling in rocky substrata of the reef (Hutchings 1974). The same is the cause of deficiency in the quantifications of the biomass of reef benthos, and especially of its meyo- and microcomponents. Much easier is apparently to sample quantitatively benthos found in separate dead coral heads. They are covered by plankton mesh and then separated from the base and crushed. This way of sampling gave extremely high biomass values (Grassle 1973). The sampling is much easier and quantitative in the soft-bottom biotopes. But even for these sites the biomass of benthos had been mostly underestimated because of meyo- and microbenthos missing in many of the benthos quantification studies. Inexactness in biomass estimations was aggravated by the variation in units of data presentation. Either the wet, the dry, the total or only the tissue weight unit were used. Thus the available information on the reef benthic biomass is still insufficient for its real evaluation.

5.1 Molluscs

Molluscs form one of the main groups of reef benthos as regards species number and also the biomass. They represent all the sizes in the trophic spectra, and display numerous ecological and morphological adaptations, which make for their extremely high density in most of the reef biotopes and for a high degree of use of trophic resources. Their sizes vary between several millimetres in micromolluscs and 1 m, and weights of up to 100 kg (tridacnides). Among them we may find species which represent all trophic levels ranging from symbiotic primary photosynthetic producers to plant feeders, omnivores, detritovores, filter-feeders, predators, coprophages, commensals, and parasites. Thus their communities are able to form themselves into a complete food-web (Taylor 1984). Molluscan species may occupy all the main bottom econiches: depths and the surfaces of sediments and rocks, dead coral heads, living corals or seaweed's thalloms.

5.1.1 Gastropods

The molluscan communities are represented on reefs mainly by bivalves and gastropods plus several species of scaphopods and chitons. Among the gastropods the prosobranchs are most numerous. The *Opistobranchia* are mostly represented by various nudibranchs and *Aplysia* species. Several species of *Pulmonata* also could be found in supralittoral areas.

The reef chitons enlist only a few species, some of which may reach 10 cm length (*Acanthopleura gemmata*). They are nocturnal animals feeding on periphytonic algae. At the beach rocks the density of their populations might be of 10–15 sp. m^{-2}.

The reef omnivorous prosobranchs feeding on seaweeds, on periphytons and on microphytobenthos of soft sediments comprise 16 basic families. Some of these families include the micromolluscs with a shell size of less than 0.7 cm. Among them are Triphoridae, Littorinidae, and Rissoinidae. In biotopes of coarse coral sands and of rubble, which usually contain a significant amount of detritus, the micromolluscs from these families may compose a main portion of the total benthic biomass (Kay 1971; Kay and Switzer 1974). For example, at the Fanning atoll in such biotopes the micromolluscs of 1–3 mm size included 19 genera, like *Obtortio*, *Odostomia*, *Tricolia*, *Enchelus*, *Merelina*, an *Diala*. The small- and medium-sized gastropods from the families Fissurellidae, Littorinidae, Patellidae, Trochidae, Rissoinidae, and Planaxidae inhabit rocky and stony littoral or sublittoral areas with medium to strong wave stress. The density of communities, which these gastropods form there, may reach several hundred sp. m^{-2}.

On the rocks of the reef flat and on outer slopes gastropods from the families Haliotidae and Stomatellidae are common. On reef flats and on patch reefs the Turbinidae, which have a beautiful nacreous shell, could often be observed. In the trenches of the rocks at the reef-flat edge and at the beach rocks the cerithiids are often abundant. The family of Cerithiidae is among the most widespread ones in reef biotopes. Their species have a size spectrum within 1 to 12 cm. They are mostly detritus-feeders and inhabit the detritus accumulating zones on the bottom of lagoons and on outer reef slopes. Another abundant on reef detritus-feeding molluscan family is the Strombidiidae. They often form dense monospecific populations in shallow areas of flat and lagoon, and especially in lagoons of the barrier reefs, where the detritus arrives from land with the terrigenous materrial. The biotopes of reefs are inhabited numerous species of the family Cypraeidae. They feed on periphyton, detritus and also on meyobenthos. Their mass species are *Cypraea moneta*, *C. caputserpentis*, *C. annularis*, *C. arabica*, *C. mauritanica*. Cypreids are nocturnal animals. They spend the daytime in their hiding places in trenches, under the stones. Being an object of the collection boom they were eagerly collected by tourists and local inhabitants. This endangers the very existence of their populations in many reef sites.

The biotopes of the soft bottom are inhabited, together with the micromolluscs mentioned above, also by the small- and average-sized species of Rissoinidae, Stomatellidae, Columbellidae, Cerithiidae, Pyramidiidae, Strombidae, Xenophoridae and Hypponicidae. The last two families represent the species which feed directly on sediments. Among the planktovorous gastropods there are several species of filter-feeders, which belong to the families Calypteridae, Siliquariidae and Vermetidae. The most abundant among them are the genera *Vermetus* and *Dendropoma*. The shells of these molluscs have the form of a prolonged cylinder. They are abundant in the flat and in lagoon areas, being attached to the surface of dead, and sometimes of living, corals (Hadfield et al. 1972; Hadfield 1976). The large phytophagous opistobranch gastropods *Aplysia* feed on seaweeds and use the intestinal microflora to accomplish the digestion of crude plant material consumed in the same way as is done by cattle (Carefoot 1981). The phytophagous pulmonate molluscs from the families Siphonariidae and Onchidiidae could often be found in supralittoral areas of coral reefs.

The predatory gastropod fauna in most of the reefs is extremely abundant (Kohn and Levitien 1976; Taylor 1978, 1984). They comprise 20 to 30% of the total species number and the same proportion of the biomass in reef molluscan fauna, including bivalves. The predatory gastropods relate to 27 families, mostly prosobranchs. The predatory opistobranchs are represented on reefs mainly by other *Nudibranchia*. Among them three main trophic groups could be distinguished: the predators on other molluscs, the predators feeding on other benthic animals, and the corallovorous gastropods. The basic taxonomic groups of predatory gastropods, as classified by their biomass and by the number of their species, are the families Muricidae, Thaisidae, Conidae and Mitridae. The first three families are represented on reefs by 20–30 species, the most abundant of which being the thaisids *Morula uva* and *M. squamosa*, and the conids *Conus sponsalis* and *C. ebraens*. For example, on the reefs off Guam conids, murcids and thaisids comprised around 80% of the total number of predatory gastropods (Taylor

1984). These predatory gastropods, as well as the species of the related families of Tonnidae, Magilidae, Buccinidae and Vasidae, feed mainly on other moluscs and on polychaetes. Muricids bore small holes in the shells of their molluscan prey with the aid of acid secreted by a special gland.

Among the families of predatory gastropods listed above several trophic guilds could be differentiated (Reichelt and Kohn 1985). The largest of them are gastropods feeding exclusively on the errant polychaetes eunicids and spionids. On the reef-flat biotope of the Guam reefs this guilds included 16 species of gastropods from the genera *Conus*, *Drupa*, *Cantharus*, *Engina*, *Vasum*, *Iredalea*, *Bursa*. The second guild is formed by predatory gastropods feeding mainly on meyo- and microbenthic omnivorous gastropods, such as Bittum, Dendropoma, and Diala. This guild includes species from the families Muricidae, Thaisidae and Costellariidae (*Pusia* sp.). *Pusia* paralyses its preys by covering them with the poisonous mucus. The third guild comprises the gastropods *Mitra litterata*, *M. cucumerina*, *Imbricularia conovula* (fam. Mitridae), *Drupa grossularia* (fam. Thaisidae) and *Vasum turbinellus*. They feed mainly on sipunculid worms. Some of the predatory gastropods feed on sea urchins (*Bursa bufonia*), asteroids (*Charonia tritonis* and some of the Cymatiidae fam.), on small benthic crustaceans (*Drupa ricinus*, *Cantharus fumosum*). Among them there are also species with a wide food spectrum, like some species of the genera *Cantharus*, *Vasum*, *Nassarins*, *Bursa*, and *Conus*. Conids feed also on the eggs of benthic invertebrates and on corpses. Several species of predatory gastropods are living in associations with the scleractinians, octocorals and zoantharians, and feed on them. The corallovorous gastropods have toothed firm radula to scrape out the tissues from the surface of coral colonies. The *Heliacus* sp. have a proboscis, which ends in a mouth armed with the same kind of radula.

The corallovorous gastropods graze the mucus and the zooxanthellae, as well as the polyps. The prosobranches prefer the scleractinian hard corals, while the opistobranchs feed mostly on soft corals – alcyonaceans and gorgonaceans. The corallovorous prosobranchs belong mainly to the families Architectonidae (*Philippia radiata*), Epitonidae (*Epitonium ulu*), Magilidae (genera *Corallophyla*, *Quoyula*, *Leptoconchus*, *Magilopsis*, *Magilus*), Thaisidae (*Drupa cornus*, *Drupella fragum*), Ovulidae (*Jenneria pustulata*, *Pedicularia decussata*). Some Magilids (*Corallophyla*) are feeding on anipatharian corals. *Corallophyla* sp. and some architectonids (*Heliactis*) live in association with zoantharians, escaping their nematocysts, and feed on their mucus and on zooxanthellae. The *Heliactis* sp. graze also their polyps (Robertson 1980). In some corallovorous genera a fine trophic specialization was recorded. For example, *Corallophyla violacea* feeds only on the coral Porites, *C. erosa* and *C. deformis* on acroporid corals, *C. abbreviata* on Montipora, *C. monodonta* on gorgonaceans and on zoantharians, *Quoyula madreporatium* prefers the coral Pocillopora, *Q. monodonta* the corals Stylophora and Seriatopora, *Leptoconchus cumingi* the coral Goniopora and hydrocoral Millepora, and *Magilus antiguus* the corals Goniastrea and Leptoria (Demond 1957; Gohar and Soliman 1963; Robertson 1970; Ott and Lewis 1972; Harris 1975). The ovulid prosobranch gastropod, *Ovula ovum* feeds mainly on alcyonacean octocorals Lobophytum, Sarcophyton and Sinularia, and the *Cyphoma gibbosum* on gorgonaceans.

The ability to feed on soft octocorals and on zoantharians is especially peculiar to the opistobranch predatory gastropods *Nudibranchia* devoid of shells. The fauna of the reef nudibranchs is extremely rich and varied (Ostergaard 1955). Its members live mostly on the surface of soft corals and sponges, which they also feed on. The soft corals possess strong chemical defence substances and are very poisonous. The nudibranchs feeding on them are able to neutralize their poisonous terpenoid toxins, which they consume together with the food, and even accumulate them in their body, becoming poisonous, and thus preserving themselves from predation. Among them there are only some eolid species of *Pinufius* and *Phestilla* genera, which are feeding on scleractinian corals. The eolid *Aeolidopsis rausonii* feeds only on the zoantharians Palythoa, this being their specific ectoparasite. They graze the polyps with the aid of a long proboscis (Robertson 1980). Some species of opistobranchs from the order Nataspidea feed on sponges and ascidians using also the proboscis armed with strong toothed radula. Several species of eolid and glaucid nudibranchs even learned to use, as secondary symbionts, the zooxanthellae, which they consume when feeding on hermatypic soft corals. Along their body they have a line of special accessory sinuses formed by intestines. These sinuses are filled with the undigested zooxanthellae. During several days these zooxanthellae remain alive and feed their new host with the translocated products of photosynthesis. Gradually they are digested and replaced by a new portion of algae. The same procedure is used by some omnivorous opistobranchs from the genera *Hermaea*, *Oxynoe* and *Elysia*, employing as secondary symbionts the nondigested chloroplasts from the seaweed tissues they consume (Clark and Busacca 1978). The nudibranchs also learn to use another cellular element of corals – the nematocysts. Swallowing the tissues of corals they do not digest the larger nematocyst cell spyrocysts, but insert them into tissues of their own epithelium, employing them for defence, the same way as they are used by their original host corals (Edmunds 1966; Robertson 1970).

The nonpoisonous nudibranchs have no shell, but they are well camouflaged by their coloration and body forms. This makes them actually undistinguishable from the substratum, e.g., from the branches of colonies of their host corals. While experimenting with the alcyonacean *Lemnalia* the author was astounded at seeing how one of its "branches" started to crawl by the wall of the aquarium. This "branch" appeared to be a nudibranch, which was later estimated as a new species. This species is living on *Lemnalia* and is absolutely undestinguishable from the branch of this coral. The poisonous nudibranchs have usually bright coloration as a warning plackard. Many of them are extremely beautiful in their coloration and forms, being one of the remarkable decorations on the reef. In the soft sediments of lagoons Scaphopoda could be found. They have curious tooth-form shells and easily move inside the sediments. Their size is 3–10 cm. In bottom reef biotopes also various cephalopod molluscs could be met with – squids and small octopuses.

5.1.2 Bivalves

The communities of Bivalvia are also abundant on reefs both faunistically and by their biomass. They comprise 10–30% of species of the reef's malacofauna and about the same part of its total biomass. Reefs have vast spaces of rocky hard substrate for the settling of sessile bivalves and plenty of soft-bottom areas for their vagile species. Being filterers they have a good food supply from reef waters rich in microplankton and detritus. Some reef bivalves have symbiotic zooxanthellae and feed autotrophically. The rocky littoral biotopes exposed to wave-stress, as well as the surfaces of dead massive coral colonies at the windward reef-flat edges, are inhabited by oysters (family Ostrea). The basal valves of their shell is attached to the rock surface, thus firmly fixing the molluscs onto the substrate surface. Some oysters, like *Ostrea cucullata*, *O. sandwichiensis*, *Crassostrea gigas*, and *Saccostrea amasca* form on reefs of the Indo-Pacific region extremely dense monospecific populations counting hundreds of specimens per square meter. The bivalves spondilus ("thorny" oysters) and some of Chamidae prefer temperate wave-stressed places, as channels and caves.

Most of the bivalves attach their shells to the rock surface with the aid of bissus. Among them are the families Mytilidae, Pteriidae, Pinnidae, and Malleidae. They also inhabit the solid substrates in areas of temperate to weak wave stress: rocky surfaces of reef-flat, dead coral heads, walls of trenches and channels. Many of these bivalves are collected for food or for pearls and nacre. In biotopes of reef-flats, on patch reefs, and in dead coral colonies the bivalves of the families Limidae, Isognomidae, Arcidae, Nucleidae are also usual. Some species of arcids, nucleids and lucinids inhabit also the coarse sands and rubble substrates on the flat and in the littoral areas. One of the most important elements of fauna in the reef-flat and patch-reef areas are the large bivalves of the family Tridacnidae, which have symbiotic zooxanthellae. Less abundant is another relative to tridacnids is the bivalve *Corculum cardissa*. It has a fine transparent shell. The tissues of mantle under these shells also contain symbiotic zooxanthellae. *Corculum* lives in soft sediments. But it digs into the sediment so that the upper part of its shell remains exposed to sunlight, necessary for his symbionts. The ecology of tridachids is described below.

The sandy bottom of channels crossing the reef-flat and its outer slopes, as well as the bottom of the lagoon, are inhabited mainly by free-living bivalves. One of them (Pectinidae) lies on the surface flat on its side and may move in water by flopping its valves. Others dig into the sediments and move with the aid of their "foot" (Cardiidae, Corbulidae, Veneridae, Psammobilidae, Tellinidae). In soft-bottom biotopes some of the sedentary bivalves are also often abundant, especially on silty coral sands in the lagoons of barrier reefs. These are mostly Pinna, and also Codakia from the family Lucinidae. They attach to pieces of corals or of tridacna shells buried in the sand. The areas with profuse coral growth often contain dense populations of boring bivalves hidden inside the coral colonies. Their biomass remains usually uncounted during benthos quantification studies. More information about this very important group is given below.

5.1.3 Tridacnids and Their Symbiosis with Algae

The tridacnids or the gigant clams are the largest among the modern and even among the fossil molluscs. The weight of a large specimen of *Tridacna gigas* may exceed 300 kg, and its length 1 m. (Yonge 1975). The size of *Tridacna deresa* – the second largest tridacnid – is up to 50 cm, of the *Hyppopus hyppopus* 45 cm, of *T. maxima* and *T. squamosa* 40 cm, *T. crocea* 20 cm. Below the lock connecting the valves of their shells tridacnids have a hole. Through this hole a mighty bissus passes which is used by the mollusc for attachment to the solid substrate. *Tridacna crocea*, by making rolling movements around her bissus, can bore into the rocky ground. Old specimens of the large tridacnids *T. gigas*, *T. deresa* and *Hyppopus* usually lose their bissus and lay free on the ground, being fixed to their places by their own weight. In the daytime tridacna opens its valves and exposes to the light her magnificently colored mantle. The tissues of the mantle contain a mass of zooxanthellae cells. The products of their photosynthesis these symbiotic algae translocate to the cells of the host (Muscatine 1967; Goreau et al. 1973). Also, an alternative mechanism for the use of algal photosynthesis products by the host exists. The excess of the zooxanthellae cells, which are permanently accumulating in mantle tissues, are consumed and digested by the amoebocytes, which travel in the blood stream. The nondigested substance of zooxanthellae is excreted by amoebocytes in the liver of the mollusc, which is extremely large. Thus, in the blood stream of tridacna an additional digestive system is readily functioning, which increases the efficiency of the use of photosynthetic production of algae by the host. The faeces of tridacna contain a mass of nondigested zooxanthellae. They comprise in *T. maxima* up to 80% of their mass (Ricard and Salvat 1972).

The symbiotic zooxanthellae of tridacnids, as well as main their physiologial features: light adaptations, translocation of photosynthates to the host, are practically the same as in zooxanthellae symbiotic to the hermatypic corals (Taylor 1973a; Goreau et al. 1973; Scott and Jitts 1977).

The degree of autotrophy of tridacnids could be evaluated from the ratio of their daytime photosynthesis to respiration (P_t/M_t). It varies within 1.2 to 3.4 in clams of 1 to 10 kg weight depending mainly upon the size of the Tridacna. In 24 h this balance corresponds to P_t/M_t ratios of 0.6–1.7 (Munro and Gwyther 1981). In the medium-sized specimens of *T. maxima* the P_t/M_t ratio in 24-h balance was 1 down to the depth 10 m (Jaubert 1977). Thus the tridacnids can cover a large portion of their daily energy expenditures with the aid of their algal symbions, the same as the hermatypic corals. The calcification function in *Tridacna* and the very ability of growing of their gigantic shells are provided by its coupling with the photosynthesis, again the same as in corals. Besides, the tridacnids have also a normal filtering apparatus, which separates even small phytoplankton cells of 10–12 μ size as well as the zooxanthellae, which often could be found in plankton over the coral reefs (Goreau et al. 1973; Ricard and Salvat 1972). But there is evidence that the plankton and the particulated matter, taken by tridacna by filtration, are poorly digested by the mollusc. The latter authors concluded that these molluses did not need heterotrophic sources of feeding at all, being actually autotrophs by their total photosynthesis – respiration balance. This thesis does not seem to be acceptable, not only in relation to the tridacnids, but for the hermatypic corals either (Sorokin 1984a; cf. 9.3). It is highly possible also that among the functions of the filtering apparatus in tridacna besides respiration there is also an inorganic nutrients supply from the water. Because of their filtration activity the tridacnids readily accumulate heavy metals in their kidneys, e.g., Zn, Cu, Cd, especially in areas of water pollution. Therefore, they could be used as a very sensitive indicator of such kinds of pollution of coral reef waters (Khristoforova et al. 1979).

The tridacnids are hermaphrodites. The eggs and the sperm are excreted into the water column where the fertilization soon proceeds. The spawning of one specimen within their population triggers the spawning of neighbors, thus increasing the possibility of cross-fertilization, which gives healthier generations. So the success of their breeding depends upon the density of their populations. On the untouched reefs their populations are usually dense and often are composed by one or two species (Ricard 1981). In places of their extensive collection their population became rare, because of the above-mentioned features of their propagation. The catch leads often to their gradual extermination. The veligers of *Tridacna* settle after 1–2 weeks of planktonic life and become tiny clams of 0.3 mm size with a well-developed leg. With its aid they crawl until they find an appropriate place for the attachment. The young tridacnids have a relatively high growth rate (Mc Michael 1974). *Tridacna gigas* during the first year of life reaches a size of 10 cm and attains the weight of 40–50 g. After 10 years this species grows up to 60 cm. Another *Tridacna T. deresa*, has a growth rate of nearly 5 cm year^{-1} (Munro and Gwyther 1981). *T. maxima* attains its maximum weight at the age of 11–13 years (Ricard 1981). The annual coefficient of

P/B in natural populations of *T. maxima* is estimated as 0.2. In the lagoon of the atoll Takapoto the population of *T. maxima* includes around 14 million specimens, which produce annually about 120 t of pure meat, having a basic stock of biomass of 530 t in some units. High growth rates, large size and an easy harvest make tridacnids a promising object for rational development of reef resources via aquaculture. Moreover, their dry meat costs about US$100 per kilo (Yamaguchi 1977; Munro and Gwyther 1981).

5.1.4 Boring Clams

The boring bivalves are extremely widespread in areas of the rocky flat and especially in areas of active coral growth. They compose there a remarkable portion of the total benthic biomass (Evseev and Silina 1982). They bore into living scleractinian corals as well as dead coral heads and also the rocks of the reef flat. Thus they are one of the main agents of bioerosion processes on the reef (together with boring sponges), and their activity is an important

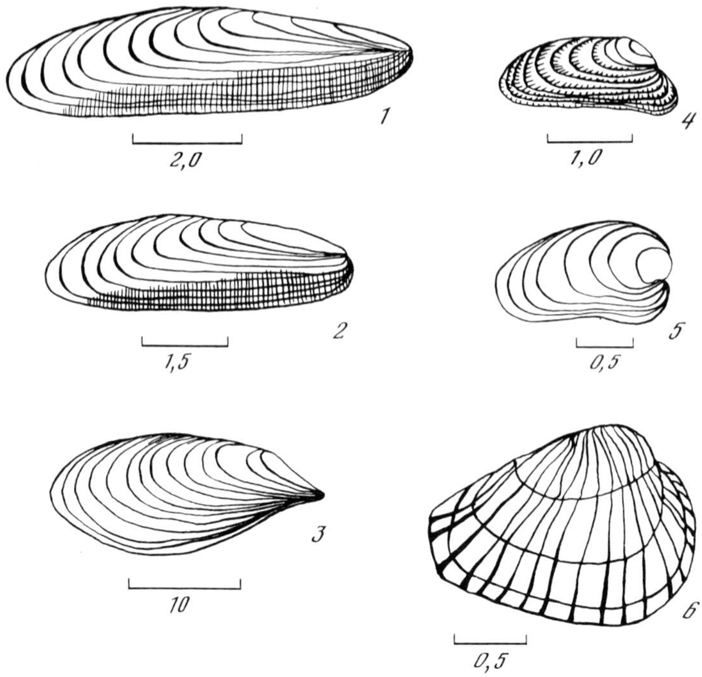

Fig. 5.1. Boring bivalves: *1 Lithophaga nigra*; *2 L. teres*; *3 Gastrochaenia impressa*; *4 Venerupis maer*; *5 Botula cinnamonea*; *6 Petricola lithophaga*; *Bar* in cm. (After Appucuttan 1972)

factor in the life of reef corals (Goreau and Hartman 1963; Connell 1973; Evseev 1981; Trudgill 1983). Among the boring molluscs there are six families of bivalves and one of gastropods (Soliman 1969; Appukutan 1972; Hadfield 1976; Kleeman 1980). The shape of their shells are shown in Fig. 5.1. The dominating taxa of boring clams are the representatives of mytilids and belong to the specialized genus *Lithophaga*, which numbers about ten species. *Lithophaga* have a finger-like shell 0.5–3 cm long. Some species attain lengths of up to 5–9 cm.

The boring molluscs are distinctly specialized in relation to the kind of preferred substrate (Soliman 1969; Evseev and Silina 1982; Peyrot-Clausade et al. 1989). Some five to six species of *Lithophaga* live only in dead colonies of scleractinian corals (*L. laevigata*, *L. malaccana* and *L. teres* are among them). Other species prefer the flat rocks or the "nigger heads" (*L. obesa*). A group of species live only within colonies of living corals: *L. cumingiana* and *L. lessepiana* in Stylophora, *L. hanleyana*, *L. lima* in Cyphastrea and in Montipora, *L. kuehnelti* in Acropora and in Stylophora, *L. simplex* in Favia, in Lobophyllia and in Symphillia. One of the causes of this fine specialization could be seen in the diversity of the mechanisms for boring in lime material, used by different species of borers. They bore mechanically, chemically or combining these two techniques (Gohar and Soliman 1963). The most widespread are the latter (Barthel 1981). The veligers of boring molluscs settle on the surface of living or dead coral colonies, selecting substrates peculiar to a given species. After their metamorphosis the young clam bores into the colony lime skeleton or into the rock. It forms then a pear-like channel with a narrow exit. Through it the clam shows outside its syphones. The distal end of the channel is gradually cemented by the clam. If the rate of growth of the host coral exceeds the rate of boring the clam begins to bore the channel backward to prevent the closing of the exit hole (Soliman 1969; Barthel 1981). The rate of boring of solid lime material by *Lithophaga* ranges within 2 to 4 mm year^{-1} in lime rocks of flats.

The clam *Fungicava* bores his holes into the central part of the corallite of the solitary coral *Fungia*, and thrusts out his syphones close to the gastral cavity of its polyp. Thus the clam filters water enriched with zooxanthellae excreted by the polyp (Goreau et al. 1970; Hadfield 1976). The boring clam *Gastrochaena* lives only in dead coral heads. With the use of X-radiographic techniques it was established that from the massive corals of Caribbean reefs the boring bivalves excavated 15–20% of their total skeletal mass (Highsmith 1980), using only the mechanical mode (Soliman 1969). The density of populations of boring molluscs is usually very high. For example, 14 colonies of living corals from six genera collected at the edge of the reef flat on one of the Swane reefs a 20 × 30 cm piece of lime rock and one tridacna shell alone contained 600 specimens of boring clams. On the reefs off Barbados the mean number of boring bivalves was three for each coral head (Highsmith 1980). The degree of the infestation of corals by these borers seems to be directly dependent on the level of planktonic productivity in reef water, which itself depends on the nutrients input. The more nutrients, the more borers, and the more damage to corals. Thus, a high nutrients content is not healthy at all for the reef ecosystem, which has been established also experimentally (Kinsey and Davies 1979b). Enhanced planktonic productivity induced in areas of anthropogenic pollution, while improving trophic conditions for the filtering, boring bivalves, drastically increases the level of infestation of corals by borers, and so increases their mortality.

Some clams are commensals of ramulose scleractinian corals, such as the *Seriatopora*. They are "nesting" in their ramules. Among them there are some species of the genera *Barbatia*, *Pedium*, and also the tridacnid *T. crocea*. The veligers of this clam settle on the dead patch of the colony or upon its living surface which then dies. The surrounding living colony grows around the attached clam and finally the clam becomes "nested" into the colony. This form of commensalism of boring clams with their scleractinian hosts is profitable mostly to the clams, saving them from predation. As for the corals, this commensalism at first sight seems to be only harmful for the host corals. Indeed, the clams bore numerous holes in their branehes decreasing so their stability against the wave-stress. But in fact the destruction of branches of ramose corals, by breaking their parts damaged by the borers, promotes their vegetative multiplication and their distribution. The filtering activity of boring clams results in excretion of large amounts of pseudo-feces, which improves the trophic conditions for their host corals.

5.1.5 Community Structure, Roles in Reef Ecosystems

The communities of coral reef molluscs are extremely diverse. The total number of their species in one reef system may attain 100 to 550 (Maes 1967; Ricard and Salvat 1972; Zlatarsky and Estalella 1980; Taylor 1984). Their fauna was described as found on the Aldabra atoll (Taylor 1976), the Chagos archipelago (Sheppard 1984), the Phantom shallows (Evseev 1982), the Scott reefs (Sirenko 1980), the New Hebrides reefs (Budin et al. 1980), the Canton atoll (Kay 1978), the Micronesian reefs (Demond 1957), the Fanning atoll (Kay 1971; Kay and Switzer 1974), the One Tree atoll (Hutchings 1974), and on the reefs of the eastern Pacific (Cantera 1989). The number of gastropod species comprised on an average 50 to 70% of the total molluscan fauna. This percentage varies in different reef biotopes (Kay 1971). The share of gastropods in benthic communities is usually less in the flat biotopes and increases in the biotopes of littoral and lagoon. In the communities of reef molluscs the following trophic groups could be named:

1. Phytophages feeding on thalloms of seaweeds and leaves of sea grasses (*Aplysia*, *Turbo*, *Trochus*; see Klumpp and Pulfrich 1989).

2. Omnivorous molluscs feeding on small seaweeds, filamentous algae, periphyton and microphytobenthos, on detritus and on small benthic animals (*Trochus*, *Turbo*, *Patella*, *Littorina*, *Cerithium*, *Strombus*, some species of muricids, chytons).
3. Predatory benthophages feeding on other molluscs, polychaetes, and echynoderms (conids, muricids, *Morula*, *Cassis*, *Cymatum*, *Charonia*, *Bursa*, *Rapa, Latirus*, *Mitra*, and some other predatory gastropods).
4. Predatory corallovorous gastropods feeding on hard and soft corals (*Oliva*, *Cyphoma*, *Heliacus*, *Epithonium*, *Drupa*).
5. Filter-feeders feeding on microplankton and suspended organic matter (bivalves, vermetids).
6. Symbiotic molluscs which cover a significant part of their energy expenditures by the photosynthesis of their algal symbionts (Tridacna, Corculum). The ratios of these groups varies in different reef sites. On average, the omnivorous gastropods, phytopages and chitons comprise 30–40% of the total number of molluscan species, the predatory gastropods and corallophages 40–50%, and the filter-feeders 10–20%.

Thus, for the reef communities of molluscs there is a peculiar and significant predomination of predatory species (Kohn 1968; Kohn and Levitien 1976). For example, the taxonomic structure of communities of reef molluscs in the barrier reef off Moorea, French Polynesia, looks as follows (Ricard and Salvat 1972). On the flat covered with sand and with rear corals, gastropods were the most numerous. Of the total number of species of Mollusca (117 species) the percentages for species of the dominating taxa were: Mitridae 15%, Cerithidae 11%, Conidae 10%, Muricidae and related taxa 8%, Cypraeidae 7%, Terebridae 6%, Strombiidae 4%. But as regards the total number of specimens (not counting the micromolluscs) the Cypraeidae were most numerous (52%) followed by Cerithiidae with 16%, Terebridae 4%, and Mitridae 2%. Five dominating species alone formed 75% of the total number of Mollusca. These were: *Cypraea oblevata*, *Cerithium piperitum*, *Strombus mutabilis*, *Terebra* sp., and *Vermetus maximus*. Three species out of 117 were responsible for 90% of the total molluscan biomass: *Vermetus maximus*, *Tridacna maxima* and *Cypraea obleviata*. In the coastal lagoon of the same barrier reef, the bottom of which was covered by sand with the admixture of detritus and with rear corals, just the one *Cypraea obleviata* comprised 60% of the total numer of species. The number of this gastropod exceeded 100 sp. m^{-2}. Other dominating species in this biotope were: *Cerithium piperum*, *Strombus mutabilis*, *Cantharus fumosus*, *C. coronatum*, and *Pyramidella* sp.

The density of communities, as a rule, drastically varies in different areas of the reef. Its maximum is usually in the littoral rocky biotopes and in the wind-stressed reef edges. The vertical cliff slopes are most often densely colonized by the bivalves, while their horizontal surfaces and the pools in them are inhabited by medium-sixed gastropods (*Nerita*, *Arca*, *Cerithium*, *Littorina*, chitons), which number up to hundreds per square meter. Therefore, the biomass of molluscs may reach in these biotopes up to several kg m^{-2} (Grassle 1973; Gulbin et al. 1987). On the reef flat and the sand bottom of the lagoon the number of macrobenthic molluscs varies within 3–20 sp. m^{-2} and their biomass within 5–50 g m^{-2} (Salvat 1972; Ricard 1981). An average molluscan biomass in the lagoon of the Takapoto atoll was around 5 g m^{-2}. But actually these numbers could be underestimates because of the missing micromolluscs, the biomass of which in soft reef sediments might be at the same level as that of macrobenthic species (Hutchings 1974). Many species of molluscs are fished on reefs for food, souvenirs, collections and the handicrafts. Among them are: *Haliotis*, *Mytilus*, *Pecten*, *Cardium*, *Littorina*, *Tridacna*, *Pinna*, *oysters*, venerids and pteriids.

5.2 Sponges

5.2.1 Features of Morphology and Physiology; Symbiosis

Sponges, as an element of the sessile benthos, are second in their importance after corals and zoantharians. Moreover; the function of sponges in the coral reef ecosystem is somewhat similar to the function of the corals themselves. Many sponges have algal symbionts, and thus they add to the primary organic production. The calcareous sponges produce lime and take part in reef-building processes cementing carbonaceous material during the formation of reef-flat rocks. Scleropongia compose the framework constructions, thus functioning also as reef-builders, especially in deep zones of the fore-reef, where the coral growth is weak (Hartman and Goreau 1970). The spicule material produced by sponges is an important source for reef-bottom sediments. The boring sponges also add to bottom sediments, being a main agent of the bioerosion of reef-flat lime rocks (Goreau and Hartman 1963; Thomas 1972; Hartman 1977).

Sponges are the most ancient and primitive among the multicellular animals. Their tissues are not differentiated in ectoderm and endoderm. They are composed by groups of specialized cells. The motion of their flagella is synchronized but each of the cells feeds independently. The body of a sponge is composed of the mesoglea – the structural substance consisting of the fibers of collagen and spongin. The latter is a hornlike structural material. The fibers of the mesoglea are framed by silicate or calcareous spicules, which are taxonomically specific (Fig. 5.2). Outside, the mesoglea is covered by a layer of the eptithelium-like flagellated cells – the pinacoderm. The same layer of pinacoderm covers also the walls of numerous channels furrowing the sponge's body. The flagella of pinacoderm

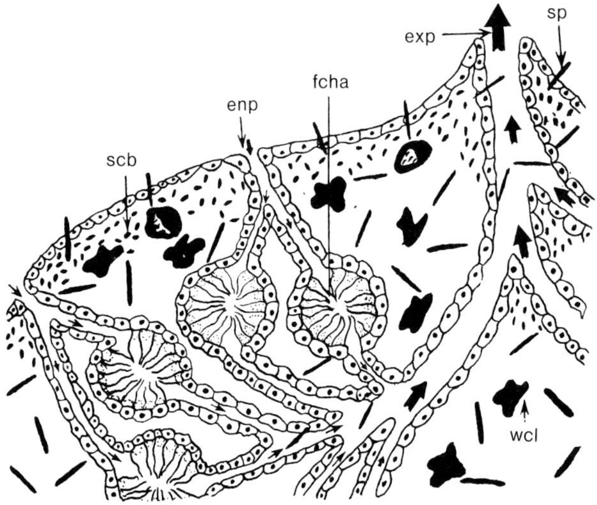

Fig. 5.2. Structure of sponge: *enp* entry pores; *exp* exit pores; *fcha* filtration chambers; *sp* spicules; *wcl* wandering cells; *scb* symbiotic cyanobacteria

cells, the pynacocytes, propel the currents in channels from outside to inside in the sponge. Along the channels filtration chambers, the so-called sinuses, are situated. Their walls are lined with choanocytes. Using their flagella, collars, and also their mucus, they catch the food particles from water which passes through the chamber. The collar of the choanocyte has the construction of a fine net (Rasmont 1968). The flagellum drives water through the collar like through the filtration funnel. A part of the food particles caught like this the choanocyte phagocytizes himself. The rest is obtained by the other cells of the sponge, mainly – amoebocytes, which then transfer it to the tissue cells of the sponge, the food wandering through the fiber thickness of the mesoglea. The choanocytes are collecting mainly fine food particles – less than 10 μ in size. The larger planktonic organisms and detritus aggregates are caught by pynacocytes, which are lining the surface of the sponge's body and its inner channels.

Sponges are among the most powerful fine filterers. Their main food source is the bacterioplankton (Sorokin 1973c, 1978b; Reiswig 1975). The filtration rate of the sponge, as estimated by direct measurement of the volume of water passed through its body, was found to be $0.5–0.8\,l\,h^{-1}\,cm^{-3}$ of the volume of the sponge's body. A volume of water equal to the volume of its body the sponge filters in 5 to 20 s. During the measurements, done by counting the rate of decrease in the ^{14}C-labeled bacterioplankton concentration in the presence of the sponge, the rate of filtration in the demosponge *Toxadocea violacea* was found to be $200–400\,l\,g^{-1}$ of its body carbon per day. It was recorded also that the sponge is able to filter out the bacterioplankton from the water down to a negligible concentration of some

$30–50 \times 10^3$ cells ml^{-1}, starting from the initial values of $1–1.5 \times 10^6 ml^{-1}$ (Sorokin, 1978a). The energy expenditures for filtration in sponges was estimated 1–4 μg $O_2 l^{-1}$ of filtered water. It is remarkable that the sponge Verongia was not able to cover the energy expended on filtration at the expense of the food so obtained, when the O_2 consumption per 1 l of water filtered was more than 4 μg. But this sponge has an additional source of energy – the photosynthesis of its symbiotic algae and the consumption of dissolved organic matter.

The respiration rates in sponges range within 0.02–0.07 ml $O_2 cm^{-3}$ of volume of their colonies per hour, or 0.4–1.5 ml $O_2 g^{-1} h^{-1}$ of dry weight (Reiswig 1974). The food ration of the sponge community in the fore-reef zone off Jamaica this author evaluates as being within 1.5–4.0 g of organic matter $m^{-2} day^{-1}$. The food of sponges includes all components of suspended matter (bacteria, algae, microzooplankton, detritus) and also the colloidal and dissolved organic matter (Stephens and Schinske 1961; Reiswig 1975; Sorokin 1978b; Vacelet 1979; Wilkinson 1978, 1983b; Wilkinson and Garrone 1980). Asymbiotic sponges (*Mycale*) may cover all energy demand by filter-feeding alone, while sponges which contain symbiotic algae only about 20% of it. The rest they get by consumption of dissolved organic matter with the aid of symbiotic heterotrophic bacteria and by using the photosynthetic products translocated by their algal symbionts. Just this symbiosis with algae and bacteria helps most sponges to inhabit the upper- and middle-depth zones of the reef and to compete successfully with other sessile fauna.

Sponges have inside their cells symbiotic blue-green algae (cyanobacteria), and also heterotrophic bacteria. The presence of bacteria inside the cells of sponges was discovered in 1930. The bacteria were supposed to be parasites of the sponges, but later it was proved that they are their real symbionts (Levi and Levi 1965; Vacelet 1971; Wilkinson 1978, 1983b; Wilkinson and Garrone 1980). These bacteria live inside the special cells of a sponge – the bacteriocytes, as well as in the mesoglea. The total number of symbiotic bacteria in sponges attains 10^7 cells g^{-1} of their wet weight. In some sponges the microbial biomass comprises up to 30–40% of the total cellular biomass of the host (Vacelet and Donadey 1977; Wilkinson 1978, 1983b). One sponge usually harbors several (up to seven) different species of bacteria. Some of these symbiotic bacteria are similar to the ordinary planktonic bacteria of surrounding waters, but several species were found in sponges collected in different geographical areas and their appeared to be specific symbiotic species (Wilkinson et al. 1981). In the mesoglea of the sponge Terpios filamentous symbiotic bacteria have been found which had blue pigment. By their morphology those bacteria were similar to cyanobacteria but they contained no chlorophyll (Santavy 1985).

The function of bacterial symbionts in sponges relates mainly to the utilization of dissolved organic matter from water that pass through the hosts body via the channels. Using it, the bacteria build up their biomass, which is then gradually digested by the sponge cells. The ^{3}H-label of alanine appeared first in the bacterial cells and only later in the cells of the sponge (Wilkinson and Garrone 1980). Other sources of feeding and biomass production by these symbiotic bacteria are the waste products, excreted by the sponge's tissues. The bacteria which live in the mesoglea consume also the collagen material and the particulated matter which penetrate into the mesoglea. Besides the production of biomass ultimately used by the sponge, these bacteria also accomplish an important function in cleaning the channels and the mesoglea, when an excess of old collagen filament and particulate matter had been consumed. Thus they support a normal filtering activity in the host sponge. Symbiotic heterotrophic bacteria are definitely found in the following common reef sponges: Pericharax, Jaspis, Neofibularia, Casterospongia. The presence of symbiotic bacteria was recorded even in eggs of some sponges. So these bacteria could be inherited through generations (Levi and Levi 1976). In some sponges the symbiotic heterotrophic bacteria are present together with the blue-green algae.

The blue-green algae are known as symbionts only in sponges and in some echiurid worms (Kawaguti 1971). They were first discovered in the cells of sponges by Feldmann in 1933. Later, besides the blue-greens also zooxanthellae and even a cryptomonad were found in them (Sara and Liaci 1964). Most of the symbiotic sponges harbor a small unicellular cyanobacterium *Aphanocapsa feldmanni* (or, more rarely, *A. raspaigelea*). They are the specialized symbiotic species relative to pelagic Synechococcus. Some sponges harbor the unicellular cyanobacteria *Stephanocapsa feldmanni* and *Synechocystic trididemnii* as symbionts (Larcum and Cox 1988). Several species of sponges (*Tethya*, *Ulosa*) contain filamentous chain species of the cyanobacterium *Phormidium spongeliae*. These algal symbionts could be found in approximately half of sponge species inhabitins the euphotic zone of a reef (Vacelet 1981; Wilkinson and Trott 1985). The sponge Dysidea harbors as symbiont the filamentous cyanobacterium *Oscillatoria spongelina* (Borowitzka et al. 1988). The symbiotic cyanobacteria are usually accumulating about 1 cm deep in the illuminated upper layer of the sponge's body, depending on the degree of light penetration into it. The unicellular symbionts are harbored by the sponge mostly inside the specialized cells-cyanocycles in quantities up to several hundred per cell (Wilkinson 1980). Some portion of them is present also in the mesoglea. The biomass of symbiotic blue-green algae attains in some sponges up to 50% of their cellular biomass. The multicellular filamentous cyanobacteria are located mostly in the mesoglea. Cyanocytes control the multiplication of symbiotic cyanobacteria inside them, as well as their metabolic rates, the same as the cells of the coral polyp's ectoderm control those in the ir zooxanthellae. They may stimulate the translocation of photosynthates by algal cells, thus receiving from them the food by the biotrophical way. Another portion they get of it by digesting the excess of the algal cell produced via the phagotrophic mechanism (Sara 1971; Vacelet 1971). The translocation was estimated to be around 12% of their net photosynthetic production, the rest corresponding to the necrotrophic way. The "infestation" of sponges with the blue-green synbionts proceeds most probably via the eggs, since they have been found also in the larvae of some sponges.

The algal symbionts are useful for sponges not only as their feeders. As many other blue-green algae, they also fix atmospheric nitrogen, thus improving the nitrogen balance of the host (Wilkinson and Fay 1979). The algae consume also the inorganic nutrients from water driven through the channels. They use the end-products of the host's metabolism. These products (mostly-nurients) are assimilated into the biomass of algae and used by the host after their digestion. So the presence of algae in a symbiotic sponge provides a semiclosed recycling of nutrients in their hosts, which is extremely important for life in nutrient-poor reef waters. Indeed, sponges which harbor the algal symbionts grow much faster in biotopes in the light than in those in the shade (Wilkinson and Vacelet 1979). The blue-greens are harbored as symbionts by the sponges from a different their taxa.

Within the sponges communities of reefs three main trophic growths could be identified: phototrophic sponges, mixotrophic sponges and heterotrophic sponges. They are characterized by the degree of participation in photosynthesis production in compensation for their respiratory losses. The latter is expressed by the ratios of photosynthesis to respiration (P_t/M_t) as measured at an average illumination typical at depths of 10–30 m, where usually most of the sponge species live ($400\,\mu E\,m^{-2}\,s^{-1}$). In the phototrophic sponges this ratio is 1.5, in mixotrophic >1.5, and in heterotrophic around zero. The maximum values of P_t/M_t ratios at noon (2.5–4) were recorded in the fan- and the cape-shaped sponges with thin transparent walls (Table 5.1). These sponges may compensate in a 24-h balance 70–120% of their respiration loss. The gross photosynthesis of the sponge associations on the reef slopes of the central Great Barrier Reef of Australia, having a density of about $300\,g\,m^{-3}$, averaged $0.1–0.2\,g\,C\,m^{-2}\,day^{-1}$. These values comprised around 2% of the gross photosynthesis production in the same biotope ($7.7\,g\,C\,m^{-2}$; Wilkinson 1983a). The optimum illumination for

Table 5.1. Ratios of surface area to wet weight (S/W), chlorophyll content (Ch, $\mu g\,g^{-1}$ wet weight), respiration (M_t, $mg\,O_2\,g^{-1}\,day^{-1}$) and photosynthesis (P, $\mu g\,O_2\,g^{-1}\,h^{-1}$) rates, and ratios of the hourly values of P/M_t at midday with illumination of $400\,\mu E\,m^{-2}\,s^{-1}$ (after Wilkinson 1983b)

Sponges	S/W	Ch	M_t	P	P/M_t
Neofibularia	2.4	116	1.09	191	4.2
Carterispongia	6.3	105	1.80	278	3.7
Pseudoaxinissa	7.4	76	2.24	317	3.4
Phyllospongia	10.5	105	1.85	331	4.3
Irchinia	1.7	65	0.60	56	2.2

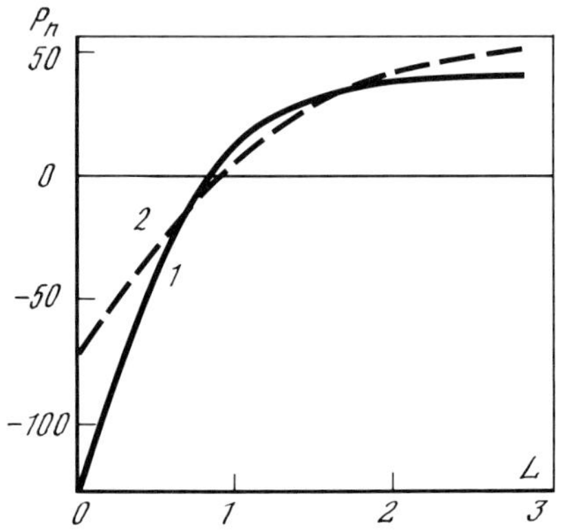

Fig. 5.3. Light dependence curves of net photosynthesis (P_n µg O_2 g^{-1} h^{-1}) in symbiotic sponges *Carterosponia* (*1*) and *Cascopongia* (*2*); *L* light intensity, E m^{-2} h^{-1}. (After Wilkinson and Trott 1985)

sponge photosynthesis ranges within a light intensity of 30–170 µ E m^{-2} s or 50–150 W m^{-2} (Fig. 5.3). This level of illumination is somewhat higher than that for corals. The same as in corals, light adaptation to low illumination was observed also in symbiotic sponges (Wilkinson 1981; Wilkinson and Trott 1985). The parameters of P_t and M_t in some mass species of reef sponges are given in Table 5.1. The chlorophyll content in sponges varied in different species from 10 to 40 µg g^{-1} of wet body weight. The pigment system of algal symbionts in sponges has the absorption maximum of 664 nm (Wilkinson 1983b, 1985).

The phototrophic sponges usually dominate in most reef-flat and outer reef-slope biotopes (Wilkinson 1985, 1989). For example, in some biotopes of the Davis reef (GBR of Australia) these sponges comprised over 80% of the total species number of sponges. In seven out of nine common species of sponges the ratios P_t/M_t at midday were around 3 (Wilkinson 1983b).

5.2.2 Community Structure and Features of Ecology

The reef benthic communities are inhabited by more than 500 species of sponges from three of their basic classes. Demospongia, Calcarea and Sclerospongia. About 90–95% of them belong to demosponges, 4–8% to Calcarea and 1–2% to sclerosponges. The demosponges have their skeleton armatures consisting of collagen and spongin fiber mass armed with the fine silicate needle-shaped spicules. The calcareous sponges are armed with 3–4-ray, star-like calcite spicules. The sclerosponges form a massive aragonite base, but their body mass is armed with silicate spicules. Some 20% of the total sponge community consists of species of the boring sponges (Pang 1973; Rützler 1975). The sponges fauna of the Caribbean reefs (down to 120 m depth) enlists about 400 species, and reefs off Madagascar 150 species (Vacelet and Vasseur 1977). The shallow biotopes are dominated by demosponges, and to a lesser degree by calcareous ones, while the deeper shadowed zones of the outer reef slopes, caves and tunnels mainly by sclerosponge species (Hartman and Goreau 1970). Among the common sponges which occupy the shallow and middle-depth biotopes of the Indo-Pacific reefs there could be named *Tethya*, *Plectronia*, *Mycale*, *Timea*, *Spirastrella*, *Ulosa*, *Higginsia*, *Pericharax*, *Jaspis*, *Mycale*, *Neofibularis*, *Asteropus*, *Haliclonia*, *Jantella*, *Phyllospongia*, *Dysidea*, *Ircinia*, *Carterospongia*, *Pseudoaxinissa*, *Scleroderma*, *Carmina*, *Reineria*, *Callyspongia*, *Cacospongia*, *Dictyonella*, *Prianos*, *Cliona*, *Psamaplysilla*. In deeper zones of the outer reef slopes and in the shadowed biotopes the sponge genera *Spirorhabda*, *Acanthotriaena*, *Scleritoderma*, *Astrosclera*, *Leucetta*, *Murrayona* are most common. At the Western Atlantic reefs in shallow and middle-depth zones the species in mass are *Tethya*, *Chondrilla*, *Haliclona*, *Gelloides*, *Iotrochota*, *Grantella*, *Cliona*, *Verongia*. Deeper, the sponges *Amphimedon*, *Ulosa*, *Ectypoplasia*, *Calyspongia*, *Niphates*, *Xestospongia*, *Chondrilla*, *Anthosigmella* become more common. In the fore-reef zone at the base of the outer reef slope the sclerosponges *Ceratoporella*, *Stromatospongia*, *Hispidoptera* and the calcareous sponge *Leucetta* are usual (Lang et al. 1975). Within the colonies of the living and dead corals as well as in the trenches and holes of calcareous reef-flat rocks, under the boulders the incrusting species of sponges *Erylis*, *Paratetilla*, *Tethya*, *Timea*, *Spirastrella*, *Ulosa*, *Higginsia* are most numerous. The sponges fauna of the Atlantic reefs include much less symbiotic species compared with the Pacific reefs (Wilkinson 1989).

On the reef-flat and on upper zones of the reef slope the spongal fauna consists mostly of phototrophic and boring species, which have small flat colonies, or of those incrusting trenched undersides of boulders and stones or covering the coral rubble. In deeper parts of the lagoons many sponges dig part of their body into the sediment or attach it to pieces of solid substratum. But the most abundant and various sponge communities inhabit the middle depths of the outer slope and especially the buttress zone and the upper part of the fore-reef, which are the areas of intensive water drive so important for the filtration activity of sponges (Wilkinson and Cheshire 1989; Wilkinson 1989). The biomass of sponges in shallow areas ranges usually within 10 to 150 $g\,m^{-2}$, and at depths of 20–40 m of outer slopes within 100 to 500 $g\,m^{-2}$. But on the reef-flat areas covered with rubble it could be also as high as 1.2 $kg\,m^{-2}$ (de Laubenfels 1950; Hartman and Goreau 1970; Reiswig 1973; Hartman 1977; Vacelet and Vasseur 1977; Bak and Luckhurst 1980; Suchanek 1983; Wilkinson and Trott 1985).

Having relatively high growth rates, sponges often overgrow corals and other colonial sessile fauna and thus successfully compete with these, especially because some sponges excrete toxins, which inhibit neighboring corals and even kill them. Many sponge species contain antibiotical substances fixed in the membranes of their cells, which prevent their fouling with bacteria (Burkholder 1973; Amede and Pesando 1985). The sponges usually are poorly grazed by reef fauna. They have low nutritional value, are armed with sharp needles, and some of them are toxic. Therefore, they are eaten only by some specialized tetradontid fishes, some species of urchins, and by nudibranchs.

5.2.3 Boring Sponges

Boring sponges are among the ecologically most important components of bottom fauna. They complete up to a quarter of the total sponge communities in the upper- and middle-depth reef zones. Their activity reduces the stability of coral communities to wave-stress. The majority of the species of boring sponges belongs to the families Clionidae, and some to the family Adocidae (Siphonodyction) and Spirastrellidae (Speciospongia, Anthosigmella). The most common species of this group belong to the genus *Cliona* (*C. lampa*, *C. copiosa*, *C. viridis*, *C. aprica*, *C. langae*, *C. laticlaviola*), *Mycale* (*M. laevis*), *Anthosigmella* (*A. varians*). The boring sponges bore holes and small caves in living and dead coral colonies, in molluscan shells, pieces of rubble, and rocks of reef-flats. Among the living corals these sponges prefer colonies with dead basal parts. The larvae settle there, grow and bore, thus weakening them and finally causing their collapse under wave or current action (Thomas 1972; Pang 1973). Their destructive activity is especially evident in coral communities on steep slopes, where whole massives of colonies of foliose corals, like Agaricia or Turbinaria, could be seen collapsing on being undermined by boring sponges (Goreau and Hartman 1966). It was also discovered that the infestation of corals by boring sponges increases on reefs influenced by pollution (Rose and Risk 1985). An affinity of some species of the clionid sponges to definite species of corals was also recorded (Goreau and Hartmann 1963).

The sponges bore the hard substrate with the aid of special "scraper" cells, they extrude the pseudopodia, which excrete the acid and thus dissolve the calcium carbonate. The pseudopodia penetrate into the lime rock like the scoop of an excavator and extract chips from it, which have a specific semispherical form, easily identifiable in the bottom sediments. The chips of boring sponges are one of the important sources of the formation of coral sand providing 20 to 30% of material composing it (Hartman 1977). The rates of boring of hard lime materials by boring sponge communities in situ varies within 0.3 to $3\,kg\,m^{-2}\,year^{-1}$. The linear rate of boring by a single sponge might be 2 to 5 mm per year (Neumann 1966; Moore and Shedd 1977). The sponges play an important role in the processes of reef genesis, and also of bioerosion (Marshall 1983). They influence even the shape of the growing reef and control the age of hard corals, not only of the ramose but of the massive kind too (Goreau and Hartman 1963). Besides their function in bioerosion they may also act as reef builders. When they develop vigorously within the masses of coral rubble in zones of moat and reef-flat depressions, they cement separate pieces of rubble into one larger aggregation attached to the rocky bottom by the same sponges. These aggregations will be then consolidated by the coralline algae and become solid lime rock, thus enhancing reef growth (Wulff 1984).

5.3 Wormlike Animals

The communities of polychaete worms on coral reefs are formed by two of their suborders: the free-living errant polychaetes Errantia and the sedentary polychaetes Sedentaria. The errant polychaetes have flat or round bodies with rows of parapodia and cirrae on their sides. Their size varies from 2 mm to 10 cm. The different taxa of Errantia, being trophically specialized, their mouth apparatus is of a specific structure (Fauchald and Jumars 1979). The sedentary polychaetes live in permanent or in temporal tube homes. Their mouth parts are transformed into a filtering organ. They feed on plankton and on suspended organic detritus.

The errant polychaetes, which inhabit reefs, belong to the following main families: Nereidae, Eunicidae, Syllidae, Polynoidae, and Amphinomidae. Less common families are: Cirratulidae, Aphroditidae, Glyceridae, Phillodocidae. The nereids are relatively large omnivorous worms. They have a flattened body with well-developed parapodia. They collect or capture their food with the aid of an outstreched pharynx and with their massive strong jaws. They prefer soft-bottom biotopes, where they build their holes, compacted inside with the aid of mucus which they are excreting. When feeding, the worm crawls out of its hole and collects all food material around it: algae, detritus, small benthic animals. After that it leaves the "home" and quickly builds a new one. Some species of nereids are even filterers (*Nereis diversicolor*). They construst the mucus nets inside their homes and pump water through it the same as chironomids do. Among them there are also pure predators (*N. grubel*) and phytophages (*N. procera*). But a main portion of the nereid species is omnivorous. Some nereids change the form of their body during the mating period. Their parapodia expand and on their heads big eyes appear. These worms in their "wedding dress" are called heteronereids. After this metamorphosis they ascend from bottom to water column. There the males and the females perform the wedding dance, and afterward shed their eggs and sperm into the water, where the fertilization takes place. This period is a feast for fish (and for people too), they graze or catch a larger portion of the "dancing" polychaetes. The mass species of the nereids are: *Nereis caudata*, *N. falcaria*, *N. jacksoni*, *N. trifasciata*, *Platynereis isocita*, *Perinereis calodonta*, *P. anomala*, *Ceratonereis mirabilis*.

The eunicids are medium-sized (1–2 cm) worms with round-shaped body, short parapodia and a fringe of gills on their dorsal sides. Most eunicid species can be found in the cryptofaunal communities, in branches of dead corals, under stones in holes and trenches of flat rocks. They can also stretch out their pharynx armed with double mandibles and with rows maxillae. Several species of this group bore holes in solid lime rocks and even in living corals (*Lyzidice*, *Palola*, *Eunice*, *Schemancephala*). The young vorms wander freely around on the bottom, while the adults build permanent tubes or live in holes bored and then abandoned by sipunculid worms. Some species make their own holes is soft or even solid bottom substrate (Day 1967). Many eunicids are commensals of sponges, molluscs, holothurians, and hydroids. Some (*Arabella*) are parasites. The eunicids also perform the wedding dource, the same as the nereids. Among the nereids there are specialized predators feeding on other worms, on ostracods and other meiobenthos (*E. aphroditois*), and phytophages feeding on coralline algae and periphyton (*E. viridis*). But most of the eunicids are omnivores animal food dominating in their food ration (Day 1967; Fauchald and Jumars 1979). Most common among the eunicids are: *Eunice atra*, *E. cincta*, *E. australis*, *E. antennata*, *E. caribbea*, *Palola siciliensis*, *Lysidice collaris*, *L. ninetta*, *Nematonereis unicornis*, *Drilonereis falcata*, *Lambrinereis etraura*, *Marphysa corallina*, *Arabella mutans*, and *A. iricolor*.

Syllids are small, flat worms with minute, fine parapodia. Their size is 2 to 20 mm. The syllids are recognized as one of the most important component of the meiobenthos often dominating the communities of flat- of rubble-bottom areas and even the soft bottom (Kohn and Lloyd 1973; Hutchings 1974). The extremely poverful outstretched pharynx of some syllids has a special suction device, the proventriculus. With the aid of it the worms can suck out the polyps of hydroids. Their mouth is armed with one large tooth and several smaller ones. Most syllids are predators. They graze hydroids, bryozoans, ascidians, sponges, zoantharians. But there are among them also omnivorous worms (genus *Eusyllis*), which live in sand and feed on microphytobenthos, on ostracods and micromolluscs. An important feature of syllids is their ability for asexual breeding by simple division of the body into several pieces. Each new piece develops head and tail at its ends and becomes a new worm. Because of this ability they may quickly colonize the bottom substrates and their populations attain large numbers and form a considerable part of the biomass (Hutchings 1974). The most common species of syllids are: *Syllis cornuta*, *S. alternata*, *S. spongicola*, *S. amica*, *S. gracilis*, *S. exilis*, *Opistosyllis laevis*, *Tryanosyllis zebra*, *Exogene longicornis*, *E. dispar*, *E. verugira*, *Brandia rhopalophera*, and *Odontosyllis ctenostoma*.

Polynoids are relatively large worms. Their body is covered by brightly colored scales. They live mostly under boulders and stones and are a remarkable element of the reef cryptofauna. They are predators feeding on other worms and eggs laid by molluscs. Their mass species are: *Hermothoe indica*, *H. dictyophora*, *Lepidonotus carinulatus*, *L. glaucus*, *L. jukesi*, *L. jachsoni*, *Dendropoma* sp.

Amphinomids, or "fire" worms, are large lazy polychaetes. They live under stones on reef flats. Their epidermis is covered by fine, brittle and very poisonous needles. They are predators. The most common of them are: *Eurithoe complanata*, *E. chilensis*, *Amphinome flave*, *A. rostrata*, *Notopygos megalops*, *Hermodyce carunculata*. The latter feeds on scleractinian corals, and lives in holes, which it bores in their living colonies (Marsden 1962).

The sedentary polychaetes are represented in reef bottom biotopes mainly by the families: Serpulidae, Sabellidae, Spionidae, Terebellidae, Capilellidae, Chaetopteridae. Most of them are fine filter-feeders and feed mainly on bacterioplankton and or small phytoplankton.

One of the most common genus among them is *Spirobris*, family Serpulidae. Spirobrises are large worms living in lime tubes incrusted into the coral colony. The diameter of their tubes measures 5 to 15 mm. The trochophore larvae of spirobris settle on dead or damaged places of coral colonies. After the metamorphosis the young worms start to construct their tubes, which grow together with the growing coral colony, but a little faster, so that the edge of the tube is permanently elevated above the colony's surface. Thus the worm escapes aggression from the host's polyps. At the head of this polychaete a sophisticated and stunningly beautiful double spiral-shaped filtering device is situated. It is formed by the mouth tentacles covered with ciliated epithelium. The spirobrises often form rather dense populations in the old colonies of massive corals. The surface of such corals seems to be covered with beautiful flowers. In fact, the spirobrises are one of the most remarkable decorations of a reef. Their common species are: *Spirobris gigantea* and *S. nautiloides*. Other serpulids construct meandering lime tubes on the surfaces of stones and rocks of sublittoral and outer slope zones. Their size is 5–30 mm. Sometimes their tubes form quite a layer on the surface of rocks. Their common species are: *Hydroides longispinosa*, *H. exaltatus*, *Pomoceros* sp., and *Serpula vermicularia*.

The sabellids are usually large polychaetes, which live in soft but very firm tubes. They inhabit the soft bottom, where they form sometimes very dense populations. Their filtering apparatus is formed by straight long feathered tentacles. On being endangered, the worm instantly hides in its tube. Their mass species are: *Megalomma vesiculatum*, *M. binoculatum*, *Sabella microphthalma*, *S. melanostigma*, *Hypsicomus phaetothaenia*.

The spionid worms construct permanent or temporary tubes or holes in soft bottom substrates. Some of them (*Polydora*) bore in the heads of dead corals. The head end of the worm is furnished with fine flexible tentacles covered with cilia. The tentacles are used by the worm to collect detritus that settles around the hole. They also may use their tentacles for filtration and for sedimentary feeding. As common species of reef spionids could be mentioned: *Malacoleros indicus*, *Prionospio cirrifera*, *Polydora armata*, *P. antennata*, *Pseudopolydora reishi*, and *Boccardia polybranchia*.

The habits of the terebellids are similar to those of the spionids. They construct their tube homes in soft bottom, under the boulders, in heads of dead corals. At the apical end of their bodies they have fans of fine ciliated tentacles. When feeding, the worm peeps out from the hole and spreads the fan of tentacles over the surface of the ground. Using the tentacles and the threads of mucus which they excrete, the animal collects the settling detritus and drives it to its mouth. With the aid of special "lips" this material is sorted out and the food particles are swallowed. Among common reef terebellids are: *Terebella ehrenbergi*, *Liomia medusa*, *Streblosoma hesslei*, *Hanchiella tribullata*, *Terebellides strommi*, *Eupolimnia nebulosa*, and *Reteterebella queenslandia*.

The capitellids build their tubes in soft bottom ground from sand particles glued together with mucus. They collect from its surface the detritus and microalgae. Among their common species are: *Stychoproctus somalus*, *Dasybranchues caducus*, *D. cadutus*, and *Pullicila armata*.

The chaetopterids also live in tubes submerged in the sand or attached to the walls of trenches in flat rocks or to boulders. They are filter-feeders. Using their ciliated epithelium they pass water through the mucus nets that are excreted by their parapodia (Barnes 1964). On reefs their usual species are: *Chaetopterus variopedutus*, *Mesochaetopterus minutus*, and *Phyllochaetopterus* sp.

Other families of reef polychaetes are represented by the following common genera: *Cirriformia, Dodecaceria* (fam. Cirratulidae), *Phyllodoce*, *Eulalia*, *Notophyllum*, *Pterocyrris* (fam. Phyllodocidae), *Armandia*, and *Ophelia* (family, Opheliidae).

Thus the polychaete worms occupy practically all the possible econiches in reef biotopes. Among them we may find the sedentary filterers, the wandering and settled detritovores and phytophages, the predators, the commensals and parasites. They inhabit all possible kinds of bottom substrates: sand, rubble, caves and trenches in lime rocks sea grasses (Ansari et al. 1991). They are also an important component of the cryptofauna, inhabiting the dead colonies of corals. For example, in a large coral head on one of the Great Barrier Reefs over 100 species of polychaetes have been found (Grassle 1973). Their grazing impact on dead colonies of corals enhance the recruitment of boring organisms, thus accelerating the bioerosion processes (Hutchings and Kiene 1989). Polychaetes are also an important part of coral reef demersal and mezozooplankton (Sale et al. 1976a; see also Sect. 3.3.2). Many small aldult polychaetes periodically rise at night up to the water column. Some of their species swarm in water during their breeding period. The dense swarms of polychaetes could be formed simultaneously by several of their species (Hutchings and Howitt 1988). The trochophore larvae of polychaetes are a usual element of reef plankton. Finally, there are among them purely planktonic forms, like Tomopteriidae.

The quantification of polychaetes in solid reef rocks, where they are among the main benthos components, is a difficult problem. To collect them there it is necessary to cut out with the aid of chisel and hammer the rock blocks. Then these blocks are soaked with formalin and crushed to wash out the fauna (Kohn and Lloyd 1973; Hutchings 1974; Vittor and Johnson

1977). Another way is to dissolve the lime blocks with acid (Hutchings and Weate 1978). Inevitably, during the precedure of cutting and breaking up the rock a large, if not the largest, part of vagile fauna may escape. For this reason, data on the taxonomy and the density of the polychaete fauna in reefs seem to underestimate significantly the latter (Fauchald 1974). One possible nondirect way to solve this problem is to place porous lime blocks of 1 dm^3 size on the bottom for 2–3 years (Hutchings 1981, 1985). During this period the block is inhabited by polychaetes from the surrounding biotope. In such blocks, kept for 2 years on the Lizard Island reef-flat, 144 species of polychaetes were discovered. This number was far greater than it was possible to find in the samples of natural flat rocks. In the blocks the polychaete fauna was dominated by the genera: *Polydora*, *Syllis*, *Hydroides*, *Ceratonereis*, and *Nematonereis*. The recruiting of polychaete fauna into these blocks proceeded mainly through the settling of their larvae from the water column. An important stage in this process was the initial appearance of the boring sipunculid worms. They bored holes in block walls which were then occupied by other polychaetes.

The total species number of polychaete fauna on a separate-standing reef was: in the western Atlantic region 100–130, on the Bahamas 46, at the Tulear reefs 118, at Guam 29, at reefs of Thailand and Sumatra 76, at the Enivetak atoll 132, and at reefs of the southern GBR about 140 (Kohn and Lloyd 1973; Hutchings 1974, 1985; Vittor and Johnson 1977; Peyrot and Clausade 1977; Devaney et al. 1987). Most numerous were the species of Syllidae and Eunicidae. These families and sometimes also the serpulids formed the bulk of the total number of the polychaete fauna. A main portion of their biomass was composed usually of eunicids, nereids and capitellids. At the reefs of the Indo-Pacific in reef-flat biotopes the number of species of polychaetes varied within 10–20 and their total number within $10–60 \times 10^3 m^{-2}$, with a domination of small syllids (Kohn and Lloyd 1973; Hutchings 1974, 1985; Peyrot-Clausade 1977; Buzhinskaya et al. 1980; cf. Fig. 5.4). On reefs of the western Atlantic, in the same biotopes the eunicids and serpulids were most numerous (Vittor and Johnson 1977). Shannon's index of species diversity of polychaete communities in biotopes of flats was usually very high: H = 1.5–2.8. In shallow soft bottom biotopes with seagrasses and seaweeds the polychaete fauna is less abundant. Nevertheless, their biomass compose a significant portions of the benthos there. Their total numbers attained $5–20 \times 10^3 m^{-2}$ and their biomass $10–100 g m^{-2}$ (Thomassin et al. 1976, 1982; Ausari et al. 1991). Among the errant polychaetes the eunicids (Marphysa), the capitellids (Dasybranchus), the polynoids (Lepidonotus), the nereids (Nereis, Perinereis) and the glycerids (Glycera) were most numerous. The sedentary polychaetes Chaetopterus, Mesochaetopterus, Terebellides, Prionospio were here also usual components of benthic fauna. The comunity structures of the polychaete communities vary on different reefs, being dependent upon the abundance of plankton and detritus on them. On reefs with an abundant

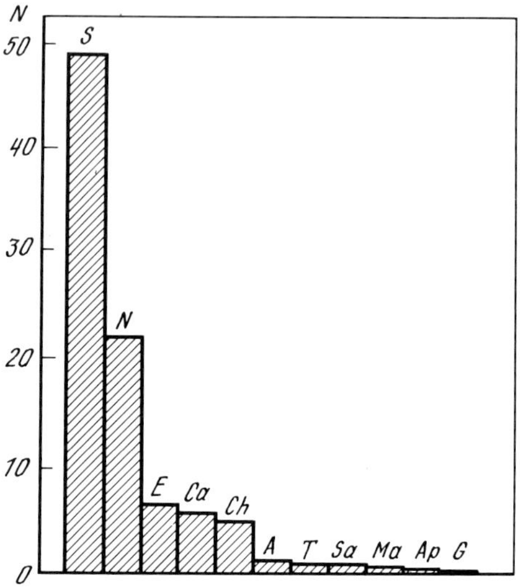

Fig. 5.4. Numbers (N, dm^{-3}) of the representatives of dominating families of polychaetes in the sample of reef-flat rock from Guam I.: *S* Syllidae; *N* Nereidae; *E* Eunicidae; *Ca* Capitellidae; *Ch* Chaetopteridae; *A* Amphinomidae; *T* Terribellidae; *Sa* Sabellidae; *M* Maldanidae; *Ap* Aphroditidae; *G* Glyceridae. (Modified after Hutchings 1974)

Table 5.2. The wet biomass ($g\,m^{-2}$) of main groups of polychaete worms at the Bahamas and the Florida reefs (after Vittor and Johnson 1977)

Groups	Areas	
	Bahamas	Florida
Eunicids	34	61
Serpulids	0.5	865
Others	13.5	105
Total biomass	48	1028

plankton and detritus in the water column the filtering sedentary polychaetes are the most numerous, and on reefs surrounded by oligotrophic waters the eunicids are dominating (cf. Table 5.2). But, on an average, in the coral reef shallow biotopes the filterers comprise about 30% of all the species of polychaetes, the detritophages and omnivores 40%, and the predators 30% (Vittor and Johnson 1977). The total number of polychaetes varies in these biotopes on different reefs within 10 to 130 × $10^3\,m^{-2}$, and the biomass within 20 to $1000\,g\,m^{-2}$ (Hutchings 1974; Kohn and White 1977; Peyrot-Clausade 1977; Vittor and Johnson 1977; Reichelt 1979; see Table 5.2).

The polychaetes use various sources of food available in reef biotopes and build up a significant biomass of a high trophical quality. Thus they play a very important role in the trophodynamics of coral reef ecosystems. They are a significant source of food for most reef fish feeding on zoobenthos and for the predatory invertebrates, especially molluscs, as well as for many planktonovorous fish (Kohn 1968; Bakus 1969; Vittor and Johnson 1977). Having a small average body size and a high growth rate and being able (syllids) to multiply vegetatively, the polychaetes compose a great part, or sometimes even the main portion, of the total production of reef zoobenthos (Hutchings 1974, 1981).

Besides the polychaetes as significant components of the reef zoobenthos, there are also other groups of worms – the sipunculids and the nematods. In some bottom biopopes the benthic fauna includes also the echiurids, the oligochaetes and the turbellarians. But the density of their populations is usually low.

The sipunculid and the echiurid worms are now recognized as a separate phyla of invertebrates. The sipunculids have a short sack- or cylinder-like nonsegmented body. Their size varies from 2 mm to 10 cm. At the end of the body they have a trunk. The mouth is situated at the end of the trunk. It is armed with horny teeth and is surrounded with a row of tentacles. They dig out holes in sediments or bore into the rocky substrates, being so an important component of the reef-boring fauna (Warme 1977; Peyrot-Clausade et al. 1989). The density of populations of these worms in reef-flat biotopes can be significant – up to $1–2 \times 10^3$ sp. m^{-2} with a biomass up to 20–30 g m^{-2}. In soft bottom areas it is less: 20–50 sp. m^{-2} (Rice and Macintyre 1982). Their communities on individual reefs include 5 to 30 species. In total, the fauna of the coral reef sipunculids lists about 300 species (Rice 1976). In the flat rocks the most numerous are the boring sipunculids, like Paraspidosiphon, Aspidosiphon, Lithacrosiphon, Siphonosoma. The worm *Aspidosiphon elegante* bores into the living corals, and *A. jukesi* lives in symbiosis with the solitary coral Heterocyathus (cf. Sect. 7.3). In soft sediments in holes live *Siphonosoma novaepommeraniae*, *S. vasum*, *Phascolosoma perlucens*, *Golfingia misakiana*.

The echiurids externally resemble the sipunculids. They live in holes, in trenches, under the stones. With the aid of their trunks similar to those of the sipunculids, they collect the detritus. The most common genera of the reef echiurids are: *Anelassorhynchus*, *Pseudolonellia*, and *Achaetobonellia*.

The reef annelid worms oligochaetes are represented in reef bottom biotopes mainly by the tubificids and to a lesser degree by the enchitreids. Their size is 5–20 mm. Some of the tubificid oligochaetes (Phillodrilus, Corallodrilus) have no intestines but have symbiotic bacteria in their bodies. Most of them feed on bacteria and on fine detritus. Besides the above-mentioned genera of tubificids Limnodriloides, Heterodriloides, Marcusaedrilus are often present in the reef meyobenthos (Jamieson 1977; Mather and Bennett 1984). They occur in relatively small quantities in the meyobenthos of the soft sediments, especially in the littoral zone of barrier and fringing reefs. But their numbers even there are usually rather low: 50–200 sp. dm^3 of sediment. This is an order less than those of the nematods. In areas of silty or of oxygen-poor sediments near big islands the importance of oligochaetes largely increases (Thomassin et al. 1982).

The round worms nematods are frequent in reef bottom sediments (Renaud-Mornant et al. 1971; Thomassin et al. 1976; Grelet 1985). Their size is 0.2–1 mm and an individual weighs only 3–10 μg. The number of nematods in coral sands varies within 0.2 to 2×10^6 sp. m^{-2}, thus they comprise a significant part of the microbenthos, attaining a level of biomass of 1 to 10 g m^{-2}. Most of them are sediment- or detritus-feeders, but some of their species are predatory. The nematods themselves are an important source of food for various benthic fauna and fish (Salvat and Renant-Mornant 1969).

The reef ribbon worms nemeretins are ordinarily present in the infaunal benthos. They live under stones, in trenches, and in coral rubble. Some of them could be very long – up to 1 m and more (Baseodiscus). Their other common genera have sizes from 1 cm (*Poseidonemeretis*) to 5–10 cm (*Gorgonorhynchus*, *Nemerites*). Most of nemeretine worms are predators. Small nemeretines often comprise a significant part of the meiobenthos (Gibson 1982; Devaney et al. 1987).

The flat worms turbellarians are mostly commensals of corals and sponges. Some of their species are criptofaunal. Most of turbellarians have sizes of 1–5 cm and a very diverse and bright coloration. Taxonomically they are represented in reef benthos by the genera *Pseudoceros*, *Stychoplana*, *Paraplanocera* (order Polycladia).

5.4 Benthic Crustaceans

5.4.1 Community Structure

Reef crustaceans are one of the most diverse and abundant groups of the coral reef vagile and sedentary benthos. Various of their taxa occupy numerous trophical niches in bottom reef biotopes, thus providing efficient use of the food sources within the benthic communities. Hundreds of species of crustaceans are commensals, symbionts or parasites of sessile and vagile benthic and planktonic reef animals. Many of their species are criptofaunal. The total number of crustacean species on the reefs of the Indo-Pacific exceeds 1200, including ca. 800 species of decapods, of which 500 species are crabs (Serene 1972; Haig 1976). Among the communities of the reef crustaceans their following ecological groups could be named: (1) the demersal crustaceans, which at night swim in the water column and in the daytime hide in bottom sediments, rubble, seaweeds, and perforated reef rocks, thus escaping grazing (mysids, cumaceans, copepods); (2) crustaceans, which inhabit soft sediments and rubble, covered with seagrasses and macrophytes; (3) the criptofaunal crustaceans; (4) commensals and symbiotic crustaceans of varions sedentary and vagile benthic animals and (Thomassin 1974; Haig 1976; Patton 1976; Tsareva 1980; Peyrot-Clausade 1981; Galkin 1982).

The lower meiobenthic vagile crustaceans are represented on reefs by small adults and by larvae, which have sizes of 0.05 to 3 mm (Thomassin et al. 1976, 1982). They are mostly cyclopids, harpacticids, ostracods, and the cyclopid nauplii. The most numerous of these crustaceans are found in coral sands, in rubble, and in the thickets of macrophytes. The number of cyclopids in coral sands of the barrier reef of Tulear was $4-15 \times 10^3\,dm^{-3}$, and their biomass $1-5\,g\,m^{-2}$. The numbers of ostracods varied there within 40 to 400 sp. dm^{-3}. In the same biotopes off Moorea Island the density of these groups was and by an order less. Most of the bottom cyclopids are the demersal predators, which at night migrate up to the water column and catch small zooplankters. But some of cyclopids are symbionts and parasites of corals. For example, cyclopids from the family Lichmolgidae are the symbionts of the scleractinian corals. They live on their surface and feed on their mucus and on their zooxanthellae, accomplishing thus the function of cleaners. The copepods Anthessius, Astroxynus, Nanaspis are the symbionts of some molluscs and echinoderms. The highly specialized parasitic copepods Xarfia live inside the coral polyps. They have worm-like bodies with the reduced legs (Patton 1976). The copepods Pandarus, Dentigryps, Caligus, and Lepeophtheris are parasites of fish. The ostracods inlude several tens of species, most of which belong to the genera *Sarsiella*, *Xestoleberis*, *Paradoxostoma*, and *Anscottiella* (Hartmann 1964; Herriott and Immerman 1979; Shornikov 1980; Jones 1984). The ostracods feed on microphytobenthos and on detritus.

The sedementary lower crustacean barnacles belong mostly to the order Cirripedia. The ordinary barnacles attach to the surfaces of the flat and of the littoral rocks, to bivalves or to turtles shells. The boring barnacle Lithotria bores into the lime rocks. The corallobiont barnacles grow into the living coral colonies so that their shells are hidden completely inside them. On the surface remains only a small hole with elevated edge, through which the animal sticks out its filtering apparatus composed by its modified legs. Most of specialized corallobiont cyrripedians belong to the family Pyrgomatidae, which list about 35 of their species. Among them there are also several species of real balanid barnacles (*Balanus cardinus*, *B. durhamy*, *Kachloria floridana*). The corallobionts are merely the commensals. But some of them lost the ability for filter-feeding and feed on the mucus and – even on the polyps of their coral hosts (Hoekia).

The higher reef crustaceans (class Malacostraca) such as mysids, stomatopods, cumaceans, isopods, and amphipods inhabit mostly the biotopes of soft sediments as well as the rubble and the solid reef-flat biotopes covered with macrophytes (Jones 1984). Most of them are demersal animals feeding as filterers or as planktonovorous predators. They go up to the water column at night for feeding. In the samples of botton sediments the mysids and the young or the small adult isopods form a significant part of meiobenthos, especially in daytime (Thomassin et al. 1982). The mysids in such samples include usually demersal species from the genera *Euchalina*, *Gastrosaccus*, *Erythrops*, and *Doxomysis*. Some isopods live permanently on the thallus of macrophytes, in sands (Exosphaeroma), in the rubble (Cirolana), or under stones (Orbinus). Their other common genera are: *Apanthura*, *Leptautura*, and *Mesanthura*. Isopods feed on periphyton, on detritus and graze thalli of macrophytes. Some isopods live at the base of colonies of corals and gorgonaceans and feed on the mucus which they shed (Kensley 1982). Many isopod species are exoparasites of fish.

The cumaceans inhabit mostly the soft sediments. Some of them are demersal as the mysids. They are predators and feed on smaller crustaceans and on nematods. The most common cumacean species belong to the genera *Bodotria*, *Discoides*, and *Sheardia*. The stomatods are relatively large benthic crustaceans, being 2–8 cm long. They live in sediments, digging in, constructing permanent holes, or hiding in seaweeds. They are also predators. Some of their species are demersal. Their most common species in reef biotopes are: *Pseudosquilla ciliata*, *Astrosquilla littoralis*, *Alima clorida*, *Lysisquilla* sp. *Leptosquilla* sp., and *Nanosquilla* sp. (Thomassin 1974). The amphipods, gammerids and caprellids, comprise the bulk of benthic communities in different reef biocopes enriched with decaying plant material and detritus. Some of them live on thalli of seaweeds. Others build there a kind of nest (Hyale, Maera, Mallacota). The amphipods Lembos, Gammaropsis, Amphithoe, Cymadusa construct the tubes from a mucous substance hardening in water, and fasten them in trenches or in branches of dead corals. The amphipod Cyparus constructs its tube home with which he can move about like the trichopteran larvae (Bernard 1976), and the Lysianassa lives under stones. The young amphipods feed on detritus, but the adult become more specialized. Among them there are filterers (Urothoe, Albunea), phytophages on benthic microalgae (Siphonoecetes), predators and a lot of commensals and parasites which live on hydroids and feed on their polyps (Podocerus), in ascidians or in sponges (Leucothoe), in the mantle of some molluscs (Cyprodinius), or in holes with polychaetes or crabs (Lizistriella). The most common reef gammarides belong to the genera *Ampelisca*, *Cymadusa*, *Gammaropsis*, *Dexamine Leucothoe*, *Listriella*, *Lysianassa*, *Lembos*, *Cerapus*, *Melita*, and *Photus*, and the common caprellids to the genera *Metaprotella*, *Dodecadina*, and *Phthilisca*.

The reef decapods include prawns and shrimps (suborder Natantia), crabs and lobsters (both from the suborder Reptantia). They include hundreds of species and comprise a major portion of crustaceans in any reef biotope by number of taxa and by their biomass. Many decapods are commensals or symbionts to various large reef animals and, primarily, to the corals themselves (Patton 1976; Bruce 1976). The reef shrimps belong mainly to the super-families Stenopodidea, Caridea and Thalassinidea. Stenopodids or the coral shrimps Stenopus, Microprosthema, Odontozona are very colorful. Some of them are commensals to actinians and corals. Caridean shrimps are extremely widespread, being an important component of the vagile benthic fauna as well as commensals to various animals: corals, hydroids, sponges, urchins, actinians, and ascidians. They belong mostly to the families Alpheidae and Hyppolithidae. Most of the free-living shrimps are crude filterers or detritophages. The ghost-shrimps thalassinds (genera *Axius*, *Scytolepthus*, *Callinassa*) live in holes which they dig out in the soft sediments. The species composition of shrimp communities varied in different reef zones. On the outer reef slope they are rare. Near the reef edge, the shrimps Brachycarpus and Odontozona are numerous. The shallows of reef flats are inhabited by shrimps and prawns from the families Pontionidae, Hyppolithidae and from the genera *Anthanas* and *Periclemenes*. The shallow parts of lagoons, covered with seagrass, are occupied by the shrimps Alpheus, Hyppolite, Thor, Saron, and the coastal beach sands by Pontophilus, Leptochela, and Metapenacapsis. The shrimps Alpheus and Callinassa build permanent holes in sand, cementing their walls with their saliva. The shrimp fauna of lagoon bottom biotopes includes commensals and symbionts of corals, alcyonaceans and sponges, as well as the free-living large detritovores Alpheus, Benaeus, Metapaenopsis, Heteropaenopsis, which are an important object of fishery (Bruce 1976).

Reef lobsters (genus *Palinurus*) are large nocturnal decapods which hide during the daytime in holes and trenches of flats and outer reef slopes. This genus of spiny and rock lobsters includes 5–7 species. Most common of them are *P. ornatus*, *P. versicolor*, and *P. penicillatus*. The last two are objects of intensive fishery (George 1974).

The reef crabs belong to the subfamilies Anomura (hermit crabs) and Brachyura (real crabs). The hermit crabs and related genera of Anomura are extremely widespread in all reef biotopes. The real hermit crabs (family Paguridae) live in abandoned shells of gastropods. They are especially abundant in benthic communities of reef flats and littorals. Most hermit crabs are active only at night, hiding in the daytime under stones. All real hermit crabs are predators feeding on polychaetes and molluscs (Thomassin 1974). The crabs Callinus and Trizopagurus live under the corals and feed on their polyps (Gillhurst 1985). Thus, they damage the tips of their branches. The mass species of pagurid hermit crabs are: *Calcinus*, *Dardanus*, *Clibanarus*, *Diogenes*, *Pagurus*, *Paguristes*, *Coenobita*, *Burgus*, and *Paguritta*. Another three families of anomuran crabs related to hermits are: Hippidae, Galatheidae and Porcellanidae. Among them there are the free-living small crabs phytophages and detritophages. Some of them are the commensals of corals (Allogalathea).

The real crabs Brachyura are the most important crustacean group in the reef macrobenthos. It lists several tens of genera and hundreds of species. Most of them are free-living crabs. They inhabit biotopes of sand and rubble, reef shallows, sandy and rock littoral zones, criptofaunal habitats in trenches and in holes of lime rocks (Taylor and Lewis 1970; Serene 1972; Thomassin 1974; Tsareva 1980; Peyrot-Clausade 1981; Jones 1984). Many brachyuran crabs are commensals, parasites, or symbionts of corals and other benthic animals. The most numerous among them are the Xanthidae. This family lists over a hundred species of reef crabs, which live mostly in cavities and trenches of reef flats, being one of the main components of the cryptofauna. Xanthid crabs feed on periphyton, on detritus and on meiobenthos. The common xanthid genera are: *Actaea*, *Paractaea*, *Eriphia*, *Leptodius*, *Euxanthus*, *Dairia*, *Atergatus*, *Pilumnus*, *Cymo*, *Chlorodiella*, *Domecia*, *Liomera*, *Caprilius*, *Xanthias*. The common commensal xanthids belong to the genera *Trapezia*, *Tetralia*, and *Phymodius*.

The spider crabs from the family Majidae inhabit mostly biotopes of flats and outer slope. Their common genera are: *Camposcia*, *Micippa*, *Tiarina*, Huenia, *Oncinopus*, and *Hyastenus*. They have a round spiny carapace camouflaged with pieces of sponges or with seaweeds. The same is done by crabs from the family Dromiidae. Their hind legs are modified, being adapted to preserving permanently the camouflage – with the colony of sponges or ascidians on their backs. The family of Portunidae includes the large swimming crabs, which are common in reef shallow biotopes. Most usual among them belong to the genera *Portunus*, *Thalamita*, *Thalamitoides*, and *Lissocarcinus*. Several species of small portunid crabs are the commensals of holothurians and alcyonaceans. The large grapsid crabs (family Grapsidae) and the box crabs from the family Calappidae (genera *Calappa* and *Matuta*) are predators and feed mainly on gastropeds. The former live under stones on reef flats, while the second prefers sand shallows, digging in during the daytime (Gibbs 1978). The same habit have the pebble crabs Phylura (family Lencosiidae). In the upper littoral zone the ghost crabs (family Ocypode) are usually numerous. They are voracious predators hunting at night on fish and shrimps and even on newly hatched young turtles. They can run at high speed and on being persued – instantly disappear, digging into the sand. Other reef crabs from the families Hapalocarcinidae, Eumedonidae, and Pinnotheridae are mostly commensals of larger benthic animals.

5.4.2 Symbiosis in Reef Crustaceans

As it was pointed out above, many crustaceans are commensals or symbionts of various reef animals. The same crustaceans are the main infaunal symbionts and parasites of corals (Mc Closkey 1970; Patton 1976; Bruce 1976; Naumov et al. 1980; Odinetz and de Forges 1985). In fact, each large colony of ramose coral Pocillopora or Acropora harbors 5 to 15 different species of symbiotic crustaceans, not to mention other infaunal animals, such as sponges, molluscs, and polychaetes. The number of specimens of crustaceans, shrimps and crabs in each such a coral colony varies within 5 to 40. Among them are the pontionid, the alpheid, and the hypolithid shrimps from the genera: *Periclimenes*, *Philarius*, *Harpilopsis*, *Fenneria*, Corallicaris, *Johaste*, *Thor*, Hymenocera, *Synalpheus*, and *Alpheus*. The infaunal symbiotic crabs belong most often to the xanthids *Chlorodiella*, *Phymodus*, *Cymo*, *Domecia*, *Trapezia* (about 20 species) and *Tetralia* (about 5 species). On the Caribbean reefs the infaunal crabs include the spider crab *Mitrax* sp., some species of the family Porcellanidae, and the crab *Quadriella*, which lives here on gorgonians (Castro 1976; Austin et al. 1980).

Three main groups can be named among the infaunal crustaceans. One of them use the coral colony as a temporary refuge, which they casually abandon, like the shrimp *Periclimenes spiniferus* and the crabs *Phymodus undulatus* and *Chlorodiella nigra*, which often inhabit colonies of the coral *Pocillopora*. The second group includes crabs and shrimps that live within the colony permanently, but do not have any specific morphological adaptations, which should distinquish them from their free-living relatives. Those are the shrimps *Periclimenes anymone*, *P. madreporae*, *Thor amboinenses*, and the crab *Chlorodiella andreoussi*. Both these groups of infaunal crustaceans include species which feed independently of their host corals, being filterers (shrimps) or feeding on mucus which the coral is shedding anyway (crabs). The representatives of the third group are specialized symbionts of corals. They have morphological adaptations for using their hosts not only as a refuge but also as a source of food. Having flattened bodies, they easily move between the coral's branches with the aid of modified legs fit for climbing over their slimy surfaces. At the end of their hind feet they have special "brushes" and "combs". Lacerating the polyps of corals with the combs, the crabs stimulate mucus excretion and then collects this mucus with the "brushes" and eats it (Bruce 1976). This symbiosis is mutualistic because the crustacean symbionts defend their hosts against the attacks of corallavorous predators (Abele and Patton 1976; Glynn 1983), as well as against the settling of the larvae of borers on the host colony. They also clean the surfaces of the colony from excess mucus and sediments. They feed only on their host's resources, eating not only mucus and zooxanthellae of the host coral, but also the tissues of coenosarc, and even snatching off the preys caught by its polyps. The symbiotic crabs form permanent couples. In one individual coral colony usually only one couple of a given species lives, which repel other crabs of the same species. The crab *Hapalocarcinus* and the shrimp *Paratypton* can control the growth of some branching corals. On their branches they form chambers in which their caged females live (Patton 1976). Their larvae settle on the tip of a growing branch, influencing then its growth so that the growing branch forms a chamber. The crab *Cryptocherius* makes "homes" in the colonies of massive corals. His larva kills polyps on settling in a coral colony and the coral stops growing in this spot. Then gradually, while it grows, a hole forms here, in which the adult crab will live as in a refuge.

The density of infaunal communities in corals appears to be dependent upon the abundance of plankton in the reef waters (Abele 1979; Kropp and Birkeland 1981). It is larger on the barrier reefs than in atoll-dwelling corals. It might depend also upon the rate of sedimentation, because the suspended matter settling on corals stimulate the production of mucus, which is a main food source of the infaunal crustaceans (Odinetz and de Forges 1985).

The biomass of infauna in hard ramose corals attains 5 to 15 g per 1 kg dry weight of the coral colony (Shirayama and Horikoshi 1982). The bulk of it compose the boring bivalves and sponges (up to 70%), while 20–50% comprise the infaunal crustaceans, and 10–20% the polychaetes. The total number of infaunal organisms in ramose corals collected at the fringing reefs off Guam Island was found to be 50 to 300 sp. kg^{-1} of coral colony. In an average coral colony (volume 0.5–3 dm^3) the number of specialized infaunal animals, according to Krasnov et al. (1982), varied within 5–30 sp. per colony, and their biomass within 0.5–4.5 g. Shannon's index (H') in the infaunal communities is within same values as found in free-living benthic fauna: 1.5–2.5. The taxonomic composition of the fauna in different species of corals is varied, even if the shape of their colonies appears to be similar (Naumov et al. 1980). But after the coral colony dies, its specialized infauna is gradually replaced by the opportunistic cryptofaunal species (Coles 1980).

Crustaceans live as symbionts or commensals with many other animals, exept corals. The crab *Anchistus custos* is the symbiont of large bivalves. He cleans their gills scraping off the pseudofeces and feeding on them. The shrimps *Alpheus*, *Hyppolite*, *Pericliclimenes*, and *Pontonia* live with sponges, hydroids, ascidians and even with nudibranchs (*Periclimenes imperator*). Several species of the shrimps *Periclimenes*, *Pedersonia*, as well as the shrimps *Hyppolyzmata grahami*, *H. californica*, *Stenopus hispida* are known as the cleaning symbiomts of fish and actinians (Limbaugh et al. 1961; Bruce 1976).

The rich fauna of benthic crustaceans is an important source of feeding for reef fish (Vivien 1973; Thomassin 1974). Of great importance as food source for various reef animals, including corals, are also their pelagic larval and young stages (cf. Sect. 9.2.1).

5.5 Reef Echinoderms

The communities of benthic echinoderms include the sea stars, the brittle stars (ophiuroides), the feathered stars (crinoids), and the sea cucumbers (holothurians). In the coral reef bottom biotopes these basic groups develop great variety while displaying some features of endemism in particular reef systems, such as the Caribbean or the Red Sea reef regions (Clark 1976). In the Indo-Pacific region they include over 1000 species, and in the Western Atlantic around 150, having only eight species common for these two zoogeographic regions. Such segregation of the echinoderm faunas in these regions is similar to the faunistic separation of their coral communities. The endemism of the echinoderm fauna of coral reefs reveals also the fact that out of 1027 of their species inhabiting reefs of the Indo-Pacific basin only 57 are common and could be found in it from one end to the other. Within a single reef system the number of species of echinoderms varies within 20–150. Thus in the Red Sea it is 48, on the Caribbean reefs about 100, on reefs of the Philippines 190, and on the Great Barrier Reef about 160 (Marsh and Marshall 1983).

The echinoderms often form dense monospecific populations in shallow reef zones, which are an important component of the reef vagile macrobenthos and play an important role in trophodynamic and nutrients regeneration (Webb et al. 1977; Huges et al. 1987). The echinoderms produce also a significant portion of lime material, thus functioning as reef builders. They occupy all the trophic niches, as filterers (ophiuroids, crinoids), detritus- and sediment-eaters (holothurians, ophiuroids), phytophages (urchins) and predators (stars and, in part, urchins and ophiuroids).

5.5.1 Sea Urchins

The urchins inhabiting coral reefs belong to four main families: Diadematidae, Ehinometridea (regular urchins), Clypeasteridae, Brissidae (irregular urchins). The total number of urchin species on reefs of the Indo-Pacific was evaluated with 140, while in particular reef systems it varied within 10 to 40: on the Great Barrier reef 26 species, on the Philippines 39. The density of their populations may attain 20–70 sp. m^{-3}, averaging 5 to 15 sp. m^{-1} (Clark 1974; Budin 1980; Fadeev and Lukin 1987; Bak 1989). Most reef urchins are nocturnal, hiding in the daytime in trenches and holes. Some urchins (Echinometra, pencil urchin Heterocen trotus) bore their holes themselves in reef lime rocks. The irregular urchins dig into sand. Most urchins are omnivores. The regular urchins feed on seaweeds and periphyton, consuming also young scleractinians, corals, and microbenthos (Bak and van Eys 1975; Sammarco 1982; Hixon 1985). Together with the periphyton and coralline algae they scrape off from the flat surfaces also the spat of corals and small gastropodes. Some diadems may feed also as sedimentary feeders from the water column. The organic particles settling on their long needles covered with mucoid film are then digested by the wandering amoebocytes, which crawl out from their porous surfaces (Thomassin 1981). The irregular heart-like urchins, which live in coral sand, feed on forams, microbenthic molluscs and worms, and also on the microphytobenthos. The daily food ration of urchin communities in the littoral area of an Atlantic reef appeared to be very large: 1–4 $g\,m^{-2}$, e.g., 1–6% of their body weight or 20–80% of a benthic plant's primary production (Ogden 1977; Hatcher 1983b).

The families of the diademid urchins and of the relative tripneustid urchins inlude the most common widespread genera *Diadema*, *Astropyga*, *Echinotrix*, *Temnopleurus*, *Mespilia*, *Toxopneustes*, *Tripneustes*, *Nudechinus*. At the Indo-Pacific reefs the most numerous species are *Diadema setosum*, *D. savignyi*, *Tripneustes pileosus*, *T. gratilla*, *Echinothrix diadema*, and *E. calamaris*. On the Caribbean reefs they are: *Diadema antillarum*, *Astropyga radiata*, *A. magnifica*, *Echinothrix calamaris*, *Tripneustes ventricosa* (Endean 1957; Clark 1976; Budin 1980; Marsh and Marshall 1983). Some mass species of regular urchins belong to other two families: Echinometridae and Sidaridae. These are: on reefs of the Indo-Pacific *Echinometra mathaei*, *E. oblonga*, *Echinostrephus aclilulatus*, *E. molaris*, *Heterocentrotus mammilaris*, *H. trigonarius*, *Eucidaris metularia*, *E. thouarsii*, and in the Western Atlantic reefs *Echinometra lucunter*, *E. viridis*, and *Eusidaris tribuloides*. The heterocentrotuses or pencil urchins have massive and extremely firm needles, which they use for boring holes in reef-flat rocks.

The irregular clypeasterid urchins and the heart-like brissid urchins have flat bodies densely covered with fine short needles. The flat body enable them to move freely inside the sediments. The common species of clypeasterid urchins on the Atlantic reefs are: *Clypeaster rosaceus*, *Encope stockesi*, *Leodia* sp., *Mellita quinquesperforata*, and *Echinoneus cyclostomus*. The latter and also *Clypeaster reticulatus*, *C. fervens*, and *Echinodiscus bisperforatus* are the most numerous also on the Indo-Pacific reefs. The mass species of heart-like urchins on the Atlantic reefs are: *Plagiobrissus grandis* and *Meoma ventricosa*, and in the Indo-Pacific: *Metalia spatagus*, *M. sternalis*, *M. dicrana*, *Brissus latecarinatus*, and *Rhinobrissus hemiasteroides*. These urchins often form very dense monospecific populations on the sandy bottom of lagoons, where 10–20 large urchins m^{-2} (Budin 1980) can be found. On sands covered with sea-grass it could be even more – up to 200 sp. m^{-2} (Fadeev and Lukin 1987).

The poisonous sharp needles defending the urchins' bodies nevertheless cannot prevent their being grazed by some specialized predators. They are eaten by diodontid and balistid fish which are steadily regulating their population densities (Ormond et al. 1989). The urchins are grazed also by large predatory gastropods (Cassis), by crabs and lobsters. The urchins, especially the diademids, consume large quantities of seaweeds, coral spat and the corals themselves, thus influencing the development and composition of whole reef benthic communities (Bak and van Eys 1975; Benayahu and Loya 1977; Ogden and Lobel 1978; Williams 1981; Sammarco 1982; Huges et al. 1987; Bak 1989).

5.5.2 Sea Stars

The fauna of reef sea stars lists over 250 species. Within a particular reef system 5 to 20 species could be found (Marsh 1974; Clark 1976). The most common reef asteroids belong to the family Oreasteridae (Budin 1980; Marsh and Marshall 1983). Among them are the blue star *Linkea laevigata*, the small motile *Linkea multiflora*, the large (over 30 cm) massive stars *Culcita coricca*, *C. schmideliana*, *C. novaeguineae*, the large stars *Pentaceraster gracilis*, *P. regulus* and *P. australis*, and related to this genus, the spiny large stars *Protoreaster nodulosus* and *P. lincki*. Some common oreasterid stars belong to the families Ophidiaster, Oreaster, Nardoa, Fromia, and Leiaster. The common asteroids Echinaster, Methrodia, Valvaster, Asterina, Luidia, Astropecten, Arhaster and well-known corallopage Acanthaster belong to several families of the order Spinulosida.

Most sea stars inhabit the hard bottom biotopes of reef flats, sublittoral and patch reefs. Some (Oreaster) prefer the lower zones of the outer reef slope (Budin 1980). Small stars could be found in cryptofaunal habitats (*Ophidiaster grandis*, *Asterina burtoni*, *Diasterina leptalacantha*). The massive large star Culcita lives on sandy shallows on the flat and in the patch-reef areas, covered with seaweeds and sea grasses.

Most stars are omnivores. When feeding, the star crawls up to the food object and extrudes its stomach and turns it inside out. The sessile organisms attached to the ground the star then envelops with its stomach and digests them on the spot (Thomassin 1976). Detritus and vagile benthic animals, forams, and gastropods the star also captures with its stomach, takes in it the food back to its place and digests it while it moves to a new feeding place. The stars *Pteraster* and *Asterina* are predators. They feed on sponges, hydroids, and ascidians. The star *Culcita*, being omnivorous, feeds also on corals prefering young colonies of species with large polyps. The damage done to corals by this star on some reefs of the Western Pacific can be quite noticeable (Endean 1971; Goreau et al. 1972). The specialized corallovorous star *Acanthaster planci* feeds only on corals. This large asteroid, called also the "crown of thorns", is armed with long poisonous thorns. These form large aggregations and cause drastic damage to the reefs. The problem of *Acanthaster* will be discussed elsewhere (cf. Sect. 11.1). The stars themselves, and especially their small-sized species, as well as their egg layings, larvae and young, are important sources of feeding for many fish, crabs and predatory gastropods.

5.5.3 Brittle Stars

The reef brittle stars, ophiuroids, are nocturnal animals. In the daytime they are hiding in their refuges: under the stones in trenches or in seagrasses. Their number in some reef biotopes attains 50 to 70 sp. m^{-2} (Finshelson

1974). The number of species of these stars varies within a particular reef system between 5 and 60. Their total number on reefs of the Indo-Pacific is close to 300, and in the Western Atlantic to 50 (Devaney 1974; Clark 1976; Hotchkiss 1982; Marsh and Marshall 1983). The brittle stars have five snake-like long slender flexible arms – tentacles lined with long ambulacral tube feet. These ambulacral feet secretes mucus and, being extremely motile, can catch particles suspended in the water column. Most brittle stars are sedimentary and filter-feeders, and can also collect detritus and dead plankton from the bottom surface. Thus they feed on plankton and hyponeuston, on detritus, and also catch small benthic animals. Their different genera have various trophic orientations. Some are pure filterers, like *Ophiothrix* and *Ophiopholis* (Roushdy and Hansen 1960). Some are feeding on hyponeustonic plankton living in the surface film, like *Ophiocoma scolopendrina*. These brittle stars live in trenches on the reef flat. When the tide is rising they anchor themselves with the aid of 1 or 2 tentacles onto the surface of flats. Another tentacle they stretch out to meet the tide current and collect detritus and plankton from the water surface. The filtering stars come out from their shelters at night and after anchoring themselves to the bottom form with their tentacles a kind of fan across the passing current, catching phytoplankton, detritus and small zooplankters (Woodley 1975). The food particles thus collected are transported along the ambulacral channel of the tentacles to the mouth (Magnus 1967; Meyer 1973a; Kissling and Taylor 1977). Some species of stars cannot filter at all, for being specialized to collect food from solid surfaces (*Ophiocomina*).

Common reef brittle stars are: Ophiocoma, Ophiocomella, Ophiomastix (family Ophiocomidae), Ophiothrix, Macrophiothrix, Ophiactis (fam. Ophiothrichidae), Ophiarachna, Ophiopholis, Ophiopeza (fam. Ophiodermatidae), Ophionereis (fam. Ophionereidae), Amphiura, Amphipholis (fam. Amphipholidae), Ophiura, Ophiolepis (fam. Ophiuridae). The trophic specialization of different species of ophiuroids makes for a zonal distribution of their taxa in reef biotopes (Meyer 1973b; Devaney 1974). Thus, on the reefs of Florida on the soft bottom overgrown with seagrass and weeds the detritovorous species *Ophiothrix erstedi* and *O. sabrisculata* were the most numerous. On the reef-flat *Ophiocoma appresum*, *O. echinata*, and *O. phiollpis impressa* were the most usual. Near the reef edge filtering stars were common: *Ophiocoma pumila*, *O. echinata*, and *O. wendtii*. In biotopes of the outer slope *Ophionereis reticulata* and in the fore-reef zone *Ophiozona impressa* and *Ophiothrix lineata* dominated (Kissling and Taylor 1977; Hotchkiss 1982). Some small ophiuroids live as commensals within the colonies of corals (*Ophiothela tigris*) or within the sponges (*Ophiocomella* sp.; Clark 1976). The largest number of brittle stars – 30 sp. m^{-2} – was recorded in the rubble biotope of an Atlantic reef (Bray 1981). Brittle stars play a quite noticeable role in reef trophodynamics, forming an important component of its biofilter.

5.5.4 Crinoids

The reef comatulid crinoids are also a component of its biofilter. Their numbers in biotopes of the outer slope could be very high – up to 50 sp.m^{-2} (Macurda and Meyer 1977). They are free-living mainly nocturnal animals. During daytime they hide in their refuges in trenches and in caves. At night, crinoids crawl out and, swimming in the water column with the aid of their mobile feathered arms, settle on some protruding element of the bottom, like colonies of ramose coral or gorgonian, or the high sponge, selecting in this way a position convenient for feeding by filtration. On being settled, the crinoid spreads its arms to form the fan or the cap, oriented across the incoming current. Some of them may turn their arms around the axis so as to adjust their ambulacral feet in position for meeting the current. With the aid of their arms they catch the detritus particles and plankton out from the passing water stream. The food particles caught are then quickly transferred along the arms to the mouth via axial ambulacral grooves (Meyer 1973b; Zmarzly and Holland 1981; Liddell 1982). Crinoids which live in areas where the current is weak, being unable to filter-feed, are adapted to collecting settling suspended matter. When feeding they spread their arms horizontally, forming a cap. The efficiency of their nocturnal feeding is enhanced by vertical zooplankton migration at night. The same manner of feeding is used by the stalked sessile deep-sea crinoids (Macurda and Meyer 1974). The number of arms in particular species of comatulid crinoids varies from 10 (*Comanthus*, *Tropiometra*, *Oligometra*) up to 200 (*Comanthus lenneti*); most of them have 20 to 40.

The number of species of crinoids in the coral-reef benthic communities of the Caribbean reefs is only eight, while in reefs of the Indo-Pacific 280. In particular reef areas of the Western Atlantic it is usually 4–5 species (Meyer 1973b; Macurda and Meyer 1977; Macurda 1982), on the Red Sea reefs up to 14, and up to 25–35 species on reefs of the GBR or of the Philippines (Marsh and Marshall 1983; Rowe et al. 1986). Most widespread are crinoids from the genera *Comanthus*, *Capillaster*, *Nemataster*, *Comaster*, *Comactima* (family Comasteridae), *Lamprometra*, *Heterometra*, *Oxymetra*, *Decametra* Family Mariametridae), *Toxometra*, and *Iridometra* (family Antedonidae). Thus at the western Atlantic reefs the common species are: *Nemataster rubiginosa*, *N. discoidea*, *N. grandis*, *Comactinia echinoidea* and also *Ctenatendon kinze* (Meyer et al. 1978). In the Indo-Pacific they are: *Lamprometra klunzingeri*, *L. palmata*, *Oxymetra finschi*, *Heterometra savignyi*, *Capillaster multiradiatus*, *Comanthus benneti*, *C. suavia*, *C. mirabilis*, *Tropiometra carinata*, and *Oligometra sirripinna*. Within the crinoid communities a definite zonal distribution was observed (Finshelson 1974). Thus, in the Red Sea reefs the shallow biotopes of flats are inhabited by the mass crinoids Lampromera and Capillaster, while at the outer slope Oligometra, Comataster, and Colobometra dwell. The crinoids usually form an important component of macrobenthos in these biotopes. Being an element of reef biofilter, they channel the organic matter input from passing oceanic waters to its bottom communities. The lime skeletons of crinoids correspond to its biogenic carbonate sediments. Many crinoids are hosts to various commensals: polychaetes, ophiuroids, small gastropods and crustaceans (Finshelson 1974).

5.5.5 Holothurians

The holothurians are the most important element of vagile macrobenthos in lagoons and in reef-flat shallows covered with sand and rubble. Some of them live under stones and boulders. They are the second object (after fish) of reef fishery. The weight of large holothurian specimens might be 1–2 kg. On average, their biomass in these biotopes varies within 50 to 500 $g\,m^{-2}$, and their numbers within 0.2–40 sp. m^{-2} (Bakus 1968; Webb et al. 1977; Levin 1979; Shelley 1985). Their maximum number has been recorded at the shallow sandy flat of the Enivetok atoll at night: 200 sp. m^{-2} of *Holothuria difficilis* (Bakus 1973). Most holothurians are sediment eaters, able somewhat to select the food particles. With the aid of their mouth tentacles they select from sand its more nutritious components: forams, detritus, microalgae and microbenthos. Therefore, the food material consumed by holothurians is enriched with carbon and nitrogen and also with the microbial biomass, as compared with surrounding bottom sediment. The main source for their feeding is the microbial biomass of the sediments. Its assimilability by holothurians was found to be rather high: 30–40% (Sorokin 1978b; Moriarty 1982). The holothurians are active bioturbators of soft reef sediments. In areas of their high density they pass the whole upper layer of sediments through their guts during a month. For example, 5 million of the holothurians *Holothuria atra* and *H. leucospilota* pass 250 000 tons of sand per year through their guts (Bonham and Held 1963). The acid excreted within their guts dissolved about 1% of sand consumed. Thus, by their average biomass of 200 $g\,m^{-2}$ they will dissolve per day 2–4 g of carbonate sediment, which is close to 10–20% of the average rates of production of carbonaceous material by corals, forams and coralline algae (Webb et al. 1977). When passing the sediment consumed through the guts, holothurians stimulate in it the development of the benthic microalgae and control the abundance of meyobenthos (Renaud-Mornant 1977). They grow very fast. An average sized holothurian grows 0.5–1 cm $month^{-1}$, increasing its weight by 15–25 g. Thus because of their high density they are often responsible for a major portion of the total benthic production in the biotopes of soft sediments, which usually dominate on reefs of their area.

The total number of species in the Indo-Pacific reefs is about 140, while within a particular reef system it varies from 10 to 60. For example, in the GBR area it is 21, and near Madagascar 48 (Hammond et al. 1985; Conand 1985). Their fauna is composed of three main families: Holothuriidae,

Stychopidae and Synaptidae. The common species of holothuriids are *Holothuria atra*, *H. leucospilota*, *H. edulis*, *H. arenicola*, *H. argus*, *H. dificilis*, *H. scabra*, *H. hilla*, *H. albiventer*, *H. coluber*, *Halodeima atra*, *H. edulis*, *Actinopyga echinites*, *A. mauritania*, *A. miliaris*, *Bohadschia argus*, and *B. marmorata*. The common stychopids are: *Stychopus chloronotus*, *S. variegatus*, and *Thelenota ananas*. The common synaptids are: *Euapta lappa*, *Synapta maculata*, and *Synaptula recta*. In soft silty sands live, buried in the sediment, the sedentary holothurians Cucumaria and Pentacta from the family Dendrotrichidae. The animals thrust their contractible mobile mouth tentacles out and so collect food material from the bottom surface.

Holothurians are scantily grazed on reefs because they are protected by their firm tunics, and many of them are poisonous for fish. Some holothurians, like *Actynopiga parvula* and *Holothuria difficilis*, on being endangered throw out culverian threads – poisonous sticky filaments, which immobilize the attacker (Bakus 1968, 1973). Therefore, their main predators on reefs are people. Their dry tunics (trepang) are sold as a delicacy on the markets of S.E. Asia and Oceania (Bakus 1973; Conand 1981). The mainly fished holothurians belong to the genera: *Holothuria*, *Thelenota*, *Microthele*, and *Actinopiga*. The production of two edible species of holothurians, *Holothuria atra* and *Actynopiga echinites*, at reef shallows off New Caledonia was estimated to be about 1 t/ha, or about 40 kg of dry meat (trepang). They grow rather fast: the linear increase is 5–10 mm, and the weight increase 15–25 g by month.

5.6 Tunicates

The sedentary tunicates, solitary or colonial ascidians, are an important component of the reef criptofauna. They live mostly in trenches, under stones or in dead corals. Their bodies – zooids – are hidden inside the firm test or tunic. It consists of tunicin – a firm material composed by the polysaccharide fibers compacted by structuring proteins and a mass of tiny carbonate spicules. In some of their colonial species many zooids are hidden inside one big common test. They are represented by the families Polyclinidae, Polyclitoridae, Corallidae, Botrillinidae, Didemnidae. In colonial ascidians united within the families Scavelinidae and Perphoridae zooids are joined within the common test by stolones only at its base. The colonial ascidians often form incrusting flat colonies which like skin spread over dead coral heads, up the walls of trenches, or fill spaces under stones. The solitary reef ascidians, which belong mostly to the families Ascidiidae, Diazonidae, Styelidae, and Pynridae, also prefer cryptic habitats, while some of their spectacularly large species could be seen also in daylight on the surface of reef rocks.

The ascidians are crude filterers. They filter out from water phytoplankton, bacteria in flakes and detritus particles (Sorokin 1978b). The stream of water driven by a ciliated epithelium enters into the pharynx. The food particles are stuck together by mucus and then swallowed.

Besides this main source of feeding the ascidians, as many other sedentary reef benthic animals, feed also autotrophically, having symbiotic algae in their tissues (Lewin et al. 1980; Kott 1982). The symbiotic genera of ascidians belong to the most common family Didemnidae (Eldredge 1966). These are: *Didemnium*, *Trididemnium*, *Diplosoma*, *Lissoclinum*, and *Echinoclinum*. The algae are situated in them extracellularly in tissues of the gastral, gill and cloacal cavities. The algae symbiotic to didemnid ascidians initially were indentified as prokariotic blue-green chroococci (Newcomb and Pugh 1975). But later it was discovered that they have another pigment composition, different from the blue-greens. They have no phycobilin, but chlorophyll *b*. So, though prokariotic, they have a light-absorbing system similar to that in the eukariotic green algae (Alberte et al. 1986). Therefore, the symbionts of didemnids were described as Prochloron, a group intermediate between the green and the blue-green algae (Lewin 1977, 1984; Lewin and Cheng 1983; Paerl et al. 1984). They have round cells of 10 to 30 μ (Kott 1982; Alberte et al. 1986; Griffith and Thinh 1987). Recently it was found that some of the dedemnids could have as symbionts also a real coccoid cyanobacterium, *Synechocystis trididemnii*, which was found also in sponges (Larcum and Cox 1988).

The content of chlorophyll *a* in symbiotic ascidians is 2–5 $mg\,g^{-1}$ of their dry weight. The ratio of chlorophyll *a* and *b* is close to 3–6 (Lewin and Cheng 1983). The rate of photosynthesis in symbiotic ascidians at midday is 2–5 $mg\,O_2\,h^{-1}\,g^{-1}$ of dry weight or about 9 $mg\,O_2\,mg^{-1}\,h^{-1}$ of chlorophyll *a*. Their gross photosynthesis is around 70 $mg\,O_2\,day^{-1}$ per 1 mg of chlorophyll *a*. At midday, photosynthesis slightly exceeds respiration. In a 24-h balance the P/M ratio is about 0.6. Thus the photosynthesis of symbiotic ascidians makes up for nearly half of their daily energy loss (Lewin and Pardy 1981; Kott et al. 1984; Olson and Porter 1985). Some species of the symbiotic didemnids may crawl with a speed of 0.5–1 $cm\,day^{-1}$, in selecting the best-illuminated place (Birkeland et al. 1981).

The total number of species of sedentary tunicates on reefs of the Indo-Pacific exceeds 200 (Eldredge 1967). Their mass genera are: *Diplosoma*, *Tridemnium*, *Didemnium*, *Lissoclinum*, *Rhopalea*, *Eudistoma*, *Aplidium*, *Ascidia*, *Perophore*, *Polycarpa*, *Pyura*, and *Molqula*. The ascidians are significant components of the reef zoobenthos, being a food source for different bottom fish, predatory gastropods, crabs, polychaetes and echinoderms.

5.7 Bryozoans

The bryozoans are a key component of the reef cryptofauna. They dwell in sheltered habitats under stones, coral colonies, in trenches, in caves on the steep cliffs of the fore-reef zone, in rubble and in dead corals on the bottom of deep lagoons (Cuffey 1973). Bryozoans are colonial tentacular animals. The skeleton of their colonies is formed by fine but firm branches consisting of mycopolysaccharide polymer incrusted with carbonate material. The size of their colonies is 2–10 cm, and that of the animal's zooids 0.1–0.3 mm. The form of their colonies varies from bushes, fans, mushroom and funnels to crusts, fine moss-like coatings and nets (Cuffey 1977; Kubanin and Murachveri 1980). When the animal's zooid is active, it displays its complex filtering appatus – a halo of tentacles covered with ciliated epithelium. They filter out detritus and phytoplankton. Some bryozoans catch small zooplankton (Bugula). The bryozoans themselves are grazed by shrimps and crabs, by some nudibranchs and some surgeon-fish (Soule and Soule 1976).

The total number of species of bryozoans inhabiting reefs exceeds 300. Within a particular reef system it could be from 30 to 200 species: in the Hawaiian reefs around 200 (Soule and Soule 1977), in the Red Sea more than 80 (Dumont 1981), on the Bahamas 40 (Cuffey 1974, 1977). Most of the reef bryozoans belong to the families Actidae and Reteporidae (order Cyclostomata). They are the calcified bryozoans. Their colonies are incrusted with calcium carbonate (Cuffey 1973). Predominating form of their colonies are thin flat crusts. Some of them have bushy colonies with firm thick branches (*Cigliscula*, *Crisina*, *Retoporella*, *Neviakopora*) or with a fine branches (*Scrupocellaria*, *Margaretta*); mushroom-like colonies are peculiar to *Lichenopora*. The most abundant bryozoan populations were found in deeper poorly illuminated zones of outer and lagoonal reef slopes, as well as in caves and in tunnels of outer slopes and in trenches and holes within the patch reefs (Cuffey 1974, 1977; Logan 1984). There dominate the incrusting forms of Cheilostomata: *Celloporaria*, *Crassimarginata*, *Parasmittina*, *Microporella*, *Schizoporella*, *Bugula*, *Canda*, and *Epistomia*. Their incrusting colonies cover the back surfaces of sponges, the walls of holes, trenches and tunnels, and the foot parts of coral colonies. In deep dark bottom biotopes the bryozoans overgrow the surfaces of molluscan shells, dead corals, rubble, and the thalli of seaweeds. The most common taxa in such biotopes are the cyclostomes *Crisia*, *Idmonea*, *Nevianopora*, *Lichenopora*. They form mostly bush-like, moss-like or mushroom-like colonies (Jackson 1984).

The bryozoans are important reef builders. On ancient reefs they were sometimes the main agents of reef growth. The tripostome bryozoans, which had massive branching colonies, were basic frameworkers in reef-like biostrome constructions during the Early Paleozoic. During the Late Paleozoic the same function of frameworkers and also the role of accumulators of carbonaceous material were accomplished by the fenestrine bryozoans. Later in the Cenozoic, their function was taken over by the scleractinian corals. Modern bryozoans also participate in the processes of reef construction, being agents of consolidation of carbonaceous material. They excrete a very sticky polysaccharide substance to glue their colonies to the rock surfaces (Soule and Soule 1976). This glue then is quickly impregnated by carbonate and forms a solid massive crust, which consolidates the surfaces of rubble, and of dead corals, and thus cements the trenches and holes in the flat and patch-reef rocks (Garrett et al. 1971; Cuffey 1977). The branching bryozoan colonies keep and accumulate sand and rubble in trenches and caverns participating also in the processes of reef-rock consolidation. The material of bryozoan skeletons is a remarkable source of carbonaceous material in bottom sediments (Cuffey 1973).

5.8 Foraminifera and Other Benthic Protozoa

One of the most specific features of reef benthic communities is an extreme abundance of foraminifera. These shelled benthic protozoans compose a large part of the microbenthos in the soft bottom biotopes and produce over 50% of carbonaceous material of the coral sands (Renaud-Mornant et al. 1971). The sedentary forams densely inhabit also the surfaces of rubble material, the leaves of weeds and grasses (Glenn et al. 1981). Forams belong to the phylum Sarcomastigophora. The cells of these protozoans live in one-chamber or multichamber shells. The latter might contain up to 100 chambers. The shells are constructed from the chitin-like material incrusted by calcium carbonate which the cell extracts from seawater. The size of the shells of most forams varies usually within 0.5–1.5 mm, but some families include one or several species of gigantic symbiotic forams, which live only on coral reefs. Such gigantic forams, like Marginopora or Alveolinella, reach 2–3 cm, which is an enormous size for unicellular protozoans. The shells of most forams are perforated. Through their pores their pseudopodia stand out. The vagile forams use them for moving, and the sessile ones for sticking to the substrate. They also serve for collecting food and for respiration. The foraminiferans feed on bacteria and algae, as well as on other benthic protozoans, and on the eggs and larvae of meiobenthic animals. The symbiotic species feed also autotrophically (Hedley 1964; Lee 1980).

The fauna of forams within a particular reef system encompasses some 40 to 120 species (Calver and Salvat 1980; Salvat and Venec-Peyre 1981; Glenn et al. 1981). Most of them belong to two suborders: Miliolina and Rotalliina. Miliolins have nonperforated multichamber shells. This family includes the above-mentioned gigantic symbiotic forams *Marginopora*, *Soritis*, *Alveolinella*, *Peneroplis*. The rotalliins have also multichamber but

Table 5.3. Number of key species of forams (N, % of total) in soft bottom sediments of different reefs (after Hirshfield et al. 1968)

Enivetok atoll		Kwarjelain atoll		Kieta Bay, Bougainville	
Genus	N, %	Genus	N, %	Genus	N, %
Marginopora+ Sorities	40	*Calcarina*	50	*Calcalina*	50
Calcarina	30	*Amphistegina*	20	*Marginopora+Sorities*	20
Amphistegina	10	*Marginopora*	20	*Amphistegina*	15
Cymbaloporetta	5	*Heterostegina*	10	*Baculogipsina*	10
Small Rotaliidae	5	Other Miliolidae	10	*Elphidium*	5

perforated shells. This group includes the most common reef forams: *Calcarina*, *Elphidium*, *Homotrema*, *Amphistegina*, *Heterostegina*, *Rotalina*. Some of these genera also contain the gigantic symbiotic forams, like *Amphistegina lessoni*, *Heterostegina depressa*, and *Calcarina* sp. (Röttger 1972; Ross 1972; Plante-Cuny 1973; Smith 1977). A single reef biotope is usually dominated by two–three common species which comprise 80–90% of the total number of forams (Table 5.3). In biotopes with seaweeds or rubble the sedentary species are more numerous. In sands large vagile symbiotic forams are dominating. Some of them may move inside the sediment. They crawl in daytime up to the light, necessary for their algal symbionts, while at night they migrate into the sediment to escape the grazers (Glenn et al. 1981). Thus, on the sandy bottom in the lagoon of the atoll Takopoto the most common forams were *Peneroplis*, *Sorities* and *Quinqueloculina*, while in sandy shallows *Amphistegina*, *Homotrema* and *Calcarina* abounded (Sournia 1976b). In the lagoon of the Scilly atoll (Tuamotu) in the biotopes of seaweeds and of sandy bottom there were found 24 and 25 species of forams respectively, of which only three species were common for both of them (Salvat and Venec-Peyre 1981). Among the common species in this lagoon were *Textularia*, *Reussella*, *Sorties*, *Elphidium*, *Bolivinia*, *Purgo*, and *Quingueloculina*. At reefs off Moorea Island 120 species of foraminifera were recorded, but about 90% of the total number of foram's cells comprised the mass species *Cymbaloporetta squamosa* (Calvez and Salvat 1980). Many common forams were found in cryptofaunal habitats in trenches, holes and caves: *Acrevalina*, *Planorbulina*, *Miniacina*, *Planogypsina*, *Homotrema*. Some forams dominated among the sediment recruits, which occupied surfaces of the lime plates fixed on the reef (Vasseur 1977; Palmieri 1985). The leaves of seagrasses are often densely inhabited by large symbiotic sessile forams, like *Amphisorus hemprici*, *Amphistegina lobifera*, and *Sorties orbiculus*. Their number may attain 100–200 sp. dm^{-2} (Zohary et al. 1980).

Most of the symbiotic reef forams belong to the families Rotaliidae and Soritidae. Their common species are: *Marginopora vertebralis* ("sand dollar"), *Amphistegina lessoni*, *Calcarina calcarea*, *Archais angulatus*, and *Spirolina arietina*. The symbiotic algae accomplish in foraminifera the same function as they do in corals. They translocate a larger portion of their photosynthates to the host foram cells. This was proved by observations of an active growth of the symbiotic foram *Heterostegina depressa*, kept in the light without any food (Röttger 1972). The symbiotic algae take part also in the process of calcification in forams, thus helping them to construct their shells. The curves of dependence on rates of photosynthesis and of calcification in the symbiotic foram *Archais angulata* were similar to those in corals (Duguay and Taylor 1978). This explains why all the gigantic forams have algal symbionts. A striking peculiarity of foraminiferan symbiosis is the wide taxonomic diversity of their symbionts. They can belong to different taxa even within the same foram family (Lee 1980). For example, in symbiotic forams from the family Soritidae there were found: four species of symbiotic dinoflagellates (*Symbiodinium microadriaticum* – zooxanthella, *Gymnodinium vertebralis*, *G. rotundatum*, *Amphidinium* sp.), two species of green algae (*Chlamydomonas provasolii*, *C. hedrei*), three species of diatoms (*Amphora* sp., *Nitzschia* sp., *Fragillaria* sp.), and even the unicellular red alga *Porphiridium*. Moreover, even the same foram species, while living in different regions, may harbor different symbionts. Thus, the foram *Homotrema depressa* from the Red Sea reefs has as symbiont the diatom *Nitzschia panduriformis*, while the same foram from the Hawaiian reefs harbor another diatom, *Nitzschia valdestrigata* (Schmaljohann and Röttger 1976; Lee et al. 1979b). In the same foram species *Amphistegina lessoni* from different regions there were found diatoms which belong even to different genera: *Nitzschia* and *Fragillaria*. Similarly, the foram *Maginopora vertebralis* contains one (out of two) species of dinoflagellates or *Chamidomonas* sp. Among forams there are also symbiotic species that harbor cyanellae – the cells of the small blue-green *Chroococcum* (Hedley 1964). Most of them live in temperate waters, but some could be met with also in reef biotopes.

It is just symbiosis which is the main factor in providing for the appearance of such fantastically sophisticated and large-shelled protozoans, which, moreover, are able to live in symbiosis with such taxonomically different species as unicellular algae (Ross 1977). The construction of their shells is extremely complex and resembles those of a greenhouse. The flat shell has on its surface hundred and more chambers with very thin walls. The roof of such a chamber is formed by the needle crystals of calcite. They are oriented along the light-stream and perform actually the function of light conductors. Each chamber harbors 10–20 algal cells. When a part of the chambers is shadowed, the algae from them migrate to the illuminated ones (Ross 1972; Lee et al. 1979a). *Marginopora* has also another kind of chambers on its surface – the nurseries. In these chambers the young cells of foram produced parthenogenetically grow. The rate of movement of *Marginopora* is 1–3 cm h^{-1}.

The symbiotic forams are one of the main primary producers of organic matter in coral sands, e.g., in the biotope, which comprises the larger part of the total area of the reef (Sournia 1976b, 1977). The chlorophyll *a* content in coral sands with abundant symbiotic foraminiferan populations is high and varies within 100 to 900 $mg\,m^{-2}$, and the primary benthic production is 1.5–4.0 $g\,C\,m^{-2}\,day^{-1}$. Per 1 cm^2 of the shell's surface, measured by the rate of photosynthesis varied from 5 in *Marginopora* to 90 $\mu g\,O_2\,h^{-1}$ in calcarines (Smith 1977). The composition of photosynthetic pigments in *Marginopora* resembles that in scleractinian corals. The assimilation number in coral sands dominated by Marginopora is 0.8–1.2 mg C/mg Chl.*a* day^{-1}, and the content of Chl.*a* is 0.3 $mg\,g^{-1}$ of dry sand.

The large symbiotic forams, being extremely efficient producers of calcium carbonate, are recognized as third in importance as a source of biogenic calcite, after corals and calcareous algae (Cushman et al. 1954). They have a very high growth rate. In young *Marginopora* it will reach 0.4 $mm\,week^{-1}$ (Ross 1972). The total rate of calcification in sand foram communities on shallows of flats or in lagoons was estimated to be within 150 to 500 $g\,CaCO_3\,m^{-2}\,year^{-1}$ (Muller 1974; Ross 1977). This is, on an average, 10–20% of the values in coral or calcarine algal communities (Smith 1983). At the Great Barrier Reef they produce 10–15% of the total biogenic carbonate. Coral sands themselves consist 50–90% of the shells of forams (Maxwell 1968; Sournia 1976b). Their content in coral sands may attain $0.6–5 \times 10^5\,m^{-2}$ with their diameter averaging 2–6 mm (Odum and Odum 1955; Sakai and Nishihira 1981). They are also one of the most important sources of food for reef benthic animals, and especially of the holothurians, benthic omnivorous fish, irregular urchins, polychaetes and shrimps.

Another important group of reef benthic protozoans is the ciliates (Devaney et al. 1987). The same as the forams many of them harbor algal symbionts. The most numerous among them are the gastrotrichs and the predatory suctorians. Available data are still scarce, nevertheless their probable role in the trophodymamics of reef benthic communities is quite obvious (Fenchel 1969; Di Salvo 1973). The taxonomic composition of ciliate communities in coral sands and also in detrital sediments was studied on the reefs of Belize and of Enivetok atoll (Devaney et al. 1987). The common genera of benthic ciliates were: *Trachelocerca*, *Tracheloraphis*, *Euplotes*, *Stylonychia*, *Prorodon*, and *Uronichia*. Some specialized peritrich and suctorian ciliates live on the surface of corals or in their vicinity and feed on mucus and on zooxanthellae, which they excrete. Moreover, the ciliate does not digest instantly the swallowed zooxanthelae, but preserves them for some time, using them as temporary symbionts, the same as do some nudibranchs (cf. Sect. 5.1.1). We have found such a commensal, the peritrich ciliate *Paramecium* sp., which lives in colonies of alcyonacean *Xenia* and feeds on its zooxanthellae. In experiments with a culture of these ciliates grown on grounded tissues of Xenia and literally packed with the living cells of zooxanthellae, we have concluded that they assimilate $^{14}CO_2$ in the light: after 1 h of exposure to light the radioactivity of ^{14}C consumed was 15 cpm $\times$ 10^2 cells, and in the dark only 4 cpm.

5.9 Insects

Among the bottom fauna of the coastal shallows and of the biotopes of reef flats covered with seaweeds the larvae of mosquitoes – chironomids – and the larvae of butterflies are often present. Their number may attain several thousands per m^2 (Robles and Cubit 1981; Cheng 1985). On exposed rocky surfaces of reef flats and of littoral areas the sea bugs Aepophilis and Omania, as well as the beetles Hyphalus could be met with (Britton 1971; Cheng 1976). In the surface film over the reef shallows the beetles water-glider Halobates and water-tiger Corixis are usually numerous (Chieng 1985). These insects and their larvae may play a significant role in the trophodynamics of the shallow's reef habitats, but data in this field are still scarce (Chieng 1985).

5.10 Distribution, Communities Structure and Abundance

The bottom fauna of coral reefs is characterized by a high degree of specific variability. It is especially varied in biotopes of the rocky flat and of the outer slope. This proves the truth of the statement that one of the main factors causing this high variability in reef fauna is the abundance of refuges which prevent grazing (Kohn and Levitien 1976). The most abundant and various benthic fauna was recorded usually in cavernous reef-flats perforated by numerous burrows and holes, and covered with stones and rubble (Sirenko 1980). The second important factor that controls the abundance of fauna is trophic conditions. Both these factors are optimal in the areas of reef flats covered with seaweeds and periphyton that are not exposed at low tide (Bakus 1969). The degree of species diversity within separate families appears to be inversely proportional to the density of their specific populations in reef biotopes. For example, in dense communities of Conidae in rocky littoral zones the species diversity was much less than in those in the biotopes of reef flats where the density of these gastropods was one order less (Kohn 1968).

The total number of species of macrobenthos (excluding corals) counted within a particular reef system varied within 250–500. But in fact it could be much more because a significant portion of the cryptofauna is underestimated. In reef benthic communities their individual groups occupy all possible econiches and use all possible sources of food: plankton, suspended

Table 5.4. The econiches of the main groups of reef zoobenthos

Mode of life, habitat	Trophic groups			
	Symbiotic phototrophes	Filterers, sedimentary feeders, planktovores	Bottom plant-eaters, detritovores and omnivores	Predators and parasites
Incrusting animals	Sponges, ascidians, forams	Bryozoans, sponges, ascidians	Forams	None
Sedentary animals	Sponges, ascidians, actinians, hydroids, tridacnis	Bivalves, sponges, ascidians, polychaetes	None	Actinians, hydroids
Borers	Sponges	Sponges, bivalves, barnacles	Urchins, sipunculids	None
Criptofaunal animals	None	Sponges, bryozoans, brittle stars, polychaetes	Decapods, gastropods, holothurians, polychaetes, sipunculids, ehiurids	Stomatopods, decapods, gastropods, polychaetes, nemeretines
Symbionts and commensals of other animals	Nudibranchs	Polychaetes, shrimps	Amphipods	Decapods[a], planarian worms[a], nudibranchs[a]
Animals buried in sand	Bivalve Corculum	Bivalves, polychaetes	Decapods, various worms, urchins, holothurians	Polychaetes, nematods, gastropods, urchins, crabs
Free-living animals	Forams	Crinoids	Ostracods, decapods, asteroids, holothurians, forams, gastropods	Decapods, asteroids urchins, gastropods

[a] Feed on mucus excreted by host.

and sedimented detritus, dissolved organic matter, various plant materials – seaweeds, grass, microphytobenthos and periphyton. A significant part of benthic fauna is represented by predators, as well as by parasites, symbionts or commensals living with other animals (Gibbs 1978), Numerous benthic animals are symbiotic with algae. A tentative classification of the main components of reef benthic communities in relation to their econiches is given in Table 5.4. It demonstrates among them an abundance of groups using external sources of energy: light and food sources which arrive to the reef from the surrounding oceanic waters, such as plankton, detritus and dissolved organic matter. Besides the involvement with external energy, benthic communities by using these external food sources introduce into the coral reef ecosystem also external nutrients – nitrogen and phosphorus – providing thus their positive balance (cf. Sect. 2.2.2). And the fact of an abundance of filter-feeders within the benthic communities themselves is an indirect proof of the abundance of plankton in reef waters (cf. Sect. 3.4).

Above, the drastic influence of grazing on the composition of reef benthic communities was also mentioned. Indeed, its stress is very high there because of a high density of predators – fish, polychaetes, gastropods, stars, urchins. In the course of evolution benthic animals developed numerous morpho-physiological and behavioral adaptations to escape grazing (Bakus 1969). These adaptations and also the availability of refuges enable them to escape grazing and also to preserve a high specific diversity (Grassle 1973). Among these adaptations there could be mentioned:

1. A high motility and a quick reaction in many free-living invertebrates and in those living in shells and tubes, which help them to escape the predators' attack.
2. Protective coloration, mimicry (nudibranchs, gastropods, crabs) and warning coloration (nudibranches, holothurians).
3. Coloration polymorphism, which prevents the imprinting of a permanent image of the prey in the predator.
4. Large size of animals decreasing the possibility of being grazed (tridacnids, holothurians, asteroids, large gastropods).
5. Defence of the body with the aid of nematocysts or poisonous culverian filaments (actinians, hydroids, holothurians, nudibranchs).
6. Firm outer skeleton (molluscs, crabs, urchins, asteroids), firm tunics (ascidians, holothurians) internal sharp spicules (sponges, octocorals), brittle body (brittle stars, crinoids).
7. A bad test, bad smell or toxicity of tissues and of covers of the body.

Of the latter kind of adaptations toxicity is extremely widespread among the reef fauna (Bakus 1969, 1981; Cameron 1976). Out of reef sponges investigated 75% were poisonous for man, and of coelenterates 100%, of annelid worms 60%, of crustaceans 75%, of urchins 54%, of fish about 80%. Such groups of sessile reef benthic animals as sponges, hydroids, and actinians are practically not subjected to predators' attacks, the same as many well-defended vagile animals, which even do not hide in the daytime (asteroids, holothurians, some urchins, and nudibranchs). Nevertheless, some predators learned to consume even defended and poisonous preys. Thus, some ten species of fish eat the well-defended urchins *Diadema*, and the large gastropod Charonia successfully attacks the star Acanthaster fortified with poisonous thorns. The nudibranchs graze on the poisonous gorgonaceans.

Shannon's index of species variability in main taxonomic groups of the macrobenthos is usually high: 1.5–2.5. It is highest in biotopes richest in various refuges: on reef flats, in rubble, and in rocky littoral zones (Sheppard 1984). Nevertheless, the dominance in some biotopes of one or several leading species could also be observed (Rostomov 1987). Regarding the distribution of the biomass of benthos, generalizations are still difficult, not only because of the methodological obstacles mentioned above and of a general lack of data, but also because of the patchiness of its distribution and the big differences in it by day and at night. The number of different groups of benthos, such as urchins, brittle stars or gastropods, in different biotopes of the same reef or in the same biotope of different reefs may fluctuate within 2–3 orders of values (Tables 5.5, 5.6). For example, the number of gastropods in the stony littoral biotopes will often reach $1–2 \times 10^3$ sp. m^{-2}, while on the reef flat it is usually 1–2 sp. m^{-2} (Grassle 1973). Moreover, different authors used to express their data on biomass in different units: in body carbon (Lee et al. 1975), in dry organic matter of tissues (Odum and Odum 1955), or in raw weight including the skeleton (Salvat 1971; Ricard and Salvat 1972; Rostomov 1987). In Table 5.7 the

Table 5.5. Biomass (g m^{-2}, wet weight[a]) of main components of macrobenthos in reef-flat biotopes of the Enivetok atoll (modified from Odum and Odum 1955)

Components	Biotopes		
	Reef crest zone, rubble	Rocky windward flat, coralline algae	Deeper backward zone of windward flat, sand, corals
Urchins	187	0.5	31
Holothurians	0.5	1	200
Stars and ophiuroids	0.5	17	18
Molluscs	0.5	5	0.7
Worms	81	40	117
Crustaceans	58	3	0.5
Actinians	0.5	22	0.5
Sponges	0.5	0.5	0.5
Total benthos	328.5	89	368.2

[a] Calculated from dry weight given in the original paper, taken as composing 20% of wet weight.

Table 5.6. Biomass ($g\,m^{-2}$) of components of macrobenthos in rocky littoral zone of reefs off islands near the coast of southern Vietnam (modified from Gulbin et al. 1987)

Components	Namsu Islands	Thu Islands
Sponges	3.6	29.2
Worms	4.8	1.0
Crustaceans	21.2	55.2
Molluscs	796.2	20.7
Echinoderms	8.9	–
Total macrobenthos	834.0	106.1

available data on the biomass of macrobenthos, expressed in units of wet weight, are presented. Its lowest values (3–30 $g\,m^{-2}$) were recorded in biotopes of soft bottom – coral sands. The low biomass here is largely a consequence of the steady grazing of medium-sized animals by predators, by fish, and by gastropods and urchins. In soft bottom biotopes with rubble and dead corals, where the latter could serve as a substratum for sedentary fauna of sponges, bivalves, and hydroids, the biomass of benthos is 5–10 times more and could reach 1 $kg\,m^{-2}$ (Rostomov 1987). But in the same areas influenced by intensive terrigenous siltation the benthic biomass is usually low because of the inhibition of the bottom fauna by the sulfate reduction developing in such reef sediments (cf. Sect. 4.1). In soft bottom biotopes overgrown by seaweeds and grasses the biomass of benthos is larger, moreover coral sands and rubble contain an abundant meiobenthos, which includes small crustaceans, nematodes, polychaetes, and micromolluscs (Riddle 1988). Its specific production rate is one order more than in communities of macrobenthos: 15–30 $year^{-1}$ opposed to 1.5–3 $year^{-1}$ in macrobenthos. The biomass of benthos in the soft bottom biotopes of atolls is less than that in barrier reefs because of deficiency of detritus in the former (Salvat and Renand-Mornant 1969; Salvat 1972; Villiers 1988).

While their standing stock of benthic biomass is relatively low, soft bottom biotopes are still a main place of feeding for many bottom fish. They can graze up to 3% of it day^{-1}. But because of the dominance of medium-sized and small animals, which grow fast, grazing losses appear to be quickly compensated here. Thus, the production of benthos in soft bottom biotopes is high, and it could be even more than in biotopes of reef flats where the standing stock is high but the slow-growing cryptofauna dominates.

Relative to the biomass of meiobenthos data are still scarce. Studies of meiobenthos in coral sands of the Davis reef, GBR, performed by Hansen et al. (1987) within the program "Mercor", showed that the most abundant meiobenthos dominated by harpacticoid copepods and nematodes could be found in sediments of reef-flat and patch-reef zones. The number of meiobenthic organisms ranged there between 100–150 × 10^3 sp. m^{-2}. In lagoonal coral sands it was about five times less. According to Villiers

Table 5.7. The biomass of coral reef macrobenthos

Reef	Biotope	Biomass, $g\,m^{-2}$ wet weight	Reference
Virgin I.	Sandy shallows	108–384[a]	Lee et al. (1975)
Bermuda	Sands in lagoon	2.0–6.8	
Barrier reef off Moorea I.	Sands in flat and in coastal lagoon (average)	34	Ricard and Salvat (1972)
Enivetok atoll	Shallows at windward flat	89–368[b]	Odum and Odum (1955)
Namsu I., Vietnam	Rocky littoral zone	834	Gulbin et al. (1987)
Thu I., Vietnam	same zone	106	Gulbin et al. (1987)
Thu I., Vietnam	Sandy littoral zone	83	Gulbin et al. (1987)
Anthoy I., Vietnam	Crude sands in fore-reef zone, 10–12 m depth	170	Rostomov (1987)
Tohtu I., Vietnam	Rubble and crude sand, same zone	3220	Rostomov (1987)
Kondao I., Vietnam	Silty sands, same zone	95	Rostomov (1987)
Moruroa atoll, Tuamotu	Coral sands	0.04–10.0	Villiers (1988)

[a] Calculated using data on carbon content in biomass, including corals.
[b] Excluding large asteroids and holothurians.

Table 5.8. The numbers (N, sp. $60\,cm^{-2}$) of basic elements of meiobenthos in coral sands on the reefs of Tulear (Thomassin et al. 1976)

Element	N	Element	N
Nauplii	185–326	Nematods	185–326
Harpacticoid copepods	340–1085	Polychaetes	26–75
Ostraeods	1–28	Total meiobenthos	1050–2440

(1988), the dry ash-free biomass of meiobenthos in soft sediments of the lagoon in the Moruroa atoll was within 80–620 $mg\,m^{-2}$, up to 10^5 sp. m^{-2} in number. An indirect evaluation of it was based upon the biomass of nematods in coral sands (Grelet 1985), and upon the average weight and average number of the main groups in meiobenthos (Thomassin et al. 1976, 1982). If the nematodes comprise about half of the medium-sized meiobenthic animals, its total biomass in coral sands of the Red Sea reefs at depths of 3–10 m should be 2–8 $g\,m^{-2}$, or 20–40% of the biomass of macrobenthos. The same values could be deduced from the total number of meiobenthos in coral sands on reefs of Tulear: $200–400 \times 10^3\,m^{-2}$ (nauplii excluded). The production of the meiobenthos should be then 1.5–2 times more than that of the macrobenthos.

The main components of meiobenthos of coral sands are crustaceans, worms (Table 5.8), and also the micromolluscs (Kay 1971). In contrast to most macrobenthic animals, the meiobenthos consists of nonpoisonous organisms which have a high nutritive value, and so are an important source of feeding for many reef fish (Bakus 1969; Chardy et al. 1988).

The communities of the rocky littoral zones are less stressed by grazing. Their biomass will often reach hundreds of grams m^{-2}. They are dominated by crabs and molluscs. On sandy littoral biotopes the biomass of benthos in usually less: 10 to $100\,g\,m^{-2}$ (Table 5.7). In the rocky biotopes of the reef flat and the outer reef slopes a large portion of the benthos hides in the cryptofaunal refuges, and so even a tentative quantification of it is actually very difficult. A possible approach to solving this problem might be the calculation of the total biomass based on estimations of its average values in particular microhabitats with parallel evaluations of their relative areas (Brander et al. 1971; Hutchings 1974). The biomass values thus evaluated in these biotopes varied within 100 to $400\,g\,m^{-2}$. Taking into account the methodological problems mentioned, even these high values might be underestimates. Among the benthic fauna the dominating groups here were most often urchins, polychaetes and decapod crustaceans. The latter two are an important source of food for nocturnal benthophagous fish (Bakus 1969).

6 Coral Reef Fish

6.1 Introduction, Diversity, Taxonomy

The degree of species diversity as well as the density of communities of reef fish are unknown in any other aquatic ecosystem of our planet (Marshall 1965; Ehrlich 1975, Ehrlich et al. 1977). Coral reefs harbor about a third of all species of bony fish. The biomass of fish on reefs could attain $2\,t\,ha^{-1}$. This raises the question how does the reef ecosystem feed such a large mass of fish, which, being a final trophie link, have a low efficiency in using the consumed food for growth (some 0.1–0.15). It is quite obvious that these phenomena are extremely important from the point of view of general biology and general ecology (Odum and Odum 1955; C.L. Smith 1978).

The group of coral reef fish comprises such species as permanently live on the reef or in its vicinity and use its trophic resources (Goldman and Talbot 1976). The ichthyofauna includes also some fish that are spending on reefs only a part of their life cycles and could be found there only as juveniles (Vivien 1973). Some species of reef fishes could be also met with outside of reefs areas in coastal biotopes among rocks and macrophytes, which give them refuge. The extremely high species diversity of reef fish was the result of their long evolution, which proceeded together with the evolution of the reefs themselves. The reefs of the Mesozoic were inhabited by ancient acanthopterigian fish (Smith and Tyler 1972). The differentiation and specialization of the bony fish actually started after the appearance of scleractinian corals with the beginning of their building up large and diverse reef constructions (Gosline 1971). The most ancient among reef fish descended from the family Beryciformes. An intensive evolution of reef fish went on during the Eocene (50×10^6 years B.P.) on the basis of the development of ancient perches (order Perciformes). The families thus formed are still dominant among reef fish, comprising about 60–70% of their total species. The main genera of reef fish also existed 15×10^6 years B.P. (Patterson 1964). The most ancient among modern reef fish are the representatives of the families Holocentridae, Gobiidae, Pomacentridae, Serranidae and Chaetodontidae. Other families of reef fish appeared in reef environements relatively recently. Being pelagic, they changed their morphology, size and coloration and adapted to living on reefs and to using

Table 6.1. Number of taxa of reef fish in samples collected on various reefs

Location	Total number of species	Number of families	Average number of species per family	Reference
Society Island	255	43	5.9	Smith (1978)
Bahamas	181	46	3.9	Smith C.L. (1978)
Virgin Island	125	44	2.8	Smith and Tyler (1972)
GBR, Australia	850	84	10.0	Sale (1980)
Marshalls	233	56	4.2	Hiatt and Strasburg (1960)
Barrier reef, Tulear	231	52	4.4	Vivien (1973)

local food sources. Among such newcomers are representatives of the families Clupeidae, Bothidae, Synodontidae, Antennariidae. The signs of evolutional adaptations of reef fish to the specific reefal environments are clearly visible in their coevolution with other reef organisms, with which they are interconnected by relations of "prey-predator", of commensalism, or of symbiosis (Losey 1972; Ehrlich 1975; Reese 1977).

The results of the process of evolution of reef fish is displayed in their breathtaking taxonomic diversity. Indeed, samples of fish fauna, obtained with the aid of rotenone poisoning of some part of the reef, could contain 150–250 species (C.L. Smith 1978). The total number of species of reef fish is within 6000 to 8000, which is about one-third of the total number of known species of bony fish. The average number of species per family in their communities is between 3 and 10 (Table 6.1) while in other marine ecosystems it exceeds 50. Compare these numbers (3 to 10) with those of butterflies which resemble reef fish in their diverse morphology and coloration: 3000 species per family (Ehrlich 1975). The above numbers reveal an impressive measure of species diversity in reef fish.

From Table 6.2 it is seen that the total number of reef fish species is largest in the central part of the Indo-Pacific region – in reefs off the Philippines – the same, by the way, as that of corals (cf. Sect. 7.1). The degree of taxonomic diversity of reef fish fauna does not depend upon the latitude of the place (Sale 1980; Talbot and Gilbert 1981). All in all, it includes over 100 families out of a total of 400 families of bony fish. One of the most specific features of reef fish is their small size – 5 cm on the average. In most of their species it varies within 1 to 10 cm (Smith 1978). The leading most common families of real reef fish include 30 to 100 species. Among these are: Chaetodontidae (ca. 100 species), Scaridae (ca. 80 species), Apogonidae (ca. 30 species), Pomacentridae (ca. 70 species), Acanthuridae (over 40 species), Gobiidae (over 30 species), and Serranidae (over 30 species). Among the leading taxa of specialized reef fish there are representatives of several families which live also outside of reefs. Among these are: Labridae (60 species on reefs), Muraenidae (over 20 species on

Table 6.2. Total number of species in the ichthyofauna of coral reefs in different geographic regions; after: Gosline (1971); Goldman and Talbot (1976); Sale (1980)

Region	Total number of fish species
Philippines	2200
New Guinea	1700
GBR, Australia	1500
Seychelles	880
One Tree Reef, southern GBR	850
Marshalls and Marianas	670
Alligator reefs, Florida	600
Bahamas	510
Venezuela	490
Hawaii	450
Barbados	310
California bay	260
Mafia Arch, eastern Africa	240
Fanning atoll	230
Johnston atoll, Pacific Ocean	186
Western coast of Florida	130

Table 6.3. Scale of relative abundance of fish specimens from 12 most common families of reef fish out of a total of 52 families discovered in samples collected at the reef flat of the Great Reef of Tulear, Madagascar (after Vivien 1973)

Family	Number of species in family	Total number of specimens of a given family found in samples
Muraenidae	14	144
Serranidae	10	65
Apogonidae	12	1036
Chaetodontidae	10	137
Pomacentridae	28	2256
Labridae	31	382
Scaridae	6	165
Eleotridae	10	104
Gobiidae	10	108
Salariidae	6	82
Acanthuridae	6	187
Scorpaenidae	13	71

reefs), and Scorpaenidae (over 15 species on reefs). Many fish families are represented on reefs by one or by several species.

An example of the domination hierarchy of fish families in collections from the Barrier Reef of Tulear, Madagascar, is given in Table 6.3. It shows

Table 6.4. Number of species in common families of reef fish found in explosion samples collected at a section across the central part of the GBR (after Williams and Hatcher 1983)

Family	Number of species in family
Pomacentridae	57
Labridae	46
Chaetodontidae	28
Gobiidae	21
Serranidae	22
Apogonidae	17
Holocentridae	16
Scaridae	14
Acanthuridae	10
Rest of 37 families	92
Total number of species recorded	323

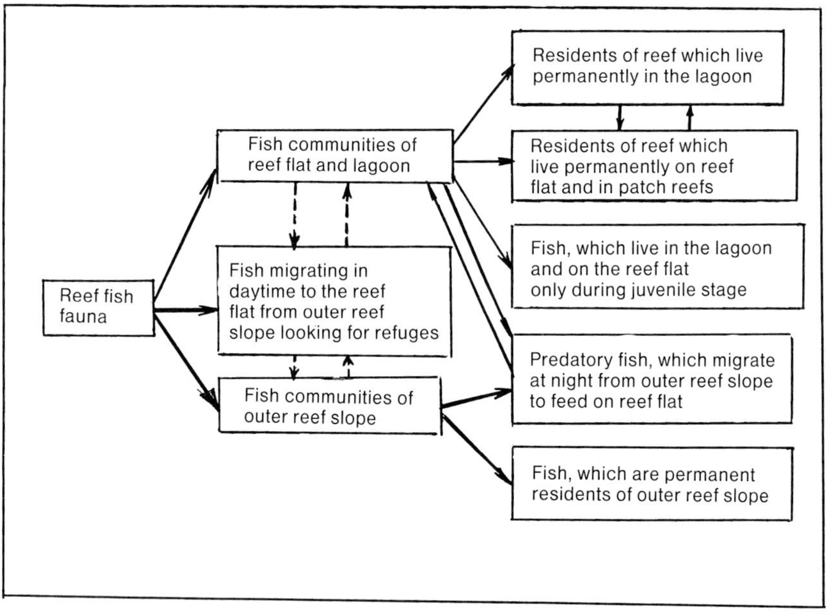

Fig. 6.1. Scheme of relationships of main ecological groups of reef fish: *solid arrows* feeding migrations of fish; *dotted arrows* daytime migrations of fish to their refuges. (After Vivien 1973)

that the dominating ones were the Pomacentridae and the Apogonidae – two families of real reef fish. The total number of specimens from these three families was three times more than the joint number of fish from the remaining 50 families. In the fauna of fish from the central GBR more than 70% of the total was represented by species from nine families of typical reef fish (Williams and Hatcher 1983). The total number of species found there at one cross section alone was 323 (Table 6.4). The communities of reef fish were subdivided by investigators into several ecological groups united by common features of feeding and behavior (Vivien 1973; Ehrlich 1975). The scheme (Fig. 6.1) shows the composition of such groups, their relationships and dynamics, caused by migrations. Such a subdivision of reef fish communities into ecological or trophic groups helps to understand better their complex structure (Hiatt and Strasburg 1960; Randall 1967). The list given below shows the taxonomic composition of such trophic groups (Table 6.5).

Table 6.5. Taxonomic composition and food spectra of the main trophic groups of reef fish

	Family	Common genera	Food spectrum
Planktonovorous fish	Atherinidae (hardheads)	*Dentatherina, Hypoatherina, Stenatherina, Atherinomorus, Pranesus*	Zooplankton, small zoobenthos
	Clupeidae (herring)	*Harengula, Jenkinsia, Sardinella, Allanetta, Spratelloides*	Zooplankton, small zoobenthos
	Hemirhamphidae (garfish)	*Hemirhamphus, Hyporhamphus, Euleptorhamphus*	Zooplankton, phytoplankton, other marine plants
	Cirrhitidae (hawkfish)	*Amblycirrhites, Cirrhites, Paracirrhites*	Zooplankton
	Caesionidae (banana fish)	*Caesio, Pterocaesio*	Zooplankton
	Syngnathidae (pipefish)	*Corythoichthys, Doryrhamphus*	Zooplankton
	Congridae (garden eels)	*Nytstactichthys, Taeniconger, Gorgasia, Haliophis*	Zooplankton
Benthophages	Labridae (wrasse)	*Cheilinus, Halichoeres, Epibulus, Pseudocheilinus, Stethojulis, Thalassoma, Labroides, Pseudolabrus, Bodjanus, Coris, Macropharingodon*	Zoobenthos, forams, algae
	Sparidae (silver breams)	*Acanthopagurus, Chryzophrys, Argyrops, Rhabdosargus*	Zoobenthos
	Apogonidae (cardinals)	*Apogon, Apogonichthys, Cheilodipteris, Fowleria, Neamia*	Zoobenthos, demersal zooplankton fish
	Pempheridae (sweepers)	*Parapriacanthus, Pempheris*	Zoobenthos, demersal zooplankton
	Holocentridae (squirrel fish)	*Holocentrus, Myripristis, Adioryx, Neoniphon, Sargocentron*	Zoobenthos, demessal zooplankton
	Mullidae (goatfish)	*Mulloidichthys, Parupeneus, Upeneus*	Zoobenthos
	Eleotridae (gudgeons)	*Valencienna, Xenisthmus*	Meiobenthos
	Balistidae (triggerfish)	*Balistes, Rhinecanthus, Melichthys, Xantichthys, Abalistes Balistapus, Sufflamen, Odonus*	Zoobenthos, demersal zooplankton

Table 6.5. *Continued*

	Family	Common genera	Food spectrum
	Diotontidae	*Diodon, Tragulichthys*	Zoobenthos
	Tetraodontidae (pufferfish)	*Arothron, Canthigaster, Lagocephalus, Sphaeroides*	Zoobenthos
Bottom omnivore fish with wide food spectrum	Pomacentridae (damselfish)	*Aboefduf, Chromis, Microspatiodon, Pomacentrus, Eupomacentrus, Dascyllus, Stegastes*	Zoobenthos, coral, algae, periphyton, zooplankton, small fish
	Gobiidae (gobies)	*Dobiodon, Signilobis, Paragobiodon, Amblyeliotris, Ctenogobius, Bathygobius, Callogobius, Eviota, Priolepis, Valencienella, Gnatholepis*	Zoobenthos, coral, algae, periphyton, zooplankton, small fish
	Chaetodontidae (butterfly fish)	*Chaetodon, Chelemon, Forcipiger Heliochus, Parachaetodon*	Corals, periphyton, zoobenthos
	Pomacanthidae (angelfish)	*Centropyge, Chaetodonoplus, Pomacanthus, Pygoplites, Holacanthus*	Corals, periphyton, zoobenthos
	Zanclus (moorish idols)	*Zanclus*	Zoobenthos, corals, macroplankton
	Ostraciontidae (box fish)	*Ostracion, Triorus, Lactoria, Lactophris*	Macrobenthos, periphyton, algae, detritus
	Hemiscyllidae (carpet sharks)	*Chilosyscyllum, Hemiscillus*	Zoobenthos, dead animals
	Blennidae (blennies)	*Meijacanthus, Ecsenius, Aspidonotus, Salarius, Pteroscriptes*	Zoobenthos, small fish, periphyton
Herbivorous fish	Acanthuridae (sturgeon)	*Acanthurus, Prionurus, Naso, Zebrasoma, Paracanturus, Ctenochaetus*	Macrophytes, periphyton, zoobenthos, gelly zooplankton
	Scaridae (parrot fish)	*Scarus, Sparisoma, Hemiglyphiodon, Cetoscarus, Hipposcarus, Calatomus*	Macrophytes, periphyton, coralline algae, corals
	Mugilidae (mullets)	*Crenimugil, Liza, Myxus*	
Piscivorous scavengers	Carcharinidae (whaler sharks)	*Carcharinus, Triaenodon, Negaprion, Rhizoprionodon*	Fish, dead animals
	Orectolobidae (wobbegong)	*Orectolobus, Eucrossorhinus*	Fish, zoobenthos
Wandering piscivores	Ginglymostomatidae	*Nebris, Gynglimostoma*	Fish
	Stegostomatidae (zebra sharks)	*Stegostoma*	Fish
	Lethrinidae (emperors)	*Lethrinus, Gymnocranius, Monotaxis*	Fish, crustaceans
	Haemulidae (grunters)	*Plectrorhynchus, Diagramma,* Pomadasys.	Fish, crustaceans
	Luthjanidae (sea perch)	*Luthjanus, Aprion, Gymnocarnis, Scolopsis, Symphorus*	Fish, zoobenthos, zooplankton (*Caesio*), phytoplankton (*Macolor*)
	Sphiraenidae (barracudas)	*Sphiraena*	Fish
	Aulostoniidae (trumpet fish)	*Aulostomus, Fistularia*	Fish
Wandering pisciuores	Carangidae (jacks)	*Trachuropsis, Caranx, Elagatis, Oligoleptis*	Fish

Table 6.5. *Continued*

	Family	Common genera	Food spectrum
	Ogcocephalidae (batfish)	*Ogcocephalus*, *Platax*	Fish, zoobenthos
	Priacanthidae (bullseyes)	*Priacanthus*	Fish, zoobeuthos
	Belonidae (needlefish)	Tilosurus, Strongylura	Fish
Bottom piscivorous ambushers	Serranidae (groupers)	*Promicrops*, *Cephalopholis*, *Plectropomus*, *Epinephelus*, *Mycteroperca*, *Anthias*, *Rainfordia*	Fish, zoobenthos, zooplankton
	Scorpaenidae (scorpion fish)	*Pterois*, *Alabys*, *Cottopistes*, *Erosa*, *Scorpaena*, *Scorpaenopsis*, *Synahceia*, *Scarpaenoides*	Fish, zooplankton
	Synodantidae (lizard fish)	*Synodus*, *Saurida*	Fish, crustaceans
	Bothidae, Soleidae (flounders)	*Bothus*, *Aserreggodes*, *Pardachirus*, *Arnoglossus*, *Pseudorhombus*	Fish
	Muraenidae (moray eels)	*Muraena*, *Echidna*, *Lucodontis*, *Gumnothorax*, *Rabula*, *Uropterygius*	Fish, zoobenthos (*Echidna*)
	Ophichthydae (snake eels)	*Myrichthus*, *Leucoranus*, *Ophichthys*	Fish, zoobenthos

6.2 Communities Structure, Abundance, Distribution

Among the main features of community structure of reef fish there should be mentioned their exclusively high taxonomic diversity and density (Ogden and Ebersole 1981). An increase in species number in a highly productive ecosystem leads to an increase in the ways of using primary production in the food web, increasing thereby the ecological efficiency of utilizing food sources. In an ecosystem with higher species diversity secondary production will be higher by same level of primary production (Margalef 1968; Goldman and Talbot 1976). Just this is the situation we find in coral reef ecosystems, where the biomass of fish alone attains 100–200 $g\,m^{-2}$ (cf. Table 6.7).

The degree of species diversity in a fish community is usually expressed by means of the Shannon-Winer index N′ (C.L. Smith 1978; Gladfelter 1980). In fish communities of most known types of reef biotopes it is extremely high. Moreover, its values seem to be similar in similar types of biotopes of geographically quite different reefs, even when the density of fish communities differs in them several times. For example, this index was equally high in fish communities of patch-reef biotopes at the Bahamas (H′ = 3.4–5.2) and of some of the Society Islands in the Pacific, while the total numbers of species in them differen 1.5–2 times (C.L. Smith 1978). These results proved the conclusion on a high degree of predictability in reef fish communities because of the existence of common regularities which influence their formation and composition. Experiments with local extermination of reef fauna on particular patch reefs with the aid of ichthyocide rotenone demonstrated that after their subsequent colonization by first fauna its composition was almost completely rehabilitated. These results were used as a proof of a deterministic character of the community structure of reef fish ("order theory"; Smith and Tyler 1972; Gladfelter et al. 1980).

But the theory of "order" has its opponents (Sale 1978b, 1980, 1989; Sale et al. 1984; Talbot et al. 1978), who insist that the structure of communities of reef fish has stochastic character (theory of "chaos") and therefore it is far from being predictable. They propose that a high mortality caused by grazing and other factors permanently renews the living space in reef biotopes for its recruitment of young fish. This process of recruitment can have only stochastic character, depending on the availability of accidental fish larvae. Detailed analysis of both theories, employing data on turnover in fish populations (Bohnsack 1983), did not prove the "chaos" theory, which appeared to be "working" only on relatively short time scales (Hunte and Côte, 1989). The theory of "order" was proved by the results of long-term observations on community structure and its temporal fluctuations (Thresher 1977), as well as the re-results of a 20-year observation on the colonization of the artificial reef constructed among natural reefs off one of the Virgin Islands in 1960 by J. Randall. Within 2 months at this "reef" a fish community had been formed which did not change much during the following years either taxonomically or quantitatively. In 1962 it was inhabited by 39 species of fish from 14 families, and in 1979 the respective numbers were: 40 species from the same 14 families, showing the same level of species diversity and the same number of dominating families (Ogden and Ebersole 1981).

The deterministic character of the community structure in reef fishes results from their high degree of specialization in the use of various reef resources, provided by their fine specific morphological and physiological adaptations. This factor controls the survival of recruites and thus provides the relative similarity of succesive ones in similar reef biotopes on a long time-scale (C.L. Smith 1978; Anderson et al. 1981). The main factors influencing the species composition of fish communities and their density include space and the morphometry of a biotope which make it available for potential refuges (Thresher 1977; Sale 1978b; Greenfield and Greenfield 1982) and also its trophic resources (Williams and Hatcher 1983). The above described determinism in the formation of reef fish communities could surely be influenced by stochastic elements during the formation of fish fauna in some reef biotopes at in different seasons of the year. This may cause their fluctuations. One of the causes of these fluctuations could also be the participation of a less specialized and more opportunistic species in the formation of fish communities. It happens especially often in biotopes of the outer reef slope where the reefal fish communities come in contact directly with those of the open sea. The recruitment of a less specialized and more opportunistic fish species could involve definite stochastic features in the formation of reef fish communities, which was detected by the author of the "chaos" theory. Some stochastic elements are introduced as a consequence of an intensive grazing of reef fish fauna, resulting in its high-speed turnover in some reef biotopes – and especially in biotopes of lagoon and outer slopes (Talbot et al. 1978).

The species composition of different biotopes within the same reef differs much more than that in similar biotopes of different reef systems (Hiatt and Strasburg 1960; Hobson 1974; C.L. Smith 1978; Williams and Hatcher 1983). It is largely determined by the landscape of the bottom, by depth, by their zonal position within the geomorphological system of the reef, and by the abundance of living corals (Chave and Eckert 1974; Goldman and Talbot 1976; Smith 1977; Galzin 1985). Most species of reef fish are living only within definite types of structural reef elements (Sale 1978a). For example, out of a total number of fish species inhabiting eight different biotopes of the One Tree Reef only 7% was common. But only about 5% out of all fish species found in each of these biotopes was met with in this given biotope and was absent in others (Goldman and Talbot 1976). The changes in composition of reef fish fauna in the GBR in different kinds of biotopes appeared to be much larger than between the same types of biotopes, situated between 11° and 22° S (Table 6.6). On one of the cross-sections out of the total number of 323 species 60% were recorded on only one out of three reefs, and only 7% was found on all three reefs (Williams 1983). Species diversity was increasing along the cross-sections from the coast to the outer reef slope. Significant changes in fish fauna could be also observed along the perimeter of the ring reefs of the GBR on their windward and leeward parts (Done 1982).

These observations demonstrated that the local communities of reef fish are largely isolated and practically do not mix, each being associared with a definite type of reef construction and reef zone. This feature of settling is peculiar only to the communities of reef fish, and is absent in coastal fish communities outside of reefs. On reefs each of their structural elements has its owner (Smith 1978). Thus it could be deduced that the density of reef fish communities is controled largely by space, which appears to be a more important factor than trophic resources and grazing pressure. The availibility of space within reef constructions regulates the conditions of hunting, hiding and propagation of fish. The species diversity of reef-fish communities therefore depends upon the complicacy of reef constructions (Hobson 1974; Hatcher 1981). For example, in reef areas covered with ramose corals the number of species is much higher compared with places of poor coral growth (Harmelin-Vivien 1977; Carpenter et al. 1981).

Table 6.6. Parameters of species diversity, total number of species and biomass in fish communities inhabiting three different biotopes. Data were obtained by analyses of explosion samples containing over 28 000 fish specimens (after Williams and Hatcher 1983 and Williams 1983)

Biotope	Species diversity		Trophic structure (% of total biomass of fish)		Parameters of average weight (W), number (N) and biomass (B)		
	Total number of species	Shannon's index of diversity (H')	Planktonovorous fish	Herbivorous fish	W, g sp.$^{-1}$	N, sp. m^{-2}	B, g m^{-2}
Coastal reef	110	1.56	9	46	12.7	7.2	92
Patch reef in the lagoon	201	2.38	11	71	27.8	5.6	156
Outer reef slope	163	2.98	29	43	48.4	4.9	237

Because of relative stability of space and trophic sources in reef biotopes the local fish associations, and even their diurnally migrating shoals, have a constant composition. The resident adult fish, which occupy permanent places in the biotope, enjoy a preference before newcomers and actively attack them. These strict residents provide the stability of species composition in fish communities in a particular reef biotope. Among them there are species from the families Pomacentridae, Eleotridae, Gobiidae, Muraenidae, and Seorpaenidae, which accomplish the function of "core species" in the community. Less strict residents, like labrides, scarids, serranids, fistulariids, and acanthurids, may use various kinds of substrate as their permanent or temporary refuges. These fish create a diversity of communities in similar reef biotopes.

High species diversity and fine specialization in the use of trophic resources by reef fish communities result in their unusually high density. In fact, as it is shown by the still not very abundant data summarized in Table 6.7, their biomass is expressed by values of 0.3 to 2.3 tonnes ha^{-1}, or 30 to 230 g m^{-2} (Stevenson and Marshall 1974). These values exceed by one order or more corresponding values in the temperate coastal ecosystems. The largest numbers were obtained by rotenon poisoning on windward reefs of the GBR: 1.95–2.37 tonnes ha^{-1} (Table 6.7). The basic stock of biomass is most often built up by common species of several planktonovorous, herbivorous and omnivorous fish from the families Caesionidae, Acanthuridae, Scaridae, and Pomacentridae (Table 6.8). The total number of fish in reef biotopes is also extremely high, attaining 7–140 sp. m^{-3} or 30–600 sp. m^{-2} (Table 6.9). On an average, it is usually within 5 to 50 sp. m^{-2}, with an average individual weight of fish within 2 to 25 g (Table 6.10). On artificial

Table 6.7. Biomass of fish (B) on coral reefs in different regions

Location, biotope	B, kg ha^{-1}	Reference
Virgin I.; fringing reefs	1590	Randall (1963)
Virgin I.; artificial reefs	17400	Randall (1963)
Bermuda; patch reefs	500	Bardach (1959)
Red Sea, fringing reefs	350	Goldman and Talbot (1976)
Hawaii; patch reefs	1850	Brock (1954)
Enivetok atoll; patch reefs	425	Odum and Odum (1955)
One Tree I. reef, GBR; outer windward slope	500	Goldman and Talbot (1976)
One Tree I.; reef flat	350	Goldman and Talbot (1976)
One Tree I.; fore-reef zone	1950	Goldman and Talbot (1976)
One Tree I.; leeward slope	1400	Goldman and Talbot (1976)
GBR, central part, windward slope	2370	Williams and Hatcher (1983)
GBR,; coastal area	920	Williams and Hatcher (1983)
GBR,; lagoon	1560	Williams and Hatcher (1983)
Southern California; reef flat	350	Bakus (1969)
Mauritius; reef flat	210	Cushing (1971)
English Channel; temperate coast	25	Carpenter (1977)

Table 6.8. Stocks (S, $kg\,ha^{-1}$) of reef fish in their most common families found on reefs of the GBR in central part of the lagoon (Zone A and on the outer slope (Zone B) (after Munro and Williams 1985)

Zone A		Zone B	
Family	S	Family	S
Caesionidae	1203	Caesionidae	323
Pomacentridae	438	Acanthuridae	247
Scaridae	130	Pomacentridae	215
Serranidae	87	Holocentridae	133
Labridae	71	Hemiramphidae	131
Acanthuridae	61	Chaetodontidae	53
Holocentridae	57	Lutjanidae	41
Chaetodontidae	57	Serranidae	32
Haemulidae	40	Lethrinidae	30
Lethrinidae	28	Haemulidae	30
Rest	202	Rest	220
Total	2374	Total	1561

Table 6.9. Density of communities of reef fish of patch-reef zones in different regions. (Data by Sale 1980)

Region	Reef	Density of fish community	
		sp. m^{-3}	sp. m^{-2}
Bahamas	Near underwater lab "Bimini 1"	27	53
	Near underwater lab "Bimini 2"	8	119
Virgin I.	Near underwater lab "Tektite A"	16	23
Bay of California	Average for 9 natural reefs	2	8
	Average for 9 artificial reefs	69	600
One Tree Reef, GBR	Average for 9 natural reefs	72	434
	Average for 8 artificial reefs	40	238
Red Sea	Reef "A"	36	50
	Reef "B"	145	302

reefs 1–2 years after their colonization by reef fauna, the density of fish populations was found to be even higher than on nearby natural reefs (Table 6.9).

The zonal distribution of fish biomass was investigated at One Tree Island ring reef. The highest values were recorded in biotopes of the outer reef slope (Fig. 6.2), also as to the total number of species (Goldman and Talbot 1976). A high biomass in this zome was created by the abundance of small planktonovorous fish and of predators feeding on them. An intensive growth of corals and coralline algae provides here food sources for corallovorous

Table 6.10. Trophical structure of the reef-fish community in the central part of the GBR according to results obtained by processing 10 explosion samples. (Data by Williams and Hatcher 1983)

Trophic guilds	Sources of food	Number of species	Percentage of total weight of fish sampled	Number of specimens, % of total in samples	Average individual weight, g
Herbivores	Detritus sediments, microphytobenthos, periphyton	16	9.4	–	–
	Seagrass, macrophytes, crusts of coralline algae	32	7.0	–	–
	Total herbivorous fish	48	16.4	6.6	8.7
Planktonovores	Mesozooplankton, phytoplankton and suspended detritus	60	20.6	–	–
	Gelatinous macrozooplankton	4	36.6	–	–
	Total planktonovorous fish	64	57.2	85.3	2.3
Benthovores	Bottom invertebrates	156	16.2	–	–
	Corals	15	3	–	–
	Zoobenthos, plants (omnivores)	14	1.5	–	–
	Total benthovorous fish	185	20.7	7.0	10.3
Piscivorous predators	Fish	11	2.2	–	–
	Fish and benthic invertebrates	15	3.5	–	–
	Total piscivorous predatory fish	26	5.7	1.1	19.0

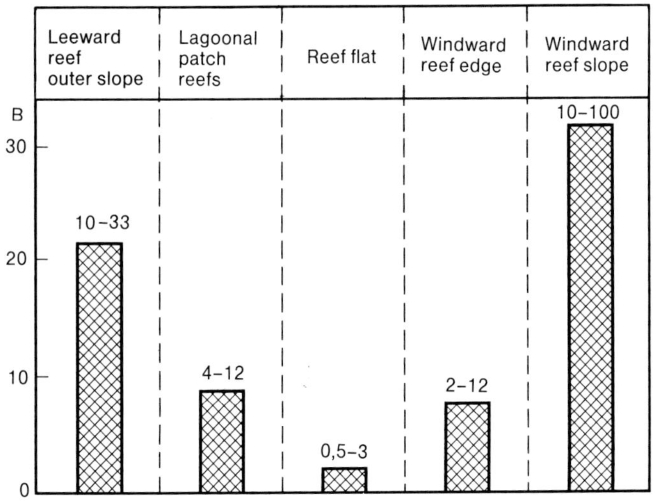

Fig. 6.2. Distribution of total ichthyomass (B, kg per explosion sample) in main biotopes of the One Tree ring reef, GBR; the *numbers* above the column correspond to the range of individual weights of fish in the sample. (Data by Goldman and Talbot 1976)

and herbivorous fishes. On the outer slope the large predatory fish are also accumulating, these come to the reef from the open ocean for cleaning. The cleaning "stations" are usually situated at the outer reef slope (cf. Sect. 6.4).

Data on fish production are still scarce. Bardach (1959) estimated it to be around 220 kg ha^{-1} $year^{-1}$ on some Bermuda reefs, accepting that the biomass of fish was 550 kg ha^{-1}. Thus the coefficient P/B per year should be 0.44 or 0.01 day^{-1}. The above value of this coefficient seems to be underestimated. The above author calculated it without taking into account the production of youth and larvae, which could even exceed the production of adult fish. The larger part of it is grazed by predators. The growth rate of large predatory reef fish like *Plectrorhynchus pictus* appears to be very high: 3.5 cm $year^{-1}$, and the increase in weight 43 g $month^{-1}$, their body weight being 200–300 g. In Acanthurus, the weight of specimens being 50–100 g, the monthly increase in weight was 21 g (Alcala et al. 1981). Thus the coefficient P/B (specific production rates) per year should be in them about 3, which is seven times more than the number taken by the above-mentioned author for the calculation of fish production. A realistic value for populations of reef fish could be evaluated as 1.5–2 $year^{-1}$. So with an approximate average biomass of 600–800 kg ha^{-1} (60–80 g m^{-2}) for reef fish, their annual production should be around 1 tonne ha^{-1}, which is equal to 100 g m^{-2} wet weight, or to 100 k.cal.m^{-2}, comprising some 1–2% of the annual primary production.

6.3 Feeding of Reef Fish

The study of feeding of reef fish had led to a separation within their communities of trophic guilds in order to simplify the analysis of their trophic structure (Hiatt and Strasburg 1960; Randall 1967; Colette and Talbot 1972; Colette and Earle 1972; Vivien 1973; Davis and Birdsong 1973; Hobson 1974; Goldman and Talbot 1976; Harmelin-Vivien and Boucshon-Navaro 1981; Williams and Hatcher 1983; Tables 6.10, 6.11, 6.12). These authors established some basic differences of reef fish fauna from that of the temperate coasts. One of the main differences is the abundance of herbivorous fish, which form on reefs one of the largest trophic guilds both by their total ichthyomass (Table 6.11) and by number of species (Table 6.10). In temperate waters, where the biomass and the number of species of fish are 5–10 times less than in reef biotopes, real herbivorous fish are actually absent. The share of plants in fish food ration does not exceed there a few percents (Bakus 1967; Choat 1989), while in coral reef biotopes it could be several tens of percents (Table 6.13). It is obvious that a high share of herbivores shortens the food chain and so increases about twice the general efficiency in the use of basic trophic sources on coral reef ecosystems at higher trophic levels.

The second feature of the trophic structure of reef-fish communities is a relatively high abundance of planktonovorous fish feeding on zooplankton. This trophic guild includes the small bottom fish that hide in refuges of flats and outer reef slope, the small shoaling planktovores, and the fish larvae.

Table 6.11. Percentage ratios of the numbers of species of main trophic guilds of reef fish in different geographical zones, and in different zones of the same reef. Data by Bakus (1969); Vivien (1973); Williams and Hatcher (1983)

Location	Total number of species	Percentage of total number of species		
		Herbivores (plants, detritus)	Omnivores (plants, detritus zoobenthos, zooplankton)	Predatory (zooplankton zoobenthos, fish)
One Tree Reef, GBR	449	30	10	60
Central GBR	323	15	20	65
Oahu I., Hawaii	107	12	8	80
Southern California	45	7	44	49
Red Sea	56	23	23	54
Great Reef of Tulear; buttress zone	210	8	18	74
Same, reef flat	227	9	20	71
Same, patch reef	193	2	26	72
Same, coastal lagoon	63	2	26	68
Puerto Rico	100	11	15	74

Table 6.12. Number of species in basic trophic groups of reef fish in collections from patch reefs of the atoll Enivetok, calculated per patch reef. Data by Gladfelter et al. (1980)

Trophic groups	Number of specimens	N, percentage of total	Approximate number of species
Herbivores	156	25	37
Daytime planktonovores	151	24	20
Nocturnal planktonovores	48	8	8
Daytime benthovores	208	34	83
Nocturnal benthovores	24	4	11
Piscivorous predators	31	5	19

Table 6.13. Components of gut contents of 142 specimens of reef fish collected at different times of the day at the Great Reef of Tulear, Madagascar. Data by Harmelin-Vivien (1981)

Components of food consumed	Weight of components of gut contents, % of total	
	In daytime	At night
Algae	25.6	0.2
Seagrass	1.4	0.9
Sponges	3.5	0.9
Coelenterates (corals, hydroids)	5.7	0.2
Polychaetes	4.4	12.5
Molluscs	6.2	1.9
Zooplankton, demersal crustaceans	26.8	62.1
Echinoderms	4.6	0.1
Ascidians	2.8	0.1
Fish	17.7	20.6
Miscellaneous	1.3	0.7

Their joint biomass is commensurable and sometimes even exceeds the biomass of benthophages (Table 6.10). For example, in explosion samples of fish from central GBR more than half of their total biomass and 85% of the total number of specimens consisted of small planktonovorous fish of 2–3 g average weight. As it is seen from Table 6.10, at some other reefs of the GBR their number comprised about a quarter of the total specimens and 15% of the total number of species. In the central part of the GBR the biomass of planktonovorous caesionid fish comprised 20–40% of the total (Table 6.8). At the Enivetok atoll, the share of planktonovores was 32% of the total number of fish collected, not to mention fish larvae and the young which feed almost exclusively on zooplankton (Table 6.12). The planktonovorous fish pick up zooplankton from oceanic waters passing over

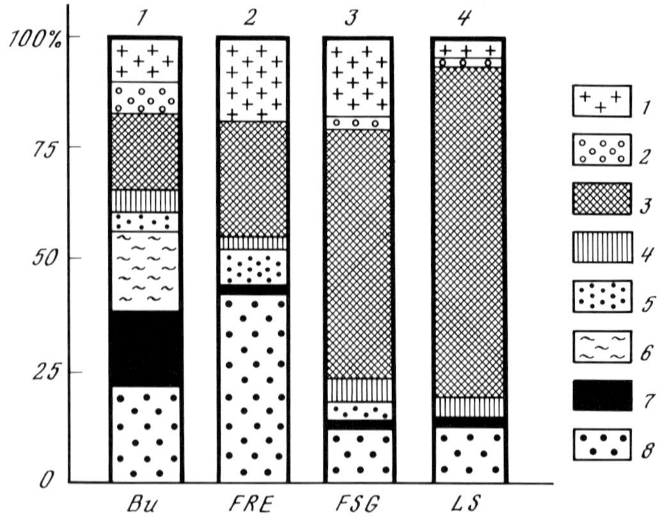

Fig. 6.3. Composition of food (% of total weight of gut contents) in fish caught in different reef zones of the Great Barrier Reef of Tulear; *1* fish; *2* echinoderms; *3* plant material; *4* molluscs; *5* corals and other Anthozoa; *6* polychaetes; *7* sponges; *8* crustaceans, mainly planktonic. Reef zones: *BU* buttress; *FRE* reef flat close to windward reef edge; *FSG* reef flat close to the leeward reef edge with patches of seagrasses; *LS* lagoon with sandy bottom and seagrass. (Data by Harmelin-Vivien 1977)

the reef. Thus they provide the input of external energy and nutrients into the reef ecosystem and so support indirectly the existence of many other reef fish. Their feces is an important source of food for corals and for other benthic animals. The share of zooplankton in food rations in some communities of reef fish often exceeds 40% (Fig. 6.3).

Another important feature of reef fish is a fine trophic specialization not only inside their families but even inside the genera (Bakus 1967). This specialization is one of the main factors ensuring the existence of dense fish populations in reef biotopes, and increasing also the efficiency of utilization of various trophic resources of the reef ecosystem by fish communities (Paine 1966; Smith and Tyler 1972). This specialization is displayed in morphological, physiological and behavioral adaptations. The most various types of specialization in the use of food sources of reef benthos and plankton we can see inside the families Pomacentridae, Chaetodontidae and Gobiidae. For example, among the pomacentrids there are: herbivores feeding on macrophytes or on periphytic turfs, omnivores feeding on zooplankton, benthovores, and cleancers. Each of these specialized trophic groups of pomacentrids have their corresponding morphology of mouth parts, appropriate form of the body and its coloration, and also appropriate mode of behavior (Emery 1973; Harmelin-Vivien 1989). Within the genus *Chaetodon* some species feed only on polyps of scleractinian corals, others prefer polyps of alcyonaceans. There are also omnivorous species of this genus, their species feeding on zoobenthos, and also herbivores feeding on macrophytes and periphyton. And to each kind of trophic specialization corresponds the appropriate construction of the mouth parts, body forms etc. (Harmelin-Vivien 1981; Reese 1981; Cox 1988). A more detailed description of the main trophic guilds of reef fish is given below.

6.3.1 Herbivorous Reef Fish

Herbivorous reef fish form one of the most important trophic guilds, which includes some 10–20% of the total species and 15–50% of total fish biomass (Tables (6.10–6.13). Sometimes their biomass comprises over 50% even of the total (Harmelin-Vivien et al. 1988). Plant food (macrophytes, periphyton) is also used by omnivorous fish as an additional food source. As to the conditions favoring the flourishing of herbivorous fish in reef environments, a main one of these could be seen in the domination on reef of plant material that is readily edible and easily digested by fish, like periphyton, microphytobenthos, coralline algae, especially the articulated kind, and mini-thallomic macrophytes (Bakus 1969). Another cause is the high temperature of reef waters, enhancing the function of enzymes which participate in the digestion of plant material by fish. Herbivorous fish are the latest branch of evolution of reefal bony fish. Being descended from ancient acanthopterigian fish of the Late Eocene (Hobson 1974), they acquired a small mouth with powerful jaws and firm massive teeth, which later in some of them fused together and formed plates of "bills" (Scaridae). Among other adaptations acquired by herbivorous fish are the throat teeth for the grinding of crude plant material, extremely elastic stomachs, which could contain a large mass of plant food, and long intestines. But most of the herbivorous fishes (exept mugilids) do not have the enzyme cellulase that dissolves the walls of plant cells. Perhaps they use for this purpose the cellulase produced by their intestine's microflora.

By their mode of feeding herbivorous fish could be subdivided into browsers, grazers and suckers. The first feed on macrophytes and seagrasses, the second on scrape periphytonic turfs and crusts of coralline algae, and the third suck in or gently collect the films of detritus and the fine turfs of microphytobenthos from the surface of soft bottom sediments. The fish of the firsh of these groups have a fine-walled but elastic stomach, and of the second a powerful thick-walled stomach which resembles the muscular stomach of birds, being destined for crushing coralline algae crusts. The long intestines and the long time of digestion make for the high efficiency of herbivorous fishes in digesting plant material. Its assimilability was evaluated to be within 30 to 60%. Some fish, which have a low assimilability compensate it by a high feeding rate (Ogden and Lobel 1978). In stomachs of acanthurids the presence of agarolythic bacteria was recorded, which participate in the digestion of macrophytes by fish (Sutton and Clemens 1988).

The trophic guild of herbivorous fish embrace a wide spectrum of taxa. It includes four complete important families of reef fish: Acanthuridae, Scaridae, Siganidae and Kyphosidae. The most numerous are the species of the first two which comprise even a significant part of the total ichthyomass. Herbivorous fishes in which plant material as food source is dominating are also present in many other families of reef fish, like the Pomacentridae, Gobiidae, Chaetodontidae, Blennidae, and Balistidae. Mugilids and acanthurids feed on microphytobenthos and on periphyton in which plant material dominates. Numerous herbivorous and omnivorous fish feed also on the feces of predatory fish, thus increasing the efficiency of energy utilization in reef food webs (Robertson 1982) Some hemiramphids (*Hemramphus dussumeri*, *H. brasiliensis*) feed on phytoplankton and on flakes of macrophytes and seagrass leaves, collecting them from the surface film. Among the omnivorous fishes in which plant material comprises a large portion of food are the angel fish *Pomacanthus arguatus* and *P. paru*, the puffer *Canthigaster rostrata*, and the balistids *Monocanthus ciliatus* and *M. chinensis* (Randall 1967; Conachear et al. 1979; Polunin 1989).

Herbivorous fish are strictly specialized with respect to the kinds of plant food they consume. In experiments with the scarid *Sparisoma radians* it was found that the scale of preference for consumed plants approximated the following: (1) seagrass *Thalassia* covered with epiphytes, (2) bare seagrass *Halodule*, (3) bare seagrass *Thalassia*, (4) the grass *Syringodium*, (5) the alga *Dictyota*, (6) the alga *Enteromorpha*, (7) the alga *Caulerpa*, (8) the alga *Halimeda*, and (9) the alga *Penicillus*. The scale of preference depends on the nutritional quality of a given plant species and on its relative toxicity for a given fish. Among those most preferted by fish are the seagrass *Thalassia*, the algae *Enteromorpha*, *Ulva*, and *Codium* (Randall 1961; Earle 1972; Hobson 1974; Ogden and Lobel 1978; Lobel and Ogden 1981). The grazing pressure induced in plants the evolutional acquisition of defence adaptations. Among these is the adaptation of some plants to grow in turbulent surf zones, where their grazing by fish is disturbed (coralline algae, filamentous algae, seagrasses). The thalloms of many algae are heavily loaded and armed with calcareous material. This decreases their food quality for fish (*Turbinaria*, *Penicilius*, *Halimeda*). Other algae are overloaded with tannin impeding their digestion by fish, or are soaked with toxins which repel their grazing. Out of 48 species of algae, which inhabit the shelf off some islands south of Japan, 38 have toxic substances in their thalloms (Hashimoto and Fusetani 1972). The thalloms of *Sargassum* contain tannins, the alga *Asparagopsis* ketones, *Caulerpa* the toxin caulerpin, *Laurencia* chlorides, and *Dictopteris* the toxin dictyopterin. The toxicity of many blue-green algae is well known.

The grazing activity of herbivorous fish is one of the most powerful ecologial factors controlling the structure, recruitment processes and productivity of the bottom algoflora (Randall 1961; Bakus 1967; Ogden and Lobel 1978; Hatcher and Larkum 1983; Hatcher and Rimmer 1985; Polunin 1989). The controlling action of grazing by fish on the growth of benthic plants was demonstrated experimentally by placing cages on the bottom, preventing so the grazing of plants by fish (Earle 1972; Wanders 1977; Nelson and Tsutui 1981). A more intensive growth inside the cages was recorded even after several weeks of experimenting (Vine 1974). But inside the cages the growth of corals and coralline algae was inhibited by accelerated development of macrophytes and especially of filamentous algae, overgrowing through the avoidance of grazing (Wanders 1977). The pressure of grazing of the algoflora by fish could be restricted naturally inside the areas defended by territorial pomacentrid fish (Vine 1974; Potts 1977).

An important function of the herbivorous fish scarids and acanthurids in reef systems could also be seen in their bioerosional activity and in the production and transport of carbonaceous material (cf. Sect. 1.3). When scraping the periphyton and coralline algae from the reef rocks and rubble they swallow a mass of calcareous material including pebbles which they use for the crushing of plant material in their throat "mill" the same as is done by grain-eating birds. The swallowed calcareous material is milled during digestion and is excreted by fish as fine carbonaceous sediments. The population of the parrot fish *Sparisoma viridis*, of average density, living on one of the Bermuda reefs, excretes about 2 t of carbonaceous sediments per year, which is equal to the erosion rate of reef flats of 1 m/100 years (Gygi 1975; Bardach 1961).

6.3.2 Planktonovorous Fish

Within this guild the groups of primary and secondary planktonovorous fish could be distinguished. The group of primary planktonovores includes several species from families of typically pelagic fish of the open sea, which adapted to reef environments. Among them clupeid and dussumerid fish are most common. Their main adaptation to reef life appears to be their small size. They are small shoaling fish, which in the daytime hideng with their schools near the reef edge, in lagoons or in caves. At night they disperse and feed in the water column on abundant nocturnal reef zooplankton. The group of secondary planktovores includes representatives of different families of typical reef fish most of which belong to the order Perciformes or are related to it (Davis and Birdsong 1973; Hobson and Chess 1978; Williams and Hatcher 1983). Planktovorous species could be found in most families of benthos-feeding predatory fish: Labridae (*Clepticus parrai*, *Microlabrichthys pascalus*), Lutjanidae (genus *Emmelichthyops*), Serranidae (*Schulzea beta*, *Serranus tortugarum*, *Paracanthes furcifer*), Pempheridae (*Pempheris schomburgki*, *Ociurus chrisurus*), Pomadasyidae (genus *Hae-*

mulon, *Eguetus acuminatus*), Pomacentridae (Genus *Chromis*, *Pomacentrus coeletus*, *Amblygliphiodon curacao*, *Dascyllus reticulatus*), Gobiidae (genera *Joglossus*, *Ptereleotris*, *Cottogobius*), Chaetodontidae (*Chaetodon sedentaris*), Apogonidae (*Apogon gracilis*, *A. novaeaquinea*, *A. savayensis*, *A. binotatus*), Holocentridae (*Myripristis pralinus*, *M. jackobius*, *M. violaceus*), Cyrrithidae (*Amblycirrithes pinos*), and Congridae (*Nystactichthys haila*, *Gorgasia sillineri*, *Taeniconger halis*). The families of reef planktonovorous fish Atherinidae and Caesionidae descended from the perchiform fish. The genera *Aterina*, *Allanetta*, *Pranesus*, *Caesio* and *Pterocassio* are among the most common reef fish feeding on zooplankton (see Tables 6.8, 6.10, 6.13).

The morphological adaptations in the secondary planktonovores are manifest largely in the changes of morphology of mouth parts. Their jaws expanded and became tube-like. On opening, the mouth sucks water inside as a pipette together with the zooplankters which the fish is hunting for. Their teeth became tiny, the gills denser, the cranium bones thin, and the eyes enlarged. The size of their body reduced and assumed a spindle-like form. In some of these fish the morphological changes connected with transition to plankton feeding were so deep, that they could easily be accepted as representatives of another family, such as the serranid *Schulzea beta*. There morphological changes were connected also with the transition from single to schooling life, because within the schools the pressure of grazing in the water column decreases (Davis and Birdsong 1973). With this, too, is connected the coloration of the daytime planktonovores, which are subjected to predators' attacks, like some aterinids. These fish have a bright coloration with a horizontal stripe which should disorient the predator attacking these fish in the shoal. The same fish which do not form permanent shoals have bright blue-white or greenish masking coloration (genus *Chromis*).

The behavioral features of planktonovore fishes depend on their individual hunting strategy. Daytime shoaling fish like *Chromis* and Dascyllus form small shoals near the reef edge close to a possible refuge and feed on zooplankton brought by the current (Hobson and Chess 1978; Hammer et al. 1988). On being attacked, the shoal instantly goes into hiding. At night, the shoals disperse and the fish spend the night in their individual lodgings. Aterinids (*Haemulon striatum*) form permanent large shoals. They feed in the water column, escaping attacks by instant reaction of the whole shoal. Some daytime planktonovores remain immobile behind their refuges and catch zooplankton preys which are driven by the currents behind them. These are for example the congrid garden eels. These snakelike fish live in holes. Exposing from their holes half of their body length, they catch zooplankton from the waters passing by (Randall 1967). Some planktonovores use ambushing tactics, stay masked near their holes or on a branch of coral and by blitz-attack hurls catch their prey. Among them there are some species of holocentrids, gobiids and cirrithids.

6.3.3 Benthovore and Piscivore Fish

The guild of fish feeding on various groups of zoobenthos is the most numerous by number of species (Tables 6.10, 6.12). The total number of species it includes (counting in the omnivorous bottom fish) comprises 40 to 80% of all communities of the reef ichthyofauna. But their share in the whole fish biomass is less (20–30%), because their species populations are less numerous and their trophic specialization is higher in comparison with other guilds (Goldman and Talbot 1976; Harmelin-Vivien 1981; Williams and Hatcher 1983). The share of their food ration varies from 15% of the total ration of the fish community in lagoonal biotopes to up to 50% in those of reef slopes. This guild could be subdivided into three main groups, depending on the level of their trophic specialization and on their preferred food: (1) fish feeding on large sessile benthic animals: ascidians, sponges, hydroids, zoantharians, sessile polychaetes; (2) fish feeding on large vagile fauna with firm exoskeleton: large crabs, barnacles, gastropods, bivalves, urchins, asteroids, holothurians, chitons; and (3) fish feeding on small vagile zoobenthos of soft sediments and on the infaunal benthos of the hard rocks and rubble: bottom crustaceans, small crabs, ostracods, stomatopods, amphypods, polychaetes and other worms. Fish from the first two groups consume also significant amounts of plant material, and those from the last group also small fish. Inside each of these group each species is specialized to feed on definite kinds of prey, which reflects the morphology of its mouth parts (Hiatt and Strasburg 1960).

Within the group of fish feeding on sessile benthos there are species which feed mostly on sponges and ascidians, like the balisitid *Cantherines*, the puffer *Arothron*, the ostraciontid (boxfish) *Acanthostracion*, the chaetodontids (angelfish) *Holacanthus*, *Pomacanthus*, and the butterfly fish *Chaetodon ephippium*. The large wrasse *Lachnolaimus* and the ogcocephalid *Ogcocephalus nasutus* prefer the barnacles, the balistid *Alutera scripta* feeds on strongly defended hydrocoral *Millepora* with poisonous nematocysts, while the sergeant fish pomacentrid *Aboefduf saxatilis* is specialized for feeding on zoantharians. Many species of the butterfly chaetodontid fish also feed on polyps of the scleractrinian, corals, octocorals and zoantharians and also pick up the filtering cones of serpulid polychaetes (Birkeland and Neudecker 1981; Wylie 1988). Some of them have special "sucking" construction mouth parts to suck out the polyps of scleractinians and octocorals (Reese 1981). Among the pomacentrids there are species adapted to scraping off the coral mucus they feed on from living colonies of scleractinians. Among these are the pomacentrids *Chromis caeruleus*, *Pomacentrus sufflavus* and the butterfly fish *Chaetodon vagabundus* (Benson and Muscatine 1974). Corals are grazed directly by some scarids: *Scarus quacamaia*, *S. croicensis*, *S. coelestrinus*, *Sparisoma aurofrenatum*, the *surge* on-fish *Acanthurus chirurgus* and the pomacentrid *Microspathodon chrysurus*. Their stomachs contains the swallowed pieces of hard corals (Randall

1974). These fish bite off the tips of coral branches and scrape coenosarc from the colonies of massive corals together with calcareous material. The balistid *Alutera scripta* and some chaltodontids graze also the octocorals.

Among fish of the second trophic group feeding on vagile benthos with firm outer skeleton there could be mentioned the ray *Aetobatus narinari*, the grunters *Haemulon plumeri*, *H. macrostomus*, the wrasses (genera *Bodianus*, *Halicoeres*, *Hemipteronotus*, *Thalassoma*), the balistid *Balistes vetula*, the box fish *Lactophris trigonus*, most tetraododontids, diodontids and ogcocephalids. They have massive craniums and jaws. Thus they are able to crush the testes of crabs, urchins and sea stars, the shells of gastropods and chitons with the aid of their jaws and throat teeth. The smaller vagile macro- and meiobenthos are consumed by rays, sand sharks, moray eels, holocentrids, by many groupers (genera *Aphestes*, *Cephalopholis*, *Epinephelus*, *Myctoperca*, *Pterometopon*), by wrasses (genera *Bodjanus*, *Halicoeres*, Talassoma), by some cerrithids and grunters (genera *Haemulon*, *Anisotremus*, *Rhonciscus*, *Macolor*), by pomacentrids (*Pomacentrus planifrons*, *P. partitus*), by goatfish, flounders (genera *Botus*, *Achirus*), by some chaetodontids, scorpion fish, and box fish (*Lactophrys triqueter*). Main components of their food ration are polychaetes, shrimps and small crabs, which comprise over 70% of their gut contents (Table 6.13). The total amount of ash-free food material consumed by this group of fish per day attains on reef flats 2–3 $\mathrm{g\,m^{-2}}$, and in lagoons 0.5 $\mathrm{g\,m^{-2}}$. Thus these benthic fish graze out the bulk of primary and secondary production of bottom biotopes (Hatcher 1981).

The guild of piscivorous predatory fish includes 10–20% of the total number of species of reef ichthyofauna (Tables 6.10, 6.12). These numbers are only tentative because the fish, and especially their young and larvae, compose a significant part of food ration also in many benthovorous and planktovorous fish (Randall 1967). The share of fish as a food in the total food ration of reef fish communities varies in different reef zones: from 5% in the lagoon up to 20% on the reef slope (Fig. 6.3). The biomass of obligate predatory fish usually does not exceed 5–10%. Among the latter the most numerous are groupers, caranxes, moray eels, barracudas, emperors (fam. Lethrinidae). This group inludes also some lutjanids (*Lutjanus cyanopercus*), mackerels (genus *Scomberonorus*), belonids, aulostomids and fistulariids, snake eels, synodontids, and reef sharks (genera *Carcharhinidae*, *Orectolobidae*). Besides fish prey the obligate piscivores hunt cephalopods – squids, octopuses, cattle fish, and shrimps and crabs. The grazing activity of predatory fish is an important factor controlling the composition and the density of fish communities (Sale 1980; Talbot et al. 1978).

6.4 Aspects of Behavior

Features of behavior are among the most important ecological adaptations of animals to environmental conditions. They are formed during their evolution and are fixed in their genetic apparatus the same as their morphological and physiological characteristics (Wickler 1971). The behavioral reactions of animals are interconnected with their morphology and physiology. Among these kinds of ecological adaptations the behavioral ones are the most mobile during evolutional changes of species adapting to changes in environmental conditions. This is most spectacularly displayed in reef fish, which form exclusively dense and varied communities, including a multitude of interacting species (Barlow 1975).

One of the important behavioral adaptations of reef fish is the rhythms of diurnal activity, which is peculiar to most of their species. During a 24-h day, fish shift from a period of active feeding to a period of relaxation. Among them there are the daytime fish. These fish are active within the light period of the day, while nocturnal fish are active only at night. A smaller part of reef fish are active the entire 24 h, and only a few of them have indefinite diurnal rhythms (Helfman 1978). With the aid of this adaptation the reef fish fulfills two main aims of their life strategy: to use optimally the trophic sources of environments and to escape grazing by predators. The first aim they achieve by their diurnal migrations to places of foraging and the second by the formation of schools and by hiding in refuges (Hobson 1973). From the viewpoint of evolution the species of nocturnal fishes are more ancient compared with the daytime ones, as they are closer to their perchid ancestors. Among them we find representatives of the most common reef-fish families Apogonidae, Lutjanidae, and Holocentridae. The daytime fish are of later origin. They are more specialized in aspects of their morphology, in their behavioral features and in food preferences. Their predatory species, like chromises, caesionids, atherinids, gobiids, and cirrithids, which are adapted to feeding on zooplankton, acquired small body (Table 6.10). A large part of daytime fish turned to feeding on sessile plants and animal organisms, like periphyton, macrophytes, corals, and sedimentary polychaetes, and also on large vagile zoobenthos, which do not hide in daytime. These kinds of food objects are more accessible to grazing during the daytime. Among the latter the herbivores Scaridae, Acanthuridae, Pomacentridae, the corallovores chaetodontids, the feeders on large benthos Ballistidae, and some of chaetodontid species could be mentioned. The appearance of the daytime and the nocturnal fish greatly increased efficiency in the use of available refuge space on reefs. At night, these are used by daytime fish which feed early in the morning. These refuges then are ready to be occupied by their successors – the nocturnal fish which use them for their daytime relaxation (Smith and Tyler 1973; Fig. 6.4). Among fish without any apparemt diurnal rhythm the large piscivorous

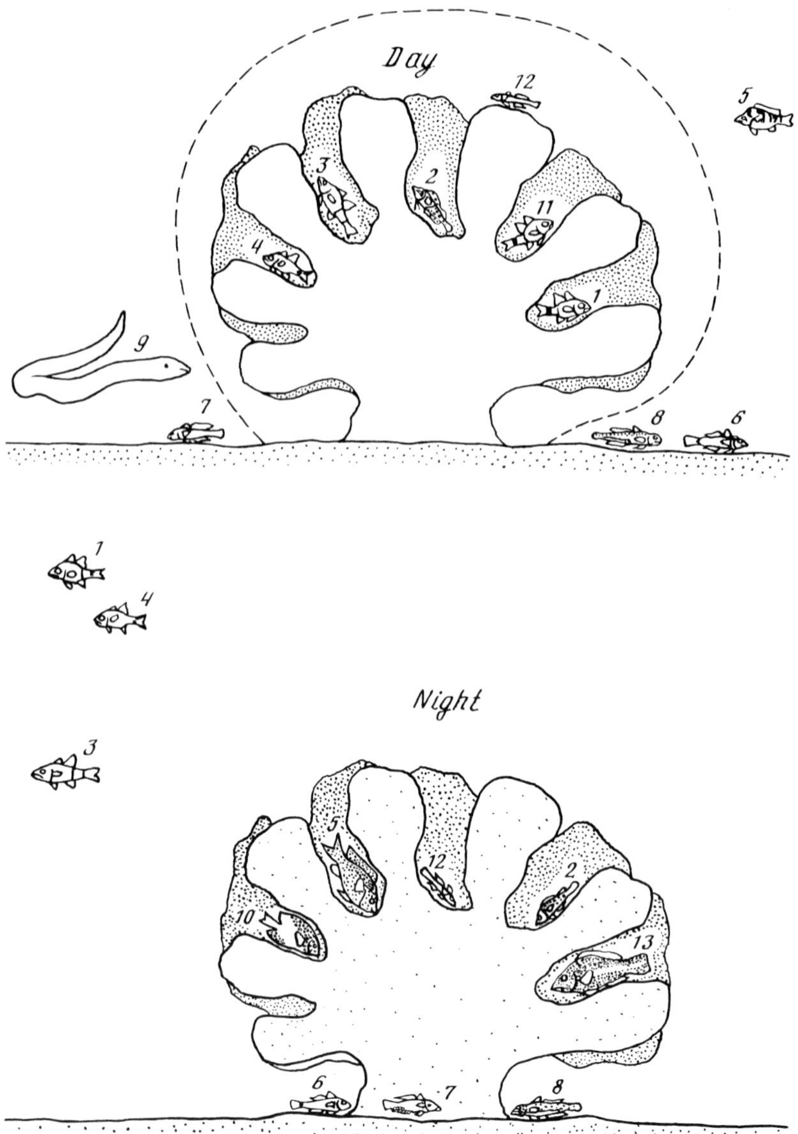

Fig. 6.4. Daytime and nocturnal fish residents in the colony of the coral *Montastrea annularis*. The permanent residents in the *upper figure* are placed inside the *dotted line*, the visitors outside. Fish species: *1*, *2* and *4* the cardinals *Apogon townsendi*, *A. binotatus* and *Phaeoptyx conklini*; *2* blenny *Labrisomis haitiensis*; *3* perch *Hypoplectrus puella*; *6*, *7* and *8* gobies *Corynopterus eilodon*, *C. glaucofraenum*, and *Gnatholepis thompsoni*; *9* moray eel *Muraena milliaris*; *10* chromis *Ch. cyaneus*; *11* damsel fish *Pomacentrus partitus*; *12* goby cleaners *Gobiosoma evelynae*; *13* parrot fish *Sparisoma aurofrenatum*. (After Smith and Tyler 1973)

Table 6.14. Diurnal activity of coral reef fish within their main families[a]

D	N	M	I
Pomacentridae	Holocentridae	Congridae	Muraenidae
Labridae	Apogonidae	Serranidae	Scorpaenidae
Scaridae	Pempheridae	Carangidae	Scombridae
Blennidae	Diodontidae	Mullidae	Serranidae
Acanthuridae	Clupeidae	Sphyraenidae	
Balistidae	Lutjanidae	Bothidae	
Chaetodontidae		Muraenidae	
Cirrithidae			
Gobiidae			

[a] D, All or dominating number of species in a given family are active only in the daytime; N, correspondingly active only at night; M, some of the species are active only in the daytime, and some only at night or during the entire 24 h; I, some species of the family do not have a definite diurnal regime.

predators like barracudas, serranids, fistulariids, scorpaenids, mooray eels, and carangids dominate. In Table 6.14 a tentative classification of reef fish by their diurnal activity is given.

The shift in fish from relaxation to activity and back occurs at dawn and at dusk. At these periods fish are especially vulnerable to a predator's attacks, because at this time their migration schools are collecting or dispersing. Safety during the period of relaxation of the fish is ensured by two different types of behavior. One of these is hiding in their mostly permanent, individual refuges. The second (most often for the small ones) form relaxation shoals, within which attacks of predators against them are less successful. Most often these shoals are not large. They stay usually in permanent places near the outer edges of the reef, near the patch reefs or near cave entrances, where they could instantly hide on being attacked (Hobson 1972). But places good for foraging and convenient for relaxation with available refuges, as well as spaces convenient for shoaling, are usually situated in different zones of the reef. It is the main cause of the diurnal migrations of reef fish, which are performed by many of their species. Some of them during these migrations cover distances of up to several kilometers (Hobson 1973). Small daytime planktonovorous fish perform diurnal vertical migrations, spending the night in bottom refuges and rising in the morning to the water column to feed on zooplankton (the pomacentrids *Chromis* and caesionids). The same migrations but in the evening is performed by the cardinal fish (*Apogon*) and the squirrel fish (*Myripristis*). For the nocturnal predatory fish feeding in the water column on fish and on demersal meso- and macrozooplankton, as well as for some nocturnal benthovores feeding on soft bottom invertebrates, which go out onto the bottom surface at night, peculiar are evening migrations from the refuges to the water column up the outer reef slope, or to the bottom biotopes of fore-reef and lagoon. Among these fish are the caranxes (Selar), sea perches (some of Lutjanidae), and atherinids (Pranesus).

Many of the daytime herbivorous, corallovorous and benthovorous fish migrate diurnally between different reef biotopes, e.g., between the biotope where they forage and the biotopes where they relax – the same as people diurnally move between home and work-place. Among these fish could be mentioned the scariids, acanthurids (*Acanthurus*, *Ziebrasoma*), chaetodontids, some pomacentrids, mugilids, mullids, and some of the labrids (Hobson 1972, 1973; Buckman and Ogden 1973; Ogden and Zieman 1977; Rublee and Zeeman 1984). These fish emerge from their night hiding places at dawn and accumulate within schools, which then start to move to their foraging places – to the thickets of macrophytes and ramose corals or to the lagoon. In the evening they return to their same nocturnal refuges (Starck and Shroeder 1985; Fig. 6.4).

The migratory schools most often are monospecific, but some fish migrate within schools containing several species. These are, for examples, the migratory schools of the Caribbean parrot fish *Scarus croicensus*. This fish, as many other scarids, spends the night "sleeping" in a mucous cocoon hidden in crevices and holes of reef slopes (Ogden and Buckman 1973). At dawn they come out from their refuges and form large schools. Often their school is joined by fish from other species: groupers, sturgeons. The schools move to the flat or to shallows covered with seeweeds or with grass for foraging. Thus they pass sometimes up to several kilometers following a complex but permanent route. The routes of schools are selected by migrating fish observing the position of "cleanear stations", where on the way to underwater meadows the fish can devoid themselves of parasites on their skin and gills. The sanitary work is done by fish cleaners *Labroides dimidiatus* and Gobiosoma sp. The formation of schools for the diurnal foraging migrations is an important behavioral adaptation of reef fish. It helps them not only to escape predators, but also to find the routes of migrations because the schools include old experienced specimens which know them and lead the shoal. Thus the information about the best routes and best foraging places is accumulated and preserved within a given local fish population. It seems that the migrating schools during their move wake use of the topographic features of the bottom landscape for navigation (Hobson 1973). Migrating fish transfer nutrients from foraging places (seagress beds) to places of relaxation (coral thickets), thus playing an important role in trophic interrelations between reef biotopes (Meyer et al. 1983).

The exact times of the morning and evening migrations in populations of different species are strictly kept apart to avoid disordered swarms, which are so convenient for hunting predators (Domm and Domm 1973). Some nocturnal fish even change coloration during their shift from daytime relaxation to nocturnal activity (Starck and Schroeder 1965). Thus some nocturnal sea perches (fam. Lutjanidae) and grunters (fam. Haemulidae), change their bluish or yellow coloration ("sleeping gowns") to silvery white with black spots. The cardinals (fam. Apogonidae), relaxing within small shoals under steep rocks of the reef edge, are bright-red, but on going to hunt at night they change their color to palish-pink. The bright coloration of these nocturnal fish in the daytime has the same adaptative importance as the permanently bright, remarkable coloration of daytime migrating fish, like parrot fish or sturgeons.

Many species of reef fish form monospecific or interspecific permanent territorial groups. The numbers of these groups reveal a complex social behavior. The same is true of the symbiotic or commensal groups of fish. These types of social behavior are peculiar mostly to species of real reef fish like pomacentrids, chaetodontids, gobiids, sturgeons, and parrot fish. The features of territorial and social behavior in these fish, just as in terrestrial animals, are actually genetically fixed ecological adaptations. They help these fish to cope with the task of defending their breeding and feeding areas from intruders and competitors. The relationships of symbiotic and commensal species with their hosts help both partners to survive. Special experiments have shown that the members of social groups know each other "by face" (Ehrlich et al. 1977), as also members of monogamic couples of fish easily recognize their mates among similar equally colored specimens (Reese 1975). Territorial fish perfectly know the landscape and the borders of their territories. To defend them, they drive out from their plots first of all their direct food competitors, perfectly recognizing them (Mahoney 1981). This kind of interspecific antagonism is also part of the complex system of social behavior in reef fish (Buckman and Ogden 1973; Sale 1980). It is well developed in territorial fish, like the sergeant fish (fam. Pomacentridae) and in sturgeons. The abundance of fish that demonstrates a settled mode of life and territorial behavior is one of most remarkable feature of reef-fish communities (Colette and Talbot 1972; Sale 1980).

Among the most common kinds of territorial behavior of reef fish the following can be mentioned: (1) defense of the territory around their holes (lizard fish, moray eels); (2) collective guarding by the permanent groups of pomacentrids of their territories only from the intrusion of their competitors – the herbivorous fish *Pomacentrus flavicanuda*, *Eupomacentrus dorsopunctatus* (Mahoney 1981); (3) collective defense of private "pasture" by a group of owners – pomacentrids *Pomacentrus jenkinsi* – attacking and repelling all intruding fish including strangers of the same species, whom they recognize "by face"; (4) permanent occupation by monogamic fish couples or by small polygamic parties of extensive bottom space, which serves as pasture and as breeding territory, and which they guard only from intrusions of other males or other couples, especially during the spawning period (sturgeons, chaerodontids, wrasses); (5) permanent living of a single fish, of a couple or of a small group near some element of reef construction which serves as a refuge, without their attempting to defend the place from other fish (pomacentrids *Dascyllus arguanus*, *Cromis* sp., some morays, many serranids, sturgeons and chaetodontids).

The most developed features of territorial and social behavior possess the damsel fish from the genera *Pomacentrus* and *Eupomacentrus*. These small and brave fish very successfully familiarize themselves with and defend their benthic pasture lots, repelling much stronger and well-armed competitors like sturgeons the urchins *Diadema*, and even ward off Acanthaster (Glynn and Coldan 1988). The damsel fish *Pomacentrus lividus*, which inhabits the submerged areas of the flat of most Red Sea reefs, live in large territorial colonies comprising up to 1000 specimens (Vine 1974). The members of a colony attack intruders jointly, even large parrot fish, biting them and sometimes even gnawing their fins.

The social behavior of reef fish includes a rather complex system of intraspecific and interspecies relationships, which are served by various kinds of signalization: light, sound or chemical. The bright sophisticated coloration is one of the main types of light signalization. Coloration serves as an important feature of adaptation for fish with territorial or schooling behavior. With its aid they inform other fish about their claim or right to a given lot of territory. The bright specific hues of the species enable them to recognize their own from strangers even in unclear coastal water, and so escape violent confrontations. Fish that have such a bright "placard" (K. Lorentz) coloration do not change it according to the coloration of the landscape, like fish with a protective one do. For example, the chaetodontids and zanklids could be pointed out, which are one of the most spectacular reef fish. Coloration serves also many other behavioral reactions connected with various ecological adaptations. It is used: (1) by butterly fish, by gobiids and by some damsel fish (fam. Pomacentridae) as the "placard" mentioned above; (2) by monogamous fish like some gobies, chaetodontids and zanklids, for mates to signal their location to each other so as not to get lost in complex landscapes of the reef; (3) by diurnally migrating parrot fish and sturgeons, for orientation of individuals inside the moving school; (4) by some schooling fish like atherinids or some grunters, for the desorientation of predators attacking the school; (5) by some butterfly fish for desorientation of a predator attacking an individual fish, provoking it to get at the fish from the caudal side (false eye near the tail and real eye masked by black stripe); (6) by some butterfly-, angel- and idol fish, as well as by scorpaenids, cerrithiids, ballistids and box fish, for warning the predator about the good defense, poor food quality or the possibility for the attacker of being poisoned, and so prevent useless attacks; (7) by pelagic fish (*Chromis*, *Caesio*), to camouflage their bodies against the bright background of the water column or water surface, or against the vertically striped background of thickets of ramose corals; (8) by some blennies and small groupers, for mimicry to resemble the fish-cleaning wrasses: making use of the peculiarities of the cleaners they are not endangered by predators, because predatory fish never eat the cleaners. But simulating the cleaners, they may also get close to the clients and gnaw their fins and scales.

The squirrel fish (fam. Holocentridae) and grunters (fam. Haemulidae) use sound signalization, producing clicking or grunting sounds (Winn et al.

1964). The sounds of these territorial nocturnal fish are related to their territorial behavior. They live within colonies that spend the daytime in their permanent lots. Upon returning from their nocturnal foraging voyages to their permanent lots on the reef bottom for daytime relaxation, the colony of squirrel fish produces permanent noise, which asserts the claim of the colony to a given territory. When the intruder approaches, the fish in the colony produce staccato sounds warning it of imminent defeat.

One sign of the high maturity of coral-reef ecosystems is the widespread development of various forms of commensalism and symbiosis. This feature applies to the communities of reef fish as well. They display many complex forms of social behavior and adaptation within this kind of relationship either between the fish themselves, or between fish and other reef animals. Thus, the fish *Caparus* lives in the coeloms of holothurians, small cardinal fish live in the mantle of the gastropod *Strombus gigas*, the fish *Lepidogaster* sp. lives among the needles of sea urchins or amid the rays of crinoids. The same is done by schools of small gobies or damsel-fish (chromis), using the needles of the sea-urchins *Diadema* or branches of corals as their refuge. Symbiosis is well known between a small reef fish (mostly gobies) and shrimps. Shrimps dig out the holes and clean them, while the fish sharing the hole with its owner give him remains of their dinners (Limbaugh 1972). But the relationship of symbiosis is demonstrated the most clearly by the clown fish living with the actinians (Mariscal 1970). The group of clown fish is represented by various pomacentrid species from the genera *Amphiprion* (12 species), *Dascyllus* (2 species) and *Premnas* (2 species). They are small brightly colored fish with clear-cut placard outlines: white cross-stripes on a bright green or red background. They live as families in the tentacles of large reef actinians. The family includes two to three adults and several babies (Ross 1978). The fish seek refuge in the well-armed tentacles of the actinian, feed on mucus excreted by the host, and on its dead tissues and remains of his food. They accomplish for the host the sanitary function of cleaning it and even feed it sometimes, bringing to the mouth of the host small fish caught by them in the vicinity. The clown fish bravely attack any fish approaching the host. Although they have very poisonous nematocysts, the host actinians never damage their symbiont fish.

Another extremely widespread kind of fish symbiosis is cleaning, performed by several specialized species of cleaner fish. They clean their fish clients from ectoparasites, from microbial overgrowths, and remove dead tissues. The cleaner fish feeds on this scraped material, thus contributing to the health of their clients. Reef fish live in warm waters and are readily attacked by numerous parasites, molds and bacteria. Parasite density is among the important factors regulating the density of their fish hosts (Adlard and Lester 1988). Therefore, cleaning is one of the most vital of their needs and life functions (Feder 1966). The fish pay close attention to this procedure and even select their migration roots depending on the position of the "cleaner stations", where the cleaner fish live and "work".

An important role of cleaning symbiosis was proved experimentally. On a small reef all cleaners had been cleared out. Several days later the fish started to leave the reef. Within 2 weeks most of the fish, except the strictly territorial ones (living in nests or holes) abandoned this reef. Among the fish that remained the percentage of ill specimens drastically increased. Cleaner fish are introduced into big marine aquaria to preserve the good health of its inhabitants. The ill fish spend at the "cleaners stations" as much time as they spend in feeding, going there several times a day. At the stations lines "of fish waiting for service" often occur and between the "clients" quarrels break out (Feder 1966). It was discovered that good places for fishing on the reef often happen to be precisely the cleaner stations, where many fish wait for the cleaning procedure. So the cleaners could be considered a key species in reef-fish communities. Without them their very existence would be endangered.

The function of cleaners on Caribbean reefs is accomplished by the most widespresd their species, the neon goby *Elacantinus oceanops* and the wrasses *Bodjanus rufus* and *Microspathodon chrysurus*. On reefs of the Indo-Pacific the most common cleaners are the wrasses *Labroides dimidiatus*, *L. phthirophascus* and *L. bicolor*. They live in small groups of three to ten specimens at the "stations", situated most often at the outer reef slope near some remarkable element of the landscape. The cleaners have a well detectable specific placard pattern which informs their "clients" of their profession. This is usually comprised of while and blue stripes along the greenish or dark-violet body. The cleaners swim near the "station" doing special moves – "gestures" – inviting the clients. If their "client" is a pelagic fish, it will float with its curled body in the water column, spread its fins, and open wide its mouth. If it is a bottom fish, it stands on a stone, and also opens its mouth and spreads its fins. The cleaners approach the client and start to "work". One of them cleans the skin scales and fins, sometimes attaching them to its body with the aid of a special sucker (neon goby). Others swim into the mouth and clean the gills, taking off parasites and gnawing swollen tissues. When the "client" feels satisfied, or when it is endangered, it signals the cleaners to leave by clapping with its mouth a couple of times. The "clients", though mostly predatory, never swallow the cleaners; they have never been found in their stomachs (Randall 1967).

6.5 Propagation of Reef Fish

Above it was stated that the main strategy of the evolution of reef fish was a fine specialization increasing the efficiency of making use of trophic resources. But the progress of specialization inevitably leads to an increase in species diversity and in the total number of species in the community, causing a decrease in the number of fish specimens per one species and so the decrease in the specific density of their populations. These developments during the evolution of reef fish created definite complications in the process of their breeding. Their different species solved these by acquiring specific adaptations ensuring success in their multiplication (Breder and Rosen 1966). Among these adaptations the following can be mentioned: (1) formation of permanent monogamic couples in some gobies and chaetodontids; (2) hermaphroditism and the ability in adult fish to change sex; (3) selection of spawning sites hidden from predations; (4) taking care of the eggs and larvae by some gobies, cardinals (*Apogon*), damsels and syngnathids (Hudson 1977; Johannes 1978a; Shapiro 1977, 1979; Sale 1980; Tresher 1984).

The most remarkable features of multiplication of reef fish among those listed above is the widespread hermaphroditism among them and the ability of adult fish to change sex. This ability is peculiar for just to the typical reef fish which form very rare populations. It could be found most often in gobies, groupers, wrasses, parrot fish and chaetodontids. The change of sex is a part of their complex social behavior purporting to even out the disproportion during the formation of spawning couples. (Sale 1982). The mechanisms of the sex changes are various. For example, in normal hermaphrodites like the grouper *Hypoplectrus* most adult fish in the population, or all of them, have well-developed milts and ovaries simultaneously. Depending on the ratio between males and females in a given spawning school they might function as males or as females (Barlow 1975). In the synchronous hermaphrodides (grouper *Serranus subligarus*) both glands may function simultaneously accomplishing self-fertilization. But most often the case in hermaphroditic fish is protogyny, i.e., the change of sex by the adult females which, if necessary, can function as males (Reinboth 1973). In local populations of these fish the females dominate. Rarer is the case of protoandry, i.e., the change of sex by males to females. It happens in hermaphroditic fish, which have a predominance of males in local populations (Robertson 1972).

The change of sex by protogynie females to males is accompanied by drastic changes in their coloration and even in the form and size of their body. These secondary males, transformed from females, are so distinct from the real ("primary") males that they happened to be described as a separate species. The primary males in a population relate by numbers to females ~1 to 20. Other males necessary for normal spawning of couples come into being through the change in sex by females. In some species of protogynic fish the "primary" males are completely absent, and all of the males which participate in spawning, are of secondary origin (Shapiro 1979). As an example of protogynic hermaphrodite fish the grouper *Anthias squamnupinnus* could be mentioned. The primary males of this species have a specific painting which drastically differs from the coloration of females: females are an even golden-orange color with a reddish stripe. The "primary" males have a reddish-violet head, and the dorsal-caudal part of the body is pale-pinkish with a big orange spot. When changing sex the female acquires a new coloration resembling that of the "primary" males (Shapiro 1981). The populations of the Caribbean parrot fish *Scarus croicensis* include two differently colored forms. One is of monotonously colored fish, which could be either females or "primary" males. The second form – brightly coloured larger fish – are the "secondary" males transformed from females which changed their sex.

The trigger mechanism which starts the change of sex by females of protogynic fish is the deficiency of males in a given local group. It was proved experimentally by removing several males a the group of typical protogynic hermaphrodites. This action inevitably caused the change in some females in the group to males, so that the balance of sexes was quickly restored (Finshelson 1970; Robertson and Choat 1981). The same was observed in the wrasse *Thalassoma bifasciatum*, the clown fish *Amphiprion melanopus*, and the angel fish *Centropyge interruptus* (Shapiro 1979). The change in sex by protogynic females could also be induced by injection of testosterone (Ogden and Buckman 1973). If the testosterone was injected into the "primary" males, they developed into specimens similar to "secondary" males. In family groups of the cleaner wrasse, *Labroides dimidiatus*, which consist of one leading "secondary" male, several females, and several young fish, "primary" males are completely absent and only females hatch from the eggs. The "boss" of the harem – that "secondary" male guards the territory of the station, suppressing even its own females. The feeling of being suppressed prevents their transformation into males. But when the "boss" was withdrawn from this group, 1–2 h later the largest female in the group took over his duty and assumed his behavior. After 2–3 weeks she changed her coloration and sex and became the "secondary" male (Robertson 1972). These examples prove the interconnection of adaptive social behavior and the physiological mechanisms of sex regulation in hermaphroditic fish. Besides the above-mentioned, the following taxa of her maphroditic reef fish could be mentioned: the groupers *Hypoplectrus chloronotus*, *H. pulella*, *Serranus subligarus*, the chaetodontids from the genus *Contropyge*, *Holaranthus tricolor*, the parrot fish from the genera *Scarus* and *Leptoscarus*, the wrasses of the genera *Halichoeres*, *Labrus*, *Bodjanus rutus*, *Thalassoma duppery*, and some others-totally over 30 species (Ross 1981; Robertson et al. 1982).

The reef fish resolve the method of propagation employing complex social and territorial ecophysiological adaptations. These adaptations are seen in the variability of the composition of spawning groups. The fish use all possible combinations: temporary or permanent monogamic couples, hurems, and collective spawning. As to the temporary couples, they are formed by many medium-sized and big single predatory fish, like some groupers, balistids, puffers, and reef sharks (Barlow 1975). To attract the sexual partner for spawning these fishes often utter a kind of "mating call" like birds, making curious curling swims with spreaded fins in an open place off the outer reef slope with a good visibility. Permanent monogamic couples are known in some species of gobies, chaetodontids and damsels (Reese 1975, 1981). Their "matrimonial" mode of life helps them in everyday life to defend their territory or to construct and to clean the common hole. It greatly increases as well successful multiplication of these fish, which have very rare specific populations. The harems are common in hermaphroditic reef fishes, as we have seen above.

The guarding of eggs is a usual mode of behavior of bottom reef fish, which lay their eggs on the solid substrate. They guard the spawns until the larvae get out (Johannes 1978b). Among these fish there are representatives of the families Apogonidae, Blennidae, Pomacentridae, Gobiidae, Tetraodontidae, Syngnathidae. The damsel fish *Acanthochromis polycanthus* and *Chromis dimidiatus* guard also their hatched larvae until they grow up (Robertson 1973). Some small bottom fish construct the nests or holes where they guard their spawn and later the larvae (Hudson 1977; Thresher and Moyer 1983).

Most reef fish, even the bottom-dwelling ones spawn in the water column and produce pelagic eggs (Mc Farland and Ogden 1985). The strategy of such spawning is destined to decrease the grazing losses in spawn and larvae. These fish select places off the upper reef slope, where the currents are directed to the reef, so that their spawn should be driven back to the reef (Johannes 1978; Sale 1980, 1985). The pelagic larvae of reef fish live in the water column of the lagoon and in the near coastal waters, feeding on zooplankton (Leis 1981). To escape the daytime grazing many of them migrate in the daytime to the bottom, as do the demersal plankters. The length of the larval period takes in most species 1–3 months (in moray eels up to 10 months), but in case of absence of an appropriate place for recruitment it could be longer (Johannes 1978b; Goldman et al. 1983; Williams et al. 1984). The recruitment of larvae has a seasonal character. Its rate varies in different years (Talbot et al. 1978; Williams 1983; Schroeder 1985). This results in definite fluctuations of the composition of fish communities on reefs (Sale et al. 1984). Nevertheless, as we have seen before, during a long period of time they are surprisingly stable. One of the factors providing this stability could be seen also in the grazing out of the larvae of foreign fish species by the residing adult fish, which thus promotes the recruitment of their own (Russel et al. 1974). After the shift from pelagic life, the recruited young fish adapt themselves to the settled life in the benthic biotope, entering local groups or shoals. An average life time of reef fish was evaluated to be 4–5 years (Ross 1978).

6.6 Conclusion

The share of fish communities in the use of trophic resources of the coral reef is very large, being commensurable with that by the zoobenthos. Their high density with a biomass of up to $200\,g\,m^{-2}$, as well as the great diversity of their trophic guilds, enable us to recognize them as the most important among the components of the coral-reef ecosystems. This share of reef fish in metabolism and productivity of the ecosystem is much larger than that of the ichthyofauna of temperate coastal ecosystems, not only because of their high density but also their dominance in herbivorous and planktovorous fish. Herbivory results in the shortening of the food chain and increases the efficiency of utilization of plant primary production by reef heterotrophs, while the second factor – the abundance of planktovores – promotes the input of additional external energy and nutrients into the reef ecosystem from the surrounding oceanic waters through the consuming of zooplankton by fish. An extremely important function of the fish communities in reef ecosystems could be seen in their participation in processes of nutrients regeneration as well as in bioerosion and production of soft carbonaceous bottom sediments (cf. Sects. 1.3, 2.2. 4).

The communities of reef fish attain the most extreme levels of density and diversity ever known in biological communities, either terrestrial or marine. They reach the highest degree in making use of the living spaces in natural biotopes, and succeed in using most efficiently the trophical resources. The scientific results and theories explaining these phenomena could be regarded as significant in sights into general ecology and theory of evolution (Sale 1984). The result of studies on reef-fish communities and especially the results of experiments with artificial reefs appear to be important for developing a scientific background for aquaculture (Russell et al. 1974; Nelson and Tsutui 1981). The problems of reef fisheries will be discussed elsewhere (cf. Sect. 11.5).

7 Communities of Corals in Reef Ecosystems

Most reef corals belong to the class Anthozoa. Only two of their families are related to another class of coelenterates–Hydrozoa: Milleporidae and Stylasteridae. The class of Anthozoa includes the two subclasses Hexacorallia (or Zoantharia) and Octocorallia, which are different in origin as well as in their morphology and physiology. The function of reef construction is mostly performed by the reef-building (or hermatypic) corals, which form massive calcareous (aragonite) skeletons. The group of hermatypic corals is represented mostly by the order Scleractinia (subclass of Hexacorallia). Two species in this group belong to the order of Octocorallia (*Tubipora musica* and *Heliopora coerulea*), and several species to the class Hydrozoa (hydrocorals *Millepora* sp. and *Stylaster roseus*) (see Fig. 7.1). The hermatypic corals harbor the algal symbionts zooxanthellae, which greatly accelerate the processes of calcification, thus enabling their host corals to construct massive colonies (cf. Sect. 1.2). The hexacorals from other orders of the subclass Hexacorallia: Corallimorpharia, Antipatharia, and Ceriantharia, including some species of the order Zoanthidea as well as most octocorals of the subclass Octocorallia, being colonizing animals, also produce hard skeletons or hard elements of their soft skeletons from the calcareous material, and thus participate in the production of crumb lime material. According to their functions in reef building (hermatype–ahermatype) and, connected with this function, their possession of algal symbionts (symbiotic–asymbiotic), all corals could be subdivided into the following groups (guilds) (Kühlmann 1985; Schuhmacher and Zibrovius 1985):

1. *Hermatypes–symbionts*. This group includes most reef-building scleractinian corals, octocorals and hydrocorals (cf. Fig. 7.1).
2. *Hermatypes–asymbionts*. These slow-growing corals can construct massive lime skeletons without the aid of zooxanthellae, in which they are able to live in dark environments–in caves, tunnels, and deep parts of the continental slope. Among them are the asymbiotic scleractinians *Tubastrea* and *Dendrophyllia*, and the hydrocoral *Stylaster rosacea*.
3. *Ahermatypes–symbionts*. Among scleractinians there could be included into this group the small fungiids, like *Heteropsammia* and *Diaseris*, and also the coral *Leptoseris* (fam. Agaricidae), which exist as small single

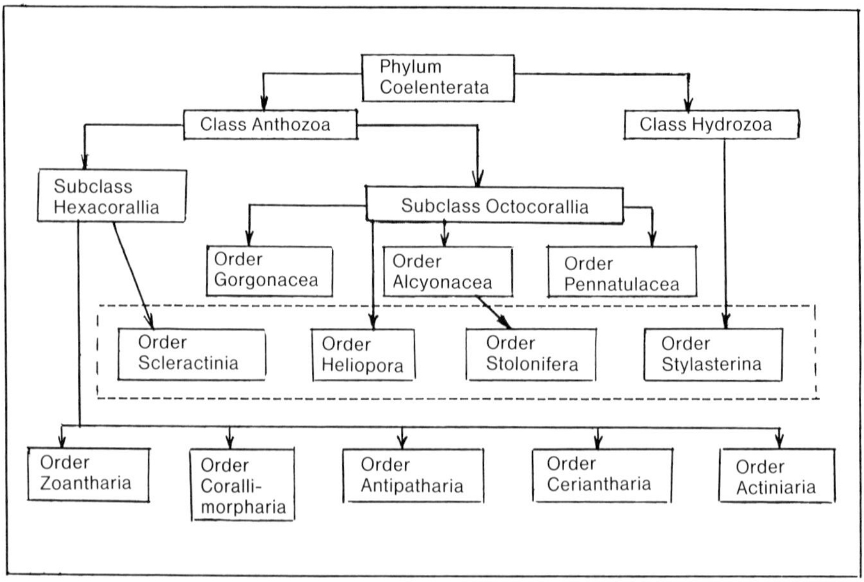

Fig. 7.1. Corals in the system of phyum Coelenterata; the taxa of hermatypic reef-building corals are placed inside the frame of dotted line

polyps or as small colonies, and so cannot be included as reef builders. This group comprises also most octocorals-alcyonaceans and gorgonaceans that have algal symbionts but do not build massive lime colonies.

4. *Ahermatypes–asymbionts*. To this group belong among scleractinians some species of the genera *Dendrophyllia* and *Tubastrea*, which have small polyps. It includes also hexacorals from the orders Antipatharia and Corallimorpharia, and asymbiotic octocorals.

In most cases the reef-building (hermatypic) corals are simultaneously the symbiotic ones. Therefore, in literature the term "hermatypic" is accepted as synonymous with the term "symbiotic". Sometimes it is not strictly correct, because we have a group of symbiotic but ahermatypic corals. But the use of these terms as synonyms became so widespread that we are using it also in this book.

7.1 Coral Communities: Composition and Formation

7.1.1 Scleractinian Corals

The communities of scleractinians inhabiting coral reefs of the World ocean are extremely various (Potts 1979). The total number of their taxa approaches 800 species which belong to ca. 110 genera. They live in warm waters in areas where the water temperature does not drop below 18–19 °C in winter, and down to a depth of 80–100 m, limited by light which they need as symbiotic animals. The composition of the fauna of sclerectinians and also the degree of its diversity vary in different reef areas and regions. The total number of their genera attains its maximum in the region of Indonesia-Philippines – North Australia, where it is over 70, with a total number of species of 250–350 (Porter 1972; Veron 1985, 1986). The less diverse fauna of scleractinians inhabits the reefs of the Atlantic basin, including only 20 genera. This unevenness in their distribution is the consequence of the relatively recent origin of this group of corals. The processes of formation of scleractinian fauna are still in progress on the reefs of the World Ocean (Veron 1985). One of the proofs for this assertion is the separation of their faunas into the Indo-Pacific and the Atlantic basins. In the young Atlantic Ocean the fauna of scleractinians is three to four times poorer compared with that of the Indo-Pacific. Only the following six genera out of 60–70 in the latter could be found on the Atlantic reefs: *Acropora*, *Cladocora*, *Favia*, *Madracis*, *Porites*, *Siderastrea*. Moreover, among the species of the above-listed common genera only one is common to both regions – *Siderastrea radians*.

Among the great variety of modern taxa of hermatypic scleractinians three groups could be distinguished. The first two have contrasting life strategies: they are the *r*- and the *k*-strategists, and the third group is intermediate between these two (Connell 1973). The first group of opportunistic corals (*r*-strategists) live in small- or medium-sized colonies, have a determined growth, reach early sexual maturity, and spend a larger part of their energy in breeding. Most of them have a monthly cycle of sexual multiplication. They have a life of short duration and a high growth rate. The success of their survival is achieved by their intensive breeding, which by their recruitment increases their chances in competition for the hard substrate. They widely use for this purpose also the vegetative multiplication by pieces of branches (cf. Sect. 7.4). The opportunistic corals may survive different kinds of stress, like exposure, low salinity overheating, pollution, and hot and turbid alkaline waters of reef shallows. Among these opportunistic corals the most common on the Indo-Pacific reefs are *Stylophora pistillata*, *Psammocora contigua*, *Pocillopora damicornis*, *Seriatopora histrix*, and many species of the genera *Montipora*, *Acropora*, and *Pavona*.

The group of conservative corals (*k*-strategists) use their energy mostly for their own growth and metabolism. Their growth is practically unlimited. Their old colonies can have diameters of 1–3 m. Correspondingly, they use less energy for propagation and cope with the hard substrate available through the formation of large colonies and thanks to their longevity. They may live for tens and hundreds of years. Their cycles of breeding have an annual periodicity. As examples of typical *k*-strategists, the massive corals *Porites* and *Montastrea* could be mentioned.

Most other corals belong to the third group with an intermediate life strategy between the contrasting types. This gives them a special lability which enables them to familiarize themselves with diverse kinds of environments and various types of hard substrates. These corals are less specialized, and the polyps of many of them are active also in daytime. Phenotypically, they are labile, forming in various reef environments a multitude of adaptive ecomorphs. Among them are the most common species of the large genus *Acropora*, most of the faviids, the genera *Pavona*, *Hydronophora*, *Galaxea*, and *Goniopora*. The communities of deep-reef biotopes with stable environmental conditions include a lot of specialized corals living there only, like some agaricids and some species of the genera *Turbinaria*, *Ehinophyllia*, *Leptoseris*, and *Diaseris*.

The taxonomy of reef corals is fraught with many problems and raises points for discussion (Boushon 1981). Most of their species are morphologically extremely versatile. Living in diverse environments they have a very high level of phenotypic variability, and form mumerous ecomorphs on different reefs or even within the same reef in its different zones. These morphs are diverse in the form of their colonies, coloration, and the size of corallites (Veron and Wallace 1984). These ecomorphs have often been described as separate species of corals. Therefore, the number of their nominal species significantly exceeds the number of the real ones. For example, the genus *Acropora* includes 365 nominal species and about 90 real ones. The genus *Montipora* includes, correspondingly, 211 nominal species against only 50 real ones, the genus *Pocillopora* 10 real species out of 35 nominal ones, *Stylophora* 4 real for 35 nominal, *Turbinaria* 10 real out of 80 nominal, etc. Small wonder then that the system of scleractinians has been revised many times (Veron et al. 1977; Ditlev 1978, Veron and Wallace 1984; UNESCO 1985). Below short descriptions of the ecological characteristics of the most important families of scleractinians are given.

Family Acroporidae. This is the largest and most important family of reef-building corals. It includes only three genera: *Acropora*, *Montipora* and *Astraepora* but about a third part of their total species belong to this family: 150 out of a total of 520–530 real species of scleractinians. The genus *Acropora* comprises around 90 real species. Together with corals from related genera, like *Montipora*, *Pocillopora*, *Stylophora* and *Seriatopora*, the genus *Acropora* dominates on most reefs both by the relative number of colonies and by the relative area covered. These corals are especially common in shallow surf zones of reef flats and of upper outer slopes. They are numerous also in lagoons and in submerged areas of reef flats. Acroporids are the main reef builders on most reefs (Kühlmann 1974; Latypov 1979; Randall 1981; Veron 1986). They grow faster than any other, even ramose, corals: 15–20 cm year^{-1} (cf. Sect. 1.2). The large colonies of staghorn *Acropora* species compose a basic framework during the growth of reef. When they live in the surf zone and have powerful colonies, several common species of this genus absorb the main energy of wave-stress and thus preserve the reef from destruction. This function on reefs of the Atlantic is performed by *Acropora cervicornis* and *A. palmata*, and on the Indo-Pacific reefs by various *Acropora* species, like *A. formosa*, *A. humilis*, *A. hyacinthus*, *A. digitifera*, and *A. palifera*. Some species of *Acropora* form the crustose colonies (*A. palifera*). Many acroporids are opportunists and may survive severe stress conditions: heating, exposure, siltation. Their species *A. squamosa*, *A. pulchra*, and *A. hebes* inhabit shallow reef zones exposed during low ebbs, or the coastal areas of the barrier and of the monsoon reefs subjected to terrigenous siltation (Ditlev 1978).

The Genus Montipora is one of the key taxa of corals on reefs of the Indo-Pacific. It includes about 50 real species, being second after *Acropora* in this respect. Many species of *Montipora* are among the most opportunistic corals and may survive most unfavorable conditions which could be met with on reefs. (This author has seen thickets of *Montipora* sp. on muddy bottoms of coastal bays of the Vietnamese coast.) They often form in such stressed environments large monospecific populations. Among the most widespread corals of this genus *M. erythraea*, *M. verrucosa* and *M. hispida* could be mentioned. The colonies of most *Montipora* species are foliose, ramose or columnar.

Family Pocillopordae. This family is related to *Acropora* and includes three genera of ramose corals most common on reefs of the Indo-Pacific: *Pocillopora*, *Stylophora* and *Seriatopora*, which include about 20 real species. Together with the acroporids these corals are the main components of scleractinian fauna in the upper 20-m reef zone, at reef flats, at outer reef slopes and at patch reefs. The species of this family, *Pocillopora damicornis*, *P. verrucosa*, *Stylophora pistillata* and *Seriatopora histrix*, are among the most widespread and opportunistic reef corals. Their colonies are present in large quantities at any reef of the Indo-Pacific and could be found practically in all its zones. Their polyps are open all day around. They grow quickly and form medium-sized (20–50 cm) colonies, which well survive stress situations. The fourth genus of this faimly, *Madracis*, lives mostly in deep zones of reef slopes.

Family Poritidae. On most of Indo-Pacific and Atlantic reefs this family ranges third in importance as component of the scleractinian fauna. The main genus of this family, *Porites*, comprises 120 nominal species. Some species of poritids have large massive colonies attaining sizes of 2–3 m and ages of several hundred years. Other species having columnal or ramose-like colonies attain the same large size. Poritids are an important component of coral communities in the surf zones of reefs, on outer slopes and in the zone of lagoonal patch reefs. On the submerged reef platforms and on shallows of fringing and barrier reefs large colonies of *Porites* form micro-atolls (Rosen 1978). Another common genus of this family, *Goniopora*, forms massive round colonies with large far-expanding polyps. This coral is opportunistic, the same as many species of *Porties* and inhabits shallow areas, patch reefs and exposed zones of reef flats.

Family Faviidae is the fourth in importance among reef coral communities. Several genera of faviids inhabit reefs of all three oceans. It includes about 150 nominal species and over 15 genera. The most common of them are the genera *Favites*, *Goniastrea*, *Montastrea*, *Cyphastrea*, *Platygyra*, *Leptastrea* and *Echinopora*. Most of these genera form massive round colonies with medium-sized or large polyps. Some of them (*Cyphastrea*, *Leptastrea*) may form flat crustose ecomorphs when living in a surf zone or in deep poorly illuminated areas of reef. Most often faviids could be found in shallows of submerged flats and in lagoonal patch-reef zones. *Montastrea* is not a key species in the surf zone and on the reef slopes of many Atlantic reefs.

Family Siderastridae is most widespread on reefs of the Western Atlantic, where it is represented by one genus (*Siderastrea*) and four real species. One of them, *Siderastrea radians*, is the single scleractinian species inhabiting both Atlantic and Indo-Pacific reefs. In the Atlantic, *Siderastrea* is an important coral on reef slopes (Goreau and Goreau 1973). On reefs of the Indo-Pacific besides *S. radians* another coral of this family, *Coscinarea columna*, is common. Its columnal colonies could be often found at reef slopes and on lagoonal patch reefs.

Family Agaricidae is represented on reefs of the Indo-Pacific by three genera: *Pavona*, *Leptoseris* and *Pashyseris*, which are very distinctive in the appearance of their colonies and their ecological features. All three of them play a key role in different reef biotopes. *Pavona* and especially *P. cactus* is one of the most common elements of coral fauna in surf zones, in zones of submerged flats, in lagoonal shallows and on patch reefs. It easily survives wave-stress, high turbidity, heating and siltation. The genus *Leptoseris* is common in the Indo-Pacific and Atlantic regions, being represented there by different species. In the Indo-Pacific, *Leptoseris* is represented by 14 species which inhabit mostly deep parts of outer reef slopes down to the lower boundary of distribution of hermatypic scleractinians, at depths of 60–90 m (Kühlmann 1983). Some of the species of *Leptoseris* could be found only deeper than 20 m. The same features are peculiar to another genus *Pashyseris*; which is common in coral communities of deep reef slope biotopes (Fricke and Schuhmacher 1983). Both these genera of scleractinians have special morpho-physiological adaptations to low illumination (cf. Sect. 8.4). In the Atlantic, this family is represented by several common species of the genus *Agaricia*. They have mostly plate-like colonies and inhabit the deep fore-reef and buttress zones of outer slopes (Goreau and Goreau 1973). Their colonies often form there continuous covers. The coral *Laptoseris* is also common in the same biotopes of Atlantic reefs, being represented there by the single species *L. cucculata*.

Family Oculinidae includes a coral very common on the Indo-Pacific reefs, *Galaxea fascicularis*, which forms small and medium-sized crustose or globe-like colonies. In its ecological features it resembles the coral *Pavona*, mentioned above. Another oculinid coral, *Achrelia horrescens*, forms extremely beautiful but very spiny ramose colonies. Several species of oculinids are living also on the Atlantic reefs.

Family Merulinidae have representatives only on reefs of the Indo-Pacific. Some of them are rather common, like *Merulina ampliata*. This coral has foliose or columnar colonies and inhabits reef flats and the upper zone of reef slopes, the same as another coral, Hydnophora exaesa, frequent on Indo-Pacific reefs, which has massive or columnar colonies.

Family Mussidae includes two genera common on reefs of the Indo-Pacific, *Lobophyllia* and *Symphillia*. Their large massive colonies are formed by separate big corallites 3 cm in diameter and 20–30 cm long. They inhabit usually the outer slope zones down to depths of 30 m, and the bottom of deep lagoonal patch reefs. In the Atlantic, this family is represented by six genera. Most common among them are *Mussa*, *Mycenophyllia* and *Scolimia*. They form massive or crustose colonies and prefer biotopes with low or medium turbulence.

Family Fungiidae unite the solitary scleractinian corals, which live only on reefs of the Indo-Pacific. Their single free-living polyps, which attain in some species 30–40 cm, hide within massive round or ellipsoid corallites. Their average size is 10–20 cm. They inhabit the bottom of pits and channels covered with rubble and sand in turbulent zones of reef flats, reef edge and moat on outer slopes. Often fungiids form large aggregations on bottom areas of several square meters. They are shifted by surf, some of them also moving actively with the aid of their long tentacles, and are able to live on the surface of sandy bottoms in lagoons. Among them the genera *Fungia*, *Heliofungia*, *Halomitra*, *Diaseris*, *Cycloseris*, and *Poliphilla* are the most common (Pichon 1974; Gill and Coates 1977).

Family Dendrophylliidae unite scleractinian genera with various structures of colonies, hermatypic and ahermatypic. It includes solitary corals (*Heteropsammia*), corals with colonies structured as a bunch of separate corallites (*Dendrophyllia*) and corals which form real colonies, either ramose (*Tubastrea*) or falt, funnel-like (*Turbinaria*). Most of the dendrophylliids are ahermatypic (asymbiotic) except the genera *Turbinaria* and *Heteropsammia*, which harbor algal symbionts. Large colonies of *Turbinaria* could be met with usually in poorly illuminated biotopes of reef slopes, on walls of channels and caves, on reefs with turbid waters and low water transparency situated on coastal reefs of continental coasts. Other hermatypic genus is *Heteropsammia*, which includes a single species *H. cochlea*, and has, besides the zooxanthellae, another remarkable symbiont – the sipunculid worm *Aspidosiphon*, which moves these small (1–3 cm) free-living solitary corals over the sandy bottom of the lagoon or at the foot of reef slopes (cf. Sect. 7.3). Their number in such biotopes is often extremely high, up to hundreds m^{-2} (Latypov 1985). The ahermatypic dendrophylliids *Dendrophyllia*, *Balanophyllia*, and *Tubastrea* are components of the sciophylous scleractinian fauna in poorly illuminated or even dark biotopes: in caves, trenches, channels, and in deep zones of the reef (Sheppard 1981b; Dinesen 1982).

7.1.2 Zoantharians and Octocorals

Zoantharians like scleractinian corals form colonies by the budding of polyps. Most of them do not have skeletal elements. Only the genus *Palythoa* forms hard massive crustose colonies of firm chitin-like material incrusted with calcareous globules. This group of *Anthosoa*, though taxonomically poor, nevertheless is very widespread in shallow reef biotopes. They form large colonies on rubble, on rock surfaces of flats, on dead, and even on living, colonies of corals (Goreau 1959; Herberts 1972; Karlson 1983). The most common reef zoantharians are represented only by three genera: *Zoanthus*, *Isaurus* and *Palythoa*. They include both in the Atlantic and in the Indo-Pacific the species *Z. sociatus*, *Z. pacificus*, and *Z. sandwichiensis*. On the Atlantic reefs also common are *Polythoa caribaeorum*, *Isaurus duchassainngi* and *I. spongiosus*, while in the Indo-Pacific *P. tuberculosa*, *P. caesia* and *P. vestitus*.

The role of octocorals in the formation of coral communities on reefs is often just as important as those of the scleractinians. By numbers of species, cover of bottom surface and by their biomass they are comparable on many reefs. Often the octocorals even dominate over scleractinians. They are pioneering in the recolonization of reefs which suffered from catastrophic events: after hurricanes, or after *Acanthaster* plagues (Nishihira and Yamazato 1974; Nishihira 1981; Yoshioka and Yoshioka 1989). Octocorals harbor many different invertebrates – commensals, symbionts, and parasites. They are also effective producers of biologically active substanses and calcareous material (Cary 1931; Crossland 1938; Bayer 1961; Velimirov and King 1979). Nevertheless, in most studies on coral communities they were most often simply ignored, apart from the fact that the description of their fauna is rather difficult because of their insufficiently developed

Table 7.1. List of common genera of Octocorallia

Indo-Pacific reefs		Western Altantic reefs	
Order Gorgonaceae		*Order Gorgonacea*	
Echinogorgia	*Pexauroides*	*Plexaura*	*Pterogorgia*
Acabaria	*Rhabdoplexaura*	*Rhabdoplexaura*	*Pseudopterogorgia*
Bebryce	*Isis, Rumphella*	*Plexaurella*	*Gorgonia*
Mopsella	*Juncella, Parerythropodium*	*Muricea*	*Elisella*
	Briareum	*Muriceopsis*	*Nicella*
		Eunicea	*Swiftia*
		Pseudoplexaura	*Lephogorgia*
			Thesea
Order Alcyonacea			
Lobophytum, Xenia, Efflatouniaria			
Sarcophyton	*Heteroxenia*	*Briareum*	*Erythropodium*
Alcyonium	*Anthelia*	*Iciligorgia*	*Diodogorgia*
Sinularia	*Cespicularia*		
Nephthea	*Capnella*		
Dendronephthea	*Cladiella*		
Stereonephthea	*Asterospicularia*		
Lemnalia	*Siphonogorgia*		
Paralemnalia	*Nephthyigorgia*		
	Telesto		
Order Stolonifera			
Tubipora	*Clavularia*		
	Pasyclavularia		
Order Helioporaceae			
Heliopora			
Order Pennatulaceae			
Virgularia	*Cavernularia*		
Veretillum			

systematics and the complexity of the procedure of taxonomic identification itself (Opresko 1973). The system of octocorals was revised several times (Bayer 1961, 1981; Cairns 1977). In Table 7.1 a list of common genera of reef octocorals is given.

Octocorals flourish on most of the Caribbean reefs and on many reefs of the Indo-Pacific (Loya 1972; Kinzie 1973; Grasshoff 1976; Benayahu and Loya 1977; Lasker and Coffroth 1983; Benayahu 1985; Yoshioka and Yoshioka 1989). They are very successful in competition for the hard substrate with scleractinians and other sessile fauna and plants owing to their fast growth (Sheppard 1979; Benayahu and Loya 1977, 1981). To suppress the growth of competitors and to repel predators, they excrete various toxic and allelopathic metabolites (Bakus 1981; Coll et al. 1982; Tursch and Tursch 1982; Sammarco 1983). Some alcyonaceans, like *Xenia macrospiculata* and *Nephthea brasilica*, are even able to move their colonies and so capture new lots of hard bottom (Benayahu and Loya 1981; La Barre and Coll 1982).

Most alcyonaceans and many gorgonian octocorals have in their polyps the algal symbionts zooxanthellae, the same as scleractinian corals (Kinzie 1974). Among common symbiotic alcyonaceans there could be mentioned the genera *Alcuonium*, Lithophyton, Lobophytom, *Sarcophyton*, *Sinularia*, *Capnella*, *Cladiella*, *Lemnalia*, *Pralemnalia*, the stolonifera coral *Tubipora*, the coral *Heliopora* (fam. Helioporaceae), and genera of the family Xeniidae. Among the gorgonacean soft corals are: *Isis*, *Rumphella*, *Eunicea*, *Pseudopterogorgia*, *Plexaura*, and *Plexaurella*. Asymbiotic alcyonaceans and gorgonians often could also be found in well-illuminated zones of the reef, though most often they prefer deeper dark biotopes, caves or channels. These are the alcyonaceans *Nephthea*, *Dendronephthea*, *Clavularia*, *Pashyclavularia*, and the gorgonians *Bebryce*, *Acabaria*, *Mopsella*, and *Juncella*.

7.2 Communities Structure of Reef Corals

7.2.1 General Considerations

Since World War II, studies on the structure of coral communities and their zonal distribution received an impulse after the introduction of scuba diving and the development of underwater photography. During these studies, the transsect method was most often used, i.e., the recording of species of corals along the rope or the chain laid across the reef by scuba divers. This method is more effective if the recording of coral colonies proceeds within squares placed on the bottom along the transsect (Loya and Slobodkin 1971; Loya 1972; Stoddart and Johannes 1978; Kenchington 1984). To study coral communities in deeper zones of the reef, submersibles have been employed (Lang 1974; Fricke and Schuhmachez 1983; Reed 1985). The dynamics of coral communities in time has been observed with the aid of photogrammetric monitoring (Done 1981; Bright et al. 1984). Parameters of species diversity and distribution of coral communities are calculated using most often Shannon-Winer's Index H′ (Loya 1972; Pichon and Morrisey 1981), cluster analysis (Done 1982) and the methodology of modeling (Abel et al. 1983).

The space distribution of the corals' taxa in bottom reef biotopes could be taken as the static reflection of the structure of their communities, because it is the result of both the stochastic processes of recruitment, growth and survival of individual corals, and of the regular socio-ecological relationships between their specific populations, and between them and the environment as well (Dana 1976). Among the factors controlling the space distribution of corals the following could be mentioned: (1) the vector gradients of parameters of physical stress, like waves, currents, tide height, nutrients concentration, illumination, turbidity of water; (2) social factors: formation of monospecific, interspecific, or reproductional bisexual groups; (3) interorganismic relationships: commensalism, symbiosis, antagonism, prey–predator; (4) various stochastic events including the extreme ones, like hurricanes, floods, Acanthaster plagues; and (5) anthropogenic influence.

Most reef corals dispose of a wide scale of ecological adaptations, being able to colonize practically all zones of the reef slope starting with the shallows of the reef flat (Figs. 7.2, 7.3). Their species diversity manifested by the total numbers of species does not significantly change within the scale of depths from 8 down to 60 m either on reefs of the Atlantic or the Indo-Pacific (Fig. 7.4). Only on reef shallows does the number of species in

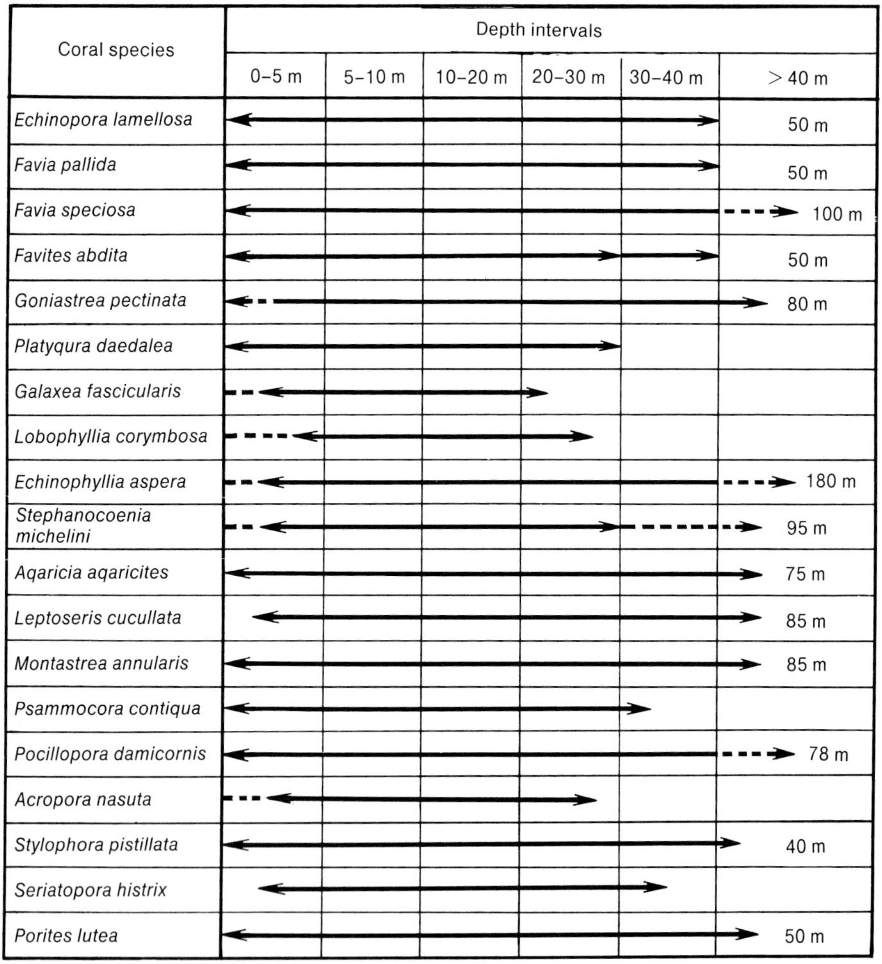

Fig. 7.2. Distribution of some common corals on the outer reef slopes of the Pacific atolls. (Data by Kühlmann 1983)

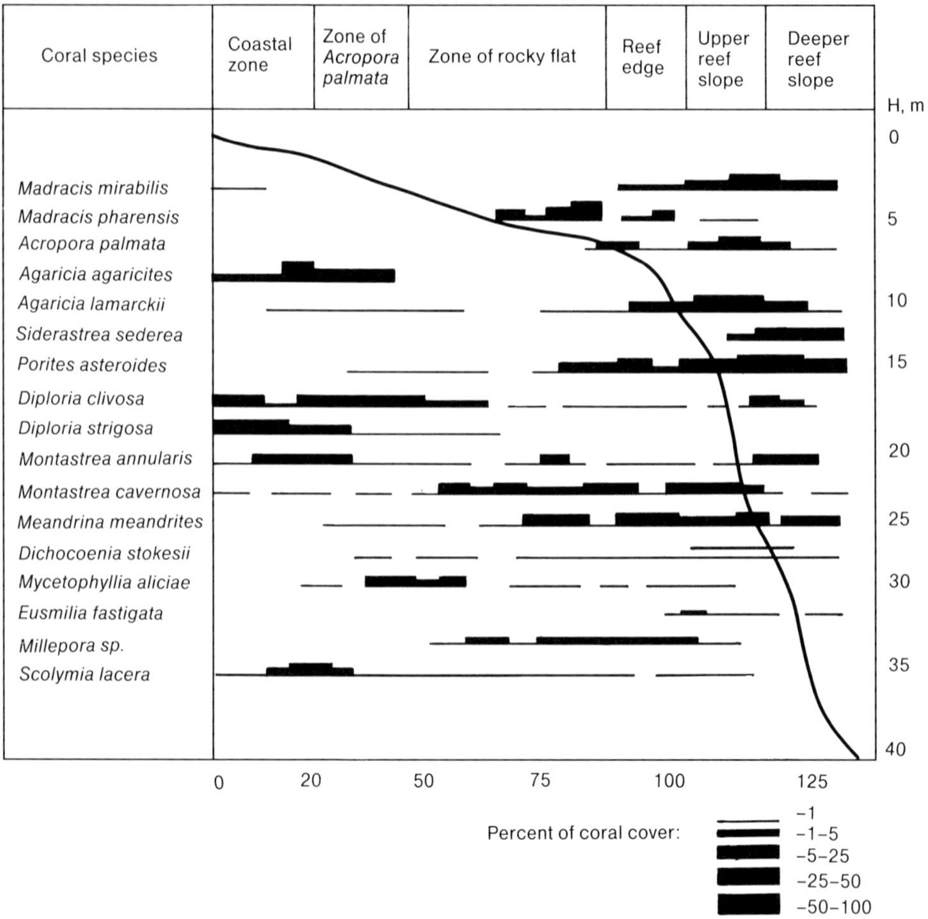

Fig. 7.3. Percentage of cover of bottom by common corals on the section across the reef in Piscadera bay, Curacao; the *curve* represents the bottom; profile H_m depth; L distance from the shore, meters. (Data by Bak 1977)

coral communities decrease. In spite of their ability to inhabit various reef zones, competition for the hard substrate is so severe that it results in a zonal distribution of coral taxa in reef biotopes and in the appearance of dominating species of corals within their communities (cf. Sect. 7.2.2). Correspondingly to the zonal changes of environmental conditions from shallows via reef-flat down the reef base the structure of the coral communities changes (Kühlmann 1974, 1983). These changes manifest themselves in their species composition, types of species ecomorphs, degree of taxonomic diversity (index H′), percentage of projective cover, and in the size and age of the colonies (Glynn 1976; Loya 1976c; Bradbury 1977; Ditlev 1978; Sheppard 1980; Morrisey 1980; Bradbury and Young 1981; Done

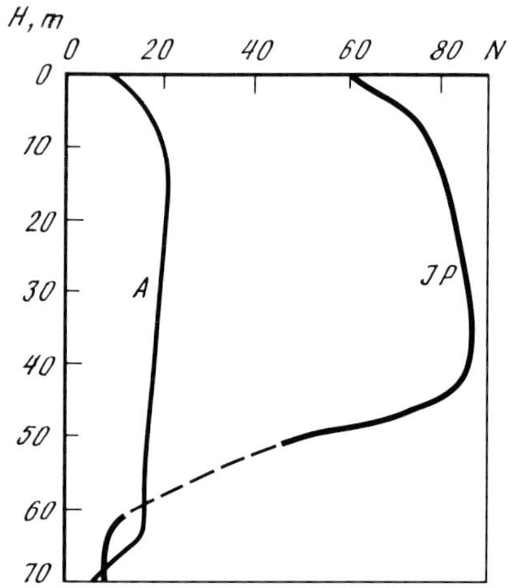

Fig. 7.4. Changes in total number of coral species on vertical profiles at the outer slopes of the Indo-Pacific (*IP*) and the west Atlantic reefs (*A*); *H* depth. (Data by Kühlmann 1983)

Table 7.2. Dominating taxa of corals in different reef zones and the percentage of cover of the bottom by corals at the cross-section through the fringing reef off one of the Namsu I., Gulf of Siam (Data by Latypov 1985)

Zone Reef	Dominating coral taxa	Percent cover	
		Total (S_t)	Cover of bottom by the two first genera listed, % S_t
Coastal lagoon ("boat channel")	*Pocillopora*, *Zoanthus*, *Psammocora*, *Acropora*, *Goniastrea*, *Millepora*	10–40	to 70
Backward part of reef flat	*Acropora*, *Galaxea*, *Lepticia*, *Montipora*, *Millepora*	20–100	30–80
Seaward part of flat	*Porites* (*murrayensis*), *Galaxea*, *Turbinaria*, *Acropora* (*millepora*), *Montipora*, *Goniastrea*	60–100	50–90
Reef slope	*Porites* (*lutea*), *Montipora*, *Acropora* (*millepora*), *Pavona*, *Favia*, *Platygyra*, *Goniastrea*	20–90	20–60
Declined plat-form near the base of reef, covered with soft sediments and seagrass	*Leptoseris*, *Goniopora*, *Leptastrea*, *Symphillia*, *Heterocyathus*, *Heteropsammia*	3	–

1982). The change in composition of coral communites at cross-sections via reef to outer slope and reef to lagoonal slope to the bottom of the lagoon is given in Tables 7.2, 7.3, 7.4. The tables demonstrate that different reef zones are inhabited by different taxonomic groups of corals with the dominance of two to four species within them (Kühlmann 1974, 1975). The dominance is especially remarkable in the wave-stressed environments, where several dominant species may occupy up to half the total surface of the bottom, covered by corals. Many of them form various ecomorphs (Morton 1974; Preobrazhensky 1975; Pichon 1978; Glynn 1982). Species which form in the upper illuminated and in wind-stressed zones massive or

Table 7.3. Percentage of cover of bottom surface (S) by corals and their dominating species in zones of lagoonal reef off Noumea, New Caledonia, subjected to medium wave-stress. (Data by Faure et al. 1981)

Zone, depth, m	Dominating corals	S, %
Submerged flat; 0–0.5 m	*Acropora digitifera*, *A. humilis*, *A. millepora*, *A. variabilis*, *Pocillopora*, *Goniastrea*	80
Reef edge; 0.5–2 m	*Favia favus*, *F. rotumana*, *F. pallida*, *Montastrea hirta*, *Leptoria phrigia*, *Platygyra daedalea*, *Hydrophora microcones*	60–80
Reef slope; 2–8 m	*A. digitifera*, *A. humilis*, *A. florida*, *A. splendida*, *A. clathrata*, *A. hyacinthus*, *Pavona*, *Mycedium*, *Pashyseris*, *Lobophyllia*, *Echinopora*, *Pectinia*, *Montipora*	80–100
Base of slope; 8–10 m	*Porites lutea*, *Platygyra*, *Leptoria*, *Stylophora*, *Acropora formosa*, *A. palifera*, *Favia*, *Leptastrea*	50–60
Bottom of lagoon covered by rubble; 9–15 m	*Turbinaria*, *Hydnophora*, *Goniopora*, *Fungia*, *Cyphastrea*, *Astraeopora*, *Montipora*	~5%
Same, bottom covered by sand; 15–25 m	*Trachyphylla*, *Heteropsammia*, *Heterocyathus*, *Cycloseris*, *Pocillopora*, *Seriatopora*, *Montipora*	<5%

Table 7.4. Characteristics of structure of coral communities in different zones of the reef Apo-Ato, Mindoro I., Philippines (Data by Ross and Hodson 1981)

Parameters of community structure	Reef zones					
	Leeward part of flat	Flat	Reef crest	Reef edge	Upper zone of reef slope	Buttress zone
Percent of cover of projective bottom surface	41	33	41	35	46	16
Number of species	17	25	44	70	90	35
Shannon's Index of species diversity (H′)	1.6	2.1	2.7	3.7	3.7	3.1

columnar colonies, when living in calm and poorly illuminated deeper zones acquire the tendency to form flat or incrusting colonies (cf. Sect. 7.3).

The total number of species and the index of species diversity also vary in various zones. The lowest numbers of species were usually recorded on reef shallows and during the ebbs exposed parts of reef flats. In zones on reefs of the Indo-Pacific the number of coral species varied between 10–30, and on reefs of the Atlantic between 3–5. Most often the numbers of species were maximal in zones of buttress and apper fore-reef, and on outer reef slope at depths of 10–20 m with illumination of 10–20% of PARS (penetrated surface photosynthetically active radiation). In these zones on reefs of the Indo-Pacific the number of species in the communities was within 40–100, and on the West Atlantic reefs within 10–30 (Sheppard 1980; Ross and Hodson 1981; Sy et al. 1981; cf. Figs. 7.5, 7.6). Below 30 m the number of species in coral communities gradually decreases down to 10–40 species in the Indo-Pacific, and to 5–10 in the Atlantic. Sometimes the maximum number of coral species could be observed also on submerged flats of Pacific reefs – up to 100–120 (Wells 1954). A high species diversity was also recorded even in the wind-stressed biotopes of windward reef flats. As examples, the Heron Island reef, GBR (Bothwell 1984), and the reefs off Mindoro Island, Philippines, could be mentioned (Ross and Hodson 1981). The indexes of species diversity H' (Shannon-Winer) in coral communities are extremely high and reach their maxima (3–4) in biotopes of the outer slope, as also in diverse communities of coral reef fish (Table 7.4). But the corals are not likewise specialized in relation to their trophic sources. Moreover, they are apparently similar in this respect (cf. Sect. 9.2.1). So this cause of high diversity, which seems to be true for fish does not apply in corals. This could be seen in the high degree of repetition of catastrophic events to which coral reef communities are subjected. These events are

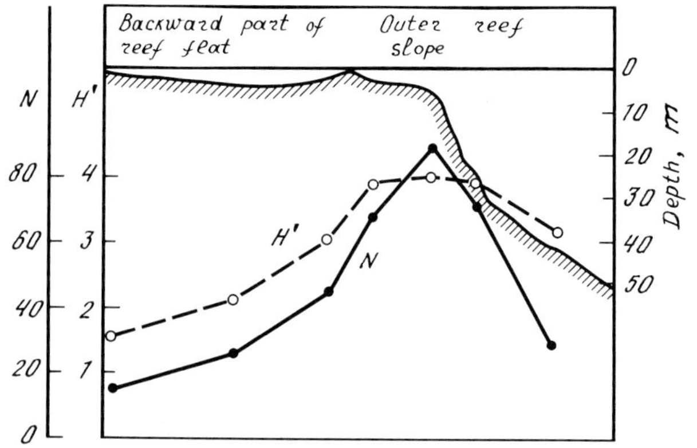

Fig. 7.5. Total number of coral species (N) and species diversity index (H') in different zones of the Apo reef, Philippines; *upper curve* profile of bottom. (Data by Ross and Hodson 1981)

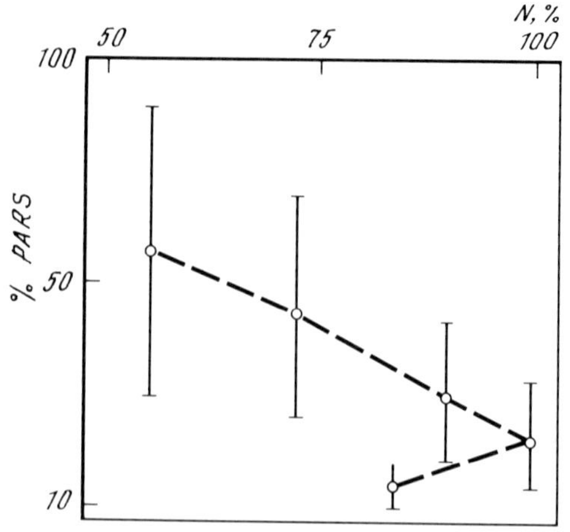

Fig. 7.6. Dependence of total number of coral species (*N*) upon light intensity (*PARS*) on slopes of reef off Tohthu I., Vietnam. (Data by Titlyanov et al. 1988c)

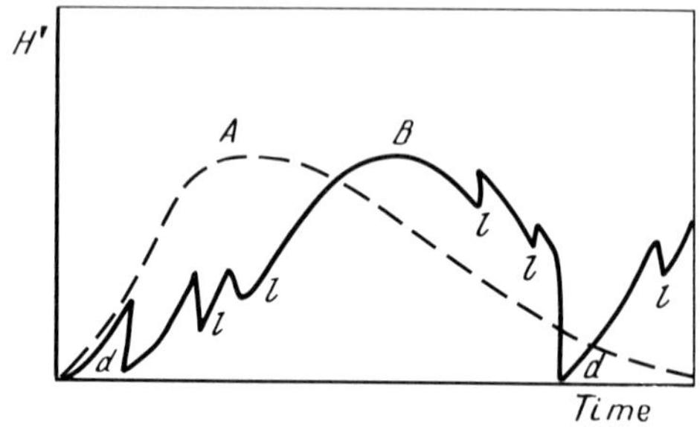

Fig. 7.7. Model of successional changes in the index of species diversity in a coral community: *A* with undisturbed gradual development, and *B* with development periodically interrupted by moderate (*l*) and destructive (*d*) stress impacts

stochastic physical and biological impacts, such as hurricanes, extremely low ebbs, long exposures, floods of cold or fresh waters accompanied by massive siltations, Acanthaster plaques, epidemic diseases of corals, etc. (Dayton and Hessler 1972; Goreau et al. 1972; Loya 1976c; Connell 1983; Bothwell 1984). These disturbances result in unselective deaths of corals and thus make space for resruitments of new species (Colgan 1981). This hypothesis is illustrated by the scheme of Fig. 7.7. The curve *A* predicts the changes of species diversity in the coral community when the above-mentioned disturbances are absent. In accordance with this theoretical prediction, at the late phase of normal succession the species diversity as well as the total number of species decrease. The curve *B* hypothetically describes the possible changes of species diversity during the succession of a community

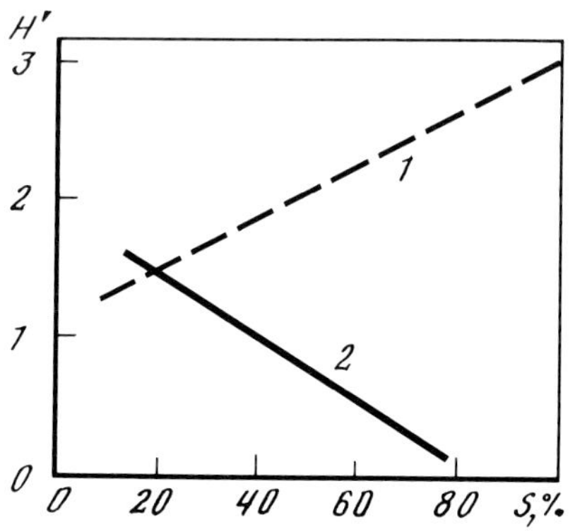

Fig. 7.8. Correlation between the species diversity index (H') in coral-reef communities and the percentage of bottom cover by corals (S): *1* at Caribbean reefs, and *2* at eastern Pacific reefs near the coast of Panama, demaged by *Acanthaster* plagues. (Data by Porter 1974a)

that is periodically disturbed by a medium (e), strong (m), and catastrophic (d) external impact.

An example of normal succession in natural nonstressed reef environments on reefs off the Pacific coast of Panama characterized by a low wave-stress was given by Porter (1974b). He used as an indirect indicator of the phase of succession the percentage of cover of bottom by corals. In this place he observed a reverse correlation between the percentages of cover and the index of diversity H′ (Fig. 7.8). This means that at the latest phase of normal unstressed succession the level of diversity decreases. But on the Caribbean reefs, which are subjected periodically to strong physical stress, like hurricanes, surf, and wave resuspension of sediments, species diversity increased with the increase in the percentage of cover. The same was shown in reefs of the GBR (Bothwell 1984). This hypothesis of disturbances is now widely recognized (Connell 1978, 1983; Grigg 1983). Nevertheless, it should be mentioned that it seems to be acceptable only for communities living in the upper, stressed reef zones. With regard to the nonstressed communities of deeper parts of the outer reef slope, this hypothesis is deficient, because we observe there a high diversity at low stress. Thus, in these biotopes a high species diversity might be the consequence of a high level of fine ecophysiological specialization, as predicted by theory (Slobodkin 1968). These examples lead to the conclusion that the level of species diversity in coral communities might be regulated in different reef zones by some of these mechanisms or by their common action (Harriott 1983).

7.2.2 Communities of Scleractinian Corals

Scleractinian corals are a dominating component of sessile zoobenthos in hard bottom biotopes, covering there 70–100% of the total area colonized by sessile benthic animals (Table 7.5). The largest percentage of bottom cover by scleractinian corals was recorded in biotopes of submerged flats and outer reef slopes on reefs of the Indo-Pacific. It often attains there

Table 7.5. Percent of cover of the bottom at slope of fringing reef off Curaçao Island by main components of sessile benthos (Data by Bak and Luckhurst 1980)

	Percent cover		Dominating corals	
Depth, m	By total sessile benthos	Only by corals	Species	Area covered by given species, % of that covered only by coral
10	41	34	*Montastrea annularis*	50
20	46	35	*Meandrina* sp. *Agaricia agaricites*	12
40	31	25	*Agaricia lamarckii*	19

Table 7.6. Percentage of projective cover of bottom surface by corals in different reef zones of several reefs

Location	Zone, depths			Reference
	Flat and reef edge (0–1 m)	Upper part of reef slope (1–3 m)	Middle part of reef slope (3–20 m)	
Guam I.	49	59	50	Colgan (1981)
Central GBR	50–100	80–100	25–60	Wallace and Bull (1981)
Cuba	73	58	36	Kühlmann (1974)
Mindoro I., Philippines	35	46	16	Ross and Hodson (1981)
Antille J.	25	35	18	Bak (1977)

Table 7.7. Biomass of main groups of benthos and fish on one of the Puerto Rican reefs with dominating association of coral Porites. (After Glynn 1973a)

Groups of benthos	Biomass, m^{-2}	
	Dry weight (kg)	Protein (g)
Macrophytes	1.02	32.5
Foraminiferans	0.13	1.8
Scleractinian corals	12.02	285.5
Zoantharians	0.001	0.4
Other zoobenthos	0.38	87.2
Bottom fish	0.006	3.6

100% of bottom surface (Wallace and Bull 1981; Latypov 1985; Tables 7.3–7.8). On reefs of the Atlantic the percentage of cover is usually less and comprises on submerged flat and on outer slope only 20–60% (Bright et al. 1984). In areas with a level of 30–60% cover the biomass of corals (dry weight together with the skeleton) varied between 3 to 20 $kg\,m^{-2}$, the

Table 7.8. Percent of cover of bottom surface by colonies of dominating coral species (S_t) and perentage of total number of colonies (S_n) on section across the fringing reef off Reunion I., Mascarenes. (After Boushon 1981)

Biotope	Dominant species	S_t,%	S_n,%
Submerged flat	*Pocillopora damicornis*	30	15
	Montipora circumvallata	18	31
	Acropora pharaonis	9	11
	Porites nigrescens	11	8
	Total for 4 species	68	65
Reef crest	*Acropora pharaonis*	13	28
	Montipora circumvallata	22	22
	Pavona divaricata	18	21
	Total for 3 species	53	71

Table 7.9. Size composition of corals on reef off Guam I. (Data by Colgan 1981)

Size class of colonies, cm	Number of colonies of given size class, % of total	
	In upper zone of reef slope	In buttress zone of reef slope
2–5	19	36
5–15	50	49
15–25	16	12
25–46	12	3
46	3	0

content of protein in them between 60–400 g m^{-2}, and the content of organic carbon between 20–260 g m^{-2} (Glynn 1973a,b; Lee et al. 1975; Table 7.8).

The size composition of corals in their communities is characterized usually by the dominance of medium-sized young colonies of 5–15 cm (diameter) with a projective area of 200–600 cm^2 (Loya 1972; Sheppard 1980; Colgan 1981; Table 7.9). In different reef zones the size composition of populations of the same coral species might change (Loya 1972). In upper reef zones the size of colonies is larger than in deeper zones of reef slopes (Fig. 7.9), but at depths lower than 10–20 m their size again increases. In reef shallows the mortality of corals and of their recruites is higher than in deeper zones. Therefore the turnover of their specific populations in upper reef zones proceeds faster than in deeper zones, where the stress is less (Huges 1983). An average age of corals in communities of reef slope biotopes for fast-growing species of acroporids and *Psammocora* could be estimated to be within 4–7 years, and that for the slow-growing poritids and faviids within 7–11 years (Table 7.10). In the populations of some massive *k*-strategist corals, like *Montastrea* and *Porites*, old, large colonies 1–2 m in

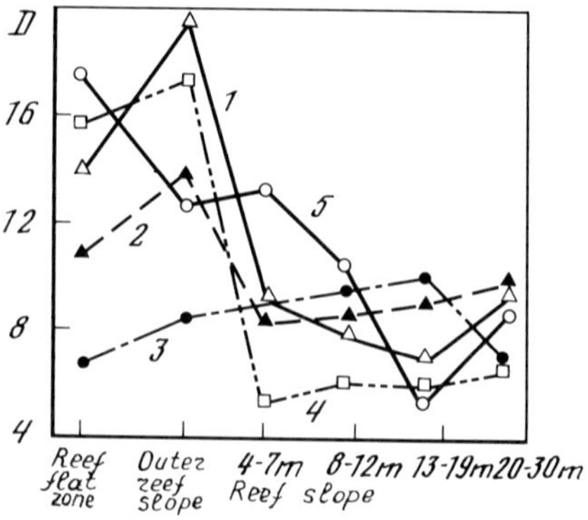

Fig. 7.9. Average size of common corals (*D* diameter in cm) in different zones of fringing reef off Eilat, Red Sea: *1 Acropora hemprichi*; *2 Montipora lobata*; *3 Stylophora pistillata*; *4 Porites lutea*; *5 Echinopora lamellosa*. (After Loya 1972)

diameter and several tens of years of age often dominate (Isdale 1984; Cameron and Endean 1985). The mortality rates of corals in their communities also vary in different reef zones depending upon the level of physical stress (Connell 1973). In young colonies with their projective area below 40 cm^2 it is the highest – up to 10–20% $year^{-1}$. In average sized colonies (40–80 cm^2) it is 3 to 10%, and in larger colonies 1–5% $year^{-1}$.

The average number of coral colonies per meter varies between 20–40 sp. with the percentage of bottom surface covered by them being 40–80% (Fig. 7.10). On reef flats this number is usually less: 5–20 sp. Maximal numbers of colonies could be found in deeper zones of reef slopes – up to 60 sp. m^{-2}. Most often a large part of the colonies in communities belong to 2–4 dominating species, which might occupy up to 50% of the total area covered by corals (Table 7.8). On many reefs a significant or even larger part of coral colonies in communities could be represented by octocorals, which may comprise there 60 to 90% of their total number (Jaap 1985).

Table 7.10. Age of coral colonies on some reefs in the GBR area; N – total number of colonies counted; A – their average age, years; G – their average growth rate (diameter), cm, $year^{-1}$ (After Connell 1973)

Corals	G	Position of reef			
		Heron Island		Low Isles	
		N	A	N	A
Acropora	4.6	32	6.5	7	2.5
Montipora	6.8	12	4.5	7	2.5
Porites	2.0	27	9.5	5	11.5
Pocillopora	6.7	4	5.5	–	–
Psammocora	4.6	–	–	3	4.5
Missive coral species (average for 11 sp.)	1.3	–	–	73	7.5
Sum (average) for all species studied	–	75	6.5	96	7.5

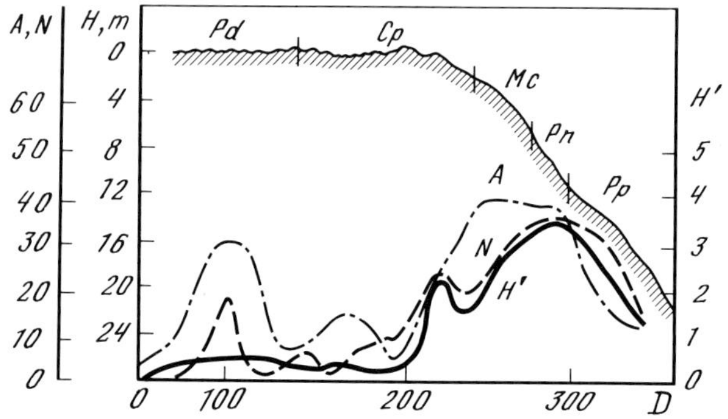

Fig. 7.10. Parameters of structure of coral-reef communities in different zones of reefs off Mactan I., Philippines: *A* number of colonies m^{-2}; *H′* index of species diversity; *N* total number of species m^{-2}; *H* depth; *D* distance from the shore, m. Letter indexes up the profile of reef correspond to zones of dominance of the following corals. P_d *Pavona dicussata*; C_p *Cycloseris patelliformis*; M_c *Montipora compressa*; P_n *Porites nigrescens*; P_p *Pavona pruetoria*. (Data by Sy et al. 1981)

As mentioned above, most scleractinians may live along all levels of depths, from shallow zones of the reef down to its base. Nevertheless, in each zone their communities are dominated by one or by several key species (Tables 7.2, 7.3). This domination is usually more pronounced in shallow zones, subjected to physical stress, where the coral communities are restricted in number of species, with the dominance of two to four of them (Connell 1973; Kühlmann 1971; Glynn 1976). In wavestressed biotopes these dominating corals belong mostly to the opportunistic, ramose, fast-growing species, which well with-stand the surf (Figs. 7.11, 7.12). Among

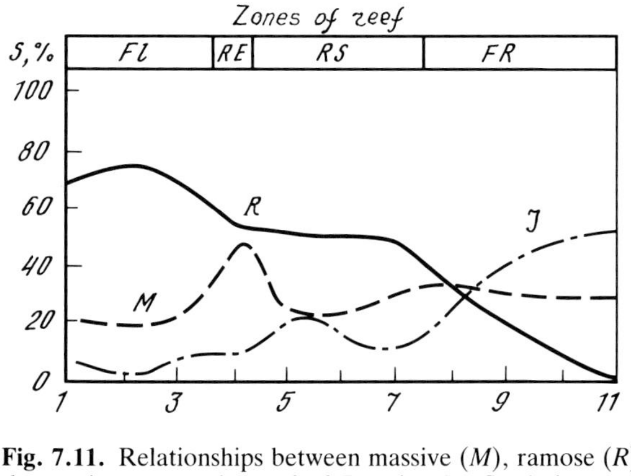

Fig. 7.11. Relationships between massive (*M*), ramose (*R*) and incrusting (*I*) corals along the section across the reef of Reunion I.: *S* relative area covered by corals of a given group, % of total area covered by corals; on *abscissa* numbers of quadrats at the transsect. Reef zones: *Fl* reef flat; *RE* reef edge; *RS* outer reef slope; *FR* fore-reef zone. (Data by Bouchon 1981)

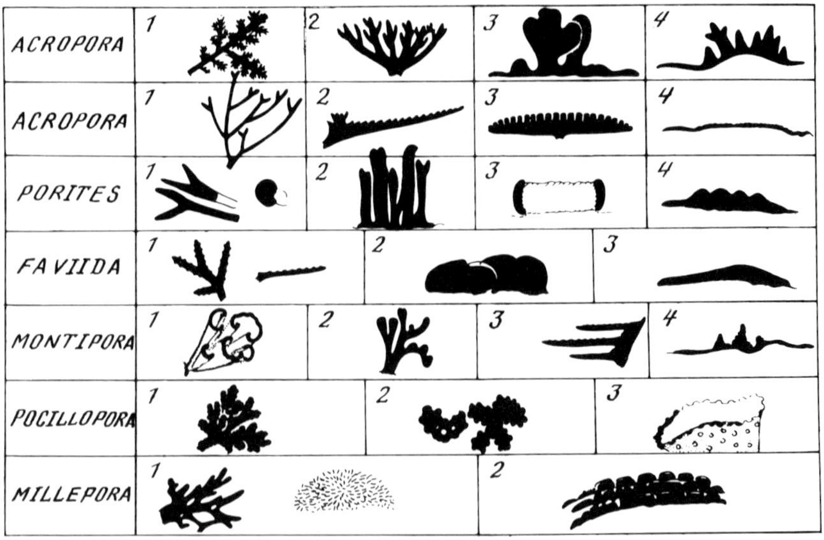

Fig. 7.12. Distribution of basic ecomorphs of some opportunistic corals on reefs off Solomon I. depending upon the level of hydrodynamic stress: *1* leeward zones, protected from wave action; *2* zones of moderate wave stress in lagoon, on reef flat and in the buttress zone on outer reef slope; *3* zones of moderate to strong wave stress of windward submerged reef-flat platforms; *4* zones of maximum wave stress on the exposed windward reef flat. (After Morton 1974)

the key species on reefs of the Indo-Pacific there are corals from the genera *Pocillopora*, *Millepora*, and the common acroporids *A. hyacinthus*, *A. cuneata*, *A. humilis*, *A. digitifera*, *A. leptocyathus*, *A. hebes*, *A. formosa*, and *A. palifera* (Done 1983). On reefs of the west Atlantic such opportunistic, dominant corals are *Acropora cervicornis*, *A. palmata*, *A. prolifera*, and *Millepora alcicornis* (Kühlmann 1975; Glynn 1976; Done 1983; Bright et al. 1984). Among the leading species in shallow surf zones there are also several massive and incrusting corals well opposing wave stress. On the Indo-Pacific reefs they are represented by *Porites* sp., faviids (genera *Favia*, *Goniastrea*), the corals *Galaxea*, *Pavona*, *Psammocora*, and *Platygyra*. On reefs of the Atlantic these are: *Diploria*, *Montastrea annularis*, and *Porites asteroides*. Some of the leading species in shallow surf zones and on exposed flats, such as *Stylophora mamillata*, *Echinopora gemmaceae*, *A. variabile*, and *A. pulchra* cannot withstand well the surf but are adapted to surviving periodical exposure – up to 60 h in total per month with each period of lengths between 1 to 4 h (Ditlev 1978). An especially enduring coral among them seems to be *Acropora puchra*, which often lives on reef flat biotopes exposed during low ebbs. It endures complete exposure for 2–4 h and does not die subjected to sun and rain (Fig. 7.13). The corals *Pocillopora* and *Montipora* endure well also siltation, high water temperature, and drops and rises in salinity (Loya 1972; Smith and Jokiel

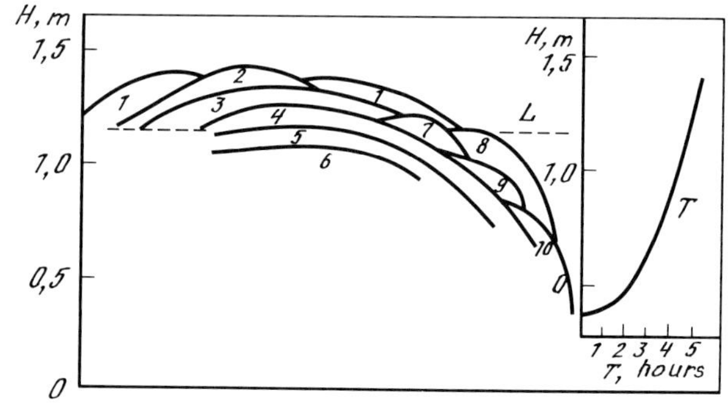

Fig. 7.13. Space distribution of the subassociations of corals on the exposed flat of the Ko-Phuket reef of Thailand; subassociations: *1 Goniastrea pectina – Coeloseris mayeri*; *2 Acropora hebes – A. pulchra*; *3 Montipora ramosa*; *4 Porites lutea – P. compressa*; *5 Favia* sp. – *Fungia fungites*; *6 Echinopora lamellosa – Hydnophora rigida*; *7 Psammocora contigua – Porites convexa*; *8 A. survulosa – A. humilis*; *9 Pocillopora damicornis*; *10* species of corals inhabiting the slope biotopes; *L* water level at reef flat during ebb tide; *H* height of the sea level above its position in the surrounding sea; *T* duration of the exposure of reef flat during ebb tide. (Based on Ditlev 1978)

1975). But if the siltation causes the formation of H_2S via microbial sulfate reduction, these corals may die (Ditlev 1978).

On submerged flat which is not exposed during ebbs and faces moderate wave stress, species diversity may often be high. For example, at this kind of flat at the Heron Island reef (GBR) within an area of $120\,m^{-2}$ 40 species were recorded with a cover percentage of less than 10%. Most of them were massive coral species (Cameron and Endean 1985). On many reefs the highest number of species in coral communities was recorded precisely in this zone (Done 1983). Dominating among them were the massive, columnar or foliose corals from the genera *Porites*, *Goniastrea*, *Galaxea*, *Favia*, and *Montipora*. The ramose corals were represented in this zone by the genera *Acropora*, *Pocillopora*, *Seriatopora*, and *Millepora* (Taylor 1968; Bak 1977; Grigg 1983; Potts et al. 1985). On the outer reef slope, in conditions of weak physical stress, species diversity and the total number of specias are usually at their maxima. Most numerous in this zone are corals with massive, columnar or foliose colonies (Figs. 7.11, 7.12). In the Indo-Pacific, in this zone the common corals are *Porites*, *Pectinia*, *Symphillia*, *Coscinarea*, *Merulina*, *Montipora*, *Goniastrea*, *Hydnophora*, and *Echinopora* (Done 1983; Sheppard 1980, 1982), and on reefs of the western Atlantic *Montipora*, *Siderastrea*, *Agaricia*, *Millepora*, and *Dichocoenia* (Kühlmann 1974, 1975, 1982, 1985; Abel et al. 1983; Lewis 1989). The ramose corals *Pocillopora*, *Stylophora*, *Seriatopora*, and several species of acroporids, also could be met with in this zone on the Indo-Pacific reefs.

In shadowed biotopes situated in shallow zones of the reef: in caves and in canyons under the peaks of the steep reef edge the hard substrate is occupied largely by coralline algae and by gorgonians. The coral communities there are usually impoverished. Nevertheless, in the GBR reefs these biotopes are inhabited by ca. 150 species of corals, most of which are hermatypic and contain symbiotic algae. Among them are *Montipora*, *Cyphastrea*, *Goniastrea*, *Pavona*, *Stylophora*, *Turbinaria*, *Echinophyllia*, *Favites*, and *Agaricia*. Besides these for other reef zones common corals, the shadowed biotopes are inhabited by really scyophylic hermatypic scleractinians, like *Leptoseris hawaiensis*, *L. glabra*, *L. mycetoserioides*, and *Pashyseris speciosa*. Among the ahermatypic corals usually the there are sciophylic scleractinia: *Tubastrea*, *Dendrophyllia*, *Phyllangia*, *Astrangia*, *Madracis*, and *Balanophyllia* (Vasseur 1974; Jackson 1977; Jaubert and Vasseur 1974; Rogers 1979b; Dinesen 1983; Kühlmann 1983).

In communities of corals in deep zones of outer reef slopes the species diversity and the number of species often remain high even at very weak illumination: 3–10% PARS. For example, at reefs of Eilat their number in this zone was three times higher than in the medium-depth zone of slope and five times higher than in the lagoon (Mergner and Schuhmacher 1974). Even by this weak light at the base of the outer reef slope over 90% of all coral species consist of the common hermatypic scleractinians, adapted to a low level of illumination (Titlyanov and Latypov 1991). Only about 8–10% of obligate sciophylic hermatypic or ahermatypic corals, which live only in deep zones of the reef. Among the first most numerous group of corals which inhabit the deep fore-reef zone on reefs of the Indo-Pacific common are *Stylophora mamillata*, *S. kühlmanni*, *Porites australiensis*, *Leptastrea transversa*, *Asteropora myriophthalma*, *Pavona varians*, *Echinopora lamellosa*, *Fungia granulosa*, *Pocillopora solida*, *Synarea convexa* and some species from the genera *Podabacia*, *Coscinarea*, and *Mycedium* (Dinesen 1977; Bouchon 1981; Titlyanov et al. 1981, 1988c; Kühlmann 1983; Dubinsky et al. 1984; Fricke and Meischner 1985). When living in this zone these corals build very fine and thin, flat and incrusting forms of colonies adapted to intercepting weak light and needing less calcareous material for their construction (Fricke and Schuhmaher 1983; Titlyanov 1987; Titlyanov and Latypov 1991). Physiological mechanisms of their light adaptation will be discussed below (cf. Sect. 8.4). Deep-water coral associations on the Atlantic reefs include the following corals: *Agaricia* sp., *Scolymia* sp., *Madracis formosa*, and *Montastrea cavernosa* (Kühlmann 1983). The specific deep-water coral associations are dominated here by specialized hermatypic sciophylic species of *Leptoseris* (*L. mycetoserioides*, *L. explanata*, *L. porosa*). They have a special light-absorbing system and may survive at light intensity levels as low as 1–3% PARS (Fricke and Schuhmacher 1983; Schlichter et al. 1985). This associations include also the hermatypic sciophyles *Cycloseris cyclolithes*, *C. patelliformis*, *Pashyseris rugosa*, *P. speciosa*, *Leptastrea biwickensis* and the ahermatypic corals *Dendrophyllia*, *Tubastrea*, *Astrangia solitaria*, and *Phyllangia americana*.

With the aid of submersibles the hermatypic corals *Agaricia grahamae* and *Montastrea cavernosa* were found at depths down to 113 m where the illumination was 0.15% PARS. The coral *Eusmilia* was found at 140 m. On the slopes of the Pacific atolls hermatypic corals were found ever deeper: *Pashyseris fastigata* at a depth of 200 m, *Coscinarea monilis* at 270 m, *Psammocora explanulata*, *Scolymia lacera*, *Leptoseris porosa* and *L. fragilis* – at 140–165 m (Lang 1974; Fricke and Schuhmaher 1983; Kühlmann 1983; Reed 1985; Fricke and Meischner 1985). These depths were far below the depth of the compensation point of photosynthesis – respiration, which even in corals best adapted to light deficiency in transparent oceanic waters was located at depths of less than 100 m (Fricke and Schuhmacher 1983). Among the ahermatypic corals at depths over 100 m, *Madracis interjecta*, *M. myriaster*, *Dendrophyllia miniscula*, and *Oculina varicosa* were recorded. These deep-water coral associations, which inhabit the lower zone of reef slopes at depths over 80 m, do not form three-dimensional reef constructions because of low calcification rates in the dark. On lots of bottom area colonized by corals they compose kinds of patchy bioherms. The rate of calcification there does not exceed the bioerosion of bioherm and coral colonies. The calcareous sediments and rubble flow down the slope, excluding so the possibility of formation of reefal construction (Fricke and Hottinger 1983; Schuhmacher and Zibrovius 1985). Similar calcareous bioherms are constructed on the continental slope down to depths of 1000 m by the deep-water ahermatypic scleractinian corals *Lophelia*, *Solenosmylia*, *Madrepora* (*oculata*), *Goniocorella*, *Desmophyllum*, and *Oculina* (Keller 1976; Zibrovius 1980; Cairns and Stanley 1981).

Specific coral communities inhabit also lagoonal bottom biotopes covered with soft sediments: carbonaceous silts, silted and pure coral sands. High rates of sedimentation and the lack of solid substrate inhibit recruitment and growth of normal sessile corals here. Nevertheless, the soft bottom biotopes often appeared to be colonized by specific coral communities in which single free-living corals dominate, such as *Fungia*, *Diaseris*, *Cycloseris* (fam. Fungiidae), *Heteropsammia* (fam. Dendrophylliidae) and *Heterocyathus* (fam. Caryophillidae). Ramose corals are also present in these communities, e.g. *Acropora*, *Pocillopora*, *Stylophora*. Their planulae settle on the tridacna shells or on large sections of the coral colonies. The corals *Pocillopora*, *Stylophora*, *Goniopora*, *Goniastrea*, *Euphillia*, *Pavona*, *Agaricella*, and *Psammocora* may live on soft bottom forming medium-sized round colonies lying free on the surface. The main factors inhibiting coral growth in these biotopes are siltation and the gradual submersion of their colonies into the sediments. A number of corals escape this using various mechanisms. Ramose corals, being rooted inside sediments with the aid of pieces of solid substrates on which they settle, have a high growth rate, and thus may overcome siltation. The round coral colonies, which lie free on the bottom, are moved periodically by fish, which seek food under them, and so escape the fate of being buried in the sediments (Glynn 1974). Fungiid free-living

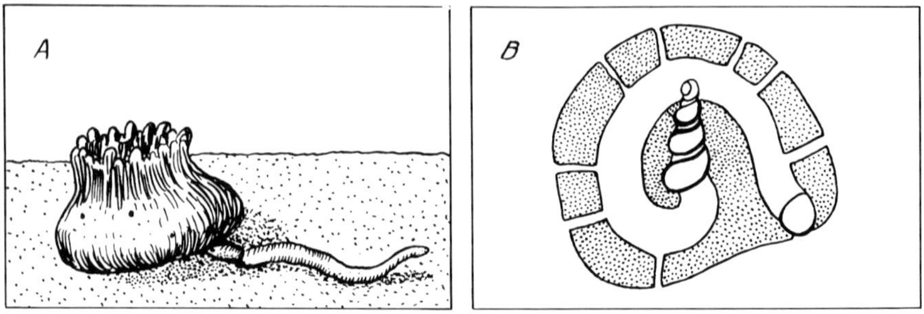

Fig. 7.14. Solitary coral *Heteropsamnia* with its symbiont, the sipunculid worm (*A*); *B* basal surface of the same coral with the gastropod shell, on which primarily the planulae of coral have settled. (After Goreau and Yonge 1968)

flat scone-like colonies are moved by waves. Some of them may become active and crawl out from the sediments with the aid of their long tentacles (Stoddart 1969; Gill and Coates 1977). But the most specialized free-living corals inhabitants of the soft bottom biotope, are the small hermatypic solitary corals *Heteropsammia* and *Heterocyathus*. They are absent in other reef zones and can live even on muddish bottom, having excellent ecological adaptations to those environments. These little corals, 1–2.5 cm in diameter, possess a very effective mechanism for shedding off sediments from the surface of their polyps, based upon the interaction of cilia with the tentacles (Fisk 1981). But their most curious feature is how they move over the surface of the soft sediment: they are driven by their symbiotic worm – the sipunculid *Aspidosiphon*. This worm drives to host coral with a speed of 1–2 m per day (Fig. 7.14). The planulae of *Heteropsammia* settle on the surface of the empty shell of this gastropod, which then is colonized by the worm. The density of coral communities in soft bottom biotopes is very high. The mass species of solitary corals may form populations of up to 1000 sp. m^{-2} (Latypov 1985).

7.2.3 Zoantharians and Octocorals in Coral Communities

Zoantharians often form large monospecific patch or belt associations on biotopes of reef flats (Goreau 1963; Kinzie 1973; Suchanek and Green 1981; Karlson 1983; Fadallah et al. 1984). They prefer shallow reef sites with intensive hydrodynamics and colonize usually rocky bottom substrates in reef-crest and reef-edge zones (*Palythoa*), rubble areas and dead corals (*Zoanthus*, *Isaurus*) and even living colonies of acroporid corals, finally killing them (*Zoanthus*). They are extremely opportunistic and withstand well all kinds of physical and anthropogenic stress (Cooke 1976), successfully competing with other reefal sessile fauna, including scleractinian corals (Herberts 1972; Walsh and Bowers 1977). Therefore, they readily colonize the destabilized or polluted reefs. We have observed this phenomenon in Kaneohe Bay, Hawaii, where large areas of dead corals killed by pollution were covered by carpets of the zoantharians *Zoanthus pacificus* and *Palythoa vestitus*. They often inhabit also the patch reefs, where there are plenty of dead coral heads. The most zone, which is usually filled with dead corals and rubble, is colonized on many reefs by *Zoanthus sociatus*.

All zoantharians possess zooxanthellae and also catch zooplankton (cf. Sect. 9.22). The rate of their lateral growth is rather high – up to 1–1.5 cm $month^{-1}$ (Karlson 1983). Thus, they could easily overgrow and force out other competing sessile organisms. They multiply mostly vegatatively, being also capable of sexual reproduction. Their mortality is high, caused by surf action and grazing pressure. They are consumed by some urchins and fish insensitive to their toxins (Cooke 1976). But the fast-growing zoantharians may quickly repair the damage (Karlson, 1983). Some zoantharians contain biologically active toxic substances of interest to pharmacology (Moore and Scheuer 1971; Kimura et al. 1972). Fish, when feeding on zoantharians, might also become toxic (Hashimoto et al. 1969).

The octocorals are a common component of coral-reef fauna, especially on destabilized, damaged or monsoon reefs subjected to terrigenous siltation, where the growth of scleractinian corals is inhibited. They grow faster with standing well siltation and changes in salinity. Among such habitats are many Caribbean reefs, reefs off Florida, off the Vietnamese coast, and reefs around high islands in lagoons of the GBR. The octocorals are often leading components of coral communities that are recovering after attacks by *Acanthaster* or after hurricanes (Nishihira and Yamazato 1974; Nishihira 1981; Yoshioka and Yoshioka 1989). They occupy mostly biotopes with low or moderate turbulence, while the ahermatypic species prefer shadowed niches or dark deep zones (Tursch and Tursch 1982). In these sites they cover 20–50% of bottom areas occupied by coral communities. On regenerating reefs it could be even more – over 90%. An important role of octocorals in reef benthic communities then seems to be obvious. Nevertheless, in descriptions of the structure of coral communities they have been often ignored. In part, this was a consequence of their indefinite situation in the taxonomic system and of the difficult procedure of their identification. Therefore, most researchers include them in the taxonomic lists of coral communities at the level of genera (Dinesen 1983).

In the Red Sea, among octocorals, the alcyonaceans largely dominate. Although not very diverse in this basin, they still play a significant role in the formation of reef bottom fauna (Finshelson 1973; Schuhmacher 1973; Verseweldt 1974; Mergner and Svoboda 1977; Benayahu and Loya 1977). On reefs of Eilat, where these studies have been mostly made, the most abundant growth of alcyonacean octocorals was recorded in shallows of submerged flats and on reef slopes. Their communities in shallows of reef flats are formed mostly by the common alcyonaceans, *Lithophyton*,

Sinularia and *Parerythropodium*. On the slopes expansive associations of *Xenia* colonies and *Lithophyton arboreum* are usual. The same large monospecific settlements are formed by the symbionic alcyonaceans *Sinularia flexibilis*, *S. gibbosa*, *Lobophytum pauciflorum*, *Sarcophytum glaucum* and *Litophyton viridis*. They may occupy areas up to several square meters on flats and at slightly inclined outer slopes. Single colonies or medium-sized associations of *Cladiella*, *Stereonephthea*, *Paralemnalia* (*eburnea*) as well as of the stoloniferans *Tubipora* and *Clavularia* could be met with in all reef zones at medium depths. Deeper down the reef slopes usual are the asymbiotic octocorals *Nephthya*, *Dendropephthya*, *Stereonephthya*, as well as the gorgonians *Acabaria*, *Bebryce*, *Gorgonella*, *Juncella*, and *Anthogorgia* (Stiasny 1940). The deep soft bottom biotopes at the base of reefs are occupied mostly by associations of the pennatulacians *Virgularia* and *Vitellum*. The total number of octocoral species on the reefs of Eilat could be evaluated at 18–20. The percentage of bottom area covered by them varied in different zones between 2 to 17%. More than half of this area is shared by two to three leading species. Among them, in different zones, are usually *Litophyton arboreum*, *Heteroxenia furcesens*, *Parerythropodium fulvum*, *Sinularia compressa* and *S. leptoclados*.

Benayahu (1985), during his revision of octocoral fauna in other reef regions of the Red Sea, has recorded over 180 species, which belong to the following taxa: family Alcyoniidae 96, Nephtheidae 56, Xeniidae 27, and Nidaliidae (genus *Siphonogorgia*) 3. By number of species most abundantly the following genera were represented: *Cladiella* 30, *Sinularia* 42, *Dendronephthea* 29, *Xenia* 20, and *Sarcophyton* 20, species. Among the most common octocorals of the Red Sea reefs there could be mentioned *Cladiella pachyclados*, *C. brachyclados*, *Sinularia leptoclados*, *S. polydactila*, *S. gardineri*, *Sarcophyton glaucum*, *S. gemmatum*, *Lobophyton pauciflorum*, *L. crassum*, *Alcyonium flaccidum*, *Parerythropodium fulvum*, *Dendronephthea curvata*, *D. ehrenbergi*, *Xenia blumi*, and *X. macrospiculata*.

An insight into the structure and composition of octocoral communities of the Pacific reefs is given by their description on reefs of Leing Island, New Guinea, offered by Tursch and Tursch (1982). The octocorals inhabited on these reefs with calm waters about a half of the rocky surfaces of the bottom, covered by corals. Over 70% of the space covered by octocorals was occupied by four leading species of the alcyonacean octocorals *Lithophyton viridis*, *Sinularia guerciformes*, *S. gibberosa* and *Lobophytum pauciflorum*, which formed there large monospecific settlements. Other octocorals (Xeniids, gorgonaceahs) were rare, and stoloniferans were absent. In total, on these reefs 115 species of octocorals have been recorded, which is by the way close to the species numbers of scleratinian corals on most of the Pacific reefs. Among them there were 23 species of the genus *Sinnularia*, 10 of *Sarcophyton*, 7 each of *Xenia*, *Dendronephthea* and *Lobophytum* and, 7 new species of octocorals (Verseveldt and Tursch 1979).

In the central GBR the octocorals are also one of the main components of sessile benthos, quite comparable by percentage of bottom cover and by number of species to the scleractinian hard corals (Verseveldt 1977; Dinesen 1983). By parameters of composition and by structure the communities of octocorals are similar here those described on reefs of the Red Sea and of New Guinea. The alcyoniids, nephtheids and xeniids were dominating there too. Some difference was noted by the appearance of new leading alcyonacean genera like *Lemnalia* and *Paralemnalia* (fam. Nephtheidae), *Efflatounaria* (fam. Xeniidae) and *Asterospicularia* (same name as family). In the lagoonal part of the GBR the octocoral communities were dominated by the leading alcyonacean species *Sarcophyton tracheliphorum*, *Lobophytum crassum*, *L. pauciflorum*, *Sinularia variabilis*, *S. flexibilis*, *S. polydactila*, *Nephthea millis*, *Xenia* sp., *Heteroxenia* sp., *Clavularia* sp., *Alcyonium* sp., and *Cespicularia* sp.

Our observations on octocorals distribution and on their taxonomic composition in the GBR region showed that their share in the formation of coral communities is most remarkable in biotopes of lagoonal patch reefs, where they survived destructive attacks of *Acanthaster* (Lizard Island), and in the biotopes of flats and slopes at the fringing reefs around high islands, where they had been subjected to terrigenous siltation (Paradise Island). In such biotopes, octocorals are largely dominating in coral communities, covering 50–90% of areas of bottom inhabited by sessile fauna. On Lizard Islands some reefs previously destroyed by *Acanthaster* were totally occupied by the common alcyonaceans *Alcyonium*, *Sarcophyton*, *Lobophytum*, *Sinularia*, and *Xenia*. Slopes and the base of reefs were inhabited by dense populations of the gorgonian *Juncella fragilis*. The symbiotic gorgonaceans, *Plexauroides* and *Isis* were common also on slopes of the fringing reefs off Paradise Island. Reefs of the Capricornia group escaped the *Acanthaster* plaques and thus preserved stable coral commuhities for a long time. The octocorals were their significant components only in the calm shallow biotopes of slopes, of leeward reefs and shallow biotopes of reef flat and patch reefs. They occupied in such biotopes 5 to 20% of area covered with corals. In shallow biotopes the associations "*Lemnalia-Paralemnalia-Cladiella-Xenia*" were most comon. At the reef-flat crevices *Capnella*, *Asterospicularia* and *Tubipora* were humerous. On windward slopes within the thickets of ramose corals Acropora large settlements of *Sinularia hirta* were recorded. The gorgonaceans *Rumpella* and *Isis* were numerous on reef slopes of the leeward side. At the foot of reefs, at depths of 18–22 m among rocks and rubble, *Alcyonium*, *Sarcophyton*, *Pashyclavularia* and the gorgonian *Juncella* occurred.

We have observed unusually abundant and diverse communities of octocorals on the monsoon reefs off the coast of central Vietnam. On some of these reefs they formed large monospecific settlements which covered big coral knolls and lots of reef-flat areas like a carpet. Most often these were the species of *Lobophytum* or *Sinularia*. The fauna of octocorals here differed from the above-considered on the Pacific reefs by the relative abundance of gorgonians. Most common among them were *Rabdoplexaura*, *Plexauroides*, *Echinogorgia*, *Acabaria*, *Mopsella*, and *Bebryce*. The hard octocorals *Heliopora* were numerous on flat shallows. This coral is common on many reefs of the Indo-Pacific. Often it forms the microatolls on submerged reef flats (Zann and Bolton 1985).

The above data show that in communities of octocorals on reefs of the Indo-Pacific the alcyonaceans most often largely dominate over the gorgonaceans. Only on reefs off the Vietnamese coasts were gorgonaceans dominating, because of their ability better to withstand siltation, drops in temperature and salinity during the winter monsoons peculiar to this area.

In contrast, on the reefs of the western Atlantic gorgonaceans absolutely predominate over alcyonaceans. Moreover, they are often a leading component of the sessile bottom fauna (Bayer 1961; Goldberg 1973; Opresko 1973; Kinzie 1973; Preston and Preston 1975; Muzik 1982; Lasker and Coffroth 1983; Yoshioka and Yoshioka 1989). The number of colonies of gorgonians reaches on many reefs 20–40 m^{-2} and the biomass over 0.5 kg m^{-2}. Shannon's index of species diversity (H′) in octocoral communities often approaches that of scleractinians: 1.5–33. Its maximum values they attain in upper zones of the outward reef slope. The total number of species of octocorals in coral communities of the western Atlantic varies between 16 on the Puerto Rican reefs and 47 on reefs of Florida, which is close to that of scleractinians.

Most common reef gorgonians on reefs of this region belong to the two families, Plexauridae and Gorgoniidae. Among them there are also several species from the family. Paramuricidae (genus *Swiftia*) Elisellidae, and the scleraxonian gorgonians *Briareum*, *Erythropodium* and *Icilligorgia*. Among the plexaurid gorgonians the most common are *Plexaura flexuosa*, *P. homomalla*, *Plexaurella dichotoma*, *P. grisea*, *P. fusifera*, *Eunicea calyculata*, *E. asperula*, *E. tourneporti*, *E. pinta*, *E. succinea*, *Pseudoplexaura circus*, *P. porosa*, *P. flagellosa*, *Muricea muricata*, *M. atlantica*, *Muriceopsis potula*, and *M. flavida*. The most common eugorgonians are *Pseudopterogorgia americana*, *P. bipinnata*, *P. rigida*, *P. callos*, *P. elisabethae*, *P. acerosa*, *Pterogorgia citrina*, *P. ancepts*, and *Gorgonia ventalina*. Elisellid gorgonians are represented on the Atlantic reefs by *Elisella barbadensis* and *Nicella schmidtii*. Among seleraxonian gorgonaceans the most common are *Iciligorgia schrammi*, *Erythropodium polyanthes*, and *E. cariboraeum*.

The composition of octocoral communities changes along the cross-sections in different reef zones. On the reef flat and on the outer reef slope down to a depth of 10–20 m the symbiotic gorgonians dominate, being represented there mostly by the genera *Gorgonia*, *Eunicea*, *Pterogogia*, *Pseudopterogorgia*, *Plexaurella*, *Plexaura* and *Briareum*. The gorgonian *Pseudopterogorgia americana* densely populates even the soft bottom biotopes, attached to pieces of rubble and shells buried in the sediments (Lasker and Coffroth 1983). On the slope downward domination gradually shifts to the asymbiotic species of gorgonians, like *Elisella*, *Nicella*, *Swiftia*, and *Icilligorgia*. But some symbiotic species of gorgonians well-adapted to a low light, could be met with sometimes down to a depth of 50–60 m (Kinzie 1973). Most species of symbiotic reef gorgonians in the western Atlantic have big, firm, flexible and fast-growing colonies, which well sustain siltation, turbid water and surf (Opresko 1973). These features enable them to compete successfully with corals and other sessile fauna and plants and so to play an important role in bottom biotopes of western Atlantic reefs, most of which are situated on the continental shelf or near high islands, and are subjected to intensive terrigenous siltation.

7.3 Factors Controlling Coral Community Structure

The main factors which control the structure of coral communities on the reef are the physical and biological stresses, on the one hand, and the differentiated reaction of various coral species to disturbing action by their different potential ability to compete for the hard substrate, on the other (Bak and Powell 1988; Claerebandt 1988). Under the heading biological stress we list here the grazing of corals by fish and other animals, their bioerosion by boring fauna, and their diseases caused by microorganisms (Connell 1973, 1978; Loya 1976c; Glynn 1976). The physical stress as a controlling factor for coral communities is more important in shallow turbulent zones of reef, while in calm leeward or deeper zones the biological stress begins to prevail (Sheppard 1981b).

Among the factors of physical stress hydrodynamics, resuspension of sediments, changes in salinity and temperature, excess of light, and exposure seem to be the most important (Bradbury and Young 1981). In shallow reef zones and on windward reef flats the structure of coral communities is influenced largely by wave stress and tidal currents (Roberts 1974; Geister 1977; Jokiel 1978; Sheppard 1980; Dollar 1982). Intensive hydrodynamics in itself is a positive factor in the life of sessile fauna. It provides permanent supply of particulated food and oxygen, driving out suspended material, which arrives from the littoral zone, thus preventing its siltation (Hubbard 1974; Wainwright and Koehl 1976). But in the surf zones of windward flats and in the channels cut across them, waves and tidal currents could be so strong that they may damage or uproot coral colonies. Therefore, coral communities in zones of active hydrodynamics are dominated by species or by their ecomorphs which well sustain hydrodynamic stress. These are most often the ramose acroporids, which may form ecomorphs corresponding to the level of this stress (Kühlmann 1975), and thus survive well in this zone. The same features are peculiar also to some other ramose corals, like *Pocillopora*, *Stylophora*, and *Millepora* which successfully survive in turbulent zones. Their ability to oppose stress depends not only upon the mechanical durability of colonies, but also upon precise orientation of their branches to the direction of stress. The ability of these ramose corals to propagate vegetatively by pieces of their branches is also one of the factors making for their abundance in stressed zones. The wave-stress also controls the species composition of octocoral communities which inhabit the surfaces of windward reef flats and upper reef slopes (Morton 1974). On reefs with a week surf the zonal features in the composition of coral communities are very slight (Loya 1972; Bouchon 1981). Instead of ramose corals, mostly the massive *Porites*, faviids and also *Montipora* dominate (Grigg 1983; Latypov 1985; Potts et al. 1985; Cameron and Endean 1985). The planula larvae of opportunistic corals seem also to be adapted to settling in stressed biotopes (Connell 1973).

The influence of fluctuations of temperature and salinity on the structure of coral communities was most spectacularly demonstrated on biotopes of exposed reef flats in the Indo-Pacific area. These reefs reached sea level long ago and their surfaces are now 0.5–1 m high over the level of sea at low ebbs. Thus corals that inhabit these biotopes are periodically exposed during low ebbs for 2–4 h, suffering at night from cold, in the day time from heat, and in rainy weather from drops of salinity (Pichon 1964; Loya 1972; Ditlev 1978). Corals living on exposed reef biotopes can endure gradual changes in salinity on a large scale: 20 to 40% (Kinsman 1964). But a sharp shift in it kills them rather quickly (Goreau 1964b). The stress influences of the above listed factors on corals living in exposed reef-flat biotopes are itetconnected and have a cumulative damaging effect (Coles and Jokiel 1978). Thus the coral associations in these biotopes exist close to the limit of sustainable stress (Jokiel and Coles 1990). Even a slight general change in environmental conditions, like some global changes in water temperature caused by the El Niño current, or an increase in tide levels caused by geostrophic events, as well as anomalous weather conditions accompanied by long periods of rain, cause of large-scale mortality of corals in the biotopes of exposed reef-flat areas.

On the exposed flat areas of the Red Sea and the Caribbean high-latitude reefs, a basic factor influencing the composition of coral communities is the extreme cooling down during low ebbs at night and during the periods of breakthrough of cold waters in winter into the reef areas (Loya 1972; Roberts et al. 1975; Scatterday 1977; Hudson 1981). These events play havor even among the most opportunistic species of *Acropora*, *Pocillopora*, *Montipora*, *Millepora*, *Diploria*, and *Psammocora*, which are usually leading in the communities of exposed reef flats (Ditlev 1978).

Among the physical factors of stress inhibiting the growth of corals is also their siltation in turbid waters in areas of intensive resuspension of sediments by waves and currents, and in biotopes receiving the terrigenous, suspended matter (Durand 1989). An intensive sedimentation of resuspended sediments after storms may even kill corals (Marshall and Orr 1931). In this respect, the resuspension of sediments and the increase in their terrigenous input caused by rains, which usually accompany the storms, damage corals not less severely than the direct physical impact of waves (Johannes 1975). The same inhibition and even the death of corals could be caused by a lack of light at lagoonal bottoms accompanying an increase in turbidity during continuous storms (Dodge et al. 1974; Rogers 1979b). Increased stress of siltation has a direct impact on the physiologial state of corals. Their respiration increases through the decreasing photosynthesis rate and the P_t/M_t-ratio (Abdel Salam and Porter 1988). The siltation and the weak light inhibit also the recruitment of corals planulae (Johannes 1972a). The siltation stress stimulated the evolutional appearance of corals with an increased ability to endure siltation, and actively sift out settling suspended material (Yonge 1940; Lewis and Price 1976). These corals may survive in very turbid waters (Roy and Smith 1971; Loya 1976c). Among these are the Atlantic corals *Siderastrea siderea* (Kühlmann 1974), *Montastrea cavernosa* and *Diploria strigosa* (Lasker 1980). It is just these corals that are usually pioneering in the colonization of turbid shallow biotopes (Hubbard 1974). A high rate of sifting out sediments is peculiar to fungiids, which successfully colonize biotopes of turbulent reef zones with turbid waters.

Corals get rid of sediments by making them sticky with mucus and expelling them with the aid of cilia and tentacles (Hubbard and Pocock 1972). Their ability to expel sediments depends also upon the geometry of corallites (Hubbard 1973). Corals which actively expel sediments have many well-developed septa and relatively large and tall corallites. Among such corals are the above-mentioned *Montastrea*, *Diploria*, and also many of other faviids: *Meandrina*, *Symphillia*, *Lobophyllia*, *Fungia*, and *Galaxea*. These corals are good survivors, also living in turbid waters in biotopes with very rapid sedimentation (Roy and Smith 1971; Loya 1975; Schuhmacher 1979). In contrast, such corals as *Agaricia*, which have rare and underdeveloped septa cannot quickly expel sediment so that they can live only in clean, deep waters at steep, deep zones of reef slopes.

Light as a physical factor influences the structure of coral communities only in the upper shallow and in the deep zones of the reef. In the upper zone corals might be inhibited by an excess of light (Siebeck 1981). But the coenosarc of corals contains protective pigments and amino acid mycosporine which absorb ultraviolet Radiation (Chalker et al. 1988). The deficiency of light begins to influence species composition of coral communities only in biotopes, where illumination drops down to 3–5% of PARS, that is in caves and in other shadowed places of reef flats and of the upper part of reef slopes, and also in the deep zone of the fore-reef and reef base. In coral communities here many ahermatypic species and the specialized hermatypic corals appear, which are adapted to a low illumination. They may live without signs of bleaching at an illumination as low as 1% PARS. These are *Leptoseris*, *Cycloseris*, and *Pashyseris*. At an illumination of 2–3% PARS communities of corals include species of scyophilic hermatypic scleractinians like *Leptastrea*, *Scolymia*, *Turbinaria*, and some of the opportunistic corals that live everywhere on the reef and can also survive at this low level of illumination, such as Psammocora, *Pocyllopora*, *Seriatopora*, *Lobophyllia*, and *Alveopora* (Dinesen 1982; Kühlmann 1983; Fricke and Schuhmacher 1983; Titlyanov et al. 1981, 1988c; Titlyanov 1987).

The main biological factors which control community structure and development of sessile benthic fauna at the level of interspecific relationships of competition for hard substrate and grazing (Daine 1966). Because of the high density of reef benthic communities these competition relationships involve extreme strain (Connell 1973; Jackson 1977, 1989). A significant influence on their structure exerts competition between the corals themselves (Sheppard 1980). Corals display usually the following kinds of competition: overgrowth and the shadowing of each other (Shinn 1966; Potts 1976; Bothwell 1983b), excretion of toxic metabolites inhibiting the competing species (Sammarco et al. 1983), and the interspecific aggression of corals against their competitors (Glynn et al. 1972; Lang 1973; Dustan 1975; Richardson et al. 1979; Sheppard 1979; Wellington 1980; Bothwell 1984). In the first case, the quickly growing ramose scleractinians or octocorals overgrow the slowly growing massive corals, shadow them and suppress their growth. The use of toxic metabolites is widespread in the octocorals. Most of them produce different kinds of toxic compounds and excrete them into the surrounding water. Using this chemical weapon and the high growth rate they often replace the scleractinian corals, thus exerting an influence upon their community structure. For example, the scleraxonian gorgonacean *Erythropodium* effectively replace many scleractinian corals and even zoantharians (Karlson 1980). The toxic metabolites excreted by alcyonacean *Sinularia* inhibited in the experiments the growth of scleractinian corals and even caused their mortality (Sammarco et al. 1983).

The interspecific aggression is a common feature of scleractinian corals which have polyps, well armed with cnidocil. Their main organs of aggression are their tantacles and mesenterial filaments equipped with nematocysts. In the process of growth the coral aggressor approaches the coral-competitor and kills its polyps with the aid of nematocysts, extending polyps and tentacles to the enemy. In this way the coral-aggressor stops the growth of the coral competitor at his side (Logan 1984). The critical distance for such an attack varies in different species of corals. In acroporids and in pocilloporids, which have small polyps, this distance was 0.5–1 cm, while in corals with polyps of an average size, it was 1–3 cm (Sheppard 1981a).

The most aggressive corals usually possess special, long tentacles – sweepers, which are functionally destined to kill the polyps of coral competitors. They may kill them even at a distance of 3–4 cm. Sweeper tentacles are possessed by most of the aggressive corals with medium or large polyps, like *Diploria strigosa*, *Montastrea annularis*, *Scolymia lacera*, and *Galaxea fascicularis* (Lang 1973; Richardson et al. 1979; Wellington 1980; Chornesky 1983). These tentacles are with a powerful cnidocil, which has an increased number of large nematocysts (Hartog 1977). Also, these sweeper tentacles increase their number in polyps situated at the side of the coral approaching its competitor. The same was observed in polyps of corals that were kept in aquarium together with their potential competitors, e.g., with corals ranking below them in the "scale of agressivity". Some corals, like *Goniopora*, for example, may extend their polyps as far as 10–15 cm, and so could kill their competitors at a significant distance. In the Atlantic corals a kind of hierarchy – the "scale of aggression" – was discovered (Lang 1974; Cope 1981). The most aggressive of them belong to the families Mussidae, Meandrinidae and Faviidae. Lang (1973) denied the aggressive relationships between corals of the same species, while Bothwell (1984) questioned this statement. According to Lang (1973), this hierarchy in the coral community of Jamaican reefs, starting with the most aggressive species, appears as follows: *Mussa angulata* → *Scolymia lacera* → *Isophyllia sinuosa* → *Mycetophillia ferox* → *Meandrina meandrites* → *Mycetophillia danaana* → *Scolymia cubensis* → *Isophyllastrea rigida* → *Montastrea annularis* → *Diploria labirinthiformis* → *D. strigosa* → *D. clivosa* → *Montastrea cavernosa* → *Manicina areolata* → *Acropora palmata* → *A. cervicornis* → *Eusmilia fastigata* → *Agaricia* sp. → *Helioseris cucullata* → *Siderastrea* → *Porites* → *Madracis* → *Stephanocoenia*. On the reefs off Bermuda this hierarchy seems to be a little different: the most aggressive corals there are *Montastrea annularis*, *Diploria labyrinthiformis* and *D. strigosa*. In communities of the GBR this hierarchy within a group of the main families of corals appears as follows Mussidae → Faviidae → Acroporidae → Pocilloporidae → Poritidae (Bothwell 1984).

In some reef regions the structure of the coral communities was subjected to the influence of the *Acanthaster* plaques, which was caused by a definite selectivity evinced by this predator in relation to its prey (Goreau et al. 1972; Glynn 1976; Done 1985). Sometimes corals may escape the attacks of the predator, e.g. when they are well armed with the powerful nematocysts, like the hydrocoral *Millepora*, the octocoral *Heliopora* (Lewis 1989), and corals with massive round colonies, onto whose surface the asteroids cannot fasten their bodies, like old colonies of the coral *Pocillopora*, which are defended by their symbiotic crabs (Glynn 1981) or like corals defended by territorial fish, the Pomacentrids. The structure of a coral community after an attack by *Acanthaster* changes drastically. Its species diversity increases with the simultaneous decrease in the percentage of bottom surface covered by corals (Fig. 7.8).

Corallovorous fish like scarids, box fish and balistids damage corals, but simultaneously further their vegetative propagation by crushing the branches of ramose corals to pieces, which then may start new colonies (Randall 1974; Tunnicliffe 1981; Highsmith 1982). Being specialized with respect to which species of corals they eat and also as to their age and size, corallovorous fish significantly influence the structure of coral communities (Grassle 1973). The grazing impact on some preferred by fish corals, such as

Pocillopora, could be so strong that in the calm zone of flats at some reefs this coral desappears, remaining abundant only in communities of the shallow surf zones, where the waves repel fish (Neudecker 1979). Fish which are scraping periphyton and macrophytes from reef rocks, improve the recruitment conditions for the coral's planulae (Birkeland 1977; Sutton 1983).

A significant influence on the formation of coral communities exert the urchins *Diadema* (Sammarco 1980, 1982) which scrape even the surface of reef rock and clean the spot for recruitment of planulae. This positive impact prevails when their number is moderate (1–4 sp. per m^{-2}). But when the populations of *Diadema* become denser, they influence recruitment negatively, because they damage coral spat and the young colonies of corals (Bak and van Eys 1975). But the latter situation is rare on reefs. Thus, in general, the influence of urchins on coral recruitment seems to be positive. Among other biological factors influencing the formation of coral communities and their structure is the activity of boring benthic fauna, e.g. boring bivalves and boring sponges, which are sharply specialized with respect to their preference for living corals to be bored (Kleeman 1980; Rylaarsdam 1983; Wilkinson 1983b). The influence of microbial diseases of corals will be discussed below (cf. Sect. 11.3).

7.4 Reproduction of Corals

Just as the other Coelenterate, corals use two ways of reproduction: sexual (partially including parthenogenesis) and vegetative. During sexual multiplication, corals produce their larvae planulae which swim in the water column for some time, and then settle, starting a new colony. In vegetative propagation pieces of corals, their bailed-out single polyps, or their "buds" distributed by currents and waves, then recruit again and give birth to new colonies. The relative importance of these two modes of breeding are different in different corals, being largely dependent upon their life strategy.

With respect to their sexual breeding, all corals for a long time were regarded as bisexual and viviparous (Yonge 1973; Stimson 1978). This concept was the result of observations on several opportunistic species of corals living in the security of aquaria, which indeed are bisexual and viviparous, like *Pocillopora*, *Stylophora* and *Cyphastrea* (Atoda 1953; Harrigan 1972; Rinkevich and Loya 1979a,c). But J. Duerden in 1902 also described coral species, whose polyps dispersed sperm and eggs instead of planulae. These corals were *Manicina areolata* and *Favia fragum*. Later, the fruitless attempts to find the hatching of planulae in most corals resulted in the opinion taking root that their sexual breeding is of secondary importance in their propagation, even though histological studies invariably demonstrated the presence of gonads in their polyps. These studies also revealed the dominance of hermaphrodites among the corals investigated (Stimson 1978; Connell 1978). The explanation of these facts came only in the early 1980s when it was discovered that the viviparous corals, which breed periodically around the year, and have several cycles of gametogenesis during this period, are actually rarities in coral communities. On the Indo-Pacific reefs they comprise less than 5% of all species and in the western Atlantic less than 10%. The rest, 90–95% of corals, appears to be spawning hermaphrodites with external fertilization, an annual cycle of multiplication and a very short period of spawning, which takes only 2 to 6 days (Harrison et al. 1984; Szamant 1986). No wonder that previous researchers waiting for the expected monthly appearance of planulae easily missed these short periods of their annual spawning (Kojis and Quinn 1981).

It is recognized now more clearly that within the coral communities taxa can be found at levels, of families, genera and even species, with different types of sexual multiplication, which specifically correspond to their life strategy.

Among the variety of members of reef coral communities, taxa can be found with one of the following modes of sexual breeding: spawning hermaphrodites, spawning bisexuals, viviparous hermaprodites and viviparous bisexuals (Stimson 1978; Fadallah 1983; Harrison 1983, 1985; Richmond and Jokiel 1984; Babcock 1984; Kojis and Quinn 1982, 1984, 1985; Jokiel 1985; Wyers 1985; Babcock et al. 1986; cf. Table 7.11). According to present knowledge, some 60% of corals of the Pacific reefs are spawning hermaphrodites with external fertilization (Harrison et al. 1984). Among the corals of the western Atlantic reefs, they comprise slightly more than 50% (Szamant-Froelich et al. 1984; Szamant 1986). In some corals one further mode of multiplication was discovered – partial parthenogenesis, when part of the oocytes does not pass through the phase of meiosis. Thus, the developing larvae are genetic copies of the parent organism. These planulae are born by polyps together with the normal planulae developed after fertilization (Stoddart and Black 1985; Ayre and Reising 1986). Among these species are *Pocillopora damicornis* and *Tubastrea* sp.

Sexual maturity in corals can be estimated by the appearance of gonads in their polyps. It comes in most corals late – at the age of 3–5 years (Rinkevich and Loya 1979a; Kojis and Quinn 1981, 1985; Harrison 1985). But the main factor which controls the coming of corals to maturity is the size of the colony, not just age, though this is also of definite importance. In massive corals (*Montastrea*, *Cyphastrea*) the projective area at this period of development should be 80–100 cm^2, and in ramose corals (*Acropora*) the length of the branches has to reach 10–20 cm (Szamant 1986). In experiments, the pieces of old mature colonies of the above-mentioned massive corals after their acclimatation and the formation of new colonies lost the ability of sexual breeding, and gonads in their polyps disappeared. They recover only when new colonies attain the necessary size and number of polyps (over 30 in *M. annularis*).

In most viviparous and in spawning corals, in their polyps and in their whole population on separate reefs, gametogenesis cycles are synchronized (Richmond and Jokiel 1984; Stoddart and Black 1985; Kojis 1986). The number of these cycles in corals with monthly periods of multiplication (coming forth of planulae) is two to three per year, and in corals with annual cycles of multiplication one (Stimson 1978; Kojis 1986). In the last case, the oocytes appear in the tissues of polyps in the spring and become mature in autumn. The mature oocytes are reddish. In some species they contain the algal symbionts, zooxanthellae (Babcock et al. 1986). Their number in gonads of one small-to-average polyp varies between 10 and 120 (Stimson 1978). In polyps of large corals, like the solitary coral *Fungia*, the number of oocytes in gonads might reach up to several millions. Their size varies between 20–600 μ.

In polyps of the viviparous hermaphroditic corals *Pocillopora damicornis* and *Acropora formosa* the ovaries and the spermaries are formed in tissues of the mesentery within the same polyp (Stoddart and Black 1985). The number of oocytes in one polyp of *Pocillopora* was 18–24, and in one polyp of Acropora close to 100. The oocytes are self-fertilized inside the polyp. In *Pocillopora* some oocytes are parthenogenetic. The ova develop to the phase of planulae, which then remain to mature inside the mesentery, attached to its walls by special stalks (Kojis 1986). In bisexual viviparous corals, like *Fungia*, *Dendrophyllia*, and *Tubastrea*, the oocytes are fertilized inside the mesentery by spermatozoids which swim into the mesentarial cavity of the polyp through its mouth.

The spawning hermaphrodite corals, which dominate in reef coral communities, employ two types of fertilization: the obligate cross-fertilization or mixed fertilization, when a part of the oocytes are fertilized by spermatozoids from the same polyp or from other polyps of the same colony genetically equivalent to the former. An example of obligate cross-fertilization is given by the coral *Montipora digitate*. The oocytes of this coral "recognize" their own spermatozoids and reject them (Heyward and Babcock 1986). Inbreeding (external fertilization by spermatozoids from the same colony) is largely widespread among the spawning species, but it is not clearcut, because a part of their oocytes is fertilized by sperm from synchronously spawning colonies. Moreover, most spawning hermaphrodite corals have mechanisms restricting the possibility of inbreeding. In some of them the final fissions occur in the oocytes 15–20 min after they have been spawned into the water. Thus, they cannot be self-fertilized inside the polyp. In the coral *Goniastrea aspera*, the oocytes thrown out by polyps settle down to the bottom and become glued onto the substrate. Here they are fertilized by sperm excreted by neighboring colonies (Heyward and Babcock 1986). But in most hermaphrodite spawning corals special lumps of ca. 1 mm size are formed in the mesenteries of their polyps, which hold together the oocytes and the unmatured sperm. The oocytes in these lumps contain much fat. On being thrown out, these lumps float up and soon desintegrate in the surface film. Then during several tens of minutes the sperm matures and the fertilization proceeds. However, this is mostly cross-fertilization. Because of the synchronized spawning of all coral populations at the sea surface, a mass of eggs and sperm accumulates in the surface film itself (Harrison et al. 1984; Babcock et al. 1986). The accumulation of sexual products of corals during their synchronous spawning results even in the formation of rosy slicks on the sea surface. The spawning proceeds synchronously in all polyps of the colony. Polyps of the corals *Acropora* or *Platygyra* slowly let out the lumps of gametes within 10–50 min. Other species throw out clouds of sperm and lumps of oocytes by instant contraction of all polyps of the colony (Goniastrea, Fungia).

The process of gametogenesis and spawning are synchronized in the polyps of corals not only within one colony, but also in all colonies of the population on a given reef. In viviparous corals the synchronization of the monthly periods of hatching of their larvae is clearly connected with and triggered by the phases of the moon. In spawning corals with annual cycles, the connection between the time of synchronous breeding and both the phases of the moon and the temperature of the water has been established (Rinkevich and Loya 1979a; Harriott 1983; Fadallah 1983; Jokiel 1984, 1985; Jokiel 1985). Moreover, on the Pacific reefs the times of breeding in spawning corals are synchronized also in most members of the coral community, be it scleractinians or octocorals. On reefs of the GBR the time of mass spawning of corals within a week of the breeding period was established to be simultaneous for over 130 species – representatives of 12 families of corals out of 15 inhabiting the central part of the GBR (Willis et al. 1985; Babcock et al. 1986). The mass spawning is triggered in most species when the water temperature reaches 26 °C on the first day after full moon, which happens in this region in late October. The process of maturation of gametes is controlled by the water temperature, while the moon phase (actually the level of illumination at night) controls the exact time of spawning. The latter was proved in experiments by imitating moonlight (Richmond and Jokiel 1984). On reefs of different regions the times of mass breeding of corals appear to be different, being triggered at various levels of water temperature and by different phases of the moon for the same coral species (Loya 1972). Some corals of the GBR with a cycle of spawning of two times annually – at the end of October and at the end of November – spawn exactly at the same moon phases (*Acropora latisetella*, *Hydnophora exaesa*). Maximum rates of spawning during synchronous breeding were observed during the first half of the night, from sunset till midnight.

The synchronous spawning of tens of different species of corals is in itself an outstanding biological phenomenon (Harrison et al. 1984). But it seem to be happening only in very diverse and abundant coral communities of the GBR. This phenomenon has also been observed in the region of Fiji Island. But on Caribbean and Red Sea reefs this kind of synchronous breeding within a larger part of the coral community does not happen. It is synchronous there too, but only within the specific population, not within the whole community (Szamant-Froelich et al. 1984; Shlesinger and Loya 1985; Szamant 1986). Moreover, Atlantic corals differ from the Pacific species not only in their periodicity of breeding but also in the ratios of species with different types of gametogenesis and fertilization (Table 7.11), which could be related to their relatively recent evolutionary origin in the "young" Atlantic Ocean (Szamant 1986).

The types of sexual breeding in corals are directly related to the strategy of their survival. The quickly growing ramose or foliose radicals, *r*-strategist corals, are most often viviparous hermaphrodites, sometimes with elements of hermaphroditism for which self-fertilization is usual. Thus, they reproduce mostly their genetic copies. They breed monthly, year-round. Their viviparity ensures them better success in fertilization and decreases the time of the larval period. Their short-living planulae soon settle near their parents, decreasing the rate of their mortality from grazing. Being genetically identical to the parents, they are adapted to surviving in the biotope where they were hatched. This increases their ability to compete with stranger planules for substrate during their recruitment. These features of breeding, besides their ability for vegetative propagation, provides *r*-strategist corals good chances of survival in the physically stressed zones, where they often dominate (Loya 1972; Szamant 1986).

Table 7.11. Types of sexual breeding in common corals of the Indo-Pacific and western Atlantic reefs (Data by Rinkevich and Loya (1979a,c); Kojis and Quinn (1980); Richmond and Jokiel (1984); Szamant-Froelich et al. (1984); Szamant (1986); Babcock et al. (1986)

Type of breeding	Sexual status	Corals of Indo-Pacific	Corals of western Altantic
Viviparous	Bisexual	*Porites murrayensis*, *Dendrophyllia*, *Tubastrea*	*Siderastrea radians*
	Hermaphrodites	*Stylophora pistillata*, *Pocillopora damicornis*,[a] *P. caespitosa*, *P. meandrina*, *Acropora palifera*, *A. cuneata*, *A. formosa*, *Galaxea aspera*, *Cyphastrea ocillina*	*Porites asteroides*, *Favia fragum*, *Manicina areolata*, *Mycetophillia ferox*, *Agaricia agaricites*, *A. humilis*
Spawning	Bisexual	Fam. Fungiidea, fam. Poritidae, fam. Agaricidae, (Pachyseris), fam. Caryophilliidae.	*Montipora cavernosa*, *Dendrogyra cylindrica*
	Hermaphrodites	Fam. Acroporidae,[b] fam. Faviidae, fam. Merulinidae, fam. Mussidae, fam. Oculinidae, fam. Pectinidae	*Acropora palmata*, *A. cervicornis*, *Diploria strigosa*, *Montastrea annularis*

[a] With partial parthenogenesis.
[b] Excluding species mentioned above.

All of the long-living massive or columnar conservative corals (*k*-strategists) have annual cycles of breeding. Most of them are spawning, and only a few of their species are viviparous. They reach maturity late and are very prolific. Their planulae live long in the water column before settling (Harriott 1982; Jokiel 1984; Szamant 1986). The spawning procedure provides them a better possibility for cross-fertilization and so for the production of heterozygote offspring, which are more viable and versatile, and better adaptive to strange biotopes when settling. The synchronous spawning of tens of their species also provides conditions for intergeneric fertilization, thus promoting the appearance of new morphs and new species in this group of corals – *k*-strategists. An important biological reason for mass synchronous spawning in coral communities (and also in the populations of viviparous species) could also be seen in the decreasing mortality of eggs and planulae from grazing by fish and other planktonic and benthic predators (Harrison 1985; Westheat and Reising 1988). The predators cannot physically graze them out significantly during the short period of their massive accumulation in the water column.

In octocorals and in zoantharians sexual breeding appears to be just as important as in scleractinians, while they reproduce successfully also asexually (Yamazato et al. 1981; Fadallah et al. 1984). Among zoantharians, such spawning species are most numerous, which have external fertilization (Cooke 1976; Karlson 1981). In populations of the mass species *Zoanthus sociatus* and *Palythoa caribbaereum* there are colonies present simultaneously with only hermaphroditic polyps, colonies with polyps of both sexes, and also colonies with only monosexual polyps. Most octocorals are bisexual, either spawning or viviparous, though hermaphroditic and even the parthenogenetic species could also be found among them (Weinberg and Weinberg 1979; Yamazato et al. 1981; Benayahu and Loya 1984). Their viviparous species are represented by *Xenia*, and *Parerythropodium*, by some species of *Alcyonium*, by *Anthelia*, *Eunicella*, and *Muricea* (Theodor 1967; Grigg 1977; Benayahu 1989). The alcyonacean *Capnella*, which has bisexual colonies, takes special "care" of its offspring. The eggs excreted by polyps instantly attach to the surface of the colony. Here they are fertilized by sperm produced by polyps of a neighboring male colony. The hatched planulae drop down to the bottom and start new colonies near the mother ones, forming so a large monospecific settlement of this octocoral, which could be up to several tens of square meters large (Farrant 1985).

The spawning octocorals produce eggs, which are fertilized externally. Among them are the common alcyonaceans *Lobophytum*, *Sarcophyton*, and *Sinularia*, and most species of the genus *Alsyonium* (Yamazato et al. 1981; Babcock et al. 1986). They have bisexual colonies. The ovogenesis takes in *Lobophytum crassum* 2 years, and the spermatogenesis 2 years. Spawning occurs once a year. The number of oocytes in one polyp is 70–100, with a diameter of 0.5 mm.

The size of planulae hatched by viviparous corals varies between 0.5–1 and 0.8–3 mm. They have a large stock of steroid lipids which in the newborn cells comprises up to 70% of the dry weight. This stock of material of high caloric value serves as a source of energy enabling them to survive until they settle. Another source of energy is photosynthetic products, translocated by symbiotic zooxanthellae, which they acquire from their parents via eggs, or through infection by free-living zooxanthellae cells after hatching (Richmond 1981; Babcock et al. 1986). Hatched planulae swim in the water column, part of them in populations of the same species crawl on the bottom (Lewis 1974b). Often they come together and form aggregates. The new-born planulae have positive phototaxis which keeps them in the water column (Kawaguti 1941; Harrigan 1972). Several days later their taxis changes to negative. Then the planulae settle on the bottom and start to seek place for recruiting. They try to find shadowed hidden places of rocky ground. The planulae of different corals display definite selectivity with respect to substrates to settle on, being able to select the biotopes where the adult corals prefer to live (Yonge 1973). They seem to be able recognize the young colonies of the same species near which they prefer to settle (Lewis 1974a). The planulae of free-living fungiid corals, living in adult age on a soft or rubble bottom, settle first on hard rocks and grow up attached to the surface by stalks like mushrooms. Later, the stalks break and the polyps on being set free are transferred to their appropriate places by wave action.

Upon being settled, the planulae start their metamorphosis. Within 15–30 h the apical end of the mouth divides and six initial tentacles appear. In its mesentery new septae appear and their total number will soon reach 12. After 1–3 days the formation of the exoskeleton starts. After 4–6 days the walls of the corallite start to grow and in the course of 10–20 days the planula transforms into corallite with a single polyp in it. Within 2–3 weeks this corallite reaches the size of 2–5 mm and in its ectoderm around its edges the buds appear, which after passing through the same stages of development as the first corallite, give start to the formation of a new colony. Within a year the settled young colony reaches a size of 3–8 mm in spawning corals, and up to 10–15 mm in the viviparous corals *Stylophora* and *Porites* asteroids. The fast-growing spat of *Pocillopora* reaches the above size in 3–5 months (Coles 1985; Sato 1985).

The planulae of *Pocillopora damicornis* were found to be able to reverse their morphogenesis, which is a unique feature among the larvae of invertebrates (Richmond 1985). These planulae also settled and formed the first (larval) corallites, which on being stressed (by excess of light, for example) may abandon them and return again to the planktonic stage, whence they are able to settle again, thus escaping the possibility to die in unfavorable conditions. Among the group of planulae hatched by this coral there could be several nonsettling specimens, which pass through their metamorphosis up to the stage of polyp, remaining still in the planktonic stage. The-free living polyps thus formed spend some time in the water column or crawl on the bottom until they find a proper place and can settle. The planulae of the viviparous species settle usually within 3–5 days, and planulae of spawning corals within 2–3 weeks (Babcock 1984; Harrison et al. 1984; Richmond and Jokiel 1984). But some of them may remain in the water column even for several months, being drived by currents tens and hundreds of kilometers away from the place of their hatching (Grigg 1981b; Babcock 1988). Another mode of the spatial distribution of corals is the settling of planulae on floating objects: on driftwood, coconuts, bottoms of ships, on floating pumice, thus traveling thousands of kilometers (Roy and Smith 1971; Jokiel 1984).

The rate of recruitment of corals was studied by placing in different reef biotopes of some artificial substrate, like concrete plates or tubes, glass plates or lime plates made by cutting, colonies of massive corals. On reefs of the GBR the rates of recruitment was estimated to be within ten colonies $m^{-2}year^{-1}$ (Connell 1973; Bothwell 1983a). On the Atlantic reefs it was less: three to four colonies (Rogers et al. 1984; Szamant-Froelich 1985). On reefs of the Red Sea it was about five colonies $m^{-2}year^{-1}$, the common coral *Stylophora pistillata* dominating among the recruits (Loya 1976b). The rate of recruitment was as a rule higher in communities where the conditions of illumination and the content of nutrients were far from optimal: in poorly illuminated, deep zones of reef slopes or zones situated in reefs surrounded by waters impoverished in nutrients (Birkeland 1977). In reef zones with optimal conditions the recruitment was inhibited by the macrophytes and algal turfs flourishing there. The highest rates of recruitment showed the most common ramose corals. On reef flats of Heren Island (GBR) these were *Acropora digitifera*, *A. millepora*, *A. cuneata*, and *Pocillopora damicornis*, which were thus contributing to the high density of their specific population in the biotope (Connell 1973). The strategy of recruitment seems to be different in viviparous and in spawning corals (Bothwell 1983a, 1984). In the viviparous species *Acropora cuneata* and *Pocillopora damicornis* recruitment proceeded mostly by the settling of their planulae, while in the spawning species *A. digitifera*, *A. millepora*, and *A. pulchra* most of new recruits descended from the rooted pieces of their branch, i.e., by vegetative propagation. This strategy provides spawning species the possibility of colonizing nearby areas of their native biotope, because their eggs and larvae exist for a long time in the surface film and are driven far from the area where they were produced (Tunnicliffe 1981; Highsmith 1982). It was detected also that planulae of some species preferred to settle together in one spot, which probably might increase the chances of their survival (Goreau et al. 1981). Sometimes aggregations of recruited larvae were formed by planulae of different species of the same genus of coral.

Corals spend for sexual breeding a significant portion of energy, especially the viviparous ones, while the survival rate of their egg larvae and spat is very low (Harrigan 1972; Goreau et al. 1981; Sammarco 1983). The larvae and eggs are eaten by various planktonovores (Connell 1973).

Available lots of hard substrate for planulae to settle on are restricted, because the large reef areas are covered by soft sediments, and settling on these most larvae die (Highsmith et al. 1980). On the hard bottom biotopes recruits face high competition from other sessile benthos, and of algae as well (Highsmith 1980; Benayahu and Loya 1981; Rosen 1981). The mortality rate of a spat of corals alone is 60–90% during 3–12 months, and it is highest in the second to fourth month of its life (Loya 1976; Sammarco 1983; cf. Fig. 7.15). The mortality of planulae and eggs of gorgonaceans was estimated to be close to 90% (Theodor 1967). The main cause of mortality of spat is the grazing by fish, urchins and other benthovores feeding on periphytonic turfs and so scraping the solid surfaces (Sammarco 1980, 1982; Rylaarsdam 1983). These grazers exert a diverse impact upon the survival of young corals. Weak grazing results in denser overgrowth on the substrate, which inhibits recruitment, while by intensive grazing the recruits could be damaged. Thus optimal conditions for recruitment of corals and for the survival of their spat seem to be created by moderate grazing, which was proved experimentally (Sato 1985; Sammarco 1985). Often the planulae settle in the holes made in reef-rock surface by the urchins *Echinometra*. This help them later to escape the scraping grazers (Birkeland and Randall 1981).

As mentioned above, in many corals the larger part of recruited colonies is formed via vegetative breeding (Connell 1973; Highsmith 1982). Among the modes of vegetative propagation of corals, the following could be mentioned: (1) the fragmentation of colonies and their rooting in reef biotopes; (2) the bailing out of polyps from corallites of adult colony and their subsequent settling; (3) the production of new colonies in solitary corals by fission of their corallites or by formation of new colonies by them through budding; (4) the shedding of separate corallites functioning like buds; (5) the fission of a colony with subsequent separation of the newly formed ones; and (6) the separation of pieces of branches by "autotomy".

The propagation by fragmentation is widespread among the ramose scleractinians and among the gorgonaceans, though it happens also in massive corals (Crassle 1973; Tunnicliffe 1982; Bothwell 1983b, 1984). Main factors causing the fragmentations are storms, especially hurricanes (Highsmith et al. 1980; Woodley et al. 1981), as well as the activity of boring sponges and clams (Tunnicliffe 1978). Probably the harboring of boring animals by living corals is a kind of adaptive reaction enhancing their multiplication and thus improving their chances in the competition for substrate (Highsmith 1982). Otherwise the ramose corals should develop defence mechanisms against borers, what actually has not happened. They are extremely numerous in most of the common corals, especially in acroporids (cf. Sect. 5.1.4). Some corals produce fragments by fission of the colony through separate sections at its base (Highsmith 1980). It happens when a part of the colony has at its base been ruined by siltation or by borers and the whole colony collapses. Then its parts attach themselves to the substrate in new places and start new coral colonies. This way of vegetative multiplication happens often in massive, foliose and columnar corals, like *Porites*, *Agaricia*, *Montipora*, and *Meandrina* (Bak 1976; Hughes and Jackson 1980). In the ramose corals (*Acropora* sp.) it was observed in zones of fast sedimentation (Connell 1973).

The fragments of living coral colonies fallen onto the bottom secrete a lot of mucus, which covers it like a cocoon, and preserve polyps from mechanical damage when the fragments are driven up from the bottom by currents and waves (Bothwell 1981). Most fragments take root on the spot. After several unsuccessful attempts to root the bulk of fragments produced in the upper surf zones of the reef is driven down the reef slope. They accumulate and then root in middle and lower depth zones, helping thereby the formation of dense and varied coral communities there (Highsmith 1982). In common species of spawning ramose corals which actively use also the vegetative way of multiplication, the share of the latter in their total recruitment is high – 70–80% (Bak and Engel 1979; Tunnicliffe 1981). In some Atlantic gorgonians this percentage was over 90% (Lasker and Coffroth 1985), After long-lasting storms, when a mass of fragments ready to be recruited appears on the reef, in some of its zones the forming of monoclonal coral populations was observed (Bull 1982; Heyward and Collins 1985).

The vegetative multiplication of corals by the bailing out of polyps was discovered in one of the common opportunistic corals, *Seriatopora histrix* in aquaria under stress conditions, such as exposure, excess or insufficient illumination (Sammarco 1982). The coenosarc around polyps of a stressed colony pull into the corallites and the polyps start to fall out of them. Carried by currents and waves, they float to new places, where they settle like the planulae and give start to the growth of new colonies. If the "pregnant" polyps of this viviparous coral bail out, they then hatch the planulae that are in this naked and solitary state, enhancing so the distribution of this species.

The propagation by buds is known in the scleractinian corals *Asteros milia* (fam. Caryophyllidae), *Balanophyllia*, and *Endocyathus* (fam. Dendrophyllidae). Buds, looking like under-developed polyps, are joined with the mother colony by a fragile connection. When this breaks, the buds fall out and later settle and grow up into a new colony (Zibrowius 1985). Somewhat similarly, the buds are produced by the solitary coral *Fungia*. It separates periodically a part of its polyp together with an appropriate part of corallite. The separated part of the colony soon grows up to a normal coral (Wells 1966). The coral *Goniopora stokesi* also forms a kind of "buds", which are small (2–4 cm) round colonies, appearing on the surface of the mother colony. They then separate from it and give rise to new colonies (Sheppard 1981b). The same way of propagation is used by the alcyonaceans Xeniids. For example, *Xenia macrospicularia* periodically separates parts of its colonies, which then settle nearby, forming gradually large settlements ("carpet growths", Benayahu and Loya 1984). A similar mode is used also by another octocoral, *Briareum asbestum* (Lasker 1983). The gorgonian *Juncella* ("sea whip") produce offspring fragments by "autotomy", i.e., by the resorption of tissues at the "Whip" 5–8 cm from its end. The pieces thus periodically separated fall to the bottom and root. The efficiency of this mode of vegetative multiplication appears to be extremely high. Populations of sea whips may produce up to 300 new colonies per square meter per year.

8 Morphology and Ecological Physiology of Corals

Many important properties of coral reef ecosystems are closely connected with the morphological and functional characteristics of the corals themselves, as dominant inhabitants of their bottom biotopes. Among these properties there could be mentioned the adaptive variability of corals, their endosymbiosis with algae, and their ability for autotrophic feeding coupled with heterotrophy. Below the most important points of the ecological morphology and physiology of corals are discussed. The anatomy of polyps and the morphology of corallites and of coral colonies have been described in corresponding reviews, and we will deal with them only in part if necessary. Among these reviews should be mentioned: Hyman (1940), Bayer (1973), Coates and Oliver (1973), Mariscal (1974), Preobrazhenski (1986), and Sokolov (1987).

8.1 Features of Ecological Morphology

Colonies of corals are constructed from separate modules, corallites, connected within a united structure. They are formed by sister polyps, multiplying by budding, and thus represent the clone of genetically identical animals. By the vegetative propagation (cf. Sect. 7.4) the coral colony is able to produce other identical clones. Therefore, in the natural populations of corals clones of polyps exist, which are represented by different colonies. Being a separate monoclonal multimodular but unique organism, the coral colony itself thus became an object of natural selection (Wells 1971). This resulted in the appearance of a multitude of species of modern corals within a relatively short geological time. The statement that the coral colony "is the animal with many mouths", has been offered by Sir Maurice Yonge long ago (Yonge 1930a). To it should be added another one – it is also something like "the plant with many leaves". In fact, the colonies of scleractinian corals, and even most of the octocorals, possess the structures that provide the interconnection between the polyps and so unite them into a single organism. In colonies of scleractinians gastrovascular channels were found through which the exchange with nutritional substances takes place

(Schlichter 1982; Marphenin 1984; Gladfelter 1988). The nervous systems of polyps in a colony are interconnected and thus united within a complete system (Horridge 1956). The electrical irritation of one polyp results in contraction of other polyps in the colony. The pulse of irritation passes through the colony with a speed of several cm s^{-1}. Another fact proving the correctness of the old sentence by M. Yonge is the actual independence of the rate of respiration of corals from the size of their polyps (cf. Sect. 9.1). In free-living aquatic animals it is the reverse function of their size.

How important this preference for modular structure is in coral colonies could be seen in the possibility for them to significantly increase with the aid of this structure the light-collecting and filtering surface at the expense of increasing their size, and branching without an increase in losses of respiration, which remains stable as per biomass of a single module (Hughes 1983; Sorokin 1984a). The coral colonies composed of module corallites and armed with skeletal material also have a definite preference in their ability to regenerate, compared with free-living animals. The loss of part of from the modules, grazed or damaged, does not cause the death of the colony, which may quickly regenetate by the budding of the remaning modules (Jackson 1977). The modular construction of a colony offers the possibility of constructing from the modules a great variety of phenotypic adaptive morphs ("living forms"), which enable the same coral species to colonize various bottom biotopes within a wide scale of environmental conditions (Vaughan and Wells 1993; Goreau 1963; Veron et al. 1977; Foster 1979, 1980; Vosburgh 1982; Foster 1983).

The structure of corallites has also an adaptive character and serves the strategy of survival of the coral, connected with preferences for sources of fedding and an acquired mechanism for shedding the sediments (Hubbard and Pocock 1972; Hubbard 1973; Lewis and Price 1975; Foster 1980). The diameter of polyps in colonial scleractinian corals varies within a large scale: from 1 mm in acroporids and poritids to 3–5 cm in some mussids, and the wet weight from 1 mg to several grams. As regards the reason for this wide scale of sizes in polyps, it had been supposed that the corals with larger polyps should easier catch zooplankton, thus being largely heterotrophic, while those with small polyps (especially ramose ones) more autotrophic (Porter 1974b; Hughes 1983). But experiments failed to prove this (Sorokin 1981a, 1984a; Gli-Turnes and Corredor 1981; cf. Sect. 9.2.1). It was shown that the ability for photosynthesis and for heterotrophic feeding in different species of corals, as calculated per biomass of polyps, was not very much dependent upon the size of their polyps or on the form of their colonies. Without refusing the definite adaptive reasons for formation of the ramose corals with respect to enlarging their light-collecting surfaces, the main reason for their appearance in the course of evolution seems to be quite another. Ramose, and also cribose, corals have high mechanical endurance of hydraulic pressure, being constructed from less calcareous skeletal material, compared with the massive corals. This enables them to have a higher rate of growth by the same rate of calcification and so overgrow the massive corals and better escape siltation (Shinn 1963; Connell 1973; Vosburgh 1982; Bottier 1980). Another preference of ramose corals could be seen in their better ability to catch zooplankton, which has been proved experimentally (cf. Sect. 9.2). In fact, corals hunt most successfully at dawn and dusk, when the demersal zooplankters rise or return to the bottom, Also, they feed on zooplankton, which is driven by the current laterally. It seems quite obvious that by the experimentally proved equal ability of large and small polyps to catch mesozooplankton, their preference will be for polyps of ramose species, properly distributed in space. Ramose corals have also one more important preference – their increased ability for vegetative multiplication by fragmentation, which greatly improves their ability to compete for the hard substrate with other sessile fauna (Highsmith 1982).

Their ability, widespread among corals, to form adaptive ecomorphs, was mentioned above. For example, some massive corals, like *Montastrea*, in poorly illuminated deep reef zones form folioses, or crustose colonies (Barnes 1973; Dustan 1975; Graus and Macintyre 1976; Preobrazhensky and Ivanov 1982). The flattened colony is formed by the module corallites which grow only at the edges of young coral, while in corals that live in normal light, growth proceeds over their whole surface. The flat form of a colony is optimal for a poorly illuminated environment. Corals under these conditions have a deficiency of energetic resources for growth and direct the available resources only to the narrow zone of growth. It provides them with the ability to attain a speed of growth sufficient to escape siltation and to compete with other corals for space, moreover the flat form of a colony enables them also to construct the necessary light-collecting surface with a minimal expenditure of resources (Dustan 1975). In the upper reef zone the massive round or columnar form should be preferable. It endures better the waves, grazing, overgrowth and siltation, compared with the flat or foliose form. The ramose corals in deep biotopes also form flattened colonies, in which most branches grow laterally, being perpendicular to the direction of the light stream (Titlyanov et al. 1988c; Titlyanov and Latypov, 1991). Thus, their energy resources are used to support growth only in one plane, which maintains its normal speed at a low general rate of calcification, inhibited in deep reef zones by a deficiency of such light energy penetrating there. The lateral-form colony also avoids the self-shadowing of branches which is important in light-deficient environments.

The formation of versatile ecomorphs in colonies living in upper reef zones is connected with their adaptive phenotypic variability under the influence of various kinds of physical strees and grazing (see Sect. 7.3). The conditions of hydrodynamics influence also the form of the colonies of gorgonacean octocorals. For example, in calm waters the colonies of *Eunicella stricta* look like so many separate whips, while in the surf zones they are bush-like, and on slopes visited by permanent currents fan-like, like many other gorgonians living there (Theodor 1963). The planes of these fan colonies are perpendicular to the direction of the dominating current. Their ends effectively ward off the action of the current upon the colony. Thus they oppose well hydraulic stress (Grigg 1972). At the same time, such a position of the fans is optimal for catching food particles passing by with the current (Leversee 1976). No wonder that this form and orientation of colonies are peculiar to most ahermatypic gorgonians feeding on plankton (Muzik and Wainwright 1977).

8.2 Biomass of Living Tissues

During studies of the functional activity of corals the parameters obtained have most often been expressed per unit of surface of their colonies or per dry or wet weight (cf. Sects. 9.1, 9.2). But the bulk of their mass in scleractinians (~98%) as well as in zoantharians and in octocorals (up to 90%) comprises inert organic and inorganic structural materials: skeletal aragonite, chitine-like organic material of coenosarc and of axial filaments, wax-like protecting substances. This raises the problem how to estimate the biomass of functionally active tissues of corals, the biomass of their polyps and zooxanthellae. Some researchers have accepted as the biomass of tissues the total organic matter content in the colony (excluding the axial chord of octocorals). The content of organic matter was estimated in the decalcinated material of colonies (Davies 1980), or in their alkaline hydrolisates (Sorokin 1984b). Muscatine and Cernichiari (1969) evaluated the biomass of polyps measuring the content of protein in corals. But in their estimates coenosarc was also included into the biomass of polyps. It was necessary to find a way to isolate single polyps from the coenosarc. This problem was finally solved. Polyps were cut out one by one – with the aid of binoculars – from the decalcinated coenosarc of the coral colony, previously kept for a week in 10% formalin solution (Sorokin 1984b).

Various methods were used to measure the biomass of zooxanthelleae. Most often their total number is counted per 1 polyp or per 1 cm^2 of coral surface, and the average volume of their cells has been measured. Zooxanthellae have a spheroid form of 9–12 μ diameter and 400–600 μ^3 average volume. To count them, pieces of corals with a definite number of polyps, which had been prepared in same way as mentioned above were crushed in a mortar. Then the carbonate sediment was dissolved with the aid of HCl. The neutralized suspension was filtered on a membrane filter, which was then stained with erythrosine and examined under the microscope after clearing it with immersion oil (Sorokin 1984b). To isolate the zooxanthellae, some other methods were employed, including mechanical scraping of living tissues, washing them out with the aid of "Water-Pik" apparatus (Johannes and Wiebe 1970), or homogenization of colonies (Lasker 1977). To separate the zooxanthellae from the flurry of tissues, gradient liquids were used.

The percentage of dry weight in colonies of scleractinian corals varies usually within 70 to 85% with their specific weight being 2.5–3 $g\,cm^{-3}$. In octocorals it varies within wide limits: 10 to 60%, depending upon the content of spicular calcareous meterial in them (Lewis and Post 1982). The substance of organic matrix (mainly polyps and coenosarc) comprise in colonies of scleractinian corals 5–20% of dry weight, or 10–70 $mg\,C\,g^{-1}$ of dry weight of a colony (Table 8.1). The largest content of total organic matter was recorded in colonies of corals *Montipora* and *Porites*, living in shallow biotopes and subjected to excess illumination. The biomass of polyps in them comprises only 10–17% of the total organic matter, while in most of other corals its share is much more (Table 8.1). In octocorals the content of organic matter is between 10 to 25% and only in xeniids, which actually do not have skeletal elements, it could be over 70%. In gorgonaceans 50–80% of the material in colony is composed of the calcareous skeleton, and 6–30% of axial cord material (Lewis and Post 1982). The dry decalcinated mass of living tissue in scleractinian corals contains 10–20% ash, 20–25% protein and 10–30% lipids (Edmunds and Davies 1986).The individual wet biomass of polyps in them varies from 0.3–1 mg in colonies of the corals *Seriatopora* and *Stylophora* to several grams in *Symphillia* or *Fungia*. The content of organic carbon per one polyp in corals with small polyps (acroporids, poritids) varied within 15 to 140 μg, in corals with medium polyps (*Leptastrea*, *Pavona*, *Meruline*, *Galaxea*) within 0.6–3 mg, and in corals with large polyps (faviids, *Symphillia, Fungia*) between tens and hundreds of milligrams (Table 8.1). The biomass of the polyps of scleractinians comprises usually 20–40% of the total organic matter or 0.5–1% of dry weight of a colony. Most octocorals have small to medium sized polyps 0.5–3 mg of wet weight or 40–250 $\mu g\,C\,sp.^{-1}$ (Table 8.2).

The number of zooxanthellae in corals is rather stable: $1–5 \times 10^6$ cells per 1 cm^2 surface of their colonies (Table 8.3). Calculated per 1 polyp, small or medium size, it varied either in scleractinians or octocorals within $20–60 \times 10^3$ cells. In scleractinians the biomass of zooxanthellae was found to be 30–100 $\mu g\ C\,g^{-1}$ of dry weight of a colony, and in octocorals 0.5–1.5 $mg\,C\,g^{-1}$ (Tables 8.1, 8.2). For the evaluation of animal and plant components of comparative importance in polyp tissues it was important to estimate the ratio of biomass of zooxanthellae to biomass of polyps. First, its estimation had given very high values for this ratio – 20–40% in protein (Muscatine and Cernichari 1969). But these data seemed to be exaggerated, maybe because of the incomplete purification of zooxanthellae from the animal tissues of polyps. Later, it was shown that as to protein this ratio in polyps of the coral *Pocillopora* is close to 5% (Muscatine and Porter 1977). The same ratio of their biomasses, expressed in units of organic carbon, varied most often between 1–3%, estimated in 23 species of scleractinians and in 10 species of octocorals (Sorokin 1981a, 1984b; cf. Tables 8.1, 8.2). Other estimates gave values of this ratio with 1–5% in the corals *Acropora*,

Table 8.1. Total content of organic carbon (B_t), biomass of polyps (B_p) and zooxanthellae (B_z) in colonies of common corals of the GBR, calculated per dry weight of colonies. (Data by Sorokin 1984b)

Species of corals	B_t, mg C g^{-1}	Polyps				Zooxanthellae			
		Number of polyps sp. g^{-1}	B_p, mg C $sp.^{-1}$	B_p, mg C g^{-1}	B_p/B_t %	Number, 10^3/polyp	B_z, μg C/polyp	B_z, μg C g^{-1}	B_z/B_p, %
Seriatopora hystrix	6	190	0.015	2.8	46	25	0.5	95	3.4
Pocillopora damicornis	14	130	0.032	4.2	30	20	0.4	52	1.2
Stylophora pistillata	12	80	0.048	3.8	32	32	0.6	51	1.3
Acropora squamosa	25	110	0.065	7.2	29	56	1.1	117	1.6
Acropora formosa	19	35	0.073	2.6	13	42	0.8	28	1.1
Porites annae	32	75	0.040	3.0	9	30	0.6	47	1.6
Montipora erythraea	70	52	0.120	6.2	9	92	1.7	90	1.4
Cyphastrea seralia	26	20	0.42	8.4	32	163	3.4	68	0.8
Leptastrea transversa	13	6.6	0.60	4.0	30	320	6.7	44	1.1
Pavona cactus	22	10	0.65	6.5	30	206	4.7	47	0.7
Merulina ampliata	36	5	1.02	5.1	14	460	10.6	53	1.0
Gonippora sp.	62	11	1.15	12.6	20	260	5.2	57	0.5
Galaxea fascicularis	34	3.6	1.80	6.5	19	370	7.7	28	0.4
Echinopora lamellosa	23	2.2	4.50	9.9	43	1 380	26.2	58	0.6
Favites abdita	16	0.6	6.85	4.1	26	2 110	44.3	27	0.6
Symphilla sp.	44	0.25	45.2	11.3	25	14 100	324.3	81	0.7
Fungia scutaria	35	0.10	145.0	14.5	41	51 000	1173.0	117	0.9

Table 8.2. Biomass of polyps (B_p) and zooxanthellae in some octocorals and zoantharians per dry weight of colonies. (Data by the author)

Groups of octocorals	Species of octocorals	Polyps			Zooxanthellae		
		Number, g^{-1}	B_p, µg C/polyp	B_p, $mg\,C\,g^{-1}$	Number/polyp, 10^3 cells	B_z, $mg\,C\,g^{-1}$	B_z/B_p, %
Symbiotic octocorals	*Litophyton* sp.	520	53	26.7	36	0.48	1.8
	Alcyonium molle	96	250	24.0	66	0.19	0.8
	Lobophytum roxasi	320	43	13.8	18	0.24	1.8
	Sarcophyton glaucum	140	160	22.4	52	0.19	0.8
	Sinularia fungoides	180	73	13.1	23	0.13	1.0
	Rhabdopexaura princeps	305	48	14.6	25	0.24	1.6
	Rumphella agregata	315	67	21.1	59	0.47	2.3
	Tubipora musica	7	860	6.0	280	0.06	1.0
Asymbiotic octocorals	*Spongotes gigantea*	200	90	18.0	–	–	–
	Echinogorgia praelonga	310	80	24.8	–	–	–
	Plexauroides lenzii	410	56	23.0	–	–	–
	Bebryce indica	140	154	21.6	–	–	–
	Mopsella aurantia	680	24	16.3	–	–	–
	Acabaria nicksoni	315	42	13.2	–	–	–
Zoantharians	*Zoanthus sociatus*	8	9100	72.8	5100	0.05	1.4
	Palythoa sp.	6	610	3.7	520	0.08	2.2

Table 8.3. Number of zooxanthellae in corals, N × $10^6 cm^{-2}$ of surface of their colonies

Species of corals	N	Authors
Stylophora pistillata	0.97–2.3	Titlyanov et al. (1988a)
Stylophora pistillata	4.5	Mc Closkey and Muscatine (1984)
Pocillopora damicornis	0.9–1.1	Titlyanov et al. (1988a)
Pocillopora eydoxy	6.1	Edmunds and Davies (1986)
Porites porites	3.13	Edmunds and Davies (1986)
Seriatopora hystrix	4.8–7.8	Black and Burris (1983)
Acropora diversa	0.6–0.7	Titlyanov et al. (1988a)
Montastrea annularis	2.7–8.8	Dustan (1979)
Montastrea cavernosa	2.0–6.0	Lasker (1977)
Plerogyra sp.	9.0–14.0	Drew (1972)
Scleractinians of 10 familes (depth 0–42 m)	0.9–2.3	Drew (1972)
Alcyonia Alcyonium	1.35–2.48	Drew (1972)
Hydrocoral *Millepora*	0.72	Drew (1972)

Stylophora and *Pocillopora* (Falkovsky et al. 1984; Titlyanov et al. 1988a).

The above-mentioned stability of the content of zooxanthellae in polyps (as well as that of chlorophyll) independently of the size of their polyps (Tables 8.1, 8.2) permitted the supposition that they form one cell layer in tissues of the host polyp. Indeed, spread within one layer, 1–2 × 10^6 cells of zooxanthellae would occupy an area of about 1 cm^2; this amount was found in corals (Table 8.3). Histological analysis also confirmed this conclusion (Kawaguti and Nakayama 1973). It seems quite obvious that just this arrangement of zooxanthellae within the polyp might enable it to make optimal use of light energy, excluding the self-shadowing effect, moreover, that their exact number in a polyp depends upon its surface area in the active, expanded state (Drew 1972). For example, their highest number, exceeding by about on order its average value in most corals, has been recorded in *Plerogyra* – the scleractinian coral which inhabits the biotopes of reef slopes. Its polyps possess special bulb-like "organs" of photosynthesis which, inflated if necessary, increase the illuminated "working" surface of the polyp (Fricke and Vareshi 1982). The host polyp regulates the number of its zooxanthellae mainly through the expulsion the excess (Goreau 1964b). As alternative possible mechanisms, there could be mentioned also the inhibition of their growth by nutrients limitation or by excretion, some inhibitors on the side of the host, as well as their digestion by polyps.

8.3 Ecological and Physiological Aspects of Endosymbiosis

The coelenterates are exceptional in the animal world in their ability to live in symbiosis with algae. Most of their sessile species living in illuminated habitats have algal symbionts. Among them there are several hundred species of actinians, corals, zoantharians, hydroids, and sessile jellyfish (Muscatine 1974). All of them harbor in their tissues the same species of symbiotic dinoflagellates-the zooxanthellae *Symbiodinium microadriaticum*, Freudentahl. They were first found by K. Brandt in the cells of a planktonic radiolarian at the end of the last century. Initial studies of their functional role as symbionts of coelenterates were made in 1920–1930 (Boshma 1925a; Yonge and Nicholls 1931). The culture of this alga in artificial medium was done successfully first by Kawaguti (1944), who discovered the motile stages of zooxanthellae, which, being in host tissues, looked like round motile cells. Later it was shown that the motile form of zooxanthellae was very similar to the free-living dinoflagellate *Gymnodinium* (Mc Laughlin and Zahl 1966). They appeared in the culture mostly at midday and their appearance seemed to be connected with illumination conditions (Lerch and Cook 1984). They have a rather complex cycle of development and their cells living in the host's tissues represents one of its stages (cf. Fig. 8.1). In particular, it is one of two kinds of their immobile cysts which are capable of cell division (Mc Laughlin and Zahl 1966; Steele 1975; Schoenberg and Trench 1980). They are disposed by groups of two-three cells inside the vacuoles of the ectoderm cells of the polyp which cover its body, tentacles, oral disc and mesentery (Kawaguti 1964; Vandermeulen et al. 1972; Smith 1979). Some of them are located in the intercellular spaces (Woscoboynikov et al. 1983). In cultures, among the cells of zooxanthellae their following cellular forms could be found: (1) the immobile cysts with thick cell walls; (2) the motile flagellated gymnodinium forms; (3) zoosporangia containing one large motile zoospora; and (4) zoosporangia containing two or three nonmotile zoospores.The motile zoospores swim actively and may penetrate into young polyps of corals and into the tissues of other symbiotic coelenterates (Schoenberg and Trench 1980).

The strains of zooxanthellae, which live in different coelenterate hosts, are very similar in their taxonomic signs, e.g. the size of cells and their morphology, the composition of pigments, and their life cycle, thus belonging to the *same species, Symbiodinium microadriaticum*. Such a wide scale of hosts encompassing several hundreds of coelenterate taxa and even clams (tridacna) with different immunic-genetic characteristics and containing one single species of the symbiont is itself an outstanding biological phenomenon (Schoenberg and Trench 1980). The latter authors showed, moreover, that the one host, for example the actinian *Aiptasia*, which had expelled its zooxanthellae, might be infected by the strange strain of these algae, taken from some other coelenterates (corals) or even from the tridacna clams. Nevertheless, it seems that the morphologically identical strains of zooxanthellae, living in taxonomically different hosts differ by some of their biochemical and physiological features (Kinzie et al. 1977; Schoenberg 1980; Chang et al. 1983; Colley and Trench 1985). They might be different even genetically, so that they could be related to different

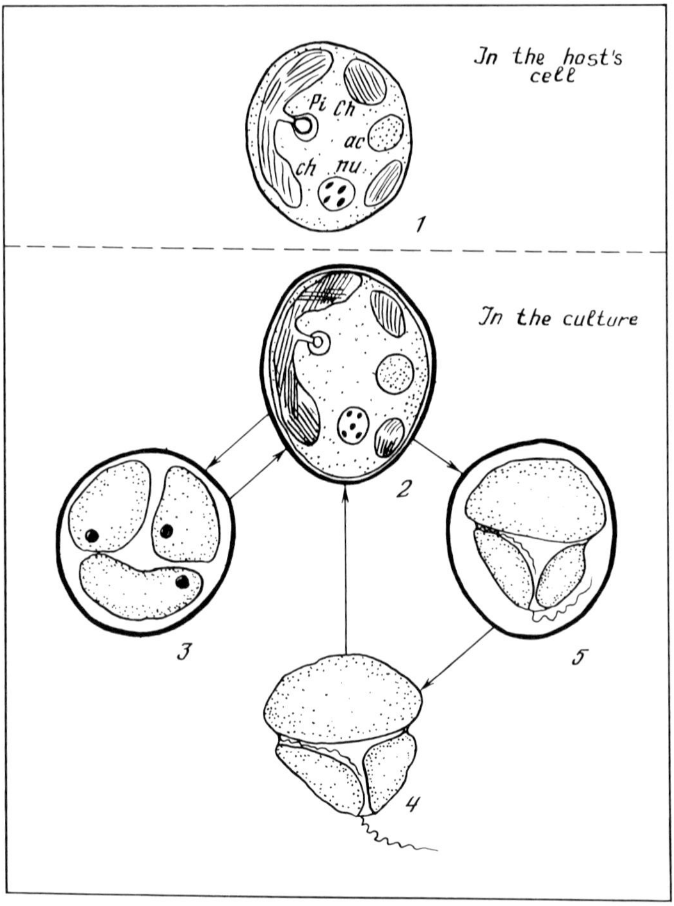

Fig. 8.1. Scheme of life cycle of zooxanthellae: *1* the cyst-like zooxanthella cell within the cell of host poly P. *Pi* Pyrenoid; *Ch* chloroplast; *ac* accumulative body; *nu* nucleus; zooxanthellae in culture: *2* cyst with a thick cell wall; *3* zoosporangium; *4* motile zoospore; *5* nonmotile tetraspore. (After Schoenberg and Trench 1980)

species (Blank and Trench 1985). For example, the symbiotic zooxanthella cell in the jellyfish *Cassiopeia* contained 97 chromosomes, in the actinians *Heteractic* and *Anthopleura* 74 and 50 respectively, in the coral Montipora 26 (Blank and Trench 1985). Later, Trench and Blank (1987) established three new species of *Symbiodinium* besides *S. microadriaticum*: *S. goreauii* (symbiont of the Caribbean Sea anemone *Ragactis lucida*), *S. kawagutii* (symbiont of the coral *Montipora verrucosa*) and *S. pilosum*, derived from the zoanthid *Zoanthus sociatus*.

The brownish-green round zooxanthella cells are separated from the cells of the host by their cell wall, the periplast, which has a complex structure consisting of three membranes incrusted with cellulose (Mc Laughlin and

Zahl 1966; Taylor 1973b; Mashansky et al. 1980). The function of the periplast is to protect the zooxanthella cell from the attack by lysosomes of the host cell (Muscatine et al. 1975). At the same time, the periplast is permeable for metabolite exchange. Some parts of the periplast have numerous protrusions and exfoliations of the outer membrane. It was supposed that through them the translocation of lipoid droplets – products of photosynthesis – proceeds from the zooxanthellae to the host cell (Patton and Burris 1983). On the side of the host, the filaments of its cell protoplasm form a kind of capsule around the cells of the zooxanthellae (Woscoboynikov et al. 1983). Thus, the existence of the complex periplast in the case of zooxanthellae decreases the difference between intracellular and interorganismic symbiosis, because their cells are largely isolated from the cell of the host, moreover, zooxanthellae preserve the ability to multiply independently of the host in external environments. But after their penetration into the host's cells an exchange with the metabolites between them and the polyp soon starts, which is regulated by the host cells, as well as by the rate of their multiplication. A remarkable feature of the morphology of zooxanthellae cells is their extremely large chloroplast, which occupies half of the total volume of their cells. The lamellar system of the chloroplasts consists of 10–12 complexes. Each of them contains three tylacoids. The pyrenoid of the chloroplast is attached to it with the aid of a stalk (schoenberg and Trench 1980; Voscoboynikov et al. 1983).

The primary "infestation" of polyps with zooxanthellae may occur in various ways. Numerous taxa of hydroids, scleractinians and octocorals, as well as their eggs or new-born larvae, contain the zooxanthellae cells (Gohar 1940; Atoda 1951; Smith 1980; Fitt 1984). In planulae, they supply them with products of their photosynthesis thus fomenting their survival (Richmond 1981). At the same time, many corals produce eggs and larvae in which zooxanthellae are originally absent (Atoda 1953; Droop 1963; Kinzie 1974; Fitt 1984; Babcock et al. 1986). Among them some octocorals (*Heteroxenia*, *Briareum, Pseudopterogorgia*) and most of the scleractinian spawning corals. The larvae hatched from these eggs could be infected by free-swimming motile zoospores of zooxanthellae, which are present in water. The "infection" takes place also at the stage where settled larvae formed a first corallite. The polyp attracts zooxanthellae, which swim into its mesenterial cavity through its mouth and thus "infect" him (Kinzie 1973). Zooxanthellae may locate the potential host polyps with the aid of chemoreception. They move along the gradient of concentration of nitrogen salts excreted by the polyp. The unmotile zoospores get into the gastral cavity of their potential polyp hosts, having been previously consumed by some filtering zooplankter, which are then captured by the polyp as its prey (Fitt 1984). The zoospores are protected by their thick cell wall from being digested during this procedure (Schoenberg 1980). The zoospores (or cysts), when they get into the gastral cavity of the polyp, are phagocytosed there by mesenterial phagocytes and then transferred by them into the polyp's tissues where they then settle (Kinzie et al. 1977; Fitt and Trench 1980). The stock of zooxanthellae in reef waters is formed through the extrusion of their excess by corals. Large amounts of zooxanthellae are released into the water column from corals, and are damaged by corallovore reef animals like parrot fish, urchins, etc. (Fitt and Trench 1980). The feces of corallovores contain a lot of undigested living cysts of zooxanthellae, which then may "infect" other hosts, the more so since the feces of fish are readily swallowed by coral polyps (Parker 1984). Large quantities of zooxanthellae-like dinoflagellates were recorded to be present in waters of reef shallows (Taylor 1980).

The composition of photosynthetic pigments in zooxanthellae is the same as in the free-living planktonic dinoflagellates, like *Peridinium* or *Amphidium* (Jeffrey and Haxo 1968; Titlyanov et al. 1980). They contain chlorophylls *a* and *c*, plus a selection of additional carotenoid pigments peridinine and dinoxanthines, specific for the dinoflagellates. The ratio of chlorophylls *c* and *a* in zooxanthellae from different hosts is rather stable (0.2–0.5). The content of chlorophyll in zooxanthellae varies between 5–12 μg 10^{-6} cells, between 15–50 μg g^{-1} of colony weight, or between 5–25 μg cm^{-2} of its surface (cf. Table 8.4). The total content of carotenoid pigments is also high in them: ca. 2–5 μg 10^{-6} cells. In corals adapted to a low illumination, the content of both groups of pigments in zooxanthellae significantly increases (cf. Table 8.6). Peridinine and other carotenoid pigments are combined in the cells of zooxanthellae with chlorophylls, a composition typical for the dinoflagellates' protein-pigment complexes (PCP). They strongly enhance their efficiency in utilizing light energy in the green range of 490 and 540 nm, thus increasing the ability of zooxanthellae for photoadaptation (Zvalinski et al. 1978, 1980; Dustan 1979; Muscatine 1980). In fact, they have a lower threshold of light saturation compared with diatoms (Scott and Jitts 1977), moreover, their large chloroplasts have a specially structured labellar system, mentioned above.

The spectra of absorption of visible light in zooxanthellae have the main maxima at 440 and 670 nm, and also additional but significant ones at 470–500 and 540 nm (Fig. 8.2). The latter are connected with the function of PCP, mentioned above (Leletkin and Zvalinski 1981). The curves of action spectra in their pigment system were estimated by measuring the rates of $^{14}CO_2$-consumption at different wavelengths of light (Scott and Jitts 1977), and those of the oxygen evolution (Halldall 1968; Zvalinski et al. 1980). They differ from the curves of the absorption spectra, and this difference is most significant just in the range of action of PSP – 470–600 nm (Fig. 8.3). It proves hour important their role is in light absorption by zooxanthellae in deep reef habitats where blue-green light dominates. The absorption maximum at 350–500 nm was especially well expressed in zooxanthellae of symbiotic gorgonaceans, which are known to be enduring well light deficiency (Fig. 8.4). The mechanism of light adaptation in zooxanthellae is connected with their ability to enlarge the size of their photosynthesis units

Table 8.4. Content of photosynthetic pigments in zooxanthellae of corals taken from illuminated (l) and shadowed (sh) habitats, $\mu g\, g^{-1}$ (Data by Zvalinski et al. (1980); Titlyanov et al. (1980))

Corals	Chlorophyll *a*		Chlorophyll *c*		Piridinine		B-carotene		Dinoxanthine	
	l	sh	l	sh	l	sh	l	sh	l	sh
Pocillopora verrucosa	5.5	7.0	2.1	2.7	1.3	2.3	0.15	0.13	0.41	0.51
Stylophora mordax	4.8	6.8	1.8	2.3	1.4	2.3	0.14	0.18	0.48	0.78
Porites nigrescens	5.4	6.7	2.2	2.5	1.1	1.8	0.10	0.15	0.47	0.51
Acropora hyacinthus	6.2	9.6	2.3	4.1	2.7	3.8	0.20	0.3	0.60	0.70
Pocillopora elegans	7.2	24.5	2.6	7.2	4.1	11.4	1.00	0.4	1.9	2.50
Acropora prostrata	18.9	34.4	7.9	12.3	7.1	12.3	2.00	1.5	1.9	3.20

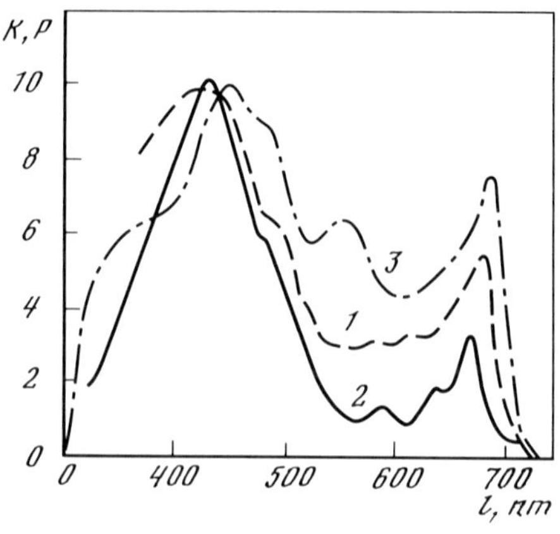

Fig. 8.2. The light absorption spectra of the photosynthetic pigments of zooxanthellae from the corals *Plerogyra sinuosa* (*1*) and *Pocillopora damicornis* (*2*), and the action spectrum of zooxanthellae from the coral *Favia* sp. (*3*); ℓ wavelength; *K* coefficients of light absorption, and *P* photosynthesis rates (relative units). (Data by Vereschi and Fricke 1986; Zvalinski et al. 1980; Halldall 1968)

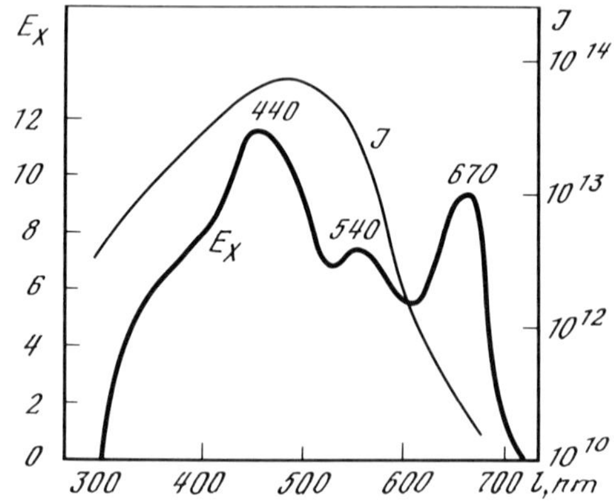

Fig. 8.3. Action spectrum of zooxanthellae from the coral *Favia* (*Ex*) and the spectral composition of light (*I*) in deep reef biotopes of outer reef slope with a level of light intensity of 10–5% PARS. *Ex* Excretion of oxygen by zooxanthellae (relative units); *I* illumination, quants $cm^{-2}s^{-1}nm^{-1}$; ℓ wavelength. (Data by Halldall 1968; Dustan 1982)

(PSU) – the protein – pigment complexes, which primarily absorb light energy and then transfer it to the reaction centers of photosystems I and II. The average size of PSU in corals adapted to a low light was 1.5 times greater than in corals living in normal light (Leletkin and Zvalinski 1981).

The light saturation curves of isolated zooxanthellae differ from those in intact coral colonies (Zvalinski et al. 1980). The saturation point of light intensity in them lies two to four times lower than in intact corals, being

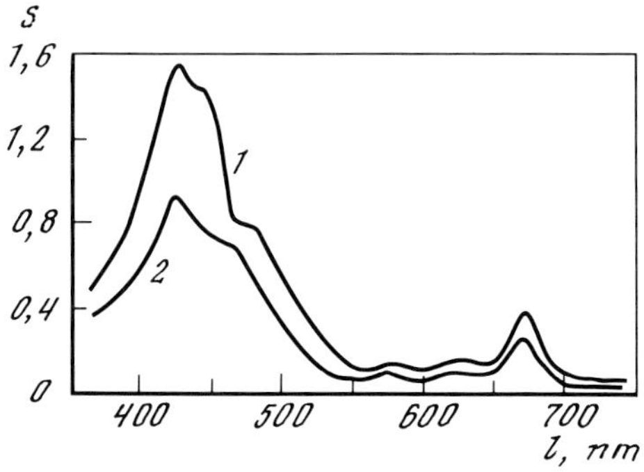

Fig. 8.4. The light absorption spectra of zooxanthellae from the gorgonaceans *Rhigidogorgia flabellum* (*1*) and *Antillogorgia turgida* (*2*); *S* optical density (relative units); ℓ wavelength. (Data by Burkholder and Burkholder 1960)

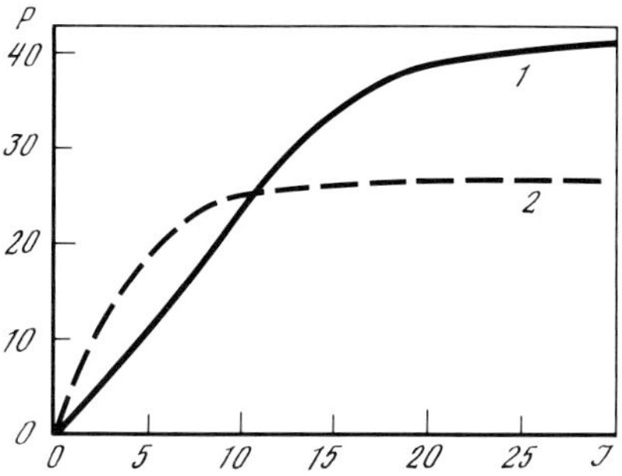

Fig. 8.5. Light dependence curves of photosynthesis in isolated zooxanthellae from colonies of the coral *Pocillopora verrucosa*, taken from well-illuminated (*1*) and from shadowed (*2*) reef habitats; *P* photosynthesis rate, $\mu l\, O_2\, h^{-1}\, 10^{-6}$ cells of zooxanthellae; *I* light intensity, $wt\, m^{-2}$. (After Titlyanov 1981)

within 20–30 $wt\, m^{-2}$ (0.5–0.7 $E\, m^{-2} h^{-1}$) (Fig. 8.5). In the light-adapted corals the zooxanthellae may have even much lower light saturation threshold – within 10–15 $wt\, m^{-2}$. Zooxanthellae also exhibit light inhibition in excessive illumination, while intact corals do not. The light saturation threshold in zooxanthellae is lower even than in free-living phytoplankton (Scott and Jitts 1977).

The rate of photosynthesis in zooxanthellae per unit of chlorophyll is somewhat similar to that in phytoplamkton. The assimilation number of chlorophyll varies in them ca. 1–5 $mg\, C\, h^{-1}\, 1\, mg^{-1}$ of chlorophyll *a* (Wethey and Porter 1976; Falkovsky and Dubinsky 1981). Average photosynthesis rates in zooxanthellae are 10–20 $\mu g\, O_2\, 10^{-6}$ cells (Zvalinski et al. 1980; Muscatine 1980; Titlyanov et al. 1988a). The respiration of zooxanthellae estimated by different methods ranges from 1% of their photosynthesis

(Muscatine et al. 1984) to 20% (Zvalinski et al. 1980; Davies 1984). Lower numbers were obtained by calculation from the ratios of biomasses of zooxanthellae and polyps, while the higher ones by direct measurements of respiration in isolated zooxanthellae; around 3–5 $\mu g\, O_2\, h^{-1}\, 10^6$ cells (Edmunds and Davies 1986). From the latter data it could be deduced that the ratio of their gross photosynthesis to respiration (P_t/M_z) should be around 5–10. But the specific growth rate in zooxanthellae is very low, being evaluated as between 0.003–0.01 day^{-1} (Muscatine 1983; Wilkerson et al. 1983; Edmunds and Davies 1986). Their own energy expenditures are also low, between 10–15%. The rest of their photosynthesis production is translocated to their host polyps.

The assimilation of CO_2 by their photosynthesis proceeds via the "C_3" cycle with the participation of the ribulose diphosphate carboxylase (Benson et al. 1978), with the posibility of involvement of the "C_4"-cycle with participation of phosphoenolpyruvate carboxylase (Trench and Blanck 1987). Being within the polyps, zooxanthellae probably have photorespiration, thus enhancing the net production of their photosynthesis (Muscatine 1980). Among the primary products of their photosynthesis are drates, phosphoglyceric acid, several simple amino acids and various lipids (von Holt and von Holt 1968; Trench 1974; Patton et al. 1977). About 50% of assimilates are soon transformed into alcohol-soluble lipid fraction (Schmitz and Kremer 1977). As regards the nutrients supply of zooxanthellae, they use the salts of phosphorus and nitrogen, formed during heterotrophic metabolism in tissues of the polyp, also using those directly consumed by corals from the sea water (Webb and Wiebe 1978; Muscatine et al. 1984; Sorokin 1990c). In cultures they use inorganic nutrients (D'Elia et al. 1983). Over 70–80% of organic P and N produced by zooxanthellae they translocate to the host polyps, providing them with the bulk of the amino acids necessary for growth (Burris 1983; Muscatine et al. 1984; Wafar et al. 1985).

Thus the symbiosis of polyps and zooxanthellae is based upon their mutualistic metabolic relationships (Muscatine 1974; Lewis 1981b). Three main forms of such relationships could me mentioned: (1) translocation of assimilates by zooxanthellae to the host; (2) consumption of inorganic nutrients – products of metabolism of host cells by zooxanthellae; and (3) acceleration of calcification in corals, which is definitely coupled with the photosynthesis of their symbionts. The host polyp stimulates the translocation of the main portion of autotrophic production by zooxanthellae. The zooxanthellae freshly isolated from coral of *Tridacna* tissues excrete over 60% of their photosynthates (Muscatine 1967; von Holt and von Holt 1968), but the addition of tissue homogenate of polyps again restores this process (Trench 1979). The main product thus excreted by the isolated zooxanthellae was glycerine plus some amounts of alanine glucose and organic acids (Trench 1971; Goreau et al. 1973). ^{14}C-acetate, which was added to the suspension of isolated zooxanthellae in light, is rapidly assimilated and transformed into the lipid fraction (Patton and Burris 1983).

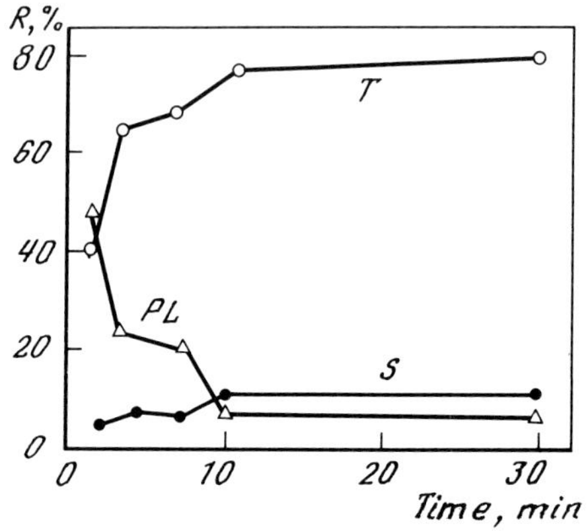

Fig. 8.6. Inclusion of ^{14}C-acetate into the lipid fraction of isolated zooxanthellae from the coral Seriatopora; *R* percent of radioactivity: in the triglycerides (*T*), in the phospholipids (*P*) and in sterols (*S*). (After Patton and Burris 1983)

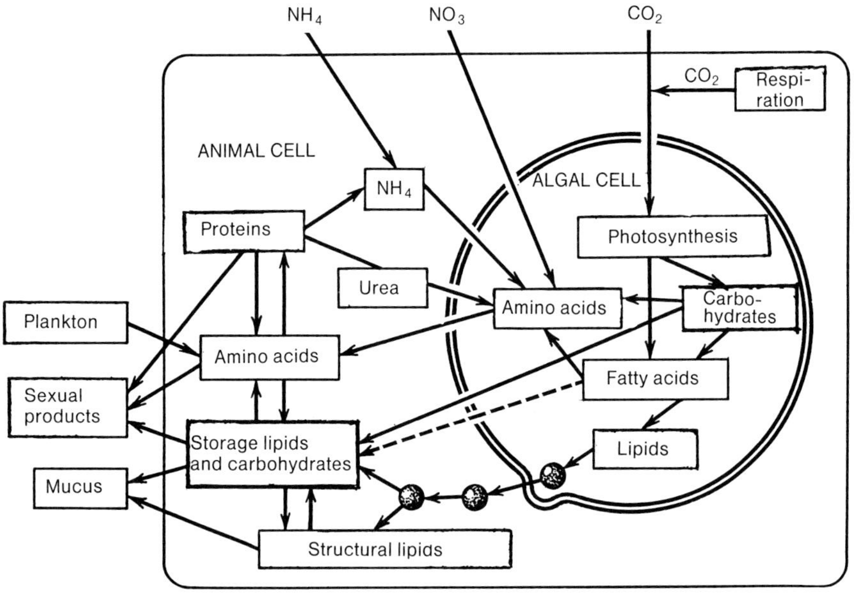

Fig. 8.7. Scheme of metabolic relationships of the cell zooxanthella with the cell of the host polyp

The same was observed in experiments with intact coral polyps, which transform the assimilated ^{14}C-acetate in light into the phospholipid and then into the triglyceride fraction (Blanquet et al. 1979; Fig. 8.6). In the intact tissues of the polyp the zooxanthellae translocate their photosynthates to the host tissues in a quite different way. The above mechanism of dissolved glycerine and amino acid excretion is of secondary importance in this case. The translocation process from zooxanthellae to their hosts' tissues proceeds as the transfer of tiny droplets of lipids via a mechanism resembling pinocytosis (Crossland 1980; Patton and Burris 1983). This process has been observed directly under the microscope. The droplet of fat, extruded from the cell, draws off the external cell membrane. Then it separates from the cell, which is surrounded by this membrane (Fig. 8.7). The droplets of fat covered with membrane might circulate via channels connecting the polyps and inside the polyps, the same as they have been observed to do in the body of the actinian *Condylactis* (Kellog and Patton 1983). They also form the lipid fraction of mucus, upon whose excretion corals lose up to 20–40% of their total photosynthetic production (Crossland 1980; Sorokin et al. 1983). The fat, which corals get mainly from their algal symbionts, serves as a main energy source in their metabolism, being also a basic stock material in polyps, which accumulates in them under favorable feeding conditions (Szamant-Froelich and Pilson 1980). The dry tissues of hermatypic corals contain 25–30% of fat and other lipoid products (Bergmann et al. 1956). Only about 25% of this fat is represented by the structural lipids (phospholipids, galactolipids). The rest is mostly the share of reserve fat and wax (Patton et al. 1977). The lipogenesis itself is localized mainly in cells of the zooxanthellae, from which lipoids thus formed are instantly transferred to the cells of the polyps and are absorbed into the total stock of fat and other lipoid materials of the coral. Besides fat, the translocation of other organic compounds, like amino acids and carbohydrates, by zooxanthellae was also observed (Mascatine et al. 1972; Burris 1983; Wafar et al. 1985). These products may be then transported through the colony to the points of its growth and calcification (Pearse and Muscatine 1971).

The translocation has been usually estimated as a percentage of the net or of the gross photosynthetic production. Estimated by a most reliable method of calculation from the values of photosynthetic production minus the metabolic and growth expenditures of the zooxanthellae themselves, it appears to be very high – within 95–99% of the net, and 90–93% of the gross autotrophic production in a coral colony (Muscatine et al. 1984, 1985).

8.4 Ecological Physiology of Photosynthesis in Corals

The course of the coral's photosynthesis rate during the day is characterized by curves with one maximum at midday (Porter 1980; Porter et al. 1984; Sorokin 1984a; cf. Figs. 8.8–8.10). In full sunlight, the midday depression due to excess light was observed only in some weakly branching (*Porites*) or foliose (*Merulina*) corals (Fig. 8.10). In most of the ramose, cribose,

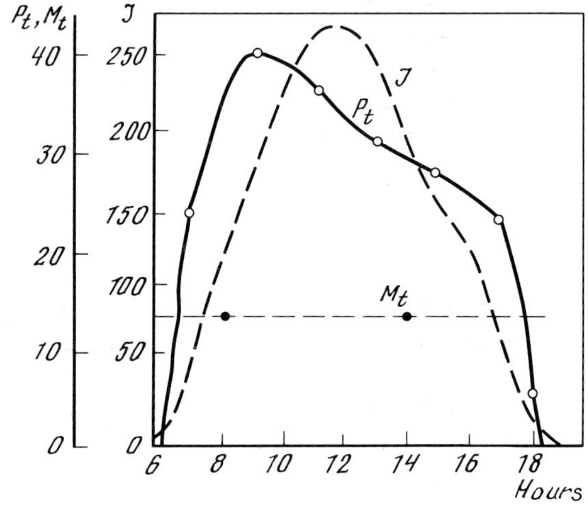

Fig. 8.8. Diurnal course of the gross photosynthesis and respiration (P_t, M_t, $\mu g\ O_2\ g^{-1}\ h^{-1}$) average in five species of common corals of the Heron I. reefs; η changes in light intensity during the day, $wt\ m^{-2}$. (After Sorokin 1986b).

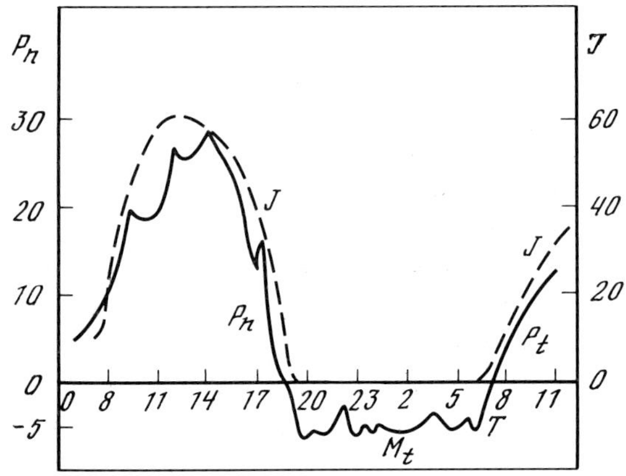

Fig. 8.9. Diurnal course of the net photosynthesis and of the respiration rates (P_n, M_t, $\mu l\ O_2\ g^{-1}\ h^{-1}$) in the coral *Acropora cervicornis* as measured in experiments by means of flow respirometers; *I* diurnal changes in illumination, $m\ E\ m^{-2}\ min^{-1}$; *T* time, h. (Data by Porter 1980)

cespitose and massive coral species this depression is absent, while the saturation light intensity ($60-120\ wt\ m^{-2}$) is several times lower than the PARS at midday: $300-350\ wt\ m^{-2}$ (Chalker and Taylor 1978). It happens in part because of self-shadowing in their branches or at the base parts of their colonies. Another factor is the possession of light-protective pigments by corals in their coenosarc, with the absorption maxima about 300–340 nm.

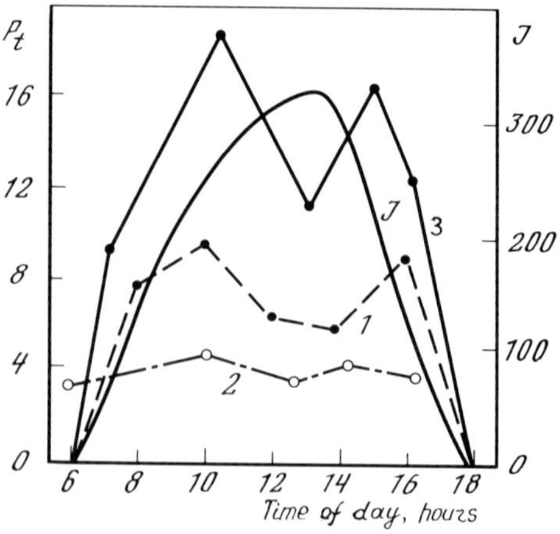

Fig. 8.10. Diurnal course of gross photosynthesis (P_t, $\mu g\,O_2\,h^{-1}\,10^{-6}$ cells of zooxanthellae) in the ramose coral *Pocillopora damicornis* (*1*), in the columnar coral *Porites porites nigrescens* (*2*) and in the foliose coral *Merulina laxa* (*3*); *I* light intensity, $wt\,m^{-2}$. (Data by Titlyanov et al. 1988c)

They protect them against the inhibiting or even destructive action of ultraviolet light, which in the tropics is extremely strong (Jokiel and York 1982). The curves of the diurnal course of photosynthesis remain equal in corals living within a wide scale of PARS: between 100% and 20%, because of their ability for light adaptation. The lower is the illumination in the native habitat of a coral, the more will the above curve resemble that of the diurnal illumination course itself (Figs. 8.13, 8.15). Also during the day changes in the functional characteristics of the photosynthetic apparatus of zooxanthellae go on. The content of chlorophyll in them decreases at midday, and their potential photosynthesis rate diminishes as well (Chalker and Taylor 1978; Titlyanov et al. 1988a).

The generalized course of diurnal changes in photosynthesis of several species of corals inhabiting shallow habitats of the Heron Island reefs is given in Fig. 8.8. This curve was used to calculate the coefficients K_p – the ratios of gross photosynthesis per whole light day to that per each hour (Table 8.5). The coefficients were used to calculate photosynthesis values per day, taking its rates measured during short-term experiments performed at various periods of the day (cf. Sect. 9.1).

The light dependence of photosynthesis in corals could be described by corresponding light saturation curves, which are rather similar in their species inhabiting shallow, well-illuminated reef biotopes (Halldall 1968; Barnes and Taylor 1973; Scott and Jitts 1977; Falkowsky and Dubinsky 1981; Titlyanov et al. 1985). The rate of photosynthesis increases proportionally to the increase in light (PARS) up to its saturation level of 90–110 $wt\,m^{-2}$ (1.5–1.8 $E\,m^{-2}\,h^{-1}$). Then the curve evens out in a plateau

Table 8.5. Coefficients K_p in corals of the Heron I. reefs (cf. text)

Period of the light day, hours.	K_p	Periods of the light day, hours.	K_p	Periods of the light day, hours.	K_p
8–9	11.8	11–12	9.7	14–15	11.8
9–10	10.8	12–13	10.2	15–16	12.8
10–11	9.1	13–14	11.1	16–17	14.0

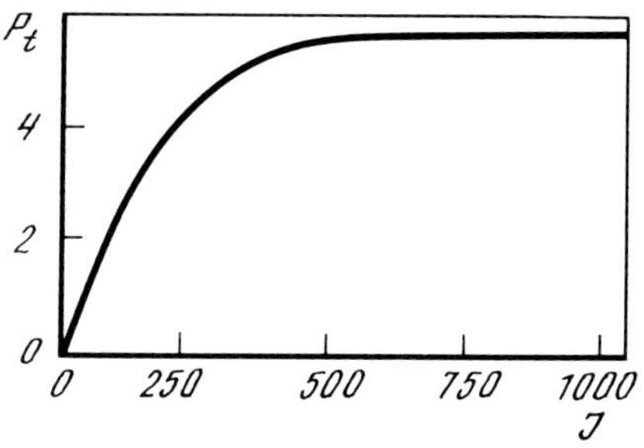

Fig. 8.11. Light saturation curve of gross photosynthesis (P_t, relative values) in the coral *Acropora cervicornis*; *I* light intensity, $E\,m^{-2}h^{-1}$. (Modified after Chalker and Taylor 1978)

(Fig. 8.11), the abscissa going parallel till the maximal light intensity peculiar to the tropical zone, 250–350 wt m^{-2}, will be reached. Thus the saturation light intensity in corals is close to 20–30% PARS. Within its limits of 90–20% PARS (in some corals even down to 10% PARS) the rate of photosynthesis is actually independent of light intensity (Titlyanov et al. 1988b). Below the light intensity of 20–10% PARS (e.g., at depths more than 20–30 m in transparent shelf waters), the their photosynthesis rate starts to drop (Drew 1973; Fig. 8.14).

The curves of light saturation of photosynthesis appear to be different in corals living in illuminated and in shadowed habitats, thus demonstrating the light adaptation of the latter (Figs. 8.12, 8.13). These curves measured in corals taken from deep reef zones or from caves or grottos and so adapted to low illumination have steeper slopes in their initial parts. These corals reach sooner the saturation point. Saturation light intensity in the corals is 40–50 wt m^{-2} compared with 90–110 wt m^{-2} in normal corals living in full light (Chalker et al. 1983; Battey and Porter 1988; Titlyanov et al. 1988a,b; Tchabardgy and Fam Van Huen 1988).

The compensation point, where the net photosynthesis approaches zero, is reached by light-adapted corals at twofold less illumination. Therefore, their potential photosynthesis rate measured in moderate light is higher than in normal corals (Zvalinski et al. 1980; Titlyanov 1981; Falkowsky and Dubinsky 1981). This increase is not accompanied by an increase in

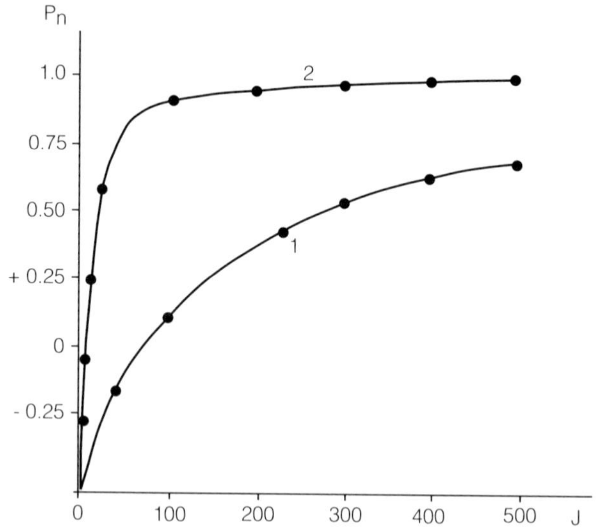

Fig. 8.12. Light saturation curves of net photosynthesis (P_n, μ mol $O_2\,h^{-1}\,\mu g^{-1}$ of protein) in the coral *Acropora digitifera* taken from 1 m depth (*1*) and in *A. divaricata* taken from 40 m depth (*2*); $\mathcal{J}$ irradiance, μ Einshteins $m^{-2}s^{-1}$. (After Chalker et al. 1983)

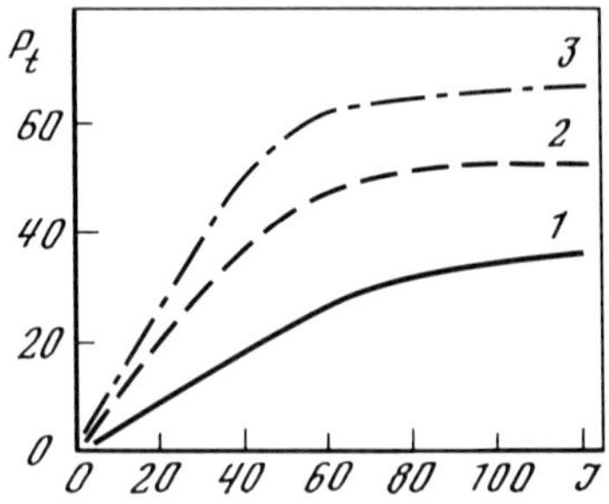

Fig. 8.13. Light saturation curves of gross photosynthesis (P_t, μg $O_2\,g^{-1}\,h^{-1}$) in colonies of the coral *Stylophora pistillata* taken from differently illuminated reef habitats: *1* from well-illuminated shallow biotope, 90–80% PARS; *2* from medium depth biotope of the reef slope, 30–20% PARS; *3* from grotto, 10–7% PARS; $\mathcal{J}$ illumination, wt m^{-2}. (Data by Titlyanov et al. 1988c)

concentration of photosynthetic units in the tylacoids of chloroplasts in corals adapted to low light, but seems to be due to the increase in the number of zooxanthellae in their polyps.

The hermatypic corals inhabit a wide scale of depths (0–140 m) within 4 orders of the scale of light intensity: from 300 to 0.1 wt m^{-2}, and with variations in the light spectrum from normal white sunlight at the surface to the blue-green light at depth (Schuhmacher and Fricke 1984) Corals possess

Table 8.6. The changes of structural and functional characteristics in corals taken from differently illuminated habitats. S/V – surface to volume ratio in coral colony. (Data by Titlyanov et al. 1988a)

Coral	Habitats		S/V	Chlorophylls $a + c$ content		Biomass of zooxanthellae $mg\,cm^{-2}$	Number of zooxanthellae $10^6\,cm^{-2}$
	Depth (m)	Light intensity (PARS) (%)		$\mu g\,cm^{-2}$	$\mu g\,10^{-6}$ cells of zooxanthellae		
Stylophora pistillata	1–2	90–80	2.0	4.7	4.8	0.16	0.97
	12–18	30–20	2.1	13.5	6.0	0.26	2.3
	1.4–5 (grotto)	10–7	3.2	10.9	5.7	0.20	1.9
Pocillopora damicornis	1.5–3	70–30	1.2	6.4	9.8	0.09	0.7
	2 (grotto)	15–10	1.9	11.5	13.5	0.11	0.9
Acropora diversa	1–2	90–80	4.3	6.0	9.8	0.31	0.6
	1.5–5 (canyon)	30–10	4.8	8.0	11.5	0.36	0.7

the mechanisms of adaptation to low light intensity and also to changes in its spectrum composition. Among these mechanisms could be mentioned: the increase in the amount of zooxanthellae and in the chlorophyll content in their polyps (Titlyanov et al. 1980; Dustan 1982; Mc Closkey and Muscatine 1984; Dubinsky et al. 1984; Falkowsky et al. 1984; Porter et al. 1984), the decrease in the rate of their respiration (P.S. Davies 1977; cf. Table 8.7), the changes (flattening) of morphology of their colonies (Titlyanov 1987), and the appearance of special organs and structures, increasing the ability of corals to use a weak light (Vereshi and Fricke 1986). Within the light intensity limits of 30–10% PARS the adaptation to decreasing illumination proceeds in corals mainly by an increase in the chlorophyll content per unit of their surface area. At light intensities down to 20% PARS the mechanism of adaptation changes. There it proceeds mainly by adequate changes in the morphology of their colonies and by an increase in the number of polyps per unit of their illuminated surface area, directed to the optimization of the intercept of light energy. The content of chlorophyll in colonies of corals living under illumination reaching at midday 30–10% PARS could increase 1.5–3 times, and sometimes even up to 7 times as compared with colonies of the same species, which live in full light. This increase results from an increase in the number and average size of zooxanthellae (Table 8.6), as well as from the growth of chlorophyll concentration in them (Zvalinski et al. 1980; Falkowsky and Dubinsky 1981; Titlyanov 1981). This growth is produced by the denser packing of the lamellae in polyps of corals adapted to a low light (Mashansky et al. 1980; Dubinsky et al. 1984), as well as by the increase in size of their photosynthetic units (Titlyanov et al. 1981). The utilization of a weak light is promoted in them by an increased content of the carotenoid pigments in their zooxanthellae (Titlyanov et al. 1980). The light-adapted corals may use the incoming light energy with an extremely high efficiency – up to 80% of PARS, with the quantum yield of photosynthesis up to 16% instead of the 1–2% in normal, unadapted corals, which in itself proves the perfection of the photosynthesis apparatus in their zooxanthellae.

Features of light adaptation were found also in permanently self-shadowed branches of coral colonies living in well-illuminated sites of the reef (Titlyanov et al. 1988b). The light intensity at lower parts of branches of the first-order in colonies of the coral *Pocillopora* growing at reef shallows, as measured at midday with the aid of a special microprobe light meter, appeared to be 9–4% PARS, while at the upper surface of branches of the second and third-orders it was 100–50% (Fig. 8.14). The light saturation curves in such shadowed first-order branches have a form corresponding to corals adapted to a low light (Fig. 8.15), and their respiration rate was about twofold twice lower than in the upper well-illuminated branches, taken from the same colony. Just because of such adaptation, the rate of photosynthesis in shadowed branches measured in the intact colony with the aid of the ^{14}C-method, was only 1.2 times less than in the upper third-order branches, while the light intensity at the surfaces of the former was by about one order less.

The decrease in the respiration rate is another adaptive reaction of corals to deficiency of light. This decrease was recorded in their specimens living in shadowed or in deep, poorly illuminated habitats (Davies 1980; Titlyanov 1987; Titlyanov et al. 1988a). It enables them to use their scarce energy resources more effectively. Therefore, the adapted corals may attain compensation points of their photosynthesis at light intensities as low as 10–5% PARS (Table 8.7; Fig. 8.15). The decrease in the respiration rate is supposedly caused by decrease in the total protein content, which results in the total volume of animal tissues decreasing in polyps, with a simultaneous

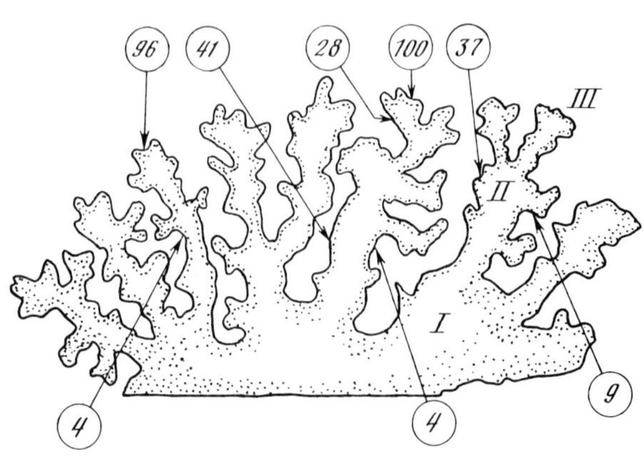

Fig. 8.14. Light intensity (wt m^{-2}) at different parts of colony in the coral *Pocillopora damicornis*. (After Titlyanov et al. 1988c)

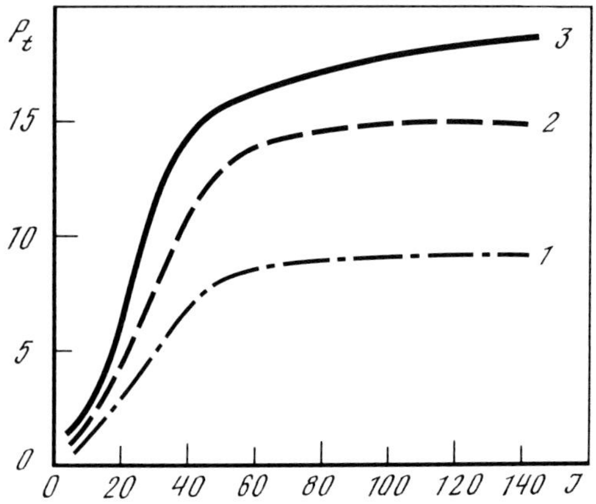

Fig. 8.15. Light saturation curves of photosynthesis in the coral *Pocillopora damicornis* as measured in strongly shadowed basal branch (*1*), in average shadowed branch of second order (*2*) and in well-illuminated upper branch of third order. (After Titlyanov et al. 1988c)

Table 8.7. Gross-photosynthesis (P) and respiration (M) in the coral Stylophora pistillata living in differently illuminated reef habitats. (Data by Titlyanov et al. 1988a)

Light intensity in habitat	P		M		P/M
	μg cm^{-2} h^{-1} (max. during the day)	μg cm^{-2} day^{-1}	μg cm^{-2} h^{-1} (max. during the day)	μg cm^{-2} day^{-1}	
Light, 50–30%	42.5	350	20.7	390	0.9
Shade, slope 15–10%	36.2	260	14.3	210	1.2
Strong shadow, grotto 7–2%	42.7	200	15.6	180	1.1

growth of the plant tissues in the zooxanthellae (Battey and Porter 1988). The period of light adaptation in corals takes only 2–3 weeks (Davies 1980; Gattuso 1985).

A lower limit of light intensity for the survival of different scleractinian species, depending upon their ability for light adaptation, varies from 10% of PARS (25–20 wt m^{-2}) in *Acropora*, down to 3–1% PARS in *Pocillopora*, *Stylophora*, and *Turbinaria* (Falkowsky and Dubinski 1981; Falkowsky et al. 1984; Titilyanov et al. 1988a,b). In corals well adapted to a low light and able to live at as low an illumination as 7–2% PARS (15–5 wt m^{-2}) their gross photosynthesis decreases only by 40–50%, while in macrophytes, for example, under the same conditions it would fall five to ten times. The P/M ratio remains in them ~1 at illuminations down to 20–10% PARS (Taylor

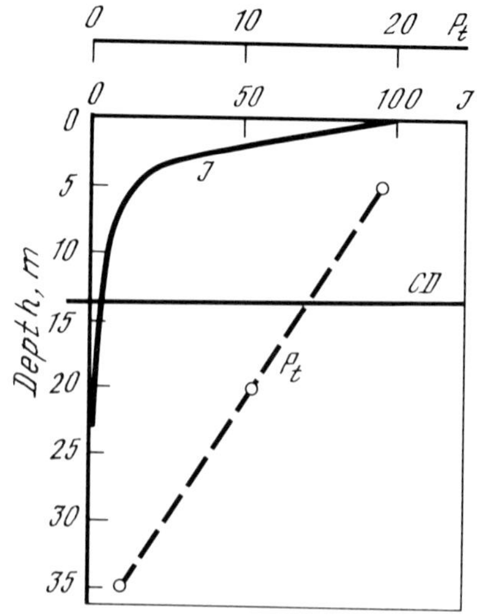

Fig. 8.16. The decrease in gross photosynthesis (P_t, $\mu g\,C\,g^{-1}\,h^{-1}$) in the coral *Acropora* sp. with depth as measured in situ; *I* illumination (relative values); *CD* depth of the compensation point of photosynthesis ($P_n = 0$). (After Drew 1973)

1973b; Jaubert 1977). In caves and in grottos of the upper reef zone the hermatypic corals, such as *Pocillopora* survive at illuminations of 3–15% PARS, which is two to three times less than their light limit in deep reef zones. This could be explained by the presence of some additional light reflected from its patches at the surface (Titlyanov 1987; Titlyanov et al. 1988b). Light adaptation takes place even in shadowed branches within the same coral colony (Titlyanov 1987; Titlyanov et al. 1988b).

The curve of changes in the photosynthesis rate in corals depending on depth, as measured in situ, is given in Fig. 8.16. The P/M ratio in this coral was over 1 down to a depth of 14 m (compensation point). In very transparent waters of the Red Sea the compensation point in some corals was observed at depths of 80–100 m (Fricke and Schuhmacher 1983).

8.5 Ecological Biochemistry of Corals

Most sessile invertebrates possess a variety of mechanisms for aggression serving their competition for the hard substrate, and also the means of defense against aggression, grazing and fouling. Among these mechanisms and remedies is their using of various toxic, repelling or indigestible metabolites and substances. The same is true of corals, but their diifferent groups make use of these on a different scale, depending on the structure of their colonies and the strategy for ensuring their survival (cf. Sect. 7.3). The best "chemists" among *Anthozoa* are the octocorals and the zoantharians. Most of them do not possess a hard exoskeleton, which could protect them from grazing. In the octocorals moreover the cnidocil is drastically reduced, and so it cannot be used for defense or for interspecific aggression, as scleractinians or hydrocorals use it.

The means of chemical defense and aggression used by octocorals are numerous and varied. They embrace a wide spectrum of antibiotic, as well as terpenoid and lipoid toxic and biologically active substances, including extremely active metabolites, such as prostaglandins. The antibacterial action of octocorals was discovered when pieces of their colonies had been placed on agar plates, infected by the test bacteria (Burkholder 1973), thus producing sterile zones. It was found that the colonies most active in this respect, gorgonaceans *Pseudoplexaura crassa* and *Eunicea mammosa*, contain the lactones crassin and mammosin, which resemble certain components of penicillin (Ciereskо 1962). These substances have an antibacterial effect and seem to be toxic also for fish, thus accomplishing a double function: preserving the gorgonaceans from fouling and repelling their predators. The gorgonians excrete antialgal agents, which prevent their fouling with microalgae (Targett et al. 1983). The numerous terpenoid compounds produced by most octocorals have also definite antimicrobial properties. The antibiotic defense of octocorals against their fouling by bacteria is moreover for most of them important, because they do not actually produce mucus, which accomplishes this function in scleractinians.

The colonies of corals are soaked with protective structural lipids, like high-molecular-weight alcohols, sterols, glyceric esters, prostaglandins, wax-like substances and hydrocarbons (Bergmann 1962; Ciereskо et al. 1968). They comprise as mush as 10 to 30% of their total biomass. The steroid lipids (including cholesterines) and the waxes of cethylpalmitate type are present in the lipoid fractions of most corals, in octocorals, zoantharians and scleractinians (King and Bergmann 1942; Ciereskо and Karns 1973; Kanazawa et al. 1976; Benson et al. 1978). These structural lipoids have a revolting taste. Some of them are even poisonous. They are also indigestible for most predators except *Acanthaster* and some fish (Benson et al. 1978). So they perform the function of peserving corals from grazing. And, in fact, corals and zoantharians are eaten only by a few highly specialized members of reef fauna.

The prostaglandins have been discovered in the lipoid fraction of the Atlantic gorgonian *Plexaura homomalla* (Weinheimer and Spraggins 1969; Bayer and Weinheimer 1974). These biologically active compounds were first isolated from sheep's testis. They exert strong polyvalent action upon the blood pressure, the muscular and nervous system, and the growth of warm-blooded animals, when injected into their blood-stream in negligible amounts of a few $\mu g\,kg^{-1}$ of their weight. Prostaglandins are now widely used in medicine and in agriculture (Pike 1974). The lipoid fraction of this gorgonian contains an enormous amount of this valuable substance – up to

15%. Its content represents in colonies of *Plexaura* up to 5–8% of their dry weight, which is 10^7 times more than in tissues of other animals, and by an order more than in sheep testis, from which it is usually extracted (Gerhardt 1984). After this discovery *Plexaura* has become a very valuable raw material and a source of hard currency for developing countries (Venkataramanusjam and Santhanam 1989). The reason for the accumulation of prostaglandins in gorgonians became clear when it was found that the injection of negligible amounts of this substance per os into fish caused the inevitable reaction of vomiting. Fish avoided the food pellets which contained prostaglandin (Gerhardt 1985), Thus, it serves as protection from predators. The same function is accomplished in octocorals also by various terpenoids, which they produce in significant amounts (Coll et al. 1977; Fenical and Harwell 1988). Before terpenoids became known as metabolites of plants, on being discovered in octocorals they had been supposed to be products of their symbiotic zooxanthellae. But recent studies have shown that they are produced by the polyps themselves (Kokke et al. 1984; Coll et al. 1985). The highest contents were found in colonies of the hermatypic gorgonians *Plexaura*, *Pseudoplexaura*, *Pseudopterogorgia*, and *Gorgonia*. The gorgonians *Eunicea palmeri*, *E. succinea*, *Pseudoplexaura pororsa*, and *P. flagellosa* contain lactone-diterpenoids, such as *crassine-acetate* (Tursch et al. 1978). All the terpenoids are toxic and have a bad smell (Cieresko and Karns 1973; Bakus 1981). Their content in both gorgonian octocorals and in the alcyonaceans appeared to be very high: 3–5% of the dry weight of the colony (Coll et al. 1985). The presence of terpenoids made them toxic or inedible. Extracts prepared from their colonies are toxic to fish and other reef fauna. Experiments showed that over 50% of octocorals inhabiting the GBR is toxic to reef fish. The most toxic of them were the common alcyonaceans *Lemnalia* and *Sarcophyton* (Coll et al. 1982). The terpenoids of alcyonaceans have antibacterial and antifungal properties (Bowden and Coll 1983). In the experiments fish avoided the food pellets containing extracts from their colonies or terpenoid preparations. Octocorals permanently excrete terpenoid toxins into the water, thus inhibiting the growth of their competitors – the scleractinian corals. Around them in calm waters the concentration of terpenoids could attain 1–5 $mg\,l^{-1}$, which was proved experimentally to be toxic for the corals *Acropora* and *Porites*. Their inhibition by neighboring octocorals has been established (Bowden and Coll 1983; Sammarco et al. 1983; Coll et al. 1985).

The zoantharians contain their own toxins – palytoxins (Scheuer 1969; Kimura et al. 1972). Therefore, they are practically left ungrazed by predators, while some of their species do not have any exosceleton. They also contain a significant amount of the biologically active substance serotonine, a stimulator of propagation in animals (Cieresko and Karns 1973).

The above information shows that octocorals and zoantharians contain a wide selection of biologically active compounds, and so they have become a valuable, medicinal raw material, which is why their main predator in the near future will be man himself.

The hermatypic scleractinian corals permanently excrete large amounts of mucus. This excretion might account for over 20% of their total photosynthetic production. The mucus is excreted also by zoantharians (Daumas and Thomassin 1977). Less is known about this function in octocorals. There were opinions that they do not secrete it at all (Lewis 1982). But we have observed some alcyonacean octocorals (Sarcophyton) using the fine mucus filaments to catch their zooplankton prey (cf. Sect. 9.2.2). The formation of mucus has been recorded also in the octocoral *Biareum* (Rublee et al. 1980).

Scleractinian corals excrete mucus in the state of a transparent fluid (Hubbard and Pocock 1972; Schuhmacher 1979; Coffroth 1985). On the surface of the colony the mucus denaturates and hardens, forming a thin cover around it (Lewis 1973; Ducklow and Mitchell 1979b). These mucus covers, which are quickly overgrown by microflora and powdered with the settling suspended matter, are soon shed by the coral, cleaning so its surface. The production of mucus by corals increases when they are damaged or stressed by heating, pollution or siltation (Bak and Elgershulzen 1976; Thompson et al. 1980), exposure, draining or by drops of salinity (Sumner et al. 1975; Daumas and Thomassin 1977). The mucus covers preserve corals from microbial fouling, draining and overheating, when they are exposed. The production of mucus was evaluated as being 2–6 mg day^{-1} of dry weight per colony of 2–3 dm^3 volume (Richman et al. 1975). With 50% cover of the bottom surface by corals these excrete some 50 mg $m^{-2} day^{-1}$ of ash-free organic matter as mucus. R. Johannes (1967) evaluated the excretion rate of mucus by corals at the reef flat of the atoll Enivetok as 20 mg m^{-2} dry weight day^{-1}. The most intensive release of mucus was recorded in acroporids. The coral *Acropora acuminata* produces over 300 µg C of mucus per 1 mg of protein of its tissue. Thus this coral excretes during the day about 30–40% of its total net photosynthetic production (Crossland et al. 1980). The content of ash in mucus is rather high – 50–70% dry weight. The caloric equivalent of mucus is 3–5 cal mg^{-1} of dry, ash-free matter (Richman et al. 1975). The mucus contains a significant amount of protein, so the C/N ratio in it is low: 4–7 (Coles and Strathmann 1973; Richman et al. 1975). The basic components of mucus are the mucopolysaccharides (40–50% dry weight), the lipids cetylpalmitates and triglycerides (30–40%), and proteins and amino acids (10%). The mucopolysaccharides provide the hydrophilic properties of mucus and are composed of amino sugar polymers. The amino acids of mucus are dominated by glycine and alanine. Proteins, peptides, and lipids form in mucus the glycoprotein and lypoprotein complexes which compose its structural basis (Benson and Muscatine 1974; Sumner et al. 1975; Daumas and Thomassin 1977; Corssland et al. 1980). The most important structural polymer among them is proteoglycan, which has in its complexes the sulfo-groups, connecting separate molecules of polymer. The number of these sulfo-groups in the mucus controls its viscosity and the process of its denaturation (Richards et al. 1983).

The mucus excreted by scleractinian corals is a significant source of food for many reef fish and invertebrates. On the mucus feed numerous infaunal animals and commensals of corals – crabs, shrimp, flat worms, ciliates (Patton 1976; Castro 1976; Bruce 1976) and some specialized fish as well (Benson and Muscatine 1977; Benson et al. 1978). The organic flakes and aggregates formed from mucus which is shed by corals are quickly overgrown with bacteria, thus increasing their food quality. At this stage these aggregates are readily consumed by zooplankton and by sessile animals (Sumner et al. 1975; Coffroth 1984). Mucus that settles on the bottom is one of the main sources of feeding for reef fauna, comprising a significant part of the organic matter in reef bottom sediments (Daumas and Thomassin 1977).

9 Nutrition of Corals

Above it was mentioned that most scleractinian corals and more than half of the taxa of octocorals inhabiting coral reefs have algal symbionts in their tissues – the zooxanthellae. The endosymbiosis enables them to use solar energy. Simultaneously, coral polyps are able to feed as heterotrophs. The autotrophic and heterotrophic nutrition of corals and the relative importance of these sources in their energy balance have been recognized long ago as key problems of biology and ecology not only of the corals themselves but also of the coral reef ecosystems as a whole (Odum and Odum 1955, 1956).

9.1 Autotrophic Feeding of Corals

The autotrophic nutrition of corals proceeds by the translocation of the photosynthates, produced by their symbiotic algae, the zooxanthellae, into the tissues of polyps. The value of photosynthetic production of corals is usually expressed in units of gross photosynthesis (P_t) or of net photosynthesis (P_n). The latter is calculted as the difference between P_t and the total respiration of the coral, including its algal component (M_t): $P_n = P_t - M_t$. The net photosynthetic production of the zooxanthellae themselves (P_{nz}) is equal to: $P_{nz} = P_t - M_z$, if M_z = respiration of zooxanthellae. The importance of autotrophic production in the energy balance of corals could be calculated as the ratio of net photosynthetic production of zooxanthellae, translocated to the animal tissues ($P_{nz} \cdot T$), to the respiration of animal tissues of the polyp (M_a). This ratio is named CZAR (contribution of zooxanthellae to animal respiration). $CZAR = (P_{nz} \times T)/M_a$, if $M_a = M_t - M_z$, and T is the part of P_{nz}, translocated to the polyp.

The rates of photosynthesis and respitation in corals were measured by the oxygen and radiocarbon methods. The latter method yields values which seem to be closer to the net photosynthesis (Cooksey and Cooksey 1972; Scott and Jitts 1977). But according to Drew (1973) it could be rather equal to the gross photosynthesis. Their measurements were accomplished in closed vessels or containers, as well as in flow respirometers. As for the reliability of both methods, it was stated that the latter gives values up to 1.5–2 times less than the former (Dennison and Barnes 1988; Tchabardgy and Fam Van Huen 1988), while there was also evideuce of a reliable comparability of both methods (Mergner and Svoboda 1977). In fact, the experiments in closed containers, being short-term (1–2 h), could not be expected to produce some artifacts, moreover, data on rates of photosynthesis obtained by different authors for the same coral species employing containers and flow respirometers are quite comparable (Tables 9.1, 9.2). For the measurements of P_t and M_t the underwater automatic flow respirometers with continuous recording of oxygen content in water have been employed in situ (Mergner and Svoboda 1977; Porter 1980; Chalker et al. 1985).

As regards the oxygen method of estimation of P_t, P_n and M_t, the P_n correspond to the net increase in O_2 content in the light, the M_t to its net decrease in the dark, and P_t to the sum of both these values: $P_t = P_n + M_t$ (Fig. 9.1). To express the results of measurements in the carbon units, the coefficients of PQ (O_2/CO_2) equal to 1.1, and the respiratory coefficients of MQ(C^{CO_2}/O_2) equal to 0.8, have been used (Muscatine et al. 1981). The respiration of zooxanthellae (M_z) was calculated from the ratio of biomasses of animal to plant tissues, accepting that the respiration of plant and animal components of polyps was proportional to their biomass (Muscatine 1980). The M_z values were also measured directly in suspensions of isolated zooxanthellae in vitro (Wilkerson et al. 1983). The values of P_t, P_n and M_t are most often expressed as their rates per hour. Their values per day could be calculated on the basis of their rates per hour, using coefficients K_p – which reflect the ratios of the values per hour (P_{th} and M_{th}), measured at definite periods of the day, to their 24-h values (Sorokin 1984a). These

Table 9.1. Gross photosynthesis (P_t) and respiration (M_t) of corals expressed cm^{-2} of their surface at optimal illumination

Coral	μg O_2 h^{-1}		μg C day^{-1}		P_t/M_t day^{-1}	Reference
	P_t	M_t	P_t	M_t		
Stylophora pistillata	40	17	137	118	1.2	Falkovsky et al. (1984)
	32	13	105	94	1.1	Porter et al. (1984)
	26–62	15–33	137	180	0.8	Titlyanov et al. (1985)
Pocillopora damicornis	9–23	7–10	52	63	0.8	Titlyanov et al. (1985)
Pocillopora eydouxi	53	11	165	79	1.6	Davies (1984)
Acropora cervicornis	37	10	119	72	1.6	Wethey and Porter (1976)
Acropora palmata	40–105	21–51	231	267	0.9	Rogers and Salesky (1981)
Acropora diversa	118	15	375	115	3.3	Edmunds and Davies (1986)
Porites porites	15–35	11–18	375	103	3.6	Titlyanov et al. (1988b)
Turbinaria reniformis	18–52	15–29	112	163	0.7	Titlyanov et al. (1988b)
Montastrea annularis	25–70	14–24	152	139	1.1	P.S. Davies (1977)
Montastrea cavernosa	7–29	7–13	58	73	0.8	P.S. Davies (1977)
Millepora tenera	25	16	81	113	0.7	Drew (1973)

Table 9.2. Gross photosynthesis (P_t) and respiration (M_t) in corals calculated g^{-1} of weight at optimal illumination

Coral	$\mu g\ O_2\ h^{-1}$		$\mu g\ C\ day^{-1}$		P_t/M_t day^{-1}	Reference
	P_t	M_t	P_t	M_t		
Pocillopora damicornis	53	26	180	188	1.0	Nair and Pillai (1972)
	117	16	400	113	3.5	Kawaguti (1937)
	44–87[a]	8–36	223	160	1.4	Tcharbadgy and Fam Van Huen (1988)
Stylophora pistillata	31–82[a]	13–26	248	115	2.2	Propp (1985)
	24–58	16–35	138	180	0.8	Titlyanov et al. (1988b)
	30–114[a]	15–48	203	226	0.9	Tcharbadgy and Fam Van Huen (1988)
Acropora cerviprnis	40	10	136	72	1.9	Porter (1980)
Acropora prostrata	90[a]	35	305	254	1.2	Propp (1985)
Acropora corymbosa	41	28	140	198	1.4	Nair and Pillai (1972)
Porites nigrescens	104[a]	42	352	300	1.2	Propp (1985)
Porites solida	12	5	41	34	1.2	Nair and Pillai (1972)
Montipora tuberculosa	147[a]	59	500	427	1.2	Propp (1985)
Merulina laxa	166[a]	46	565	322	1.8	Propp (1985)
Fungia fungites	79[a]	33	269	236	1.1	Tcharbadgy and Fam Van Huen (1988)
Cyphastrea microphthalmus	13	8	41	59	6.7	Nair and Pillai (1972)
Goniopora stokesi	27	16	92	113	0.8	Nair and Pillai (1972)
Psammocora contigua	104	30	136	216	0.6	Nair and Pillai (1972)
Millepora platyphylla	13[a]	4	41	30	1.4	Tcharbadgy and Fam Van Huen (1988)

[a] Measured in flow respirometers.

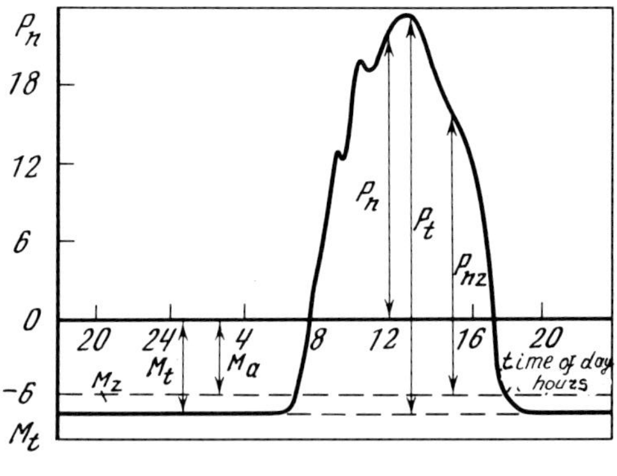

Fig. 9.1. Diurnal course of photosynthesis and respiration in the coral *Stylophora pistillata* (P_t, M_t, $\mu g\,C\,g^{-1}\,h^{-1}$) and the elements of gross and net production: P_n net (visible) photosynthesis of coral; P_{nz} net photosynthesis of zooxanthellae; M_z respiration of zooxanthellae; M_a respiration of animal tissues of the polyps themselves. (After Muscatine et al. 1981)

coefficients could be estimated by using the curves of diurnal course of photosynthesis and respiration. The value of P_t per day (P_{td}) is equal to: $P_{td} = K_p \cdot P_{th}$ (cf. Table 8.5). Most authors accepted that M_{th} was approximately constant during the 24-h cycle. Thus: $M_t = M_{th} \times 24$. The M_{th} values in this case were taken as an average of those measured in daytime and at night (Wethey and Porter 1976; Porter 1980; Sorokin 1984c). There is also an opinion that the daytime respiration of corals could be 1.5–2 times less than the nocturnal one (Porter et al. 1984; Titlyanov et al. 1988a). But the corresponding data could also be an artifice, because the daytime experiments were conducted in waters oversaturated with oxygen, which might lose part of it during the bottle experiments as bubbles.

Also, the first estimations of photosynthesis and respiration in corals made in the early 1930s by M. Yonge and S. Kawaguti showed that, as in green plants, photosynthesis outweighs respiration in diurnal balance (see Kawaguti 1937). To date we have a wide range of data on the rates of these processes, which were obtained by various methods and techniques. But the data available were presented by authors in various dimensions: per hour or per day, per cm^2 of surface, per g of weight of colony, per mg of chlorophyll or per 10^6 cells of zooxanthellae, expressed in units of O_2 or carbon. Therefore, it is quite obvious that the fluctuation of data, given by different authors, could sometimes reflect mostly differences in skeletal composition, colony forms or in the biomass of individual corals used in experiments rather than the efficient functioning of their photosynthesis apparatus. It seems reasonable therefore to express the metabolic rates measured, either of photosynthesis or respiration, in carbon units per biomass of active tissues of the corals colony. Davils (1980) and Lewis and Post (1982) have calculated them per biomass of colony decalcinated matter. But a significant part of it – more than half – is comprised of inert organic substances of coenosarc and skeleton. Therefore, it seems even more reasonable to calculate their rates per biomass of the polyps, isolated from coenosarc (Sorokin 1984b).

The data on the photosynthesis rates per cm^2 of coral colony surface, or per projective area, permit one to calculate also the metabolic rates of corals in their natural communities (Wethey and Porter 1976). An important value in this respect is the ratio P_t/M_t per day. The same ratios calculated per hour and measured at midday are much less informative for the evaluation of the importance of photosynthesis in the energy balance of corals. Nevertheless, the latter sometimes have been also used to evaluate the relative "autotrophy" of various corals and to estimate the depth of the compensation point, where $P_t = M_t$, which is obviously wrong. But even if the depth of the compensation point could be estimated correctly, e.g., with a 24-h balance of P_t/M_t, it remains largely conventional, because corals spend energy not only for respiration, but also for their multiplication, growth and mucus production. The respiration comprises only a part (60–70%) of the total energy demand. So the coral would need additional

Table 9.3. Gross photosynthesis (P_t) and respiration (M_t) calculated 10^{-6} of zooxanthellae cells, and mg^{-1} of chlorophyll *a*

Units	Group of corals	Coral	P_t	M_t	Reference
mg C h^{-1} mg^{-1} of chlorophyll	Scleractinians	*Stylophora pistillata*	3.1	0.9	Gattuso (1985)
		Acropora prostrata	3.6	1.1	Titlyanov et al. (1988a)
		Pavona praetoria	4.2	0.6	Wethey and Porter (1976)
		Pocillopora damicornis	3.2	2.9	Muscatine et al. (1981)
	Octocorals	*Antillogorgia turgida*	0.6	–	Burkholder and Burkholder (1960)
		Plexaura flexuosa	0.3	–	Burkholder and Burkholder (1960)
		Rhigidogorgia flabellum	0.4	–	Burkholder and Burkholder (1960)
μg C h^{-1} 10^{-6} cells of zooxanthellae	Scleractinians	*Stylophora pistillata*	3.4	1.1	Titlyanov et al. (1988a)
		Acropora prostrata	3.2	1.2	Titlyanov et al. (1988a)
		Merulina laxa	5.1	1.3	Titlyanov et al. (1988a)
		Porites nigrescens	2.6	0.9	Titlyanov et al. (1988a)
	Octocorals	*Sinularia* sp.	5.2	1.7	Data by the author
		Alcyonium molle	5.8	2.6	Data by the author
		Sarcophyton glaucum	3.13	1.1	Data by the author

(heterotrophic) nutrition even at the formally correctly measured compensation point.

In accordance with the above considerations about the importance of presentation of the P_t and M_t data on a 24-h basis, during our generalization of available data given per hour, we recalculated them as per 24-h, employing the ratio of these values (coefficients K_p; Table 8.5), which we have obtained in experiments. The results of such a generalization are given in Tables 9.1–9.3. The P_t and M_t data are represented in them per 1 cm^2 of colony surface (Table 9.1), per g of colony dry weight (Tables 9.2, 9.4), per 10^6 cells of zooxanthellae, and per mg of chlorophyll (Table 9.3). The corresponding data on the photosynthesis and respiration of octocorals and of the zoantharians are given in Tables 9.5, 9.6. Tables 9.4–9.6 show the ratios of photosynthesis to respiration and to the biomasses of polyps and of zooxanthellae (P_t/M_t, P_t/B_p, P_t/B_z), as well as the ratio of respiration to the biomass of polyps (M_t/B_p). The analysis of available data, given in the tables, shows that within the light intensities 100–10% PARS the rates of gross photosynthesis in most of scleractinian corals ranges within 30 to 100 $\mu g\, O_2\, g^{-1} h^{-1}$, rarely attaining 140–160 $\mu g\, O_2\, h^{-1}$ (Propp 1985). The latter values were measured in the corals *Montipora tuberculosa* and *Merulina laxa* in flow respirometers. The values below 20 $\mu g\, O_2\, g^{-1} h^{-1}$ were recorded only in some massive corals, in which the ratio of biomass to skeletal mass was lower than in ramose or foliose species. Among such massive corals with a low photosynthesis per g of colony weight were *Cyphastrea*, *Goniastrea*, massive *Porites* species, and *Favites*. In our experiments carried out under stable conditions, the range of values obtained for various taxa of corals were much narrower than could be found in the literature (cf. Tables 9.2, 9.4). In the first case, the rates of

Table 9.4. Gross photosynthesis (P_t), net photosynthesis (P_n) and respiration M_t in common corals, which inhabit the upper well-illuminated zones of the Heron I. reefs (GBR); B_p and B_z – biomass of polyps and zooxanthellae, μg C g^{-1}. (After Sorokin 1984a)

Coral	Wet weight of polyp, mg $sp.^{-1}$	Gross photosynthesis				P_n, μg C g^{-1} day^{-1}	Respiration			P_t/M_t (day^{-1})
		μg O_2 g^{-1} h^{-1} (ranges)	μg C day^{-1} (P_t)	P_t/B_p, %	P_t/B_z		μg O_2 g^{-1} h^{-1} (ranges)	μg C g^{-1} day^{-1} (M_t)	M_t/B_p, %	
Acropora squamosa	0.7	35–58	184	2.3	1.4	72	10–28	112	1.6	1.5
Pocillopora damicornis	0.4	36–93	200	4.2	3.8	72	12–32	128	3.8	1.6
Stylophora pistillata	0.5	48–55	182	4.7	3.5	54	16–22	128	4.2	1.4
Seriatopora hystrix	0.2	29–67	227	8.1	2.3	87	14–22	140	6.3	1.6
Montipora erythraea	1.2	76–101	282	4.5	3.1	82	24–30	200	4.0	1.4
Porites lobata	0.7	21–46	100	2.0	4.0	22	9–12	78	1.9	1.3
Psammocora contigua	1.5	29–44	118	2.9	4.5	30	10–15	88	1.6	1.3
Cyphastrea seralia	5.0	21–29	127	1.5	1.9	31	9–13	96	1.3	1.3
Leptastrea transversa	6.0	28–39	110	2.4	2.5	39	10–13	71	1.6	1.5
Pavona cactus	7.0	12–49	150	3.0	3.2	73	10–15	77	1.9	1.9
Merulina ampliata	11.0	29–36	105	2.0	2.0	−1	13–18	106	1.9	1.0
Goniastrea pectinata	38.4	22–28	85	1.9	1.9	9	7–10	74	1.1	1.0
Echinopora lamellosa	45.0	29–39	110	1.8	1.8	32	9–11	78	1.4	1.4
Symphillia sp.	460.0	29–37	110	1.4	1.4	10	12–15	100	1.1	1.1
Fungia scutaria	2000.0	35–41	127	1.1	1.1	39	11–13	88	1.4	1.4
Tubastrea sp.	170.0	0	0	–	–	0	5–10	50	0.3	–

photosynthesis varied within a relatively narrow scale: 20–40 $\mu g\, O_2\, g^{-1}\, h^{-1}$ in massive corals, and 30–50 $\mu g\, O_2$ in ramose or foliose species with high ratios of biomass to skeletal material.

The rates of photosynthesis cm^{-2} of surface in corals ranged from 10 to 50 $\mu g\, O_2\, h^{-1}$ (Table 9.1). Sometimes it was even over 100 $\mu g\, g^{-1}$ (Edmunds and Davies 1986). Calculated m^{-2} of surface of colonies, it will be some 0.2–0.6 $g\, O_2\, h^{-1}$ or 0.7–2 $g\, C\, m^{-2}\, day^{-1}$. The photosynthesis rates 10^{-6} of zooxanthellae cells are very stable. In different species of corals it ranges from 10 to 15 $\mu g\, O_2\, h^{-1}$ (Table 9.3). Taking their usual numbers of 2–3 × 10^6 cells g^{-1}, the rate of photosynthesis should be 20–40 $\mu g\, g^{-1}\, h^{-1}$. The rates of photosynthesis mg^{-1} of chlorophyll *a* are 10–12 $mg\, O_2$ or 4–5 $mg\, C\, mg^{-1}$ chl *a* h^{-1}. The latter value corresponds to the assimilation number of chlorophyll h^{-1}. As can be seen, it is in zooxanthellae within the same range as in phytoplankton, in which it varies between 2 to 10 mg C.

The photosynthesis rates in symbiotic octocorals, calculated g^{-1} wet weight, appears to be very similar to those in the ramose scleractinians (Table 9.6): 30–100 $\mu g\, O_2\, h^{-1}$. This reflects for them a common strategy of disposition of the zooxanthellae in polyps by one cell layer, and the disposition of polyps in the colony in such a way as to escape self-shadowing. Thus, in both groups of corals their photosynthetic apparatus reached during its development optimal power by the given morphometry of the colonies. The maximum rate of photosynthesis among the investigated symbiotic *Anthozoa*, recorded in the zoantharian *Zoanthus sociatus*, devoid of mineral skeleton was: 214 $\mu g\, O_2\, h^{-1}$ or about 2 $mg\, O_2\, day^{-1}\, g^{-1}$ wet weight (Tables 9.6, 9.9).

The value of gross photosynthesis (P_t) per day is approximately ten times more than the values per hour of the midday period (Tables 9.1, 9.2). It has been usually expressed in carbon units. Per gram of dry colony weight in ramose and in foliose corals it ranges usually from 100 to 300 $\mu g\, C\, g^{-1}$. In some species (*Merulina*, *Pocillopora*) it will reach even 400–500 μg. In massive corals it is between 40–80 μg. Calculated cm^{-2} of colony surface, the P_t values in scleractinian corals range from 50 to 300 μg C (Table 9.4), which corresponds to 0.5–3 $g\, C\, m^{-2}\, day^{-1}$ (P.S. Davies 1977). An average ratio of the active (photosynthesizing) surface area in corals to the projective area of their colonies is between 5–8 (Dahl 1976). Thus at 100% of cover of bottom surface coral communities could produce 3–25 $g\, C\, m^{-2}\, day^{-1}$. Actually, the percentage of cover is less and their gross photosynthesis production ranges from 2 to 10 g C. But sometimes it could reach 20–22 g C. Calculated by their projective area, the photosynthesis of corals even with relatively low rates of photosynthesis, like *Millepora* and *Porites*, was between 3.4–10.7 $g\, C\, m^{-2}\, day^{-1}$ (Beyer 1966). In the octocorals of Caribbean reefs the P_t values were 3.7–6.8 g C, and in some scleractinians from the same reefs: 2.7–10.2 $g\, C\, m^{-2}\, day^{-1}$ (Kanwisher and Wainwright 1967). According to Taylor (1973b), with 100% coral cover of the bottom on the Caribbean reefs, and at the rate of their photosynthesis cm^{-2} being

200–260 μg C day^{-1}, their gross photosynthetic production should be about 12–15 g C m^{-2} day^{-1}.

In the octocorals the values of photosynthesis calculated per gram of dry weight of their colonies were two to ten times more than in scleractinians (0.6–3 mg C day^{-1}), while per gram wet weight they were comparable (Tables 9.5, 9.6). This happens because their colonies contain much water and less hard skeletal material than those of scleractinians. The largest values of P_t were recorded in *Zoanthus sociatus*: 4.85 mg C g^{-1} day^{-1} (Tables 9.6, 9.9). Calculated per gram C of polyp tissues, in the scleractinian corals the P_t values ranged from 1.1 to 4.5 mg C day^{-1}, and only in the coral *Seriatopora histrix* was it 8 mg C. The highest values were recorded in small polyps, which have a large, specific surface area.

The net photosynthesis (P_n), which is equal to the difference between gross production and respiration ($P_t - M_t$), in most corals ranges at optimal illumination from 30 to 100 μg C g^{-1} day^{-1}, or 20 to 50% of P_t. It was largest in ramose and foliose corals (Table 9.4). In most corals at the upper reef zone the P_n-values calculated per m^2 of active colony surface appeared to be significant: 150–200 mg C day^{-1}, while in the coral *Turbinaria reniformis* it was even up to 850 mg C (Rogers and Salesky 1981; Falkovsky et al. 1984). The efficiency of utilization of light energy by corals at the above values of P_t in well-illuminated reef zones ranges from 0.5 to 2.5% (Beyers 1966), increasing in more shaded or deep biotopes to 5–10% (Gattuso 1985; Muscatine et al. 1985). Because of the corals' ability for light adaptation their photosynthesis remains relatively constant within a wide scale of light intensities – up to one order of their values (cf. Sect. 8.4). In Table 9.9 an example is given of P_t in corals at the surface and at depths of 20–30 m. It shows that in some of their species the photosynthesis rate could be even higher under moderate than under strong light (Taylor 1973; Wethey and Porter 1976).

Table 9.5. Gross photosynthesis production (P_t) and respiration per day in common corals and zoantharians of the Red Sea, mg O_2 g^{-1} of dry decalcinated organic matter (B_c). (Data by Mergner and Svoboda 1977)

Groups of corals	Species of corals	P_t	P_t/B_c	M_t	M_t/B_c[a] %	P_t/M_t
Scleractinian corals	*Fungia scutaria*	31	0.19	22	0.14	1.4
	Acropora variabilis	99	0.61	81	0.50	1.2
	Stylophora pistillata	88	0.75	69	0.42	1.3
	Favia pallida	30	0.54	24	0.15	1.2
	Goniastrea pectinata	60	0.37	56	0.35	1.1
Zoantharians	Palythoa mammilata	15	0.09	13	0.80	1.2
Hydrocoral	*Millepora dichotoma*	81	0.50	70	0.43	1.2
Octocorals	Lithophyton arboreum	75	0.46	72	0.45	1.0
	Heteroxenia ghardaquensis	75	0.46	62	0.38	1.2
	Xenia macrospiculata	72	0.45	50	0.37	1.2

[a] Values P_t, M_t and B_c expressed in carbon units to calculate their ratios.

Table 9.6. Total photosynthesis (P_t) and respiration (M_t) in common species of symbiotic zoantharians and octocorals of the Indo-Pacific reefs; B_p – biomass of polyps, μg C sp.$^{-1}$. (Data by Sorokin 1990a)

Groups	Species	% dry weight in colony	μg O_2 g^{-1} wet weight of colony per hour (ranges)		μg C g^{-1} dry weight of colony per day		P_t/M_t	M_t/B_p, %
			P_t	M_t	P_t	M_t		
Zoantharians	*Zoanthus sociatus*	18	134–214	60–92	3230	3070	1.1	0.76
	Palythoa tuberculosa	45	18–29	8–12	190	160	1.2	1.95
Alcyonacean octocorals	*Lemnalia rhabdota*	15	32–43	5–12	887	391	2.3	1.26
	Paralemnalia digitiformis	16	40–48	7–13	891	442	2.0	1.13
	Paralemnalia clavata	14	35–53	8–15	979	550	1.8	1.62
	Capnella sp.	44	16–22	3–6	158	71	2.2	0.33
	Cladiella humesi	18	45–58	8–10	1052	360	2.9	0.87
	Sinularia hirta	36	41–56	7–13	455	191	2.4	0.51
	Sinularia densa	32	34–45	5–11	500	240	2.1	0.58
	Sinularia sp.	38	51–75	19–29	600	250	2.6	0.48
	Sarcophyton glaucum	28	41–53	12–21	660	420	1.4	0.52
	Lithophyton arboreum	24	102–144	18–22	1675	600	2.8	3.61
	Alcyonium malle	15	120–157	46–79	3730	4860	0.8	3.03
	Lobophytum pauciflorum	39	50–69	18–30	550	416	1.3	1.30
	Xenia elongata	9	10–18	5–11	660	554	1.2	0.80
Gorgonacean octocorals	*Isis hippurus*	38	60–83	15–24	744	312	2.4	0.60
	Rumphella aggregata	31	51–68	12–19	667	456	1.5	0.67
	Hicksoniella princeps	28	96–126	23–41	1520	920	1.7	1.76
Stoloniferan	*Tubipora musica*	48	20–36	7–12	214	150	1.5	1.20
Scleractinian	*Stylophora pistillata*	76	35–61	12–19	195	190	1.0	2.80

The rate of respiration in corals (M_t) is one of the important ecophysiological parameters, which characterize their energy demand (Davies 1980; Lewis and Post 1982). As it was also mentioned above, the M_t-value is still not equivalent to the energy demand but only a part of it, which comprises after different evaluations about 60–70% of the total. The remaining energy is needed for mucus production, propagation, and growth (Muscatine et al. 1984; Sorokin 1984b; Edmunds and Davies 1986). The generalized data on respiration rates in corals are given in Tables 9.1–9.6. In the hermatypic scleractinians it is 10–30 g O_2 g of dry colonies weight, or 7 to 40 μg O_2 cm^{-2} of their surface per hour. Corals with a higher P_t have also most often higher M_t rates (Mc Closkey et al. 1978; Porter et al. 1984). Among such corals are the acroporids, *Pocillopora* and some foliose species: *Montipora*, *Merulina*, *Turbinaria*. This could be a cousequence of a higher rate of translocation of photosynthates in them by their zooxanthellae. The same could be a cause of an elevation of the M_t at post-midday time when the trasslocation is the most intensive (Mc Closkey et al. 1978; Mc Closkey and Muscatine 1984; Porter et al. 1984; Titlyanov et al. 1985). The rates of respiration in ahermatypic scleractinians as *Tubastrea*, devoid of zooxanthellae, are two to three times lower than in hermatypic corals (Tables 9.7, 9.8). Possible dependence of the respiration rate in cnidarians upon water-flow speed was also recorded – (Patterson and Sebens, per. Comm.).

The rate of respiration in different scleractinian species, calculated per organic carbon of the polyps' biomass, ranged between relatively narrow limits of 11 to 42 mg C g^{-1} (ratios $M_t:B_p$ = 0.7–4.2%). Only in *Seriatopora* with its minute polyps was it 6% (Table 9.4). In corals with average and large polyps this ratio was only about three times higher compared with corals having large polyps, while the size of polyps between these two groups varied within 3 orders of values. Thus, the respiration of corals does not actually depend much upon the individual weight of its individual

Table 9.7. Respiration of some octocorals and of the zoantharian Palythoa

Species	μg O_2 h^{-1} g^{-1} dry colony weight per hour	mg O_2 h^{-1} g^{-1} of total organic matter in colony	Author
Plexaura flexuosa	28	0.31	Lewis and Post (1982)
Eunicea tourneforti	33	0.30	Lewis and Post (1982)
Muriceopsis flavida	150	0.75	Lewis and Post (1982)
Gorgonia ventalina	121	0.76	Lewis and Post (1982)
Briareum asbestinum	36	0.15	Lewis and Post (1982)
Heteroxenia fuscescens	160	2.80	Mergner and Svoboda (1977)
Lithophyton arboreum	480	3.00	Mergner and Svoboda (1977)
Palythoa mammillata (zoantharian)	108	0.54	Mergner and Svoboda (1977)

Table 9.8. Respiration (M_t) in some asymbiotic octocorals; B_p – biomaas of polyps, $mg\,C\,g^{-1}$ of dry colony weight. (Data by Sorokin 1990a)

Species	Percentage dry weight	B_p	M_t, $\mu g\,O_2\,h^{-1}\,g^{-1}$ of wet colony weight (ranges)	M_t, $mg\,C\,g^{-1}$ of dry colony weight per day	M_t/B_p
Spongotes gigantea	19	95	41–55	2.15	2.3
Echinogorgia praelonga	22	109	34–48	1.77	1.6
Plexauroides lenzii	25	92	77–108	2.92	3.2
Bebryce indica	29	74	32–49	1.31	1.8
Mopsella aurantia	65	25	86–102	1.29	5.1
Acabaria nicksoni	57	23	61–86	1.12	4.8

polyps, while in the free-living animals the rate of respiration is a reverse function of their individual weights. This is the cousequence of the colony functioning as a whole organism. In octocorals, M_t-values calculated per raw weight of colonies are quite comparable with those in scleractinians (10–40 $\mu g\,O_2\,g^{-1}\,h^{-1}$). The asymbiotic octocorals have higher respiration rates calculated per dry colony weight, as compared with the symbiotic ones, which is reversed in scleractinians. Their values range from 30 to 500 $\mu g\,O_2\,g^{-1}\,h^{-1}$ (Tables 9.5–9.7). This wide range of M_t depends upon the variations in the percentage of skeletal material in their different species. Calculated per carbon of polyps tissues, it varies less: 25–50 $mg\,C\,g^{-1}\,day^{-1}$, which is close to its values in scleractinian polyps (Tables 9.4, 9.6). Per gram dry organic matter of colony its range in octocorals is 0.3–3 $mg\,O_2\,g^{-1}\,h^{-1}$ (Tables 9.7, 9.8). The respiration rates in zoantharians appear to be comparable with those in octocotals (Tables 9.5, 9.6). The diurnal energy demand of corals calculated from the M_t data per decalcinated biomass in scleractinian corals was estimated to be 80–200 $cal\,g^{-1}$, and in octocorals 200–500 $cal\,g^{-1}$ (Mergner and Swoboda 1977; Davies 1980; Lewis and Post 1982). But this should be an underestimate by at least 30%, because of the above mentioned energy losses other than respiration.

Among light adaptation mechanisms of corals, the decrease in respiration of their colonies living in deep reef biotopes and adapted to a low light was mentioned above (Muscatine and Porter 1977; P.S. Davies 1977; Titlyanov et al. 1985; Table 9.9). This reaction could be supposed also a consequence of the decrease in photosynthesis rates and translocation at low light (Muscatine et al. 1984). But the adaptive character of this reaction was proved by direct experiments on the transplantation of colonies from upper to deeper zone of the reef slope. The respiration rate decreased only after a 2-week period of adaptation to low light (Davies 1980).

One of the most important indices of metabolism in corals are the ratio P_t/M_t. This ratio was calculated using the corals' numbers per hour at daytime or per day. The P_t/M_t ratios per hour range in scleractinians from 2

Table 9.9. Rate of photosynthesis and respiration (P_t,M_t, mg O_2 h^{-1} mg^{-1}, chl *a*) in colonies of the coral *Stylophora pistillata*, living at different depths; I – light intensity; E, $m^{-2}s^{-1}$. (Data by Gattuso 1985)

Depth, m	I	P_t	M_t	P_t/M_t
1	318	4.0	1.3	3.1
5	193	9.1	3.2	2.8
10	158	4.9	1.3	3.8
30	111	6.1	0.8	6.4

to 5, and in octocorals 3 to 6 (Kanwisher and Wainwright 1967). Because of the corals' ability for light adaptation this ratio remains >1 down to the lower limit of illumination, at which a given species lives. For example, in the Atlantic coral *Synarea convexa* the P_t/M_t ratio was >1 down to the depth where the illumination was only 2.5% PARS (Jaubert 1977). The P_t/M_t balances per day varied even in experiments of the same author with the same coral species (Table 9.2). Even at optimal light intensity they were mostly little more than 1. In many species and in many experiments they were <1. In our estimations made at the Heron Island reef with 23 common corals living in the upper reef zones, the P_t/M_t ratios were in most corals between 1.1 to 1.4. In four species they were 1.5, 1.6, and in one (*Acropora hyacinthus*) 2.2 (Sorokin 1984a). The ratio P_t/M_t, calculated per day, decreases much faster with depth and the decrease in light than these ratios, calculated per hour, mostly because of the shortening of the light period of day at depth. For example, in the coral *Stylophora pistillata* the P_t/M_t per day at a depth of 2 m was 2.0, and at a depth of 45 m it was 0.4 (Falkovsky and Dubinsky 1981). Thus, at the depth of 45 m this coral may compensate its energy expenditures only by about 25–30% (counting expenditures additional to respiration). The rest the coral should compensate for by heterotrophic feeding. Therefore, the depth of the compensation point should be estimated on the basis of the diurnal P_t/M_t balance.

9.2 Heterotrophic Feeding of Corals

Hermatypic corals possess the ability, unique in the animal world, of occupying several trophic levels at once. Being capable of autotrophic feeding with the participation of their zooxanthellae, these corals make use of all miodes of feeding known in sedentary benthic animals. They have also corresponding morphophysiological structures serving them. Corals actively consume the dissolved organic matter with the aid of the ciliar apparatus of polyps, employing the mechanism of active transport of molecules through the cell membrane. They are also capable of sedimentary filter feeding, using for thus the ciliar apparatus and mucus nets. Lastly, they are highly specialized predators, possessing mechanisms for catching, swallowing and digesting the moving animal preys. The polyps of scleractinian corals have the largest ratio of catching area of the body to its biomass among all other aquatic animals. It seems, therefore, very possible that their apparent evolutionary success and their flourishing in biotopes of the tropical shelf are to be ascribed precisely to their capability to use most sources of energy available for sessile animals: light, plankton and dissolved organic matter. In spite of the remarkably strong competition for solid bottom substrate in tropical shelf zones, corals had been during former geological periods and are at present time among the dominating elements of the benthic fauna.

The problem of the feeding of corals has been a topic of discussion for over 60 years (Muscatine 1973). Early investigations of the feeding of actinians and corals in 1905–1910 demonstrated that coral polyps, like the actinians, actively capture animal food, such as pieces of meat or planktonic crustaceans. In has been established that in their gastral cavities food was crushed by the joint action of moving mesenterial filaments and digestive enzymes. The particles formed were then phagocytosed and transported into the polyp's tissues by wandering cells. All kinds of plant food including zooxanthellae were rejected by them (Vaughan 1912; Boshma 1925b). But also at that time Gardiner and later Gravier (1913) concluded that even if capable of predation, coral polyps got their basic nutrition anyway from their algal symbionts with which their cells were literally packed. This quite reasonable conclusion was later countered by the famous investigator of coral physiology, Sir Maurice Yonge (1930a,b). He insisted on the predatory feeding of corals as being the dominating one and did not recognize the role of the algal symbionts in coral nutrition. His arguments were that corals have a large catching surface and extremely developed structures for predation, and so the polyps supporting these without any profit would be unreasonable, moreover, that the animal tissues in polyp comprise over 90% of the total biomass. The polyps also actively rejected in his experiments all kinds of plant food. He strongly insisted on the correctness of his opinion even in the 1960s and early 1970s, when it looked already somewhat stronge, especially after the conclusive experiments of Franziskett (1970), proved the importance of autotrophic feeding of corals. The opinion of such an acknowledged expert as Maurice Yonge opposing autotrophic feeding of corals definitely retarded at that time the development of the idea about their symbiotic nature. But the appearance in the early 1970s of numerous publications about photosynthesis in corals, which proved the translocation of photosynthates from zooxanthellae to polyps, finally have led to the recognition that the energy budget in them is based upon the autotrophic production of their algal symbionts. Moreover, common opinion after Fanziskett (1970) started to reject the opposite conclusion that corals really did not need any other sources of energy at all except the one photosynthetically produced by their zooxanthellae. A kind of referee opinion on the problem autotrophy – heterotrophy in corals was expressed by Goreau et al. (1971) and then developed by Porter (1976). It said that among the variety of corals there are taxa with either the heterotrophic or the autotrophic feeding dominating in them. The preference for a type of feeding in coral species was supposed to be dependent upon their ecological adaptations and life strategy, reflected in their morphology, and first of all upon the size of their polyps. Corals with large polyps and long tentacles were supposed to be feeding more heterotrophically, and their dependence upon their zooxanthellae supposedly minimal. Porter's hypothesis seemed to be attractive. Whether correct or not, it appeared to explain the great diversity in the forms of colonies and the sizes of polyps among scleractinian corals, with a view to the background of their trophic specialization, which provides for the sharing of energy sources and for decreasing competition between them. But despite its attractiveness this theory still needed direct approval in experiments.

By the end of the 1970s it became clear that all hermatypic scleractinian corals, independently of the morphology of their colonies or the size of their polyps, make use with high efficiency of their mechanisms for predatory feeding, filter-feeding, "osmotic" feeding with DOM as well as of the autotrophic products translocated by their symbionts (Lewis and Price 1975; Lewis 1977b; Sorokin 1978b, 1981c; Falkovski et al. 1984). The next step needed was the evaluation of the shares of the above-mentioned resources in the energy balance of the different coral species. In our attempt to solve this problem we have studied feeding and metabolism of 23 common corals of the GBR at Heron Island Research Station of the University of Queensland. The results did not prove the theory of Porter regarding the subdivision of scleractinian corals into "mostly autothrophic" and "mostly heterotrophic", and they disproved as well the proposals about their trophic specialization with respect to their ability to use heterotrophic sources for feeding. Instead, an amazing uniformity of their parameters for feeding and metabolism has been established, which is independent of their taxonomic position, and the morphology of their colonies or of the size of their polyps (cf. Table 9.14).

During early studies of corals' nutrition, the investigators used the method of visual observation by binocular microscope of how polyps seized their food objects (small crustaceans or pieces of fish meat, stained with neutral dye) (Pratt 1906; Vaughan 1919; Boshma 1925b; Yonge 1930a,b). To record the predatory feeding of corals in situ, stroboscopic photography of the polyps at night was employed. The number of polyps with captured prey was then estimated in this way (Johannes and Tepley 1974). To study the coral's predation under conditions close to in situ conditions, individual colonies of corals were covered with plastic bags and the food (planktonic crustaceans or labeled organic matter) was then injected into the latter (Lasker 1981; Coffroth 1984). To attract prey, Porter (1974b, 1978) illuminated with a flash light a part of the coral colony. To study the spectrum of feeding, this author extracted in situ the contents of the gastral cavity of polyps, inserting the needle of a syringe through their mouths. The function of cilia and mucus filaments in the capturing of particulated food was studied by using the suspension of heavy particles of corund (Yonge 1930b). The rate of sedimentary filter feeding of corals was studied by recording the decrease in concentration of food particles or neutral particles in the presence of corals (Roushdy and Hansen 1961; Lewis 1976). But this method cannot yield really reliable information about the rate of feeding of corals, because in this case the amount of the suspended matter actually swallowed by polyps, remains unknown.

Really quantitative data on the feeding of corals could be most reliably obtained with the aid of radioisotopic labels. For the labeling of the potential food of corals the isotopes of sulfur ^{35}S and of tritium ^{3}H were used (Di Salvo 1971a; Coffroth 1984). But the most convenient and reliable label could be made using the radiocarbon ^{14}C. This radioisotope has been employed for the establishing of location of assimilated labeled food material within the polyp's body with the aid of radioautography (Goreau and Goreau 1960b). The radiocarbon method of quantitative study of feeding of aquatic animals (Sorokin, 1968) have been used by us also for the study of different aspects of the heterotrophic feeding of corals of the GBR, of reefs of the Indian Ocean and of the South China Sea (Sorokin 1978b, 1981c). As labeled food were used: dissolved organic matter (DOM) protein hydrolizate from labeled planktonic algae, bacterioplankton, labeled with the aid of this hydrolizate, and various labeled forms of zooplankton. The zooplankters: ciliates, veligers, rotifers, of nauplii of artemia were marked through feeding them with labeled food – bacteria or microalgae (Sorokin 1984a). The veligers were obtained by hatching eggs of the nudibranch mollusc *Aplysia*. To induce their spawning, we injected them with KCl, or serotonin solutions. Corals, selected for the feeding experiments, were first adapted to the experimental conditions. The labeled food was added at night only if their polyps were opened.

9.2.1 Feeding of Scleractinian Corals

In their hunting for animal prey scleractinian corals use the ways known in other sessile coelenterata. They may capture prey with their tentacles armed with the powerful cnidocil. The preys are immobilized by being hit by poisonous cnidae. Then they are captured by the tentacles and transferred to the mouth. The preys could be captured and immobilized also with the aid of mucus nets, formed by the polyps. Both these mechanisms interact during hunting (Yonge 1930b, 1940; Goreau et al. 1971). The activity of polyps is stimulated by amino acids which the zooplankters excrete. The polyp "feels" the presence of the prey and, stimulated, starts in pursuit with its tentacles (Porter 1974b). Even if a large strong prey, on being hit by cnidae, then brake them, it will be most often captured by the polyp anyway. Paralyzed by cnids, it starts to move chaotically, and then usually comes up against the polyp again and is captured by the mucus net or by the tentacles. The swallowing of prey is done with the aid of mucus, which the polyp sucks into the mouth together with the captured prey. The whole procedure of capturing and ingesting an animal prey takes 20–40 s. If the prey is captured by the polyp with short tentacles, which cannot reach the mouth and bring the captured prey, the polyp's mouth extends and forms a "tube", which reaches out to the prey and takes it off the tentacle (coral *Favia*). The same is done by corals which have a large exposed septal covered with the epithelium and armed with chidocill. The preys captured on the surface of this epithelium are ingested by the same procedure (corals *Hydnophora* and *Leptastrea*). Maurice Yonge (1930b) had also described the mechanism of transport of captured food to the mouth of the polyp by cilia, which were supposed to be reversing the direction of their beats in the presence of food. He erroneously concluded that the movement of corund particles to the mouth of polyps, excited by a drop of fish broth, took place through the action of cilia, which may reverse the direction of their beats in a polyp that exhibits the "food reaction". But later it was shown that it was an artifice. The cilia never reverse their beat, which is directed from the mouth to the side of the polyp, serving, as their main function, to shed the sediments (Lewis and Price 1975). But basically Maurice Yonge was right – the ciliar apparatus indeed participates in the process of feeding of coral polyps, but its function is quite different. In the inactive, closed state of the mouth the polyp lets out strips of mucus, which pass between their septae or between the tentacles. Under the action of the beating cilia the ends of these strips because whiskered and form the so-called mucus nets a bunch of sticky finest mucus filaments. The same cilia, beating in one direction, induce the ciliar currents, which pass through these mucus nets. Thus the system "ciliar epithelium–mucus net" forms a powerful filtering apparatus. The mucus nets collect the suspended food particles from the ciliar streams: slowly moving zooplankters, eggs and cysts, bacteria and pseudoplankton. Once the mucus net has collected enough food, this is then dragged by the polyp into its mouth and swallowed (Lewis and Price 1975).

The reaction of swallowing the mucus net and the activation of the feeding mechanism in the polyp, accompanied by the expansion of tentacles and the opening of the mouth, is started by the same chemical trigger – the appearance of a minute concentration of some amino acids, which interact with the chemoreceptors of the polyp (Mariscal 1971). The traces of amino acids, which are excreted by zooplankters and other living food objects, induce the "food reaction" in polyps. The sequence of actions after the start of the "food reaction" usually results in the capture and swallowing of the prey by the polyp (Lehnhof 1968; Lehman and Porter 1973; Porter 1974b; Lewis and Price 1975).

Among the majority of corals groups of species could be named which exhibit a definite preference for the use of one of the two above-mentioned ways of feeding: the catching of prey with the aid of tentacles armed with cnidocil, or with the aid of the mucus nets (Yonge 1930b, 1940; Lewis and Price 1975). The first group includes some corals with small polyps, like *Pocillopora*, *Stylophora*, and *Madracis*. They do not produce real mucus nets, but excrete mucus in the form of flakes. Another group of corals, which have small polyps with very short tentacles and flat colonies with well-developed septae (Agaricidae), catch prey only with the use of mucus nets. In corals of this group (*Pashyseris*) the tentacles are completely dwarfed and their function in capturing preys is to take the mesenterial filaments protruding from the mouths of polyps (Yonge 1930b). The most numerous is the third group of corals, which use both mechanisms of feeding. These corals are *Acropora*, *Favia*, *Syderastrea*, *Mussa*, *Isophyllia*, *Meandrina*, *Pavona*, *Hydnophora* and many other relative taxa. Most of them have polyps with a large oral disc and long tentacles. In active state, their polyps completely emerge from the corallites and expand their tentacles wide so that uder the net of the so expanded tentacles the colony remains sometimes unseen. The mouths of polyps are semi-opened, their rolled edges forming a "lip". From the oral disk out to the side of septae strong ciliar currents of water pass and from the mouths the stripes of mucus come out, forming at their ends the mucus nets. In the presence of food objects, such as little crustaceans, the "food reaction" in their polyps immediately sets in the polyps open their mouths, the mucus stripes are pulled in and swallowed. The tentacles are busily moving trying to capture the prey. The coral *Montastrea* has special long "sweeper" tentacles, which energetically "pursue" the preys.

The size of preys which corals can catch and digest actually is not strictly limited by the size of their polyps. Small polyps have several adaptations which permit them to manage large preys. In corals with small corallites, which have a column in the middle, the polyp's body may emerge from the corallite completely, and so expanded it may swallow large preys. The coral *Pocillopora* may swallow preys two to three times larger than its own size. The polyps of this coral swallow sagittas 10 mm long in 30 s, their own diameter measuring 1–1.5 mm within the corallite. Polyps of *Leptastrea* have corallites 2 mm in diameter. However, in the active state they come out and expand them 4–5 mm, being there able to swallow large pieces of meat (Yonge 1930b). Corals in which polyps do not come out in the active state, have as a rule a spacious corallites with reduced central column (*Euphyllia*, *Asteropora*). Other corals have well-developed long mesenterial filaments. With the aid of these filaments their polyps may digest large preys outside, even without swallowing them. When the prey is caught but not yet swallowed, being attached to the oral disk by mucus nets and by cnidae, these filamenta, which are actually the powerful digestive organs protruding from the polyp's mouth or through the wall of the gastral cavity, entwine the

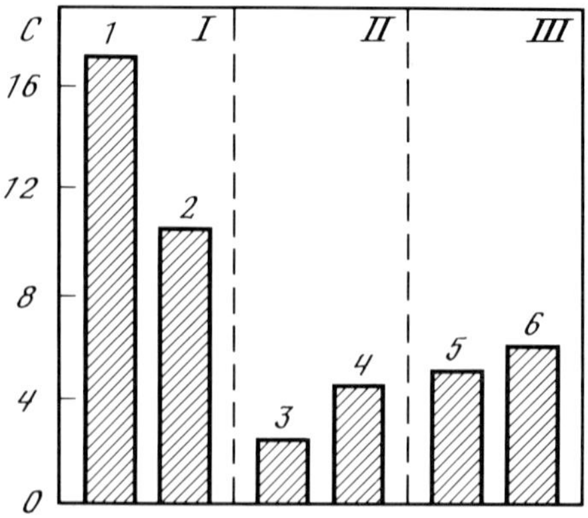

Fig. 9.2. Consumption of large ^{14}C-labeled *Artemia* nauplii of 1.4 mm size by corals with different polyp sizes: *I* corals *Stylophora* (*1*) and Pocillopora (*2*) with small polyps (0.5–0.7 mg); *II* corals *Pavona* (*3*) and *Leptastrea* (*4*) with medium-sized polyps (10–20 mg); *III* corals *Goniastrea* (*5*) and *Symphillia* (*6*) with large polyps (60–600 mg); C- relative food rations. (After Sorokin 1984a)

prey and digest it outside. Among such corals are the common scleractinians *Pashyseris*, *Montipora*, *Galaxea*, *Stylophora*, and *Merulina*. We have estimated in experiments the efficiency of feeding in corals with different sized polyps on large preys – the late nauplii of artemia 1.4 mm large, labeled with ^{14}C (Fig. 9.2). It was found that corals with small 1–1.5 mm sized polyps digested large preys even more efficiently than those with large polyps.

As active and specialized predators, coral polyps quickly digest and assimilate their preys with the aid of their digestive organs, the mesenterial filaments (Boshma 1925b; Yonge 1930a,b; Yonge and Nicholls 1930). The mesenterial filament is a very complex organ of the polyp's mesenterium. At its ends it has a ciliar epithelium, cnidocil, and the bunches of gland cells. At the base of the filament a special zone is situated, which contains phagocytizing cells. Through this zone passes also the excretion of waste products (Yonge 1940; Goreau and Philpott 1956). The process of digestion consists in two phases. First, the prey is attacked by proteases, excreted by the gland cells. The pH around the prey after its swallowing immediately drops from 8 to 7. The proteases quickly dissolve its tissues, hydrolyzing proteins to peptides. Simultaneously, the mesenterial filaments become active and break down the softened body of the prey into small particles, which are immediately phagocytosed by the wandering cells which come out from the same mesenterial filaments. They transfer then the food particles into the tissues of the polyp where they are finally digested. The undigested wastes are transported back to the gastral cavity and are then excreted. The digestion of animal preys, like small crustaceans, by polyps takes 1.5 to 4 h (Yonge and Nicholls 1930; Coffroth 1984). According to Porter (1974), polyps of the coral *Montastrea* digest the artemia nauplii in 1.5–2 h. But the time of digestion of large preys or pieces of meat swallowed by the polyp could be much longer – up to 10–12 h. We have estimated the time of digestion of labeled artemia nauplii. Corals were fed during 10 min with these preys. Then they were transferred into the beakers with unlabeled food, and the time of appearance of labeled particles of undigested remains of food was recorded. The results of the experiments showed that corals with large polyps digest the preys in 2–2.5 h, and corals with a small polyps in 2.5–3 h. The assimilability of the animal preys ingested by polyps, as estimated by the radiocarbon method in experiments with different corals, appears to be high: 70–80% (Table 9.12).

Scleractinian corals were shown to be capable of sedimentary filter feeding. They may filter out from the water column planktonic bacteria (Di Salvo 1971a, 1973; Sorokin 1971b, 1973c, 1978b), pseudoplankton and detritus (Lewis and Price 1975, 1976) and even the neutral particles, like suspended graphite (Lewis 1977a). Filtration proceeds by the interaction of ciliar currents over the polyp with the mucus nets, as described above. The particles contained in water passing through the mucus net or over the mucus strips stick to the latter's surface. In this process a definite role is played by electrostatic forces. The mucus substance contains amino groups, charged positively, while the micells of colloids, the bacterial cells or the detritus particles, are charged most often negatively, having on their surface the radicals of carbonic acids. The filtering mechanism in corals works permanently around the clock independently of the state of the coral, whether active and expanded or passive, hidden inside the corallite and closed. Even when closed, the polyps emit the strips of mucus from their mouths, and the automatically working ciliar apparatus makes at their ends the mucus nets (Coles and Strathmann 1973). The particulated matter thus caught is gradually accumulated on them. If it does not contain enough nutritive substance, the mucus nets overloaded by it are driven out from the polyp by ciliar currents thus cleaning its surface. But if the mucus net has caught something edible, the "feeding reaction" of the polyp is triggered, as described above, the polyp ingests food caught together with mucus. The polyps can manifest the "food reaction" also in closed inactive state.

Corals using this mechanism most actively include *Montastrea*, *Mussa*, *Diploria*, *Colpophyllia*, *Lobophyllia*, *Hydnophora*, *Pavona*, and *Acropora*. But some corals, like *Pocillopora* or *Porites*, which were proved to be capable of filter feeding on the bacterioplankton, did not form mucus nets at all. The flakes of mucus they excrete is used to immobilize the preys caught and to relieve the process of swallowing (Abe 1938). These corals use a different mechanism of sedimentary feeding on suspended food. Their ciliar currents are directed from the coenosarc to the tentacles and to the mouth, but not backward as in other corals using the mucus nets. These currents bring the particulated matter from outside to their oral disc where it is caught stuck to the mucus flakes, which form lumps. If they have enough edible material, the "food reaction" starts and it is swallowed, if not it is rejected and drops off the polyp. And other way of sedimentary feeding in these corals is the digestion of food particles settled on or stuck to the surface of the expanded polyp body by proteases secreted by the cells of their epithelium. The polypeptides thus formed are phagocytosed by the wandering cells (Tiffon and Bouillon 1975). The epithelial cells of the polyp are apparently able to absorb the cells of bacterioplankton settled on its surface via pinocytosis. Indeed, their surface is covered by numerous invaginations of the cell wall, which were proved to be functionally capable of pinocyctosis (Goreau and Philpott 1956; Goreau et al. 1971).

The ability of corals to take up dissolved organic matter (DOM) was first recorded by T. and N. Goreau (Goreau and Goreau 1960a) and by Stephens (1962). It was supposed that an extremely vigorous consumption of DOM by corals could be accomplished only by the active enzyme transport with participation of the permease mechanisms (Goreau et al. 1971). Later, we have proved it by direct experiments (Sorokin 1977b, 1986a). The nutritional behavior in most corals is characterized by the alternation of the periods of relaxation, when the polyps are closed and hidden in corallites, and the periods of activity, when they are opened and expanded over the corallites (Yonge 1930b; Glynn 1973b; Porter 1974b; Lewis and Price 1975; Sebens and de Riemer 1977). The adaptive reason behind this type of behavior could be seen in the corals' economy with living resources during the daytime period, when the zooplankton in reef waters is very scarce, being hidden in bottom biotopes (cf. Sect. 3.3.2). In fact, their respiration at this stage of relaxation is slower than during the period of activity (Roffmann 1968; Sebens and de Riemer 1977). At the same time, the photosynthesis of polyps during the inactive phase of the diurnal cycle does not decrease, and their activity as the filter feeders is preserved as well (Lewis and Price 1975). There are also, proofs of corals being most active in their predatory feeding at dusk, at dawn (~75% of total preys ingested) and during the night time (Glynn 1973b; Porter 1976). The diurnal rhythm of activity varies in different taxa of corals. Groups of species which are active only at night are the most numerous. They include corals with large-to-average sizes of polyps, which readily feed on zooplankton, using for hunting both the tentacles and the mucus nets. Their nocturnal activity seems to be directly connected with their success in predatory feeding. Among the corals which are active diurnally the ramose corals *Pocillopora*, *Stylophora* and some species of *Porites* and *Acropora* could be mentioned (Table 9.10). The diurnal rhythm in corals seems to be controlled by the level of illumination (Swenney 1976).

We have tried to obtain a quantitative expression of the diurnal activity in various corals by measuring in experiments the ratios of their predatory feeding at midday and midnight. The food was labeled with ^{14}C nauplii of *Artemia* (Table 9.11). The results showed that even corals with small polyps, which were active also in the daytime (*Pocillopora*, *Stylophora*, *Acropora*

Table 9.10. Diurnal activity rhythms in some common corals. (Data by Porter (1974b) and Lasker (1976), and the author's own observations)

Rhythm of diurnal activity	Corals	Size class of coral polyps
All polyps in colony are active 24 h	*Stephanocoenia*	Average
	Acropora	Small
	Pocillopora	Small
	Madracis	Small
	Porites	Small
	Millepora	Small
	Stylophora	Small
	Montastrea annularis (daytime morph)	Average
All polyps are active only at night, but some of them in the colony are active also during the day	*Cyphastrea*	Small
	Goniopora	Average
	Pavona	Average
	Hydnophora	Average
	Diploria	Average
	Eusmilia	Large
	Fungia	Large
Over 90% of polyps active only at night	*Agaricia*	Small
	Siderastrea	Average
	Montastrea cavernosa	Average
	Montastrea annularis (nocturnal morph)	Large
	Manicina	Average
	Meandrina	Large
	Mussa	Large
	Isophylia	Large
	Lobophyllia	Large
	Mycetophillia	Large
	Goniastrea	Large
	Favia	Large

squamosa), at night hunted more efficiently than in the daytime. The ratios of rates of feeding at night compured to those in the daytime ranged between 4.1–8.7, while in corals with large polyps, which were active only at night, these ratios were between 13.7–15.3. The values of these ratios were used then to calculate the 24-h food rations after their values per hour measured at night.

As seen above, the energy income in corals stems from differents sources. This raises the problem of estimating the share of each of these sources in their energy balance in the natural environments of the reef. To solve this important problem of reef trophodynamics different ways were tried. One of these could be the estimation of the availability of food sources for corals whose use by them has been established in experiments. This method could be fruitful when adequate data are available on the stocks of given food sources and on the dependence of the corals' rates of feeding on food

Table 9.11. Comparative rates of predatory feeding of corals in daytime and at night. (After Sorokin 1984a)

Size group of polyps; B-biomass of polyps, mg C $sp.^{-1}$	Corals	Rates of feeding (food rations, μg C g^{-1} per 40 min		C_2/C_1
		In daytime 3–4 p.m. (C_1)	At night, 1–2 a.m. (C_2)	
Small polyps, B = 0.015–0.15	*Stylophora*	3.80	15.74	4.1
	Pocillopora	3.23	17.18	5.3
	Acropora squamosa	2.21	15.60	7.1
	Seriatopora	1.77	10.68	6.0
	Montipora	0.42	3.66	8.7
Medium polyps, B = 0.6–1.5	*Hydnophora*	0.47	4.33	9.2
	Leptastrea	0.21	3.06	14.5
	Tubipora	0.88	9.37	10.6
	Galaxea	0.71	8.75	12.3
	Merulina	0.39	4.91	12.6
Large polyps, B = 3–40	*Goniastrea*	0.29	4.08	14.1
	Echinopora	0.31	4.26	13.7
	Favites	0.27	3.80	14.1
	Symphilla	0.23	3.51	15.3

concentration (Sorokin 1984b). These conditions were not met by most investigators. Therefore, their conclusions were often controversial. For example, Abe (1938) demonstrated that the concentration of zooplankton over the reefs off Palau Island was sufficient to provide corals with all their food needs at the expense of only predatory feeding. But other researchers concluded on the same basis that the reef zooplankton was too scarce to be a significant food source for corals (Tranter and George 1972; Johannes et al. 1970; Muscatine and Porter 1977). As we now know, the last conclusion has been invalidated because of inadequacy of the zooplankton estimates (cf. Sect. 3.3.2).

During our study of this problem we tried to adjust the above-mentioned conditions. We estimated the biomass of nocturnal zooplankton, and of bacterioplankton, as well as the stock of labile dissolved organic matter in waters over the Heron Island reefs, as potential food sources for corals. Simultaneously, in the experiments with corals the curves of dependence of their feeding rates (food rations) upon the concentration of the same kinds of potential food have been established. Comparing these curves with real food concentrations in reef waters, it was possible to calculate the most probable food rations, corresponding to the in situ concentrations of food (see below).

The composition of animal food, ingested by corals in situ has been studied by Porter (1974b). He extracted the gut contents of the coral *Montastrea annularis* in situ with the aid of a syringe. The extracted material contained copepods, ostracods, mysids, chaetognaths, appendicularians, nematods, polychaetes, small jelly fish and salps. The dominating components in the gut contents were zoea and copepods. The suspended organic material ingested by corals in situ via sedimentary filtration included bacteria, protozoa, detritus, and dead zooplamkters (Lewis and Price 1975; Lewis 1976). Corals readily ingest also the faeces of fish (Mc Closkey and Chester 1971). As regards algae and other plant material, the scleractinian corals rejected them. Even if swallowed by the polyp, this material was not digested and was soon regurgitated (Yonge 1973; Sorokin 1973a).

The quantitative characteristics of the spectrum of feeding in corals and their food preferences could be arrived at experimentally by estimating the relative rates of their feeding and assimilation of various kinds of possible food in equal concentrations. Under these conditions the rate of assimilation will reflect the food preferences of the coral (Sorokin 1968, 1984a). The rates of feeding and assimilation of food consumed in the experiments have been estimated by employing the radiocarbon method mentioned above. The results of such estimations are given in Fig. 9.3. At the optimal food concentration: $3–5\,g\,m^{-3}$ of zooplankton, $1–3\,g\,m^{-3}$ of bacterioplankton, $0.5–2\,mg\,C\,l^{-1}$ of DOM, most of the scleractinians studied, exept *Leptastrea*, were feeding largely on zooplankton. The share of predatory feeding was curiously the largest in ramose corals with small polyps, which controversed the conclusion of Porter (1974b) about their relative autotrophy. But even in these most efficient predators the share of bacterioplankton + DOM corresponded to about a third part of their food ration. In corals which actively use ciliar–mucus filtering mechanism like *Acropora*, *Pavona*, *Goniopora*,

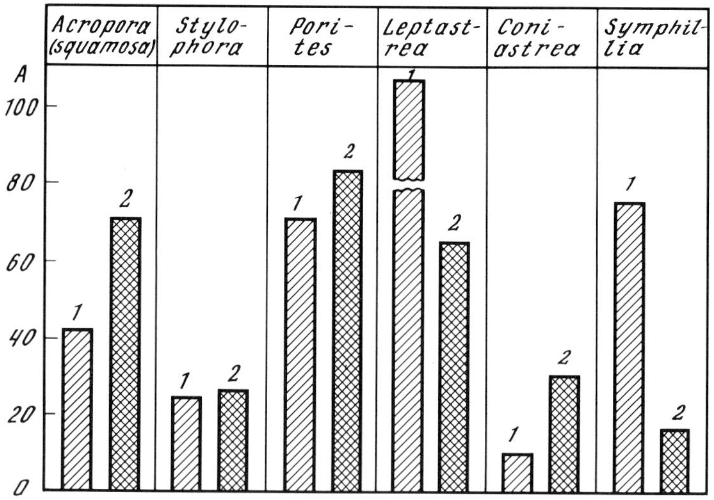

Fig. 9.3. Spectrum of feeding in common corals of the Heron I. reef; *A* relative rate of feeding with bacterioplankton (*1*) and with DOM (*2*) given as percentage of their predatory feeding rate on nauplii of *Artemia*. (Data by the author)

Favites, and *Symphillia*, the share of these food components was up to 50%, and in the coral *Leptastrea* up to 57% (Fig. 9.3, Table 9.14). These numbers reflect the potential ability of common corals of the Indo-Pacific to use various sources of heterotrophic nutrition.

The rate and the efficiency of heterotrophic nutrition in corals could be characterized by the elements of food balance: C = A + M + F, where: C is food ration, A assimilated food, M respiratory losses, F nondigested food. The ratio A:C reflects the assimilability of the ingested food, and the ratio A:M the degree of compensation of respiratory losses by the assimilated foods (Sorokin 1984b; cf. Table 9.12). In some other studies the ratio of the real or probable ration (C) to the metabolic losses (M) was also used to characterize the importance of heterotrophic feeding of corals (Johannes et al. 1970; Johannes and Tepley 1974; Coffroth 1984). But it is actually only a qualitative rather than a quantitative indicator, even if real ration values have been taken. In fact, animals respire only assimilated food, which comprises only a part (40–70%) of the ingested (i.e., of ration C). Porter (1974b) estimated the food ration in the coral *Montastrea annularis* by extraction of gastral cavities of polyps in situ. The from these data calculated A/M ratios showed that at the expense of predation this coral could compensate in situ up to 20–40% of its metabolic expenditures. The same level of this ratio (20%) was obtained by Johannes and Tepley (1974) for the coral *Porites*. They have estimated the food pation by counting the relative number of preys caught by polyps in the colony, periodically taking photographs of this at night. Glynn (1973b) arrived at about the same values of this ratio by estimating the grazing out of zooplankton from waters passing over the reef. Predatory feeding was found to be an important food source in both hermatypic and ahermatypic colonies of the temperate-water coral *Astrangia danae* (Szamant-Froelich 1981).

In our quantitative study of feeding of common corals from the GBR we have obtained the above-mentioned parameters (C, A) by the radiocarbon

Table 9.12. Elements of food balance, obtained in experiments with corals fed for 40 min with the ^{14}C-labeled nauplii 0.8 mm long. Designations: C_r – reverse specific radioactivity of food-labeled nauplii, 10^{-3} μg C cpm^{-1}; r_a – radioactivity of assimilated food (corrected for respiratory loss of $^{14}CO_2$ during the experiment), 10^3 cpm g^{-1}; r_f – radioactivity, nonassimilated part of consumed food, 10^3 cpm g^{-1}; C, A, F – consumed, assimilated and nomassimilated part of consumed food, respectively, μg C g^{-1} per 40 min; I – assimilability coefficients [(A · 100)/c]%. (Data by Sorokin 1984b)

Corals	C_r	r_a	r_f	C	A	F	I,%
Pocillopora damicornis	75	161.1	69.1	17.3	12.1	5.2	70
Stylophora pistillata	84	212.3	66.8	23.4	17.8	5.6	76
Pavona cactus	75	47.1	10.2	4.3	3.5	0.8	82
Galaxea fascicularis	84	45.1	9.2	4.6	3.8	0.8	83
Lobophyllia sp.	84	44.8	11.6	4.7	3.8	1.0	79
Fungia scularia	75	96.9	20.2	8.8	7.3	1.5	83

Table 9.13. Elements of food balance in corals ($\mu g\,C\,g^{-1}\,day^{-1}$) by their predatory feeding on nauplii of *Artemia*. C – food ration; A – assimilated food; I – assimilability; M_t – respiration; B_p – biomass of polyps, $\mu g\,C$. (After Sorokin 1984a)

Corals	Elements of food balance $\mu g\,C\,g^{-1}\,day^{-1}$			A/B_p, %	A/M, %
	C	A	I, %		
Acropora pulchra	299.3	164.2	72	2.1	106
Pocillopora	241.6	169.1	70	5.7	105
Seriatopora	155.9	107.1	69	1.7	61
Stylophora	327.8	249.6	76	8.6	156
Porites annae	55.7	40.5	73	1.85	49
Porites lobata	10.8	7.7	71	2.2	8
Montipora	68.6	47.4	69	1.1	19
Cyphastrea	35.2	25.3	72	0.4	21
Leptastrea	41.9	31.2	74	1.0	35
Pavona	55.8	45.9	82	0.9	44
Hydnophora	63.2	51.6	82	0.9	100
Galaxea	59.3	49.3	83	0.9	54
Merulina	53.8	45.8	85	1.0	34
Coscinarea	44.8	35.9	80	1.4	29
Goniastrea	69.4	58.8	85	2.0	63
Favites	26.6	22.9	86	0.6	44
Symphillia	59.1	47.0	79	0.5	38
Fungia scutaria	114.0	90.8	80	0.8	83
Tubastrea	81.5	58.1	71	0.3	92

method (Sorokin 1968), and the metabolic rates by the ordinary oxygen method in short term experiments conducted in closed dark containers. In experiments on predatory feeding nauplii of 0.7–0.9 mm size were used as labeled food. But, as it is shown in (Fig. 9.4), corals catch also larger nauplii up to 1.4 mm long. We estimated the food rations in experiments of 40-min duration at night when corals are active (Tables 9.12, 9.13). During these short-term experiments corals still did not start to loose the ingested labeled food. Because of significant variations of results obtained in experiments with different colonies of the same species, six to ten parallel estimations with each species of corals were carried out. The results of measurements of food rations in short-term experiments were used to calculate the diurnal values of amounts of assimilated food (A). To make these calculations, the values of assimilability of preys (A/C, Tables 9.12, 9.13) as well as the ratios of diurnal periodicity of the predatory feeding of corals (Table 9.11) were used. To characterize the relative importance of each source of feeding in the coral's energy expenditure, the A:M ratios, calculated per 24-h cycle, have been applied.

The results of these estimations, done with common corals of the GBR are displayed in Tables 9.13, 9.14. They show that most of the scleractinian corals studied are active predators. At an optimum food concentration,

Table 9.14. Generalized results of experimental estimations of rates of heterotrophic feeding in common corals from the Heron I. reefs with the use of the radiocarbon method. Concentrations of food in experimental aquaria: DOM – 0.3 mg C l^{-1}; bacterioplankton – 0.2 mg C l^{-1}; nauplii – 0.35 mg C l^{-1}; A – assimilated food, and M_t – respiration, both as μg C g^{-1} day^{-1}. (After Sorkoin 1984b)

Corals	Predatory feeding		Feeding on bacterioplankton		Feeding on DOM	
	A	A/M_t, %	A	A/M_t, %	A	A/M_t, %
Acropora pulchra	161.3	105	33.8	22	44.7	52
Acropora squamosa	87.5	83	36.5	35	62.5	60
Acropora palifera	38.7	35	10.8	10	33.5	30
Pocillopora damicornis	176.4	110	31.8	20	44.7	28
Stylophora pistillata	170.3	106	39.7	25	45.2	28
Seriatopora hystrix	163.6	93	37.9	22	50.9	29
Montipora erythraca	52.4	21	28.9	12	52.6	21
Porites annae	37.8	46	23.1	27	32.9	40
Psammocora contigua	32.2	29	7.5	7	23.3	21
Cyphastrea seralia	36.6	30	12.8	11	10.0	8
Leptastrea transversa	43.0	46	74.9	84	28.0	31
Achrelia horrescens	48.6	54	19.2	15	46.5	52
Merulina ampliata	32.6	24	14.0	11	23.1	17
Goniopora tenuidens	33.1	30	11.8	13	21.1	23
Galaxea fascicularis	54.6	59	17.4	19	42.9	47
Goniastrea pectinata	51.0	55	4.6	5	16.2	17
Echinopora lamellosa	56.6	56	7.0	7	24.1	25
Turbinaria danae	53.9	64	11.3	13	17.6	21
Favites abdita	37.3	72	26.0	50	16.0	31
Lobophyllia sp.	38.3	31	9.2	15	28.0	22
Fungia scutaria	70.5	64	21.5	19	26.0	24
Tubastrea sp.	176.3	280	84.3	133	44.4	70

which actually is not very far from that in waters over the reef at night, many of the corals studied may compensate over 100% of their energy expenditures. The largest values of food ration, as well as of assimilation rates and A/M_t ratios, were recorded in ramose corals with small polyps: *Acropora*, *Pocillopora*, and *Stylophora*. Their food rations per 40 min ranged within 16 to 25 μg C g^{-1}, and the assimilation values (A) within 160–250 μg C g^{-1} day^{-1}. Calculated per unit of biomass of polyps, the assimilation of preys per day ran to 2 to 8% (Table 9.14). The A/M_t ratio often exceeded in them 100%. In the common coral *Stylophora* it reached 156%. It means that by means of heterotrophic feeding on zooplankton only in concentrations of 1–3 g m^{-3} these corals could acquire the necessary energy. Less active predators among corals with small polyps appeared to be *Montipora* and *Porites*. A weak activity of *Porites* as predator has been menthoned before (Johannes and Tepley 1974).

Food rations per 40 min in corals with medium-sized polyps ranged from 3 to 6 μg C g^{-1}. By means of predation at optimal food concentration they

Table 9.15. Rate of coral's feeding on labeled ciliates. Initial concentration of ciliates in aquaria – $0.4\,mg\,l^{-1}$ (wet weight); C_r – value 0.26×10^{-3} $\mu g\,C\,cpm^{-1}$; M_t – respiration, $\mu g\,g^{-1}\,h^{-1}$. Duration of experiment 2 h; for designations see Table 9.12. (Data by Sorokin 1984b)

Coral	r_a $10^3\,cpm\,g^{-1}$	A, $\mu g\,C\,g^{-1}\,h^{-1}$	A/M_t, %
Acropora squamosa	4.35	0.65	9
Pocillopora	15.90	2.38	22
Seriatopora	7.12	1.06	8
Porites annae	23.60	3.52	31
Porites lobata	7.40	0.70	10
Psammocora	4.20	0.63	9
Tubipora	30.15	4.50	26
Leptastrea	4.96	0.74	11
Goniopora	3.22	0.48	7
Tubastrea	13.87	2.07	55

were able to compensate 25–50% of their energy expenditures. The most active predators among the massive corals studied appeared to be corals that have long tentacles, such as *Goniastrea*, *Turbinaria*, *Echinopora*, *Favites*, and *Fungia*. Their food rations per 40 min varied wthin 3 to $7\,\mu g\,C\,g^{-1}$, and the A/M_t ratios within 55–72%. But the most active predator among them was the ahermatypic coral *Tubastrea* sp. The ratio of A/M_t in this coral attained 200–280%, with the values of 40-min food rations being 13–$18\,\mu g\,C\,g^{-1}$.

The plankton of coral reefs may contain a detectable amount of ciliates (cf. Sect. 3.3.3). Therefore, the experiments were also performed so as to evaluate their probable role as food for corals. The results showed that most corals cannot feed on ciliates so efficiently as they feed on mesozooplankton, excepting the ahermatypic coral *Turbinaria* (Table 9.15). Nevertheless, even for some hermatypic corals like Porites, Tubipora, and Pocillopora there could be an additional source of feeding.

As mentioned above, to evaluate the real importance of predation in the coral's nutrition at ambient concentrations of preys, the dependence of the rate of their feeding upon the concentration of preys has to be measured. The corresponding curves are given in Fig. 9.4. The saturation points of these curves were attained by the wet biomass of zooplankton at about $1.5\,mg\,l^{-1}$. This concentration was within ranges established in reef waters: $0.5–4\,mg\,l^{-1}$ (cf. Sect. 3.3.2). In Fig. 9.5 the generalized curve is given, in which the ordinate corresponds to the ratio of assimilated food (A) at a given zooplankton concentration (biomass) to total energy expenditure of the coral (MG). The latter value includes respiration (60%) and the expenditures for mucus production, growth and multiplication (40%). In our experiments, with a food concentration of about $4\,mg\,l^{-1}$ of wet biomass, an average ratio A/MG for 23 species of corals was evaluated as to be close to 36% (at an average ratio A/M_t 60% MG = $0.6 \times 0.6 = 0.36$). This point

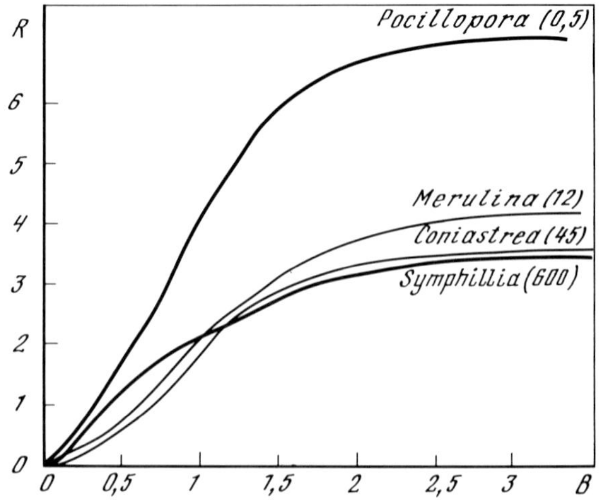

Fig. 9.4. Dependence rates of predatory feeding in corals with different polyp sizes upon the concentration of preys, nauplii of *Artemia*; *B* biomass of food, $mg\,l^{-1}$, wet weight; *numbers in brackets* wet weight of polyps, mg. (After Sorokin 1984a)

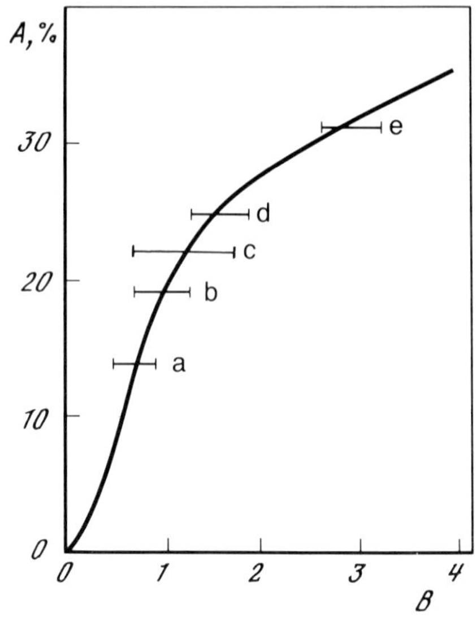

Fig. 9.5. Generalized curve of the dependence of the predatory feeding rate in corals upon food concentration; *B* wet nocturnal biomass of food, $g\,m^{-3}$; *A* rate of feeding (assimilation), % of total energy demand; *cross bars* show the ranges of ambient nocturnal zooplankton biomass in waters of various reef regions: *a* Bermuda; *b*, *c* Caribbean; *d* Philippines; *e* Heron I., GBR. (After Sorokin 1984a)

(36%) was taken as the upper point of the A/MG ratio on the ordinate. The abscissa of this curve represents the scale of zooplankton biomass from 0 to $4\,mg\,l^{-1}$. On this curve the points marked correspond to the ambient zooplankton biomass at night in waters over different reefs. Thus the shares of food have been established that were assimilated by predation, in compensation for the general energy demand of corals. Witin a real range of zooplankton biomass of 0.3–2 g m^{-3} on the reef at night, these shares (A/MG ratios), in accordance with this curve (Fig. 9.5), should be 10 to 30%. At the Heron Island reefs, where these experiments had been performed, the zooplankton biomass at night was 0.5–1 g m^{-3}. In accordance with this curve, corals by means of predatory feeding could compensate there 10–20% of their total energy demand.

The sedimentary filter feeding of corals with bacteria was first established by Di Salvo (1971a, 1973). After exposing corals for 10–12 h to the presence of suspended bacteria labeled with the redioisotope ^{35}S, he discovered 30–40% of the tadioactivity inserted into the aquaria in the tissues of polyps. Quantitative estimations of coral's feeding on bacteria labeled with ^{14}C we have done at the reefs of Kaneohe bay (Sorokin 1973c, 1978b) and of the Heron Island reefs (Sorokin 1981c, 1984a). The experiments were conducted with natural labeled bacterioplankton in concentration close to its upper level in reef waters: 1.5–3 × 10^6 cells ml^{-1}. The results of the estimation of feeding rates A and of the A/M ratios are summarized in Tables 6.14, 6.16. The common corals of the Indo-Pacific reefs appeared to be able to feed on the bacterioplankton, thus compensating 10–20% of respiratory losses. In some of the species tested, like *Leptastrea*, *Hydnophora*, and *Favites*, the rate of feeding on bacterioplankton was two to three times higher. The real assimilation of the microbial biomass by coral polyps was established by measuring the evolution of $^{14}CO_2$ by corals previously fed on labeled bakterioplankton (Sorokin 1978b, 1984b). The curve showing the dependence of feeding rates of corals upon the bacterioplankton's concentration in water are given in Fig. 9.6. It increased quasi exponentially up to the level of 0.6–0.8 mg m^{-3} of the microbial biomass in water then the curve declined. The above-mentioned concentration of bacterioplankton approaches its upper limit in reef waters (cf. Sect. 3.1.1). In accordance with the generalized curve constructed, as in the case of predatory feeding (see above), taking the ambient

Table 9.16. Feeding of corals on labeled bacterioplankton. For designations see Tables 9.12, 9.13. (Data by Sorokin 1984b)

Coral	r_a 10^3 cmp g^{-1}	A, μg C g^{-1} h^{-1}	A/M_t, %
Acropora pulchra	5.37	1.17	18
Porites lobata	2.43	0.53	13
Pavona cactus	10.21	2.21	56
Psammocora contigua	1.64	0.36	8
Hydnophora excesa	8.70	1.90	91
Cyphastrea seralia	1.38	0.30	6
Merulina ampliata	2.71	0.57	10
Leptastrea transversa	14.02	3.05	82
Favites abdita	5.90	1.28	59
Fungia scutaria	5.79	0.67	15

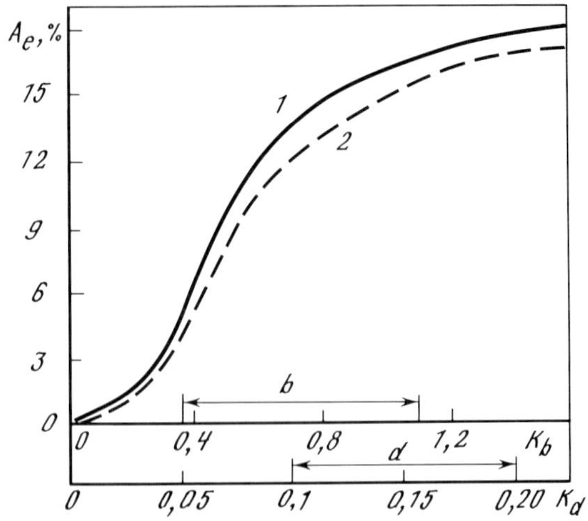

Fig. 9.6. Dependence of the rates of feeding in corals on DOM (*1*) and on bacterioplankton (*2*); *A* feeding rate, % of total energy demand; K_b and K_d concentration of food – bacterioplankton (mg l^{-1}) and *DOM* (mg C l^{-1}); *b* and *d* ranges of their ambient concentrations in waters over the reef

bacterioplankton concentrations in waters over the reef, corals should, on an average, compensate by means of feeding on bacterioplankton about 10–15% of their total energy demand (K_b) (Fig. 9.6).

The first estimations of coral's feeding on DOM, employing labeled organic compounds (Stephens 1962; Goreau et al. 1971) were rather qualitative. They did not answer the question about the possibility of corals acquiring significant amounts of food at ambient concentrations of assimilable fractions of labeled dissolved organic matter in coral-reef waters. The results of our quantitative estimations of corals feeding on DOM based on the measurements of its specific radioactivity in the experiments, are given in Table 9.17. They proved that the rate of feeding of most reef corals tested with DOM is commensurable with the rates of their feeding on bacterioplankton. At a concentration of assimilable DOM of 1 mg C l^{-1}, the rate of its assimilation by corals was 20–60 μg C g^{-1} day^{-1}, or 20–40% of respiration losses. The real content of labile DOM in reef waters was evaluated to be within 0.2–0.5 mg C. At these cocentrations, in accordance with the curve (Fig. 9.6), corals should compensate 12–18% of their total energy demand (A/MG). The assimilation of dissolved amino acids by corals was recently demonstrated by Titlyanov et al. (1988b); and by Drew-Ferrier (1991).

Summarizing the data on the quantitative aspects of feeding of scleractinian corals obtained at the Heron Island reefs, it could be concluded that at the ambient concentration of available food they, on an average, could compose their energy balance as follows: 70% of the needed energy is

Table 9.17. Assimilation by corals of DOM, labeled with ^{14}C. For designations see Tables 9.12, 9.14. (Data by Sorokin 1984b)

Coral	r_a	A	A/M_t, %
Acropora pulchra	68.4	1.64	26
Porites lobata	56.4	1.35	33
Montipora	72.3	1.73	17
Psammocora	36.0	0.86	18
Merulina	35.3	0.85	15
Pavona	37.2	0.89	22
Hydnophora	33.1	0.79	38
Fungia	41.5	1.00	21
Tubastrea	79.9	1.91	73

supplied to them by the photosynthesis of their zooxanthellae, 20% by predation, and 20% by feeding on DOM and on bacterioplankton.

9.2.2 Feeding of Zoantharia, Antipatharia, Ceriantharia and Octocorals

Data on the feeding of these groups of Anthozoa are still scarce. The zoantharians by the morphology of their polyps as well as by their symbiosis with zooxanthellae should be similar in this aspect to the scleractinians. In fact, a larger part of the necessary nutrition they get from their algal symbionts, using the other available sources of heterotrophic feeding: zooplankton, DOM, and bacterioplankton (Reimer 1970, 1971, 1972; Trench 1974; Sebens 1977). It seems that their different species may use the autotrophic source of nutrition to a different degree. Such common zoantharians as *Zoanthus sociatus* are primarily autotrophs, while the *Palythoa* are powerful predators, which may acquire a large portion of necessary energy and nutrients in this way. In accordance with our experiments (Table 9.18) they may compensate, less than scleractinians, their energy expenditures by photosynthesis: the P_t/M_t ratios in them was 0.9–1, compared with 1.2–1.5 in scleractinians. The species of *Palythoa* we have studied was established to be capable of predation, but assimilated DOM and bacterioplankton much slower. They also appeared to be able to ingest and assimilate plant material – the marine Chlorella. In this too zoantharians differ from scleractinian corals, which reject algae as a foodstuff.

The antipatharians and ceriantharians do not have algal symbionts. In the mode of their nutrition they are somewhat similar to the ahermatypic scleractinians. The main source of their nutrition is predation. The polyps of the black coral *Antiphates grandii* from the Hawaiian reefs was seen in

Table 9.18. Rates of nutrition of two zoantharian species and one scleractinian coral (for comparison) on various sources of food. For designations see Table 9.13; P_t – gross photosynthesis; P_t, A and M_t values are given as $\mu g\,C\,g^{-1}\,h^{-1}$. (Data by Sorokin 1984b)

Taxa of zoantharians and coral	M_t	Photosynthesis		Heterotrophic nutrition							
		P_t	P_t/M_t	on nauplii		on algae (marine Chlorella)		on bacterio-plankton		on DOM	
				A	A/M_t	A	A/M_t	A	A/M_t	A	A/M_t
Zoathus sociatus	5750	5330	0.9	3750	65	372	6.4	370	6.4	176	3.0
Palythoa tuberculosa	200	204	1.0	176	86	14.4	7.2	36	19.4	23	11.5
Pocillopora damicornis	138	190	1.4	155	113	1.3	1.0	35	25.0	22	16.0

experiments catching and swallowing various zooplankters, such as copepods, amphipods, and chaetognaths (Grigg 1965). The same was true for the Caribbean black corals *Cirripathes lutkeni*, *Antipathes pennacea* and *A. peduta*. They catch the food particles with the aid of mucus nets. The prey thus caught is dragged to the area around the oral disc. From there it is driven to the mouth by cilia and swallowed. Large preys which the polyp caught, but cannot swallow, may be digested outside the gastral cavity with the aid of the polyp's powerful mesenterial filaments (Lewis 1978). As it is able to use efficiently the mucus nets and the ciliar apparatus the polyp could be supposed to make use also of bacterioplankton and DOM as additional sources of heterotrophic nutrition for these corals, as do scleractinian corals.

The polyps of ceriantharians are morphologicall very similar to those in Antipatharia. Thus they too could be supposed to have the same spectrum of feeding. They have active proteases in their mesentery, which enable them to digest large preys (Tiffon and Boullon 1975).

As regards the nutrition of the octocorals, this extremely important group of Anthozoa could be characterized by a high variaty of types of feeding in their taxa. As octocorals are very vulnerable to mechanical damage and to any kind of transplantation, they are difficult objects to make experiments with (Lewis 1982). An important general feature of octocorals is an apparent reduction of most organs and organelles in them that serve the function of heterotrophic feeding in scleractinian corals and in other Hexacorallia (Kawaguti 1964a,b; Mariscal and Bigger 1977). The polyps of Octocorallia have reduced cnidocils and reduced mesenterial filaments. The cnidocil, which is so well developed in Sceractinia and is represented in them by a wide selection of various large and small nematocysts, in octocorals is represented only by tiny rear nematocysts of the same kind, which seem to be useless for hunting the mesozooplankters. Many octocorals have only rudimentary nematocysts which, moreover, are difficult to locate (Pratt 1906; Gohar 1940; Mariscal and Bigger 1977). The ciliar epithelium covers like a fur the polyps of Scleractinia and form conical structures, spirocysts, around their large nematocysts, which accomplish a key function in their feeding. In octocorals the ciliar epithelium is drastically reduced, remaining only at the oral disk of polyps and on the surface of their gullets. The conical structures around their rear nematocysts are absent sutirely. And finally, the edges of tentacles in the polyps of octocorals are fringed with pinnules, which on the tentacles of scleractinia are absent (Lewis 1982).

The above specific features of morphology in the nutritional organs of octocorals nutrition appeared already at an early stage of their evolution (Schmidt 1972). This lets us suppose that they had their own strategy of feeding and capturing preys different from that in scleractinians. Many of them have symbiotic zooxanthellae. These symbiotic (hermatypic) octocorals are just the dominating ones on the reefs. Among them are the symbiotic genera of the families Alcyonacea (*Lobophytum*, *Sarcophyton*, *Alcyonium*, *Lithophyton*, *Sinilaria*, *Xenia*), of Gorgonacea (*Hicksoniella*, *Plexaura*, *Pseudopterogorgia*, *Isis*, *Rumphella*), and of Stolonifera (*Tubipora*). The first attempts to investigate the food ingested by polyps of asymbiotic temperate-water species of *Alcyonium* – *A. digitatum* – by microscopic analysis of their gut contents shewed that it fed on small crustaceans, but much less efficiently when compared with scleractinian corals. Most of their polyps did not contain any traces of food rests (Pratt 1906). Subsequent studies have proved the opinion that the octocorals are very poor predators (Gohar 1940, 1948; Kinzie 1970). In *Xenia* for example, Gohar did not find any signs of predatory feeding at all. The guts of *Xenia* were permanently empty. This alcyonacean did not even display any "food reaction" in the presence of food, which is a most important feature of feeding behavior in Scleractinia (see above). As for the nutrition of another symbiotic octocoral, *Clavularia*, its polyps extruded periodically an excess of zooxanthellae into their gastral cavity, where these were digested and assimilated.

The conclusion by Gohar (1948) was that the symbiotic alcyonaceans did not need any predation or other sources of heterotrophic food, being supplied with the necessary nutrition by their algal symbionts. This conclusion was then supported by Goreau (1961) and Goreau et al. (1971). But there still remained the reasonable question: why have these alcyonaceans refused during evolution to use the sources of heterotrophic feeding, with the aid of which other corals got not only energy, but, what is maybe even more important, nutrients. Further research should supply the answer. But even recent studies conducted in the 1970s, produced again only somewhat indefinite results. The analyses of gut contents in polyps of *Alcyonium diditatum* confirmed early data about their ability to catch small slowly moving zooplankters, like young Oicopleura, larvaceans of ascidians and foraminifera (Sebens and Koehl 1984). In the guts of gorgonacean's polyps the veligers of Mollusca have been found (Grigg 1972). According to Kinzie (1970), most of the Atlantic gorgonians investigated did not readily feed on zooplankton, while in polyps of *Eunicella clavigera* he found swallowed planktonic crustaceans. The gorgonian *Leptogorgia* was observed catching the nauplii of artemia (Leversee 1976) and the *Pseudoplexaura* the pieces of rat liver (Murdock 1978). Nevertheless, most authors were inclined to accept the view that octocorals in general are weak predators and may ingest preferably passively floating particles of detritus of pseudoplankton, as well as foraminifera, eggs or slowly moving larvae of invertebrates. In experiments, the polyps of gorgonians caught the sephadex beads soaked with broth, or the cysts of artemia, but hardly caught the artemia nauplii. Even those which had been caught and even swallowed, most often were then again released by the polyps (Mariscal and Bigger 1977; Lasker 1981; Lasker et al. 1983). According to Roushdy and Hansen (1961), some alcyonaceans were able to feed on the phytoplankton. The alcyonacean

Heteroxenia was proved to digest the zooxanthellae extruded by polyps into their gastral cavity (Schlichter 1982). This author demonstrated also the ability of this alcyonacean to feed on DOM. Coffroth (1984) proved an important role of the mucus excreted by invertebrates (largely by corals), as food for some octocorals. In her experiments the gorgonian *Plexaura porosa* could compensate energy expenditurs by feeding only on mucus of scleractinian corals.

In his study on nutrition of a variety of octocorals Lewis (1982) offered them various kinds of zooplankton, and also investigated the gut contents of polyps in specimens taken at the reef and immediately inspected. He established that among the 24 species thus checked some did not catch any animal preys under experimental conditions. Among them were the symbiotic species *Sinularia capillosa*, *Eunicea tourneforti*, *Paralemnalia digitiformis*, *Plexaura flexuosa*, and *Gorgonia ventalina*. The same was valid for the xeniids. Nevertheless, when analyzing the guts in xeniid specimens caught in nature he found in some of them planktonic crustaceans, small polychaetes, veligers, and eggs of invertebrates. Probably, to catch food the polyps of xeniids need to be kept permanently in motion by waves on the shallows where they are living. Other octocorals studied, both their symbiotic species and genera (*Lobophytum*, *Sarcophyton*, *Sinularia densa*, *Capnella*, *Lemnalia*, *Isis*, *Juncella*, *Pseudopterogorgia*, *Plexaura*, *Telesto*, *Tubipora*), and the asymbiotic ones (*Dendronephthea*, *Rumphella*, *Briareum*, *Muriceopsis*, *Gorgonia*) were observed catching and swallowing the planktonic crustaceams and veligers.

With a view to the variability of feeding strategy in different taxa of octocorals, we have investigated the spectrum of feeding in a selection of coral species, which inhabit reefs in the Indian Ocean and the GBR, employing the radiocarbon method. To the specimens of corals acclimated to experimental conditions various kinds of labeled food were offered excluding so the influence of their physiological state on the results of the experiments. The typical results are given in Fig. 9.7. In general, they proved the basic conclusion on the ability of most octocorals for predatory feeding. Second in importance among the sources of their feeding were the bacterioplankton and DOM. The most active predators among the octocorals tested were their asymbiotic species *Dendronephthea*, *Spongotes*, *Bebryce*, *Acabaria*. The gorgonian *Mopsella* was found to be feeding also on phytoplankton.

Being more ancient and primitive in their morphology, compared with scleractinians, the octocorals use also more primitive and quite different mechanisms of catching their preys or food particles. They employ mostly direct mechanical seizure of the prey with their motile tentacles, armed by pinnules at the edges. The pinnules enlarge the catching surface of the whole bunch of tentacles. The reaction of the tentacles is very swift, being triggered by tactile contact of the prey with the tentacle or pinnules. Thus the polyps grab not only moving preys but also immobile passively floating

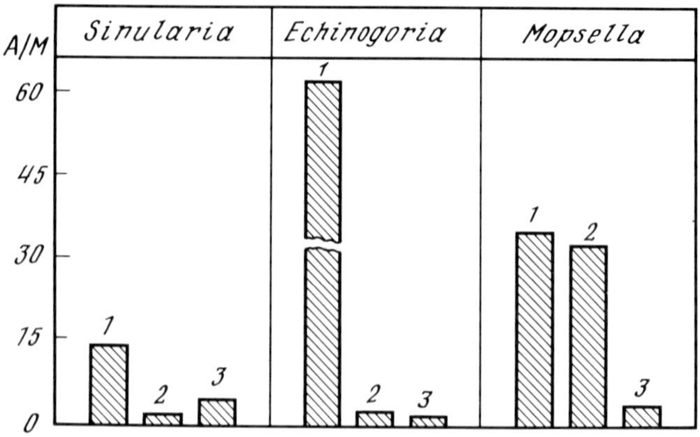

Fig. 9.7. Spectrum of feeding in some octocorals food; *1* – nauplii of *Artemia*; *2* planktonic algae; *3* bacterioplankton *A/M* ratio of assimilation to respiration, %. (Data by the author)

food objects, like cysts, eggs, detritus particles, faeces of fish, mucus flakes, etc. The food object caught is immediately transferred by the tentacle to the mouth and swallowed by the polyp (Lewis 1982). Octocorals with small polyps, like *Acabaria*, *Mopsella*, and *Leptogorgia* catch their preys with the whole bunch of tentacles, as if with a fist, and push it into the gullet. Octocorals with larger polyps and longer tentacles most often seize the prey by bending swiftly one of the tentacles, which together with the pinnules form a kind of trap, grabbing the food particle. Then the prey is pushed by the tentacle into the mouth of the polyp. This method is used by the octocorals *Lobophytum*, *Sarcophyton*, *Dendronephthea*, *Plexaura*, and *Tubipora*. In some of the asymbiotic gorgonaceans and in the octocorals *Corallum rubrum* the pinnules look like fine threads, which are wound around the preys, zooplankters or particles passively drifting around, catching them in this way (Abel 1970). The reduced cnidocil could not participate in the hunting of large preys, but probably could be used by the polyp for catching microzooplankton. The small planktonic organisms, like bacteria or microalgae, probably could be caught by the polyps of octocorals with the aid of microvilli, which cover the surfaces of their tentacles and pinnules (Mariscal and Bigger 1977).

Specific for the octocorals is also the instant reaction of a group of polyps together, when one of the polyps in this group has been triggered by contact with the potential prey. Such reaction in the polyps of scleractinian corals is absent. The adaptive reason for this reaction seems to be obvious. The contact of the polyp with the food signals its appearance in the surroundig space nearby and, moreover, that the plankton in reef environment has the tendency to form swarms and aggregations. Therefore the simultaneous reaction of the other polyps increases the probability of the neighbouring polyps catching also food accidentally.

The participation of mucus in the processes of feeding in octocorals remains disputable (Lewis 1982). But actually *Xenia* and some other symbiotic alcyonaceans are definitely able to excrete mucus. We have observed the following way of hunting the preys, early nauplii of artemia, by polyps of *Sarcophyton*. The prey, which is swimming in the vicinity of colony, when caught in the mucus net stretched out by the polyp, starts to move chaotically winding so around its body the mucus stripe, and thus inevitably approaching the place where this stripe originates, i.e., the mouth of the polyp. The process of digestion of food in octocorals is similar to that in other *Anthozoa*, but lasts two to three times longer, probably because of reduced mesenterial filaments in them. The artemia nauplii swallowed were digested by polyps of *Sarcophyton* only in 10–12 h (Lewis 1982).

Quantitative evaluations of feeding rats in octocorals have been recently missing. Semiquantitative estimations of the gut contents in polyps of some octocoral species were made by Lewis (1982). The amounts of artemia cysts or sephadex particles soaked in broth swallowed by them were also estimated (Lasker 1981; Lasker et al. 1983), and the decrease in the number of artemia nauplii in aquaria in the presence of gorgonian colonies was observed as well (Leversee 1976). These data have the slackness of their activity as predators. So they hardly could acquire enough energy this way. But at least the asymbiotic octocorals have to get all the necessary nutrition by heterotrophy only. This situation suggested the proposition that just sedimentary filtration of microplankton and other suspended matter, but not predation might be the main mechanism of nutrition in octocorals (Mariscal and Bigger 1977; Lasker 1981). Therefore, when we have started our study of the quantitative aspects of octocoral nutrition we also supposed the latter assumption as being the most probable. In our study we have used as labeled animal food the nauplii of artemia, the veligers of *Aplysia*, the rotifers *Brachionus* and the ciliates *Strombidium*. As particulated food, suspensions of labeled algae *Isochrysis* and marine *Chlorella* were used, as well as labeled bacterioplankton. The experiments were performed with the most common octocorals of the Indo-Pacific. For comparison of feeding in octocorals and scleractinians in the same experiments, the hermatypic scleractinian coral *Stylophora pistillata* and the ahermatypic species *Tubastrea* were also used. As DOM, labeled protein hydrolizate was employed.

The results showed that different species of symbiotic octocorals may ingest practically all kinds of food offered them but to different degrees (Table 9.19; Fig. 9.7). When feeding on zooplankton, the symbiotic octocorals might acquire up to 30% of their total energy demand (A/MG). The assumption on their ability to feed on the microzooplankton was not proved (Table 9.20). The rates of their feeding on veligers, rotifers and ciliates were rather low: A/MG ratios were in this case only 1–5%. Among

Table 9.19. Rates of heterotrophic feeding of octocorals (A) and the shares of various sources of food in compensation for their total energy demand (A/MG). (Data by Sorokin 1990a)

Taxa	Species of corals	Kinds of food								Total A/MG
		Zooplankton		Planktonic algae		Bacterioplankton		DOM		
		A	A/MG	A	A/MG	A	A/MG	A	A/MG	
Family	*Nephthea* sp.	34[b]	3	8	7	17	2	43	4	6
Alcyonaceae	*Spongotes* sp.	3120[a]	144	26	1	145	7	70	3	155
Family	*Echinogorgia gigantea*	1540[a]	87	28	2	42	2	24	1	92
Gorgonaceae	*Plexauroides lenzii*	757[a]	52	50	4	70	4	17	2	62
	Bebryce indica	1320[a]	100	23	2	117	9	50	4	115
	Acabaria nicksoni	1280[a]	115	36	3	82	7	27	2	127
	Mopsella aurantia	45[a]	6	220	34	110	16	26	4	60
Order Scleractinia	*Tubastrea*	176[a]	280	–	–	8	13	44	70	

[a] Food nauplii of *Artemia*. [b] Food rotifers *Brachionus*.

Table 9.20. Feeding of octocorals on microzooplankton. For designations see Table 9.13. (Data by Sorokin 1990a)

Corals	Food ciliates *Strombidium*		Food veligers of *Aplysia*	
	A	A/M, %	A	A/M, %
Alcyonium	7.4	0.7	4.1	0.4
Cladiella	16.4	3.6	0.3	0.1
Lobophytum	2.1	0.4	26.7	5.1
Sarcophyton	2.7	0.5	19.6	3.4
Sinularia	8.7	3.6	8.8	3.6
Lemnalia	2.7	0.6	27.4	5.7
Paralemnalia	5.1	0.9	20.4	3.7
Xenia	12.2	1.7	2.4	0.3
Isis	6.6	1.7	2.2	0.6
Rumphella	14.4	3.6	35.0	8.9
Acabaria	62.2	9.1	98.6	14.5
Tubipora	2.5	1.3	19.6	10.3
Pocillopora (sclaractinian coral)	42.0	27.0	59.7	38.5

other octocorals the best feeders on the microzooplankton were *Rumphella* and *Tubipora*. Bacteria were secondary in importance as food for octocorals: A/MG ratios were 3–10% in various species. The most active feeder on bacterioplankton appeared to be *Tubipora*, which might compensate in this way up to 30% of respiratory losses. The algae were consumed and assimilated in minimal quantities, but DOM seems to be a significant source of their feeding: A/MG 2 to 14%. In total, the joint heterotrophic feeding on all four of these probable sources of food tested may provide, according to our experiments, something between 10 and 40% of total energy demand in symbiotic octocorals. The rest they may acquire from their algal symbionts (cf. Sect. 9.1). The asymbiotic octocorals appeared to be much more active predators. Most of them might compensate their energy expenditures (Table 9.18). The most active predators recorded were *Spongotes*, *Bebryce* and *Acabaria*. The gorgonacean *Mopsella*, which had very small polyps, displayed a unique ability to feed on planktonic algae and bacteria, thus acquiring up to half of the necessary energy. The asymbiotic gorgonians may also actively ingest microzooplankton (Table 9.19).

The above data point to a fine trophic specialization in various taxa of asymbiotic octocorals, which is not the case with scleractinians. Some of their species with small polyps are apparently oriented toward feeding on plant and animal microplankton and suspended detritus. But most of them are preferably predatory, getting from other sourses, including DOM, only some 3–8% of the necessary energy. Comparison of asymbiotic octocorals with scleractinians (*Tubastrea*) shows that the latter use much more efficiently all available sources of heterotrophic nutrition. This conclusion applies also to symbiotic octocorals, whose abilities for all the tested ways of heterotrophic feeding were much poorer compared with those in hermatypic scleractinians. Under similar experimental conditions the latter much more efficiently fed on either zooplankton, bacteria, of DOM. This corresponds to a more developed morphology in scleractinian polyps, compared with that in octocorals. The above data enable us to suppose that the high ability of octocorals to compete with scleractinian corals in reef benthic communities could be based mostly on their chemical defence capacity and on their high growth rate, rather than on their trophic activity or on their trophic opportunism in relation to sources of heterotrophic feeding.

9.3 Energy Balance and Specific Growth Rate of Corals

The shares of autotrophic and heterotrophic nutrition of corals in their energy balance is one of the main questions of their ecophysiology (Muscatine 1980). Many observations have proved the point that hermatypic corals may completely satisfy their energy and growth needs by means of the autotrophic production of their zooxanthellae. For example, kept in the light they survived for months in water deprived of plankton by filtration (Franziskett 1970). Corals may live inside large sponges in water completely cleaned of any suspended matter by the filtering activity of their hosts (Porter 1974b), Corals isolated from light by being covered in situ with black plastic film died or bleached (Rogers 1979b). Their photosynthetic production in diurnal balance exceeds respiration ($P_t/M_t > 1$). And nevertheless, the above facts in the coral's life could not be accepted as a proof of purely autotrophic nutrition of hermatypic corals. On the one hand, their polyps cannot use all the autotrophic production of zooxanthellae, but only its translocated part. On the other hand, respiration is only a part of their energy credit, moreover the P_t/M_t ratio in corals drops down below 1 in shadowed or deep biotopes in spite of their light adaptation (Drew 1973). Thus, in these biotopes the share of heterotrophic sources of nutrition in their energy balance should be compellibly significant. Our calculations, based upon the assumption that corals spend on their respiration ~60% of their total energy expenditure, and also on the average shares of various food sources in the coral's nutrition (see above), resulted in the following evaluation of the components in the energy debet in coral communities of the shallow areas of the Heron Island reef: 70% photosynthesis of symbiotic zooxanthellae, 20% predation, and 10% feeding on DOM and on bacterioplankton. According to Lewis (1981), zooxanthellae provide around 65% of the energy debet in hermatypic corals.

Another approach to this problem could be the calculations of CZAR – the share of translocated autotrophic production of zooxanthellae in the respiration of the polyp's animal tissues: CZAR = $[(P_{nz} \cdot T)/M_a]\%$ (Muscatine et al. 1985; see Sects. 8.3, 9.1). The index P_{nz} in the above formula is the net production of zooxanthellae, T part of P_{nz} translocated to the polyp and M_a respiration of the animal tissues of the polyp. The values of P_{nz} were calculated by accepting that the ratio of respiration to biomass in zooxanthellae is the same as in animal tissues of polyps. The value of T was calculated as the ratio $(P_{nz} - G)/P_{nz}$, where G is the production of zooxanthellae used by them for their own growth. The latter was estimated by measurements of their generation time when recording the frequency of occurrence of cells in the state of mitosis in freshly isolated suspension of zooxanthellae (Wilkerson et al. 1983). The last-mentioned authors estimated in this way the energy balance in the corals *Stylophora pistillata* living at depths of 15 and 45 m in different illuminations (Table 9.2; Fig. 9.8). The table shows that in the illuminated zone colonies of this coral may complete their energy debet even from only the autotrophic source, while those living in poor light less than half, with the P_t/M_t-ratio in the latter at midday over 1. Thus, these colonies should get the other part of the necessary energy of over 50% from heterotrophic sources of nutrition. The CZAR values in this coral were estimated with 93–96% (Fig. 9.8).

The same estimations were cavried out also in the corals *Pocillopora eydoxy* (Davies 1984) and *Porites* (Edmunds and Davise 1986). But in this case the values of P_{nz} have been calculated using the directly measured values of zooxanthellae respiration (M_z) in their suspensions in vitro: $P_{nz} = P_t - M_z$. The M_z values, as measured in suspensions of zooxanthellae in vitro, were estimated to be 9–21% of P_t, whereas after the calculations mentioned above only 2–3%. Correspondingly, the obtained CZAR values were lower: 90% for *Pocillopora* and 78% for *Porites* (Table 9.22). The polyps used for their own respiration about 30% of translocated photosynthesis production. They were assumed to be using the rest on

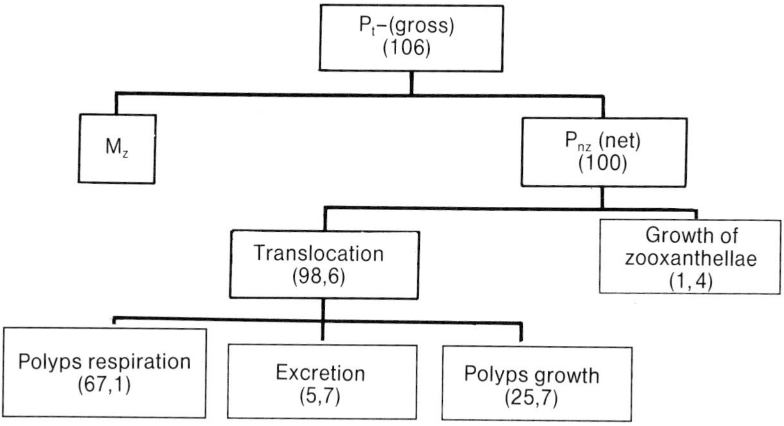

Fig. 9.8. Scheme of expenditure of photosynthetic production by zooxanthellae (P_t, relative values) in the energy metabolism of the coral *Stylophora*: M_z respiration of zooxanthellae, and P_{nz} their net autotrophic production. (After Muscatine 1984)

Table 9.21. Balance of photosynthetic production expenditures ($\mu g\, C\, cm^{-2} day^{-1}$) in colonies of the coral *Stylophora pistillata*, living under different illumination conditions. (Data by Muscatine et al. 1984, 1985; Falkovsky et al. 1984)

Components of energy balance; indices of ecological efficiency and translocation	Colony from 15 m depth	Colony from 45 m depth
Gross photosynthesis (P_t)	140.1	32.2
Respiration of zooxanthellae (M_z)	2.6	1.0
Respiration of animal tissues of polyps (M_a)	94.8	52.5
Net photosynthesis of zooxanthellae ($P_t - M_z$)	137.5	31.2
Growth of zooxanthellae (C_z)	1.2	0.7
Translocated production of zooxanthellae (T_i)	136.3	30.5
Release of organic matter (E_x)	8.2	14.0
Assimilated by polyp and stored part of T_f	33.0	–
Percentage of PARS consumed	53.0	86.0
Quantum yield of photosynthesis, %	1.9	16.8
CZAR, %	96.0	93.0
Ratio: $[(T_i \cdot 100)/(M_a + E_x)]$% (satisfaction of energy demand)	132.0	46.0

Table 9.22. Balance of photosynthetic production expenditures (calories per 300 mg of dry weight of living coral tissues) in colonies of the coral *Porites porites*. (Data by Edmunds and Davies 1986)

Elements of balance	Zooxanthellae		Animal tissues of polyps	
	Calories	% of P_t	Calories	% of P_t
Total credit of energy (gross photosynthesis, P_t)	47.6	100.0	37.2	78.0
Respiration (M_t)	11.1	21.2	12.5	26.3
Growth (G_t)	0.4	0.8	3.0	6.3
Translocation (T_i)	37.2	78.0 (CZAR)	–	–
Multiplication	–	–	0.2	0.4
Release of organic matter	–	–	21.4	45.0

mucus excretion, on reserve substances, and on growth (Davies 1989). The reserve substances could be used by the coral during cloudy days when the P_t/M_t ratio falls below 1 even in corals living in shallow biotopes. The above estimates are surely only an approximation, and have serious gaps. In the energy balance calculated for the coral *Stylophora pistillata*, for example, the respiration of the zooxanthellae has been obviously underestimated (2–3% of P_t). In free-living algae it is 15–20%. The release of organic matter by this coral seems to have been also largely underestimated in short-term experiments with ^{14}C. The release proceeds in this case mostly in form of unlabeled matter (mucus), which was not accounted for. It should be close to 30–40% of P_t (Table 9.21), but not 8%, as it is (cf. Sects. 8.4, 8.5). Therefore, the real percentage of translocation should be accepted with 70–75%, which corresponds to direct measurements of CZAR by Titlyanov et al. (1985). From this point of view the energy balance calculated for the coral Porites and shown in Table 9.22 seems to be more realistic.

The specific growth rate of zooxanthellae, expressed as G_z/B coefficient per day was evaluated with 0.01–0.04, and their generation time with 40–70 days. The specific growth rates of polyps were found to be much less: 0.002–0.004 day^{-1} or 0.7–1.5 $year^{-1}$ (Wilkerson et al. 1983; Muscatine et al. 1984, 1985; Edmunds and Davies 1986). Thus the rate of growth of zooxanthellae exceeds by much that of polyps, and so the excess of zooxanthellae should be building up. But polyps can regulate their number in their tissues by partial digestion or by expelling them (Fitt and Trench 1980; Colley and Trench 1985).

We have evaluated the probable specific growth rates in common corals using the formula: $G/B = (M_t \cdot K_2)/[B_p \cdot (1 - K_2)]$, where B_p is the biomass of polyps and K_2 the efficiency coefficients of the use of assimilated food for growth, which in most sessile animal is close to 0.3. In different coral species the G/B coefficients ranged from 0.006 to 0.02 day^{-1} (Sorokin 1984a). Lewis (1981) has obtained the specific growth rate in the community of hermatypic corals with close to 0.005 day^{-1}, and 1.8–2.0 $year^{-1}$. Such small energy expenditures in corals for their growth are easily provided for by translocation and heterotrophic nutrition. Thus, their growth rate should not be limited by the energy supply, but mostly by the rate of calcification and by the limited growth rate of the lime skeleton of their colonies.

10 Biogeochemical Metabolism and Energy Flows in Reef Ecosystems

The term "metabolism of reef communities" was first used by Odum and Odum (1955) in their prioneer work on the metabolism of communities at the Enivetok atoll, based upon the recording of diurnal fluctuations of oxygen content in waters over the reef. Later, for studies on primary production and calcification processes in reef communities, the measurements of fluctuations of CO_2, pH, PO_4-P in diurnal cycle, or downstream the waters surging up the reef from the ocean, were also employed. This approach could be called biogeochemical, and the metabolism of the whole biota of a given reef zone or biotope, the biogeochemical metabolism. It reflects the balance values of reciprocal flows of consumption and excretion of elements or compounds by different components of communities. The evaluations of rates of biogeochemical metabolism could therefore be also done by measuring them in separate components of communities, taking into account the relative areas they occupy on any given reef biotope or zone (Kinsey 1985b).

In the biogeochemical metabolism of coral-reef communities the aerobic processes predominate, e.g. photosynthesis, respiration, calcification, nutrients consumption, and nitrogen fixation. Their rates are usually expressed in units of carbon, $CaCO_3$, P, N consumed or released per m^2 of area of a given bottom biotope. In soft reef sediments anaerobic processes with the dominance of sulfate reduction and denitrification could also have taken place (cf. Sects. 1.3, 4.1).

10.1 Biogeochemical Metabolism of Organic Matter: Autotrophic Production and Heterotrophic Decomposition

Also, the first measurements of the relationships between photosynthetic oxygen evolution (P_t) and heterotrophic oxygen consumption (M_t) at some Pacific atolls testified that reef ecosystems are autotrophic in their diurnal balance, while functioning in environments poor in nutrients (Odum and Odum 1955). But subsequent studies on this balance in reef system situated near high islands, and thus receiving a large organic input from land, showed that they could be heterotrophic as well: $P_t/M_t < 1$ (Gordon and Kelley 1962; Goreau et al. 1971; Kinsey 1985b). But on an average, in most reef systems the autotrophic and heterotrophic processes are well balanced within the whole reef ($P_t/M_t \sim 1$), notwithstanding the fact that within a separate one of their biotopes (lagoons) it could be <1, or (reef flat, reef slope) >1 (Kinsey 1983; Sorokin 1986b; Zharikov 1987). In a well-balanced system, where the gross photosynthesis is in balance with the total heterotrophic respiration, the net autotrophic production ($P_n = P_t - M_t$) should be around zero. But the exploitation of reef resources by man without ruining the ecosystem could proceed only at the expense of the net autotrophic production (more correctly, the net external energy input). Therefore, the estimation of the exact size of the total net production P_n in reef ecosystems becomes extremely important from the viewpoint of their monitoring and preservation (Kinsey 1985b).

But the problem of correct evaluation of P_n does not appear to be as simple as it seems to be. As in any open system in coral-reef ecosystems the gross autotrophic production (P_t) is not actually equivalent to the total input of energy (P_g), nor is its total energy expenditure (Mg) equivalent to the respiratory losses measured (M_t). The P_g, besides P_t, includes also the energy of the external organic matter entering the reef system from the land with drainage and river waters, from mangrove thickets, from the ocean as drifting blue-green algae, or as oceanic plankton consumed by reef filterers and planktonovores (cf. Sects. 3.1.2, 6.3.2). As regards the values of Mg, besides respiration losses reef ecosystems lose a significant protion of energy as organic metter into the surrounding ocean and into bottom sediments. The level of gross autotrophic production in most kinds of reef bottom biotopes is usually very high and will often reach top level, as known in the most productive communities of our planet, like the tropical rain forests, in the thickets of macrophytes at the temperate-zone shelf in summer, or the areas of intensive upwelling in the oceans: $8–15\,g\,C\,m^{-2}\,day^{-1}$. This level of P_t becomes possible because of the multilayer or tree-dimentional structure of photosynthetic elements of the community within the illuminated zone and because the level of light saturation in them is less than the maximum day time illumination (Odum et al. 1958; Dahl 1976; Jokiel and Morrissey 1986; Table 10.1, Fig. 10.2). The basic producing agents of the reef: corals, macrophytes, seagrasses, compose the ramose multilayer structures whose functional surface of illuminated areas several times exceed the proajective area of the bottom. Among the main elements contributing to reef primary production are the zooxanthellae of symbiotic animals (corals, zoantharians, forams, clams), the benthic algae (macrophytes, filamentous algae of turfs and periphyton, microalgae of microphytobenthos, coralline and endolythic algae), the seagrasses (Lewis 1977b) and, occasionally, the phytoplankton (cf. Sects. 4.2, 4.3).

The methodology problems in measurements of reef biogeochemical metabolism and the main results were reviewed by Lewis (1977b, 1981),

Table 10.1. Ratios of photosynthesizing areas of basic functional components and their projective areas (SI) in bottom biotopes of the buttress zone at the barrier reef of Belise, Caribbean. (After Dahl 1976)

Component	SI	Percentage of total photosynthesizing area
Living corals (including surface area of the expanded polyps themselves)	13.8	53
Algae	6.0	23
Periphytic turfs	6.2	24

Sournia (1977), and Kinsey (1983, 1985b). One approach to the problem was based upon the above mentioned estimations of the fluctuations of the necessary parameters (O_2, CO_2, nutrients) in the open water over the reefs (Sargent and Austin 1954; Odum and Odum 1955; Odum et al. 1959; Kohn and Helfrich 1957; Nair and Pillai 1972; Smith and Marsh 1973; Pichon 1985). In shallow reef biotopes, where there was a permanent stream of water, the fluctuations of the above parameters had been measured at points upstream and downstream, accepting that the changes in measured concentrations resulted from the metabolic processes in benthic communities situated between the points of observations (Griffith et al. 1987). Up deeper biotopes or in areas with low currents the diurnal course of necessary metabolic parameters was measured at one point (Quasim et al. 1972; Kinsey and Domm 1974; Kinsey 1978). The measurements of δO_2 between points upstream and downstream in daytime give the net photosynthesis rate per time during which a portion of water passes between them. The same at night gives the value for respiration (M_t). Their sum yields the gross primary production P_t. Measuring P_t and M_t per hour at different times of the day and thus obtaining the curve of their diurnal course, the necessary coefficients could be found by calculating P_t, P_n and M_t values per 24-h cycle, the same as it had been described before for corals (cf. Fig. 10.1 and Table 10.2). The needed corrections on the exchange of O_2 between water and atmosphere were made employing the coefficients of diffusion as recommended by Odum and Hoskin (1958), Marsh and Smith (1978), Kinsey and Domm (1974). One of the weak points in this method could be seen in the actual impossibility to measure directly the daytime respiration of communities. It had been accepted as equal to its nocturnal level, which obviouly was not correct because of the possibility of photorespiration (Tolbert and Osmond 1976).

Besides dissolved oxygen, the parameters of pH, CO_2 and alkalinity were also used for the estimation of total reef production and metabolism (Smith and Kinsey 1978; Matcher 1990). For permanent observations of their

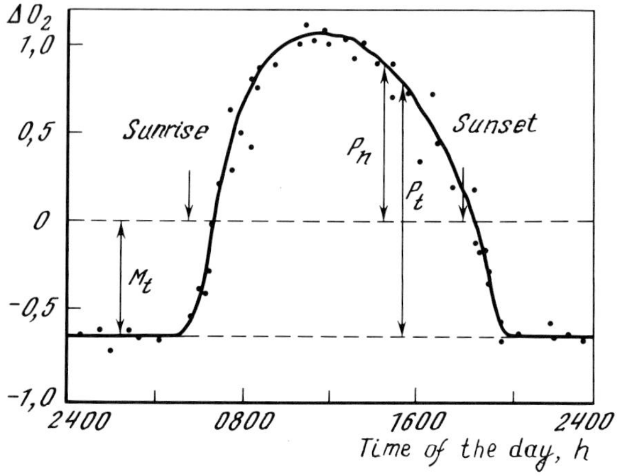

Fig. 10.1. Typical curve of the diurnal course of the change of the dissolved oxygen contents (O_2, $mg\,l^{-1}h^{-1}$) in water over reef-bottom communities, which are used for the calculations of their gross (P_t) and net (P_n) photosynthesis and also of their respiration (M_t). (After Smith and Kinsey 1978)

fluctuations monitors have been constructed, which records in situ automatically the fluctuations of the necessary parameters in water passing over the reef (Barnes 1983; Barnes and Deveraux 1984; Griffith et al. 1987). Also one of the methods employed to measure in situ reef metabolism was the isolation of part of the bottom community with the aid of plastic film (Kinsey 1972, 1978), or the construction of galleries from same film over the bottom, directed along the dominating current, and measuring O_2 changes at both their ends upstream and downstreamm, avoiding so exchange with the atmosphere (Rogers 1979a; Griffith et al. 1987).

Community metabolism as a whole has been evaluated also by measuring P_t and M_t in separate functional elements of benthic communities – in corals, in macrophtes, in pieces of hard substrates covered with algal turfs. The measurements were performed in short-term experiments in isolated ("bottle" method) or in flow respirometers, or in in situ enlosures of 10–50 l capacity, isolating lots of typical bottom biotope areas. With the aid of curves of the diurnal courses of P_t and M_t, and knowing the share of given elements or communities in the total area of a biotope, the necessary parameters of the entire biogeochemical metabolism in it could be calculated

Table 10.2. Ratios (K) of photosynthesis rates per day to those per hour in periphytic turfs on dead corals at different time intervals of a day. (Data by Sorokin 1986a)

Time intervals. h	K	Time intervals. h	K	Time intervals. h	K
8–9	10.2	11–12	7.9	14–15	13.2
9–10	8.4	12–13	8.8	15–16	16.8
10–11	7.8	13–14	10.7	16–17	24.0

(Marsh 1970; Sorokin 1971b, 1986b; Quasim et al. 1972; Sournia 1976b; Propp et al. 1983; Griffith et al. 1987). To measure the rates of photosynthesis, besides the oxygen method the radiocarbon method was also used in this way, taking into account that it gives them close to the net production P_n (Sorokin 1973c, 1986a; Bittaker and Iverson 1976). To study reef metabolism microcosms were also constructed, which included basic elements of bottom biotopes and had a volume of several cubic meters (Adey 1983; Griffith et al. 1987). The coefficients, necessary for recalculations of the gross photosynthesis values, measured during different periods of the day per whole day in periphytic turfs over dead corals, are given in Table 10.2.

High rates of biogeochemical metabolism in reef biotopes result in extremely large amplitudes in diurnal fluctuations of O_2, CO_2 and pH in waters over the reef. The amplitude of fluctuations of dissolved O_2 could reach 5–8 mg $O_2\,l^{+1}$. In Fig. 10.1 an example of the diurnal course of rates of changes in O_2 content ($\delta\ O_2\,h^{-1}$) in water is given, which has been used for calculations of P_t, P_n and M_t values per day. In Fig. 10.3 typical measurement results of O_2 content in water over the reef at different times of the day are shown. The rate of visible increase in O_2 content in water, which is proportional to the net photosynthesis (P_n) at midday, ranges usually from 0.5 to 1.5 mg $O_2\,l^{-1}\,h^{-1}$. The rates of decrease at night, proportional to M_t, are expressed by the values 0.2 to 1.5 mg $O_2\,l^{-1}\,h^{-1}$ at verage depths of 0.5–1.5 m. The largest fluctuations take place at shallows over the reef flat and over patch reefs (Fig. 10.4). The corresponding diurnal fluctuations of alkalinity in such biotopes range usually between 100–300 μg equivalents l^{-1}, and the shifts of pH – 1 to 2 units – from 7.5 to 8.5–9.5, caused by processes of photosynthesis, respiration and calcification (Kinsey 1972; Smith 1974). The dependence of reef gross photosynthesis upon illumination condition is given in Fig. 10.2.

The share of plankton in biogeochemical metabolism of carbon in reef communities could be estmiated as being between 5 to 20% of the total, depending upon the depth of the site. The rest, 80–95%, results from the function of their benthic components, taking into account that 10–30% of

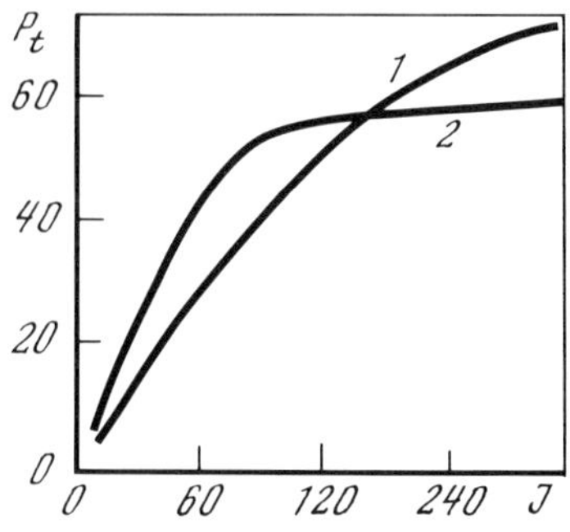

Fig. 10.2. Light dependence curves of the gross photosynthesis (P_t, mol $O_2\,m^{-2}\,h^{-1}$) in bottom reef-flat communities at the Rib reef, GBR; *I* light intensity, wt m^{-2}; *1* bottom area dominated by coralline calcareous algae; *2* area dominated by corals. (After Barnes and Devereux 1984)

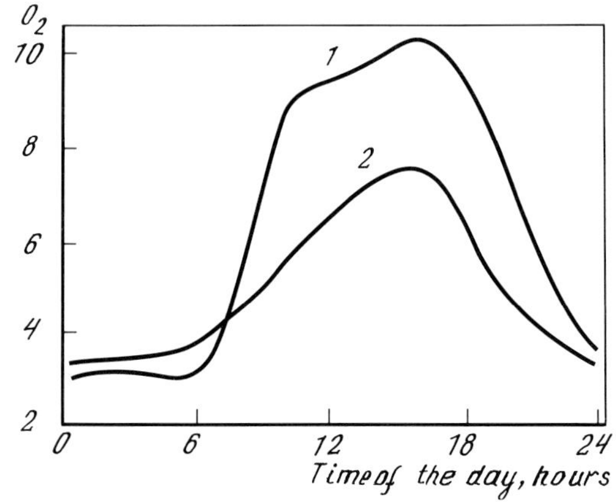

Fig. 10.3. Diurnal changes in dissolved oxygen contents (O_2, $mg\,l^{-1}$) in waters over the Lizard I. reef (*1*) and over the Kauai I. reef, Hawaii. (Data by Kohn and Helfrich 1957; Sournia 1976b)

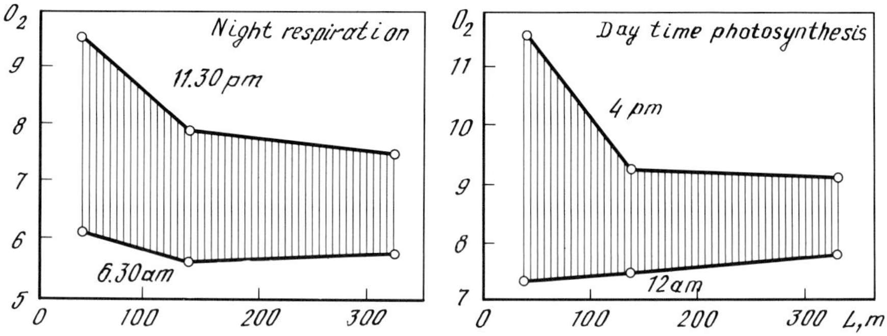

Fig. 10.4. Changes in dissolved oxygen content ($O_2\,mg\,l^{-1}$) in waters over the leeward reef off Koetivi I., Seychelles at the cross-section from the shore; *L* distance from the shore, m. (Data by the author)

the latter is the carbon flow from water to bottom communities through calcification, and the rest of 70–90% – through its photosynthetic assimilation by benthic autotrophs (Atkinson and Grigg 1984; Kinsey 1985b; Crossland et al. 1991). The level of the total carbon flow in coral-reef ecosystems is close to its maximum known in the natural biota. Moreover, the values of P_t, for example, recorded in the reef systems were sometimes close to their theoretically possible upper level in nature. The latter, in communities of ramose corals with 100% bottom cover has been evaluated to be about $60\,g\,C\,m^{-2}\,day^{-1}$ (Odum and Odum 1955), in communities of aquatic plants 40 g, and in those of dry-land plants 8 g (Larkum 1983). Maximal P_t values actually recorded on reefs in biotopes dominated by corals were as high as $34\,g\,C\,m^{-2}\,day^{-1}$, and in those dominated by

Table 10.3. Indexes of biogeochemical metabolism in reef benthic communities: P_t and p_n – gross and net photosynthetic production; M_t – respiration (all as $g\,C\,m^{-2}\,day^{-1}$); G – calcification, $kg\,CaCO_3\,m^{-2}\,year^{-1}$. (Data by Odum and Odum 1955; Odum 1956; Kohn and Helfrich 1957; Odum et al. 1959; Gordon and Kelley 1962; Nair and Pillai 1972; Quasim et al. 1972; Kinsey 1972, 1978, 1983; Smith Marsh 1973; Kinsey and Domm 1974; Marsh) 1974; Hillis-Colinvaux 1974; Smith 1974, 1981; Smith and Jokiel 1975; Sournia and Ricard 1976a; Wanders 1976a; Connor and Adey 1977; Kinsey and Davies 1979b; Rogers 1979a; Sournia et al. 1981; Vooren 1981; Adey 1983; Propp et al. 1983; Murphey and Kremer 1983; Atkinson and Grigg 1984; Barnes and Devereux 1984; Pichon and Morrissey 1985; Pichon 1985; Jokiel and Morrissey 1986; Griffith et al. 1987)

Type of community	Location	P_t	M_t	P_n	P_t/M_t	G
Mixed communities of reef flat including corals, coralline algae, and periphytic turfs	Rongelap atoll	6.0	6.0	0	1	–
	Enivetok atoll	9.6	9.6	0	1.0	–
	One Tree ring reef	7.5	6.8	0.7	1.1	4.6
	Kawaratti atoll	6.2	2.5	3.7	2.5	–
	Kauai reef, Hawaii	7.9	7.2	0.7	1.0	1.0
	Lizard I., GBR	7.8	8.9	−1.1	0.9	4.6
	Rib reef, GBR	8.8	7.6	4.2	1.2	3.5–10.0
	Paletoai reef, Moorea I.	8.8	4.4	4.4	2.0	9.6
	Galeta Point, Panama	21.4	10.0	11.4	2.1	–
	Piti reef, Guam	7.2	3.3	3.9	2.2	–
Communities of flat and lagoon dominated by macrophytes and seagrasees	One Tree I. ring reef, GBR	9.0	7.5	1.5	1.2	4.0
	Lizard I., GBR	7.0	6.7	0.3	1.0	2.4
	Jamaica (Halimeda)	4.0	1.7	2.3	2.3	–
	Virgin I.	30.0	12.8	17.2	2.3	–
	St. Croix	11.1	2.2	8.9	5.0	–
	Kaneohe bay, Hawaii	19.6	4.3	15.3	4.2	–
	Curaçao I.	9.5	4.2	5.3	2.3	–
	Florida (seegrass)	13.6	9.6	4.0	1.4	–
	Enivetok atoll	11.6	6.0	5.6	1.9	–
	Kawaratti atoll (seagrass)	12.9	9.5	3.4	1.4	–
Communities of patch reefs and rich growth of ramose corals in patchr-eef zones and at upper zones of outer reef slopes with weak to moderate turbulence.	Oahu, Hawaii	14	24	−10	0.6	–
	Lizard I., GBR	14	12	2	1.2	5.0
	Abrolhos I., West Australia	21	19.6	1.4	1.0	11.7
	San Croix, Virgin I.	23	23	0	1.0	–
	Moorea I.	17	17	0	1.0	0.5
	Barrier reef, Madagascar	19	11	8	1.7	1.9
	El Mario reef, Puerto Pico	16.8	14.2	2.6	1.2	–
	One Tree reef, GBR	20	18	2	1.1	4.6
Lagoons	San Croix, Virgin I.	7.1	6.3	0.8	1.1	–
	San Cristobal, Puerto Rico	2	3	−1	0.7	–
	Phantom bank, Timor Sea	4.0	15.9	−11.9	0.4	–
	Enivetok atoll	6.0	6.0	0	1	–
	Davis atoll	5.2	5.8	−0.6	0.9	–
	One Tree reef, GBR (sand)	2.3	3.6	−1.3	0.6	–
	One Tree reef (corals)	9.0	7.5	1.5	1.2	–
	Clipperton atoll	8.6	8.0	0.6	1.1	–
	Minkoy atoll, Adamant I.	3.9	17.2	−13.3	0.2	–
Whole reef systems	French Frigate atoll	3.56	3.2	0.36	1.1	6.5
	Canton atoll	6.0	5.9	0.1	1.0	0.5
	Takapoto atoll	4.0	4.0	0	1.0	–
	One Tree ring reef	3.8	3.6	0.2	1.0	1.5
	Lizard. I. fringing reef	3.2	3.2	0	1.0	1.8

macrophytes 30 g (Gordon and Kelley 1962; Connor and Adey 1977; Griffith et al. 1987). The largest P_t values (10–20 g C m^{-2} day^1-^2) were recorded in zones of flourishing ramose coral associations in submerged leeward patch reefs and on outer reef slope with relatively low turbulence. Its high values (7–20 g C) are peculiar also to biotopes dominated by macrophytes, algal turfs, and seagrasses. In biotopes of submerged reef flat covered by mixed communities of corals, corallines, macrophytes, and turfs of filamentous algae, the P_t values are on an average level of 4–9 g C m^{-2} (Tables 10.3–10.5) In communities of the outer reef slope they vaicy between 0.6–7 g C m^{-2}, depending upon depth and the level of illumination (P.S. Davies 1977). The communities of the outer reef slope and of the reef edge, dominated by coralline algae, have a relatively low gross production – 0.5–3.5 g C m^{-2} day^{-1}. In shallow bottom biotopes of lagoons, inhabited by living corals, macrophytes, and seagrasses, the P_t values could be high, up to 10–15 g C m^{-2}. In lagoonal biotopes with the bottom covered by rubble or sand P_t they vary between 0.4–4 g C m^{-2} (Tables 10.3–10.6). The depths of photosynthesis in the column of coral sands were estimated to be between 2–3 cm (Sorokin 1986a). The primary production in coral sands is built up by the microbenthic diatoms and cyanobacteria, by filamentous algae and symbiotic foraminifera. The coral sands contain 100 to 900 mg m^{-2} of chlorophyll *a*, and its assimilation numbers could be rather high, ranging from 0.2 to 1.5 mg O_2 h^{-1} mg^{-1} of chlorophyll *a* (Sournia 1976b). While the general level of primary production in biotopes of coral sands is two to four times less than in those of the flat or of patch reefs, their share in the total autotrophic production of reef ecosystems could be significant because they occupy a high percentage of their total bottom areas, especially in the atoll systems, where it has beem estimated at 20 to 50% (cf. Table 10.9). At one

Table 10.4. Parameters of metabolism in reef bottom biotopes, occupied by relatively monotonous plant associations (g C m^{-2} day^{-1}). For designations, see Table 14.3. (Data by Bakus 1967; Marsh 1970; Littler 1973b; Smith 1974; Wanders 1976b; Kinsey 1978; Rogers and Salesky 1981; Vooren 1981; Morrissey 1985; Sorokin 1986a)

Biotope, dominating associations	P_t	M_t	P_n	P_t/M_t
Coral sand, dominated by diatoms and symbiotic foraminifera	1.40	0.61	0.78	2.3
	0.9	1.4	−0.5	0.6
	0.55	1.1	−0.55	0.5
	0.4–1.6	0.6–2.4	0.2–0.8	1.5
Rubble dominated by filamentous blue-green alga *Ostreobium*	1.34	0.75	0.59	1.8
Reef-flat rocks, dominated by coralline algae	1.50	0.84	0.66	1.8
	0.9–1.6	0.6–1.1	0.3–0.5	1.5
	0.6–3.4	0.1–1.2	0.5–2.6	3.0
Reef-flat rocks with rubble dominated by periphytic turfs	1.4–2.8	0.6–0.8	0.8–2.0	3.0
	2.15	1.50	0.65	1.4
	3.5	2.76	0.80	1.3
	2.8	0.9	1.90	3.1

Table 10.5. Parameters of metabolism in basic bottom plant and symbiotic associations of the Curaçao reefs. For designations see Table 14.3. (Data by Vooren 1981)

Parameters of metabolism	Integral values for the whole reef[a]	Associations		
		Coralline algae (Neogoniolithon)	Red thallomic algae (Wurdemannia, Gelidium)	Coral sands (diatoms, foraminifera)
P_t, g O_2 m^{-2} h^{-1}; midday	1.37	1.07	1.26	0.33
P_n, g O_2 m^{-2} h^{-1}	0.75	0.87	1.0	0.27
M_t, g O_2 m^{-2} h^{-1}	0.62	0.20	0.26	0.06
P_t, g C m^{-2} day^{-1}	7.2	3.8	6.0	1.1
M_t, g C m^{-2} day^{-1}	8.4	1.8	2.3	0.5
P_n, g C m^{-2} day^{-1}	−1.2	2.0	3.7	0.6
P_t/M_t (day^{-1})	0.9	2.1	2.2	2.2

[a] Estimated after measurements of diurnal O_2 fluctuations in water over the reef.

Table 10.6. Flows of carbon in main reef zones of the French Frigate shoals, Hawaii. Pt and P_n – gross and net photosynthetic production; M_t – total respiration (all as $g\,C\,m^{-2}\,day^{-1}$); G – calcification, $kg\,CaCO_3\,m^{-2}\,year^{-1}$. (Data by Atkinson and Grigg 1984)

Parameters of metabolism and of community structure	Reef zones					
	Reef flat		Patch reef		Lagoon	
	Winter	Summer	Winter	Summer	Winter	Summer
P_t	4.27	8.53	5.41	14.41	2.40	2.64
P_n	1.71	3.66	0.33	1.22	0.25	0.15
M_t	2.56	4.87	5.08	13.21	2.15	2.49
G	2.80	10.2	6.51	12.08	1.40	1.40
P_t/M_t	1.66	1.75	1.06	1.09	1.12	1.06
Percentage of bottom cover by plant and symbiotic primary production associations						
Coralline algae		19		70		46
Macrophytes and filamentous		66		4		4
algae		15		0		15

of the microatolls within the lagoon of the One Tree reef the share of periphytic turfs over rubble and of coral sand in total autotrophic production was estimated to be around 80% (Borowitzka et al. 1983).

In some studies on reef metabolism, besides the gross (P_t) and net (P_n) photosynthetic production, the primary photosynthetic production (P_{np}) of reef plant associations was also calculated. The latter means the net photosynthesis of the plant elements themselves, while P_n is the net production of the whole reef community including its autotrophic and heterotrophic elements ($P_t - M_t$). Thus: $P_{np} = P_t - M_p$, where M_p is the respiration of only the plants themselves. The latter value could be calculated assuming that it comprises ca. 15 to 30% of P_t, so: $P_{np} = P_t \cdot 0.85$. (Atkinson and Grigg 1984; Hatcher 1990). The values of primary production of reef plant associations, P_{np}, are of ecological importance, because they correspond to the share of autotrophic production available for use by heterotrophic components of the community. A part of this primary production is used for their metabolism (M_n). The rest of it ($P_{np} - M_n$) will correspond to the net photosynthetic production of the reef community (P_n): $P_n = P_t - M_p - M_n$. Therefore, just the value of P_{np}, but not P_n or P_t, should be accepted as the contribution of local autotrophic production to the total energy input into the reef ecosystem. From this point of view, in the comparison of productivity of different reef ecosystems or reef communities the use of the annual P_{nt} values as basic indicators seems to be most reasonable.

The generalized data on annual primary production (P_{np}), total metabolism (P_t), and net autotrophic production (P_n) in various reefs and in whole reef systems are given in Table 10.7. The annual primary production in reef systems appear to be close to its upper values recorded in the most

Table 10.7. Annual values of primary photosynthesis production (P_{np}), net photosynthesis production (P_n), and total respiration (M_t), all in $g\,C\,m^2$, in coral-reef communities

Place (zone of reef)	P_{np}	M_t	P_n	P_{np}/M_t	Reference
Enivetok atoll (flat)	3570	4200	0	0.9	Odum and Odum (1955)
Enivetok atoll (flat)	3600	2190	2040	1.6	Smith and Marsh (1973)
Kawaratti atoll	1910	880	1370	2.2	Quasim et al. (1972)
Rongelap atoll	1060	1090	160	1.0	Sargent and Austin (1954)
Moorea I. barrier reef	2230	3050	−420	0.7	Sournia (1976b)
Guam I. (flat)	5860	2600	4300	2.3	Marsh (1974)
Phantom bank, Timor Sea	2490	7000	−4070	0.4	Propp et al. (1983)
El Mario reef, Puerto Rico	3780	4100	350	0.9	Odum et al. (1959)
Long Key reef, Florida (thallassia beds)	3290	2740	1140	1.2	Odum (1956)
Jamaica (Halimeda fields)	1240	630	830	2.0	Hillis-Colinvaux (1974)
Atoll Canton (whole system)	1860	2150	40	0.9	Smith and Jokiel (1975)
Takapoto atoll (whole system)	1240	1460	0	0.9	Sournia and Ricard (1976a)
Lizard I. (whole system)	980	1070	80	0.9	Kinsey and Davies (1979b)
French Frigate atoll (whole system)	1100	1170	130	1.0	Atkinson and Grigg (1984)

productive ecosystems of our planet (Lewis 1977b, 1981). In communities of reef flat, dominated by corals, seaweeds and algal turfs, it ranged between 1.5–3.0 $kg\,C\,m^{-2}$, which is equivalent to 3.5–7.0 kg of dry organic matter, or to 20–40 $kg\,m^{-2}$of wet biomass (Crossland et al. 1991). At some reef zones it reached even 4–6 $kg\,m^{-2}$, or up to 80 $kg\,m^{-2}$ of wet biomass $year^{-1}$. These values are three to ten times more than those in the most productive pelagic ecosystems of quasi-permanent coastal upwelling areas, where they range between 0.5–0.9 $kg\,C\,m^{-2}\,year^{-1}$. The average annual value of primary production for most productive communites of reef-flat and patch-reef zones could be assumed to be close to 2–3 $kg\,C\,m^{-2}\,year^{-1}$, the values of annual net photosynthetic production of communities (P_n) being between 0–0.3 $kg\,C\,m^{-2}$, and the P_{np}/M_t ratio close to 1 (Kinsey 1983; Tables 10.3, 10.7). The average annual values of primary production when calculated per complete reef systems appeared to be 1.5–2 times less than the above value because a significant part of their area belongs to the lagoonal biotopes, where annual primary production is two to four times lower, amounting to 0.5–1 $kg\,C\,m^{-2}$. Therefore, average values of 1.1–1.4 $kg\,C\,m^{-2}\,year^{-1}$ for the total system could be considered reasonable (Kinsey 1985b; Table 10.7). The most comprehensive estimations of the gross (P_t), primary (P_{np}), and net photosynthetic production for the complete ecosystem of the French Frigate Shoals atoll gave the values of 1.3, 1.1, and 0.13 $kg\,C\,m^{-2}\,year^{-1}$ respectively (Atkinson and Grigg 1984). However, when estimating the most probable annual primary production of reef systems for the calculation of its global value, it is necessary to take also into account that the above-mentioned level referred mostly to atoll systems. But most reefs of the World Ocean belong to the barrier and fringing types, in which the relative areas of lagoons are much smaller than in atolls, and their lagoons are more productive than those of the atolls. Therefore, it would be more reasonable to accept the referee values of their generalized annual P_t and P_{nt} at the levels of 1.8 and 1.5 kg C m^{-2}.

The rate of metabolism in aquatic animals depends on the water temperature decreasing 2–2.5 times for each drop of 10 °C. In high-latitude reef systems the seasonal changes in temperature (and illumination), and thus – the metabolic rates fluctuations could be rather high (Table 10.8), and thus they should be taken into account in calculations of the annual $P_t - M_t$ balances (Table 10.9). For example, in the reef and reef-flat biotopes of the French Frigate atoll in winter the P_t and M_t values dropped 2–2.8 times and the rate of calcification 3 times (Table 10.6). But in the lagoon such a large decrease in metabolism did not take place, most probably because of the winter bloom of phytoplankton, which often occurs in lagoons, and whose production taken into account automatically when measuring the total reef metabolism, with diurnal O_2 fluctuations occurring in the water. Significant seasonal changes in metabolic rates (1.8 times) were also recorded on the reefs of St. Croix at 18° N by Abey a. Steneck (Kinsey 1985b). The P_t/M_t rates in most cases do not change with the seasons, which points to the stability of the basic trophic status of reef communities year-round (Table 10.8; Fig. 105).

The rates of heterotrophic metabolism in reef communities are also very high, attaining 10–30 g O_2 m^{-1} day^{-1}or 3 to 10 g C m^{-2}. In some cases it will reach even 15–18 g C m^{-2} day^{-1}, which corresponds to the decomposition of 40–50 g of dry, or 300–400 g of wet biomass. Because of a high heterotrophic respiration, the P_t and M_t values in reef biotopes are usually well balanced at a high level of their rates: P_t/M_t – 0.7–1.5. This balance has been most remarkable, especially in its calculation for complete reef systems that include different biotopes with either autotrophic or

Table 10.8. Parameters of metabolism in shallow biotopes at different seasons of the year: P_t, M_t, g C m^{-2} day^{-1}, G (calcification), g $CaCO_3$ m^{-2} day^{-1}. (Data by Kinsey 1978; Kinsey and Davies 1979b; Smith 1981)

Place, latitudes	Season	P_t	M_t	P_t/M_t	G
Abrolhos reefs, western Australia; 29° S	summer	21.0	19.6	1.0	50.1
	winter	12.1	14.4	0.8	12.8
Kaneohe bay, Oahu I., Hawaii; 21° N	summer	11.0	15.1	0.7	27.3
	winter	5.0	6.4	0.9	21.6
Lisard I., GBR; 15° S	summer	9.7	11.8	0.8	9.0
	winter	4.7	3.8	1.1	6.6

Table 10.9. Gross and net photosynthesis production and shares of the main reef zones in the total autotrophic production of the whole reef system of the French Frigate atoll, Hawaii, with a total area of 700 km^2. (Data by Atkinson and Grigg 1984)

Reef zones	Cross production of communities			Net production of communities			
	$g C m^{-2} day^{-1}$	$10^3 t C year^{-1}$ per whole atoll	Share of each zone, %	$g C m^{-2}$	$10^3 t C year^{-1}$ per whole atoll	$10^3 t year^{-1}$ of wet biomass per whole atoll	Share of each zone, %
Outer reef slope and fore-reef zone, 30–10 m depth	1.0	69.2	9	0	0	0	0
Reef flat and outer reef slope, 10–0 m depth	6.4	59.8	8	2.68	25.0	185	34
Lagoon	2.5	387.8	53	0.20	31.0	301	42
Patch reefs	9.9	216.4	30	0.77	0.77	175	23
Whole atoll	–	733.3	100	–	72.8	661	100

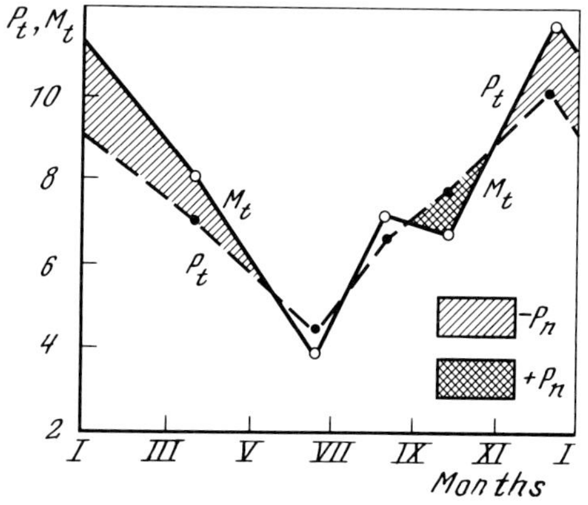

Fig. 10.5. Seasonal changes in the gross photosynthesis (P_t) and respiration (M_t) in communities of the reef flat at Lizard I., GBR; P_n net photosynthesis. (After Kinsey 1983)

heterotrophic processes dominating (Tables 10.3, 10.9). In biotopes of barrier and fringing reefs, the P_t/M_t (or the P_{np}/M_t) ratios are lower than in those of atolls, which was often below 1. Thus the P_n values were correspondingly negative (Zharikov 1987; Table 10.3). This reflects a possible additional input of organic matter, arriving from the land and used by heterotrophs. In biotopes dominated by seaweeds and seagrass the P_t/M_t ratios were usually high, 2 to 4, and thus the net production was also high, up to 3–5 $g\,C\,m^{-2}\,day^{-1}$. In biotopes with a dominance of soft sediments in lagoons, or in shadowed or deep reef biotopes where heterotrophic processes predominated, the P_t/M_t ratios were <1, and the net photosynthetic production P_n was often negative. These communities then definitely received an additional input of external organic matter as surplus to the local photosynthesis. Its main source should be just the net autotrophic production in those reef biotopes where the P_t/M_t is >1 (Sorokin 1973c; Kinsey 1983; Zharikov 1987). It comes in the form of mucus flakes, plant detritus, fish feces, and DOM. Such a transfer of organic matter from zones with excess production, i.e., from "autotrophic" to "heterotrophic" sites, while providing the energy interconnections between the reef biotopes, actually unite its separate communities into the complete ecosystem.

A comparatively recent attempt to evaluate the relative input of bottom communities of the main reef zones (biotopes) into the total balance of organic matter – energy of a whole reef system has been made at the French Frigate atoll within the EOPATH project (Polovina 1984; Atkinson and Grigg 1984; Atkinson et al. 1984; Tables 10.6, 10.9). According to these studies, which had been performed through all seasons of the year, the largest total input into the total P_t of the atoll system (700×10^3 tonnes C per year) was estimated to come from the lagoon because of its relatively large area, while the production per m^2 there was about three times less than in biotopes of patch reefs of the reef flat. The net photosynthetic production in biotopes of the outer reef slope, including the fore-reef zone, was found to be close to zero.

The biogeochemical metabolism of carbon in reef communities includes also the process of growth of the lime skeletons of benthic animals and of calcareous algae, i.e., the process of calcification. Its rates in reef communities is extremely high, ranging from 1 to 20 kg $CaSO_3\,m^{-2}\,year^{-1}$. This corresponds to diurnal values of 3 to 50 g of $CaCO_2$ or 0.2–4 g $C\,m^{-2}\,day^{-1}$, which is close to the lower level of carbon flow rates by photosynthesis processes.

10.2 Biogeochemical Metabolism of Other Elements

Both the photosynthetic production and the destruction of organic matter, as well as the calcification processes, are the energy and chemical backgrounds of the biogeochemical turnover of elements coupled with them, which are the components of the living tissues and the skeletal material of reef organisms, and also of some other elements whose turnover in the aquatic habitats is linked with the formetion of redox zones in the columns of bottom sediments (Livingston and Thomson 1971; Howard and Brown 1984). Among these elements N, P, S, Fe, Mn, Ba, Cd and Co (Shen et al. 1987; Brown et al. 1991) . The dynamics of the first four elements have been discussed above (cf. Sects. 2.2, 4.1). Here, we give only approximate calculations for them, based upon the C:N:P ratios, which in the reef ecosystem, because of the semiclosed cycles of nutrients in their plant, and especially animal symbiotic associations, looks like 100:3:0.15 by weight (Atkinson 1981a). Thus, with an average primary production in reef plant communities of 4 g $C\,m^{-2}\,day^{-1}$, the flow rates of N and P in them should be ca. 120 and 15 mg $m^{-2}\,day^{-1}$, respectively. The ratios of C:Fe:Mn in the aquatic plant biomass are close to 100:0.3:0.03 by weight. Thus, the flows of Fe and Mn coupled with the primary production in reef communities could be evaluated at approximately 12 mg and 1 mg $m^{-2}\,day^{-1}$ respectively. The flow of cobaltum could be coupled only with the microbieal biosynthesis (Sorokin 1972). With its rate in reef environments of 0.5 g $C\,m^{-2}\,day^{-1}$, the flow cobalt should be at the level of 100 μg $m^{-2}day^{-1}$.

The global evaluations of biogeochemical activity of the reef ecosystems are based upon their total area of 6×10^5 km^2 in the World Ocean which includes 15% of its shallows at 0–30 m depth, or about 0.2% of its total area (S.V. Smith 1978; Grossland et al. 1991). The primary production on this area has been evaluated, as shown above, with 1.5 kg $C\,m^{-2}\,year^{-1}$

(excluding the respiration of the plants themselves). The global annual gross primary production in reef systems should be then about 7–9 $\times$ 10^8 t C, which contributes 1.5–2% to the total primary organic production in the World Ocean. The net production of organic carton was evaluated as 20 $\times$ 10^6 tC, of which 3 $\times$ 10^6 t is buried in sediments, 15 $\times$ 10^6 t is exported to the surrounding ocean and 2 $\times$ 10^6 t can be used by man (Crossland et al. 1991). When evaluating this contribution of coral reef ecosystems to the global oceanic productivity we have to keep in mind that it is several times more than the joint primary production in the productive zones of upwelling, from where a large portion of the total world fish catch comes (Muscatine 1980). With respect to the above integral primary production, the global annual flow of elements in reef ecosystems could be expressed by the following levels: N = 2.7 $\times$ 10^7 t, P = 1.3 $\times$ 10^6 t, Fe = 3 $\times$ 10^6 t, Mn = 3 $\times$ 10^5 t.

The coral reefs of the World Ocean are the main generators of calcareous sediments on the Earth. The size of the global processes of lime sedimentation in coral reefs could be assumed to be close to 2 $\times$ 10^9 t $CaCO_3$ $year^{-1}$ Besides originating such a large-scale global biogeochemical process of calcium sedimentation from seawater, coral reef systems regulate, together with calcium, also carbon dioxide, so that they are one of the most important regulators of CO_2 content in the World Ocean. In conjunction with the now developing global "greenhouse" effect, caused by the anthropogenically induced increase in CO_2 content in the atmosphere, the CO_2 uptake by coral reefs is becoming more and more important. The progressive destabilization and destruction of reef systems may enhance the dangers from the "green house" effect which, in itself, endangers also the coral reefs' very existence, killing them by overheating, and by drowing them during the expected elevation of the sea-level. Thus, the two processes – destabilization of the reef systems and buildup of the "green house" effect – are cumulative, reinforcing each other, which could have a drastic impact on environmental conditions on which the survival of mankind, depends, unless both are stopped (Hopley and Kinsey 1988). By the middle of the twenty-first century a rise of 0.2–1.4 m in sea level is expected.

10.3 Model Studies and Energy Budget in Reef Ecosystems

One of the ways for understanding the mechanisms of functioning of coral reef ecosystems is the generalizations at the level of the schemes of trophic or energy relationships between its component communities, representing different links in the food chain. Such generalizations seem to lead to the next step – the composition of models of coral reef ecosystems as a tool for their further holistic investigation, including the influence of anthropogenic impacts. This method has been supposed to be helpful for developing practical recommendations for the optimal management of their resources (Polovina 1984).

The first of such schemes was construed by Odum and Odum (1955) for one of the reefs at the Enivetok atoll. In their classical work the authors distributed the bottom communities into separate trophic groups. Then they created a pyramid of their biomass as follows: plant primary producers – herbivores – carnivores, in a relation of 703:132:11. Their analysis of reef trophodynamics at this level resulted in developing several, still important, concepts. Among them there could be mentioned the conclusions about the definite similarity of reef systems to the ecosystems of the tropical rain forests in their maturity, the variability of sources of primary production (microphytobenthos, periphyton, macrophytes, seagrasses, corals and symbiotic foraminifera) and the complexity of their food webs. Laters other schemes appeared, revealing the relative functional roles of separate components or communities in energy and matter flows within the reef ecosystems, and between them and the pelagic systems of the surrounding ocean. In our scheme, the central role of microbial populations in the energy interconnection between separate reef communites has been pointed out (Sorokin 1973c). In other schemes (Michel et al. 1972; Geptner 1979), the role of the reef system as a biofilter was demonstrated, and with it the close relationships between the ecosystems of reef and the surrounding ocean were established at the level of energy and nutrients exchange.

The first attempt at a model study of coral reef ecosystems has been made by a special team of 40 specialists (Dahl et al. 1974). This group intended to divide its biota into several blocks – communities – which were to represent the trophical levels. Then the participants in this project set about to describe in their model the trophical links inside these blocks and between them, as well as the relationships between the blocks and environmental factors. They intended then to describe the functional role of each block and to evaluate the energy flows between them. The model so constructed was supposed to be used to locate the factors which control productivity and the metabolic processes, and thus to get a key for the monitoring and management of complex and vulnerable reef systems. But, as work progressed, this desirable and precious goal was honestly recognized as unattainable. The reef ecosystem appeared to be too sophisticated for such brave endeavors and manipulations. The number of the above-mentioned "blocks" alone was found to be as many as 104. They were united within a system by over 10^4 various interconnections and over 2×10^3 separate trophic links. Therefore, this model was never realized.

Later, ignoring this userful and instructive experience, the following attempt at such a model study was made within the ECOPATH project of the University of Hawaii at one of the Hawaiian atolls: the French Frigate shoals atoll (Polovina 1984; Atkinson and Grigg 1984; Grigg et al. 1984). This is a large crescent reef with a shallow lagoon, and 700 km^2 of the total area. As background of the future model, a simplified food-web scheme was composed, which contained 13 basic components instead of the 104 mentioned above. While this scheme contained some rather insignificant components such as, for example, turtles, it omitted several basically important ones, such as the bacterioplankton, the periphytic and bottom microflora. The system was supposed to be isolated from the surrounding ocean, but only insofar as the acquiring of additional suspended matter, DOM and plankton input. Only their probable loss was taken into account. Corals were united within one compartment with the benthic algae. Detritus and DOM were totally excluded from the food web. The herbivorous and predatory zooplankton were also united within one compartment. The same was done with the zoobenthos and reef fish. The demersal character of the reef zooplankton and its use by zoobenthos and bottom fish were ignored.

The mode of construction of the above model, from the viewpoint of the author, seems to be also too audacious. Besides an arbitrary selection of components for the main trophic groups, all field material collected at the atoll, which composed the basis of this model, was represented by results of random analyses of the gut contents of large predatory fish (sharks, carangids) and of the approximate evaluations of the standing stocks of top predators: fish, seals, sea-birds. Further necessary information for the construction of the model, including the efficiency and the species production coefficients, was taken from the literature, dealing mostly with temperate water organisms and communities. Using the standing stocks and items of information on food spectra, the food demands of the top predators were thus calculated. Then, assuming that the coefficients of grazed production in all trophic groups are uniform (0.85), all the necessary parameters of their biomass, production and food rations were calculated from the model, including then also the autotrophic ones, assured that the system is balanced. The chief final result of this model analysis was the size of the primary energy input into the system, necessary for its functioning. The primary production, so computed, was found to be $4.3 \times 10^6\,\mathrm{kg\,km^2\,year^{-1}}$, or $1.2\,\mathrm{g\,C\,m^{-2}\,day^{-1}}$. This basic value was used to verify the model by estimating its actual value in zones of the atoll at different seasons of the year. The latter was estimated to be $6 \times 10^6\,\mathrm{kg}$ of wet biomass, which is not very far from the predicted value. Another result of the model analysis was the construction of an energy-flow scheme for the whole atoll ecosystem, in which the food rations of subsequent links within the food web were expressed in $\mathrm{kg\,m^{-2}\,year^{-1}}$ (Fig. 10.6). To make this scheme more comparable to other such schemes, the above values are given as $\mathrm{kcal\,m^{-2}\,day^{-1}}$. The model was over simplified and just for this reason contained many unacceptable or a piori erroneous prerequisies. But being successfully verified it was then used to deduce some basic conclusions about the mechanisms controlling production in coral reef ecosystems. The model was recommended by the authors to be employed for the development of monitorig and management strategies concerning the use of resources of the

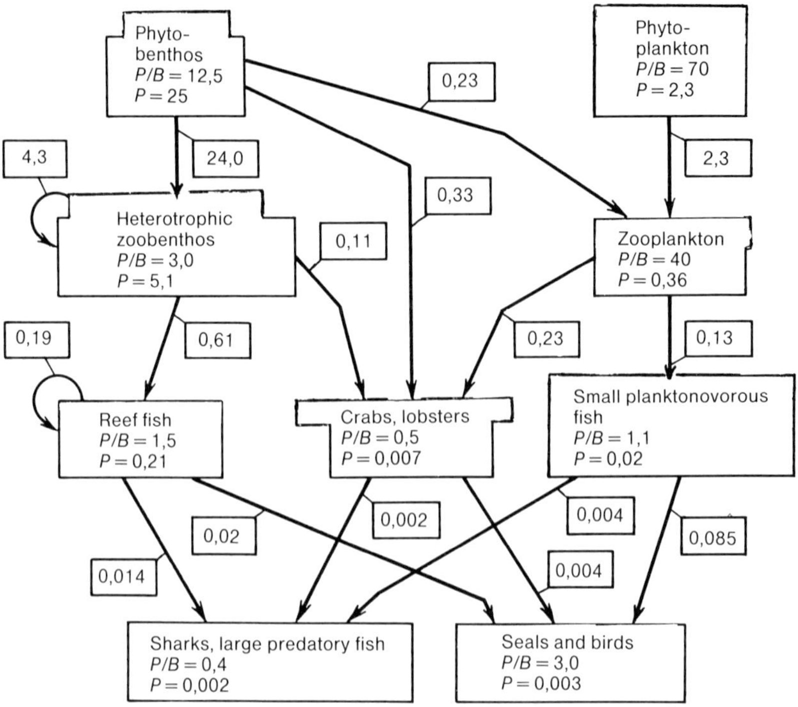

Fig. 10.6. Scheme of energy flows (K, kcal m^{-2} day^{-1}) and the *P*/*B* coefficients per day in the main components of the coral reef ecosystem at the French Frigate Shoals atoll; *P* production; *numbers in squares* energy equivalents of the food rations of the subsequent component of the food web. (After Grigg et al. 1984)

reefs. Nevertheless, this favorable outcome was also predetermined by a successful selection of basic coefficients used in the model, such as the coefficients of efficiency in the use of consumed food for growth (K_1), and coefficients of specific production (P/B) and mortality (grazing). For example, using an admittedly exaggerated coefficient of grazing in all trophic groups (0.85), the authors, with the aid of the same model, then deduced that just the grazing (but not, for example, the nutrients) controlled the production of the ecosystem, But it should be obvious that in reef biotopes with their abundance of refuges grazing could hardly be more than in the pelagic systems, where it is accepted not to exceed 0.6–0.7. By taking this coefficient to be lower (ca. 0.5), the authors obtained absurdly high primary production: 200×10^6 kg km^{-2}, which was 30 times more than that arrived at in its field estimations. This feature of the model seems to be easily explained by employing a too low K_1-coefficent (0.17 as average), and by the absence of the trophic group of microheterotrophs (bacteria). According to the model, all decomposition of organic matter proceeded via its direct grazing within the food web, while in the real reef ecosystem it was taken care of largely by microflora, which in the model was absent as a functional component. Equally absent was the detritus food chain, which had been proved to be important in real systems (Peres and Pickard 1969; Gerber and Marshall 1974; Chardy et al. 1988). Finally, as a real reef ecosystem is actually an open system, it should not be internally balanced as it has been asserted, in the model, this being one of its basic principles. For example, the reef biofilter consumes a significant portion of energy from the passing oceanic waters.

Concluding the analysis of the ECOPATH model, it could be assumed that the simplification of the coral reef ecosystem postulated in it, which, in fact, has made possible its very creation, at the same time deprirved it of the necessary adequacy, and thus of serious scientific value. While any model represents actually only a hypothesis, it should not ignore also established facts. So we have to recognize that the reef ecosystem seems to be too complex to be managed in this way by imitation modeng. However, there is a rather realistic way to generalize also established data regarding the functioning of coral reef communities at the level of a whole ecosystem, i.e., the composition of the above-mentioned schemes of energy flows. An example of such a scheme is given in Fig. 10.7. The basic data used in it are given in corresponding chapters of this book.

According to the scheme, the basis of energy metabolism in a reef ecosystem is composed of the detritus pool. It is represented by the permanently regenerating stock of dissolved and particulated organic matter. The scheme includes values of the production of the food rations, of subsequent links in the food web, and of nonassimilated and nonconsumed food. All these values are expressed in $\mathrm{k\,cal\,m^{-2}\,day^{-1}}$. The main flows of energy are expressed by thick arrows. The energy input into the reef ecosystem is composed of bottom plants from internal sources of antotrophic production of planktonic algae, and of symbiotic algae, as well as coming from its external sources, like plankton and organic matter of the oceanic waters utilized by reef biota. On barrier and fringing reefs a significant portion of external organic matter reaches the reef systems from the land and from mangroves. This external organic matter enters into its detritus pool and is then incorporated into the food web via the intermedial participation of bacteria. Despite the abundance of herbivores in reef communities, a large portion of autotrophic production created by benthic plants enters the food web via the detritus pool → bacteria. The flow of energy from the bottom plant and coral communities via the detritus pool to the reef fauna is, according to this scheme, a main way of using autotrophic production in heterotrophic processes. This was proved by field studies (Chardy et al. 1988). By its size the second energy flow passes intraorganismically from symbiotic zooxanthellae to animal tissues of their hosts – corals, clams and forams. These organisms, especially corals and *Tridacna* clams, excrete a large portion of this translocated autotrophic production as mucus and DOM, which then enter the detritus pool and are

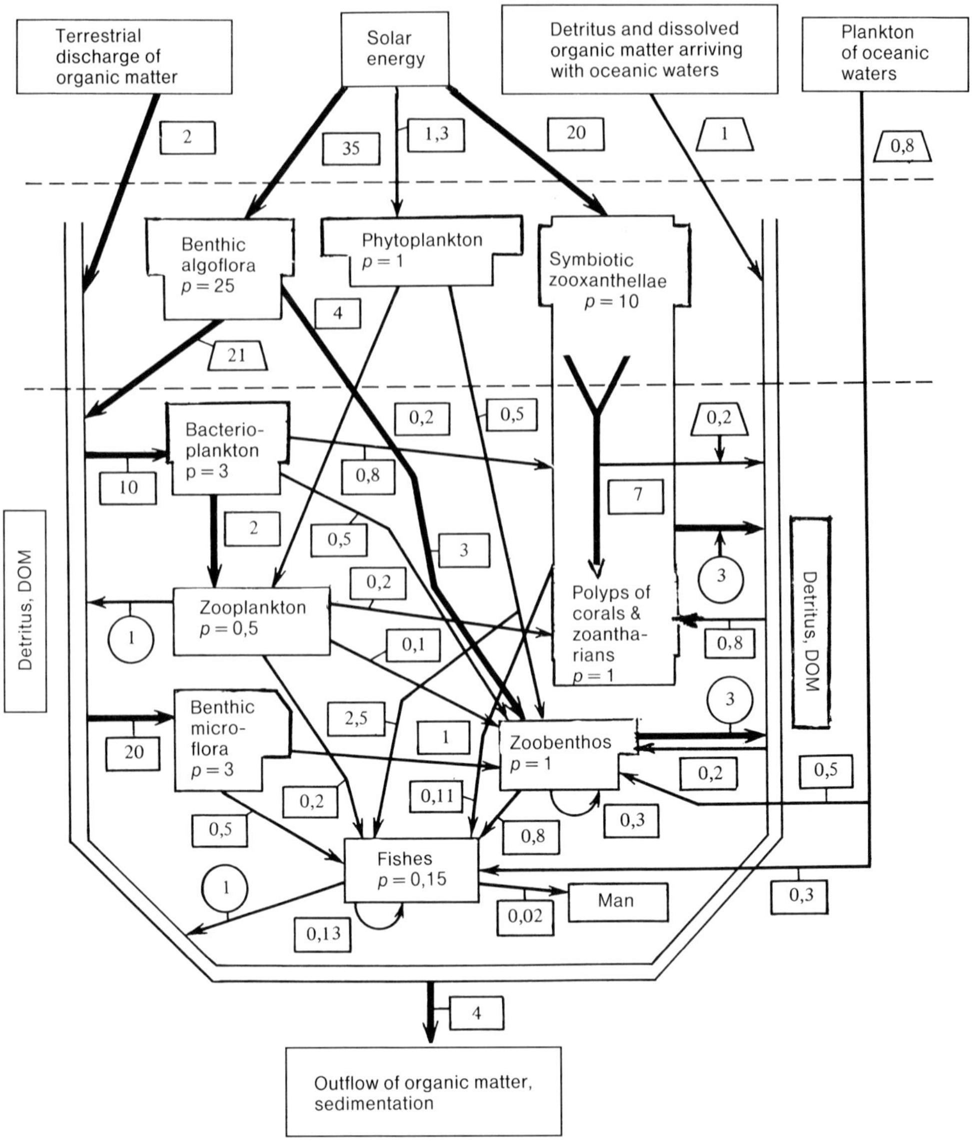

Fig. 10.7. Generalized scheme of the energy flow (kcal m^{-2} day^{-1}) in coral reef ecosystems; *P* production; *numbers in squares* energy of food ration of the subsequent component of the food web; *numbers in circles* nonassimilated food; *numbers in trapezia* nonconsumed production or the organic matter entering into the system from outside. (After Sorokin 1986b)

used through the microbial link. The microbial population in the reef ecosystem uses about 70% of the total energy flow in its metabolism, vote, that this, as well as the detritus pool, were omitted in the ECOPATH model. A significant portion of the energy flow passes via the reef herbivore fish and invertebrates which directly graze a significant part of the total plant production.

11 Destruction and Recovery of Reef Systems, Reef and Man

Coral reef ecosystems, as also the ecosystems of tropical rain forests, are among the most mature ecosystems on our planet. Such mature ecosystems are very sensitive to external impacts, be they natural or manmade, which violate their homeostasis. During the last three decades, catastrophic events affected coral reef systems, resulting in their destruction in vast areas; hurricanes, the warming of water by the El Niño, cuuurent extra low tides, *Acanthaster* plagues, and various kinds of anthropogenic stress were among the causes. The regeneration of thus degraded reef ecosystems takes decades. The close interconnection and cumulative effect of the natural and anthropogenic stress factors in processes of destruction of the reef systems caused by them appears to be clearer and clearer. For example, in polluted reef areas corals are much more exposed to the attacks of boring sponges and so they can be destroyed easier by waves. A definite causal connection between human activity and the *Acanthaster* outbreaks is also assumed.

11.1 The Acanthaster Problem

The "crown of thorns" sea-star *Acanthaster planci* is a large specialized corrallovore asteroid, which feeds exclusively on scleractinian corals. Massive infestations were first recorded in the early 1960s at some of the Indo-Pacific reefs (Barnes and Endean 1964; Barnes, 1966 Chesher 1969; Kühlmann 1971; Endean and Stablum 1973). They resulted in the catastrophic degradation of the latter. The number of reef sites infested and damaged by this star plague till the end of the 1970s drastically increased (Talbot and Talbot 1971). At the beginning of the 1980s many of the previous deteriorated and in part also recovered reefs experienced repetitions of the *Acanthaster* invasions (Moran, 1986). At reefs attacked by hordes of these stars 50 to 90% of scleractinian corals die. The subsequent recovery might take 10–40 years, depending on the degree of damage and on not being attacked by stars repeatedly. But events on reefs of the GBR region, which had been damaged earlier by *Acanthaster* in 1960–1969 indicated the possibility of repetition of attacks. This implies the threat of chronic infestation of reefs by this plaque, endangering their very existence in large reef regions (Chesher 1969). So the attacks by this sea star on reefs could be considered as an ecocatastrophe on a large scale, triggered mainly by human activity, which has increased in the postwar period (Kühlmann 1971; Endean 1974). The alarming descriptions in papers of the awful pictures preseated by the degradation of formerly beautiful reefs to a state of rubble heaps covered with ooze, attracted the attention of, government agencies, the public and researchers to the *Acanthaster* problem (Endean 1973). Special studies were organized to evaluate the scale of the disabitrous infestations and to find ways to combat them (Endean and Chesher 1973).

The genus *Acanthaster*, besides the "crown of thorn" star *A. planci*, includes also two more species: *A. ellisi* and *A. brevispinnus*. *A. ellisi* is a very rare spiny star found on Philippine reefs, which feeds on corals. *A. brevispinnus* is detritovore. All three species are genetically very close and easily form hybrids with each other (Lucas and Nash 1985). *A. planci* is supposed to be a species which appeared recently, descended from *A. brevispinnus* and acquired the ability for predation on corals. *A. ellisi* could be an intermediate form between them on the way to these evolutionary developments. The *Acanthaster* stars have very flexible motile spiny rays. By their use they are able to crawl easily upon colonies of corals and hold onto them. This lends them an advantage in competition with the other large corallovore asteroid *Culcita*, a bulky star though devoid of rays, which is able to graze only on small coral colonies or crusts. As a consequence, when the *Acanthaster* species acquired the ability to digest the protective cetyl-palmitate substances, soaking up the coenosarc of scleractinian corals, and to feed on their tissues, it dropped out of competition and was free of food limitation (Benson et al. 1978). In spite of this ability, these stars were among the rare species of large reef invertebrate predators, just like the star *Culcita*. The average size of *A. planci* is 30–35 cm, and some specimens can be up to 50–60 cm large. The dorsal side of its body is covered by large poisonous thorns 2–4 cm long. The tissues of the stars is saturated with saponin, repelling potential predators by the bad taste. *Acanthaster* has many rays: between 9 and 21. At the tip of the rays light-sensory organs and chemoreceptors are situated. With the aid of its long ambulacral legs the star moves very swiftly, which is unusual in other asteroids – up to $20\,m\,h^{-1}$, and may migrate several kilometers a month (Chesher 1969; Endean 1973). When attacking the coral head, the star crawls onto it with the aid of its rays, hiding the ambulacral legs which are sensitive to the coral's nematocysts. After fastening its body on the colony the star turns its stomach inside out. The stomach, as a white foam, engulfs an area of $20–30\,cm^2$ on the surface of the coral colony and digests tissues there for 4–5 h. Then the star pulls its stomach in, moves to a neighboring part of the colony and repeats the procedure. When the number of stars on the reef is normal ($3–4\,sp.\,km^{-2}$), they behave as nocturnal predators, being active only at night and hiding in daytime. During the night, the star "processess" in the above manner 2–3 lots of the coral's surface, most often at various colonies. But when the infestation proceeds and their papulations become dense (5–10 sp. 100 m^{-2}), the stars feed throughout the day and each eats up about 160 $cm^2 day^{-1}$ of the coral's surface. In this conditions one star eats 6–12 m^{-2} per year of dense coral cover. On their way within the "front" moving through the reef, the stars kill up to 80–90% of all scleractinian corals, leaving behind what is actually dead reef (Chesher 1969, Endean 1979).

In experiments, the stars displayed a definite preference lowards prey corals. They preferred the species of *Pocillopora*, *Acropora* and *Faviids* and avoided the poritids. But in very dense aggregations they attacked even the latter. Under natural conditions their preference depends also upon turbulence conditions. In the surf zones the stars prefer ramose colonies, on which they hold themselves better. But in calm zones they attack also the massive corals, on which they cannot stay in the surf zones (Campbell and Ormond 1970; Cameron et al. 1989). Sometimes they avoid also the large old colonies of *Pocillopora*, while in experiments this coral is among the most "beloved" by stars. It appeared that they are defended by the symbiotic crabs *Trapezia* and the shrimps *Alpheus*, which live in colonies of *Pocillopora* (Glynn 1976, 1981, 1984). The hydrocoral Millepora is avoided by stars being defensed by its powerful nematocysts (Lewis, 1989).

The stars *A. planci* are bisexual. Once a year they assemble in large aggregations for the purpose of multiplication. They cast out eggs and sperem into the water. One star produces about 60 million eggs per year. Their pelagic larvae bipinnaria feed on phytoplankton (Fig. 11.1). After 3 weeks they settle and transform into the crawling larvae brachiolaria, which feed on coralline algae (Yamaguchi 1973; Lucas 1973). Passing through their metamorpothosis, the larvae develop up to the stage of young stars with five rays, which start to feed on corals. Half a year after their hatching the young stars attain the size of 1–1.5 cm. The rate of their growth is 1–3 cm per month (Olson, 1987). In 2–3 years since their settling the star becomes mature, having a size of 20–30 cm (Kenchington 1977). The percentage of survival of *Acanthaster* larvae is very low, just as in any large predator. Their populations are represented mainly by adults. Their pelagic larvae are grazed out by zooplankton and fish. The spawn and the settled young stars are eaten by omnivorous and predatory trochid, muricid and conid gastropods (Endean 1973). The adult stars, although well defended, are eaten anyway by some large predatory gastropods, like *Charonia tritonis*, and, probably, *Cassis cornuta*, and by fish like some balistids (*Pseudobalistes flavimarginatus*), tetradontids (*Arothron hispidus*), and groupers (*Cheilinus undulatus*). The most specialized predator among them is *Charonia tritonis*, which feeds exclusively on large stars and on holothurians. When attacking the *Acanthaster*, this gastropod, with the aid of a peak at the edge of its shell, turns the star over and grazes it out from its unprotected ventral side. In experiments, one specimen of *Charonia* killed up to ten *Acanthaster* per

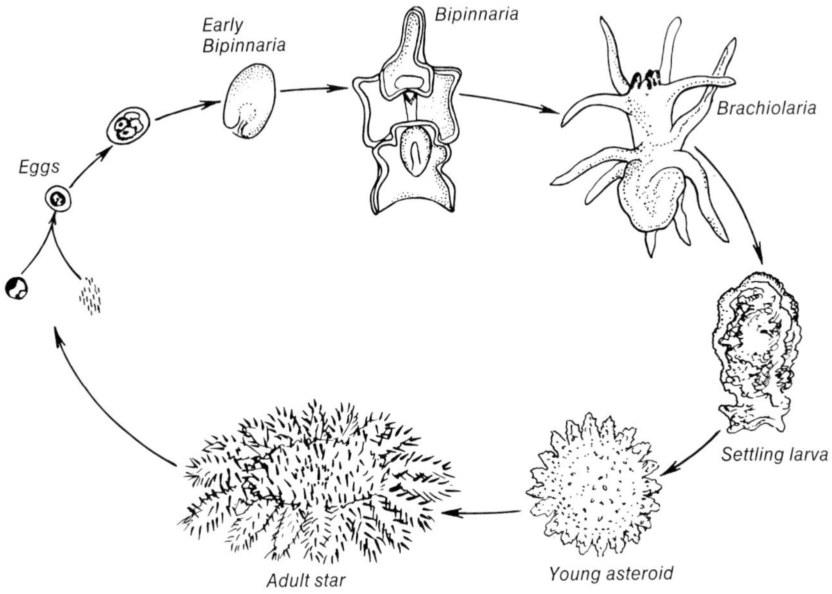

Fig. 11.1. Life cycle of *Acanthaster*

month. On reefs, where *Charonia* is present, about 50% of *Acanthaster* stars bear signs of its attack (Endean 1973). On the reefs off the eastern Pacific *Charonia* is absent. The main enemies of *Acanthaster* here are symbionts of corals, the polychaete *Heminocera picta* and the shrimp *Pericardia striata* (Glynn 1981). Often they attack the star together. The polychaetes cut a hole in the ventral side of the star. Then they both crawl into the wound and graze out the star from the inside. They attack not only adults, but also the young stars. They keep about 6% of the total *Acanthaster* population under attack. The shrimp and the polychaete are the representatives of the infauna of the coral *Pocillopora* in these reefs. Therefore, the *Acanthaster* avoid colonies of this coral. Glynn (1981) supposed that just these two small but voracious predators control the modern population of *Acanthaster* on the East Pacific reefs off Panama, keeping it at a relatively high (7–30 sp. ha^{-1}), but not catastrophic level, thus saving them from destruction.

As mentioned above, before the outbreaks of the 1960s the presence of rare populations of *A. planci* (3–6 sp. km^{-2}) had been recorded at many Indo-Pacific reefs. On an average, one star could be met with after 5 h of swimming over the reef. During surveys of the vast benthos, no denser populations of it had been detected. Only temporal aggregations of stars, during periods of multiplication, had been recorded (Campbell and Ormond 1970; Cheney 1974). The population explosion of *Acanthaster* first started on reefs of the western Pacific. Then it spread to reefs of its eastern part, its members still proceed to grow. In zones of its attack its density attained

100–1000 sp. ha^{-1}, which is 4–5 orders more than natural. During 20 min of swimming over the infested reef, 30 to 1000 stars could be recorded, which had diameters of 30–60 cm (Chesher 1969; Dana et al. 1972; Kenchington 1987). On one coral knoll near the Natan reef (GBR) R. Endean recorded over 700 stars. On one platform reef of moderate size in the same region their population could number 10^5–10^6 stars. At the same time, on some of the nearly reefs, in areas of infestation, on eight such reefs only six stars were found. On 22 reefs to the north off Lizard Island, outside the zone of invasion in 1970 not a single star had been found. But later these same reefs were severely damaged by *Acanthaster* (Cameron and Endean 1981). In their density of 150–200 sp. ha^{-1}, the stars kill a main portion of scleractinian corals on the infested reef, which looks afterwards like an abandoned, mournful cemetery (Endean 1976; Cameron 1977). At the reefs of the Eastern Pacific, which had been also infested, their density did not rise above 20–50 ha^{-1}. It was about three orders more than their natural population density, but 3–5 times less than the catastrophic level. Therefore, most reefs in this region, which were (and even now are) infested by *Acanthaster*, have not been so destroyed as some in other places of the central and western Pacific. They reached a kind of equillibrium at such a density of this predator. But the community structure of scleractinian corals is changing in the direction of large colonies of the coral *Pocillopora* dominating in them, as these are defended by their infaunal symbionts.

The most severe attacks of *Acanthaster* befell in 1957–1973 on the Indo-Pacific reefs situated along the coasts of high islands and continents. Among them were most reefs in the central part of the GBR between 14° and 20007 S, reefs off Okinawa, Guam, Saipan, Palau, Fiji, Hawaii, Eastern Samoa Islands and off the Comore Islands (Indian Ocean, Goreau et al. 1973, Wilson and Marsh 1974, Sano et al. 1987, Eahl et al. 1990). At some of the atolls and on reefs around some other high islands in the Cook, Tuamotu and Society archipelagos, near the west coast of Australia, around Mauritius and Madagascar, and off the coast of Kenya the density of *Acanthaster* increased by 2–3 orders compared with the natural one, and unmistakable signs of their attack on corals appeared (Barnes 1966; Chesher 1969; Kühlmann 1971; Dana et al. 1972, Endean and ; Stablum 1973; Endean 1973, 1974; Marsh and Tsuda 1973; Sale et al. 1976; Kühlmann 1985). On atolls, situated in the open deep ocean, the *Acanthaster* outbreaks were rarer, and as a rule they were not so destructive, except on those subjected to intensive human activity, either agricultural, constructional or military (Marsh and Tsuda 1973). On such atolls the attacks of stars were especially disastrous. Among the se there could be mentioned the Majuro, Kwargelein and Tarawa atolls. Later outbreaks occurred in the Red Sea, in Malaysia and in Vietnam (Ridzman 1988; De Silva 1988).

On the GBR the first outbreak of *Acanthaster* was observed at some lagoonal reefs. By 1966 the stars infested over 50% of reefs reviewed there between 15° and 19° S. In 1969–1971 82 reefs in this area had been reviewed out of their total number of 152 (Endean and Stablum, 1973 Ormond and Cambell, 1974; Moran et al. 1985). At 18 of them mass mortality of corals killed by stars was recorded. At 36 reefs large aggregations of them were observed and signs of their attacks on corals were visible. But at 26 reefs situated near the infested ones the stars were absent. Another review revealed that in a section of the GBR between 14° and 20° out of 240 reefs over 150 were infested. At the beginning of the 1970s large hordes of stars appeared on reefs to the north of 14° S up to 12° S, and on many of the Swane reefs in the southern-section of the GBR (Endean and Stablum 1975). Among the worst destroyed during this period were the reefs of Green Island. The number of stars on them in 1966 was estimated to be ca. 2 million. Over 2.5 years the stars practically exterminated scleractinian corals on them (Endean 1973, 1976). After they had destroyed the coral communities at most of the reefs in the central GBR region, by the middle 1970s their density started to decrease. Till 1977 it remained high only in the south at some of the Swane reefs. But in 1979–1981 reefs of the central GBR, previously also damaged by *Acanthaster*, were subjected to a repetition of its invasion, as it had been predicted by R. Endean (1977). The young corals which started to grow on them were attacked by stars again in 1977–1982 (Kenchington and Pearson 1981; Endean and Cameron 1985; Fisk et al. 1988; Reichelt et al. 1990). So a dangerous tendency of permanent infestations seems to be developing, which could result in a gradual irreversible degradation or a drastic transformation of the ecosystems in vast reef regions.

Populations of *Acanthaster*, which have a density over 150–200 sp. ha^{-1} destroy scleractinian coral communities in the space of 1 to 6 years. For example, at the fringing reef off Guam Island on a stretch 38 km long over 90% of the corals were killed by stars during the time of only 2.5 years (Chesher 1969). The extermination of corals was most severe at leeward flats, in patch-reef zones and at outer reef slopes. Corals living in the windward surf zones, where the waves impeded the stars holding on to coral heads suffered less (Endean 1973). The stars avoid the hydrocoral *Millepora*, which has powerfu cnidicil. They practically do not eat the octocoral *Heliopora* and rarely attack large, round, massive corals, which are difficult for them to hold onto. On reefs, where the ramose corals *Pocillopora* and *Stylophora* are defended by their infaunal symbionts, the stars avoid these too. Rarely are the poritids attacked by stars, which are loaded full with various structural lipids and contain toxic substances. Only 5 cm^2 of surface area on a massive coral colony damaged by the star causes, in most cases, its death (Bak and van Eys 1980). An exact account of the percentage of coral mortality at some reefs of the GBR attacked by *Acanthaster* gave its level with 60–90% (Endean and Cameron 1985). On the Brewer reef in the same region out of the young corals that had appeared after the 1970–1972 attack, after the repeated attack by stars in the middle of 1983, 90–95% were killed (Moran et al. 1985). Once all corals on a reef are eaten, the stars may move to the neighboring one advancing on the bottom between them at depths down to 60 m (Endean and Stablum 1973). At isolated reefs, after the extermination of corals the stars for a long time still remain on the reefs. Then they gradually die out.

After an *Acanthaster* invasion only white skeletons of scleractinians remain. Then they are overgrown by filamentous algae and covered with black patches of slimy colonies of blue-green algae. The coral's infauna dies. Fish abandon the destroyed reefs which present a pitiful picture to the observer. The colonies of killed corals are later attacked by boring animals and gradually collapse (Chesher 1969). Instead of a flourishing reef, 2–3 years after the stars' attack only heaps of rubble and the rare colonies of dead massive corals covered by alcyonaceans remain. Later on, such killed reefs are most often inhabited by alcyonaceans, which form on the reef flat a continuous cover. The surface of massive corals is overgrown by incrusting *Sinularia*, and amid the rubble colonies of *Sarcophyton*, *Lemnalia* and *Paralemnalia* appear in mass. Thus in the place of exterminated reef usually alcyonacean associations are forming, which continue to exist sometimes for several years preventing the restarting of scleractinian growth (Endean and Stablum 1973; Nishihira and Yamazato 1974; Cameron and Endean 1981). This author has observed this scenario on the reefs of Lizard Island (GBR), where formerly beautiful reefs were attacked by *Acanthaster* in the early 1970s But even in 1986 most of the reef-flat areas there were occupied by thickets of ugly alcyonaceans without any sign of recovery of scleractinian populations.

On the fields of rubble, which remain on the destroyed reef, the filamentous algae are later gradually replaced by coralline algae, at which point the recruitment of new scleractinian populations starts and their new colonies begin to grow. Among them the most numerous are the opportunistic fast-growing specises of *Acropora*, *Pocillopora*, *Seriatopora*, *Turbinaria* and *Montipora*. The more reef has been damaged, the later starts the recovery of scleractinian corals (Endean 1973; Endean and Stablum 1975). On one of the platform reefs of the GBR, destroyed by stars, even 7 years after the invasion the percentage of its cover by coral was only 3%, while before the invasion if had been about 30%. Most corals in this recovering community were represented by large colonies of *Porites* and *Diploastrea*, which survived the invasion, and by the newly appearing young colonies of *Acropora*, which numbered about 60 m^{-2} (Pearson 1974). About 10–15 years after invasion, in reef areas where scleractinians had managed to replace alcyonaceans, their communities were composed of average-sized ramose and foliose colonies and had a low species diversity. These regenerating communities are poor in large massive corals and are not resistant to the surf. So they were susceptible to destruction during strong storms, and also endangered by the repetition of *Acanthaster* attacks. In such permanently inhibited "ill" reef communities bioerosion could overwhelm the calcification processes, as it was pointed out by first observers of the *Acanthaster* phenomenon (Goreau 1963). Therefore, such reefs permanently inhibited by *Acanthaster* would have a tendency toward gradual destruction. Even if repeated attacks stop, a reef once damaged by *Acanthaster* could completely recover only after 20–40 years (Endean and Cameron 1985).

The actual cause of recent *Acanthaster* outbreaks on reefs of the Indo-Pacific is still disputed (Eann, 1990). Existing hypotheses are based mainly on two controversial postulates: (1) the outbreaks are a natural periodical phenomenon and happened also in the past (Walsh et al. 1971; Dana et al. 1972; Vine 1973), and (2) they are a unique biological phenomenon, the ecocatastrophe, which did not happen before, being in some manner triggered by increased human activity in reef regions (Kühlmann 1971; Randall 1972; Endean 1973, 1974, 1977; Cameron 1977; Cameron and Endean 1985). The followers of the first opinion have serious arguments of their own but, the latter position seems to be more reasonable. (It coincides also with the author's own observations.) In fact, in abundant literature on the reef bottom fauna, describing results of more than a century of observations in various areas of the Indo-Pacific, no mention was made of such catastrophic infestations of reefs by this star. Through the whole period before 1960 *A. planci* had been described as a rare species of sea stars. R. Endean (1973) stated that such allusious were absent also in the mythology of island aborigines. Cases of the presence of skeletal remains of *Acanthaster* in ancient reef rocks (Frankel 1977; Walbran et al. 1989) could not be accepted as proof of past outbreaks because of the ability of these stars to form aggregations in their breeding period, while still remaining a rare species. Also, quite reasonable is the remark that the ecosystems of Holocene reefs could not exist at all had it been periodically subjected to catastrophic damage by *Acanthaster* invasions, as happened in 1960–1970, which, as we have seen above, have a tendency to become chronic.

As for this author's own experience, during 2 weeks of swimming among reefs off Tahiti in 1961 he did not meet with a single *Acanthaster*. But in 1968, when large construction works started there in the lagoon of the Barrier Reef, in connection with future nuclear tests at Moruroa, during 1 h of swimming he came upon 2–3 stars, which were devouring corals in the daytime. As it was pointed out above, this transfer from nocturnal to daytime habit is a sign of future outbreak of invasion by this star. In 1984 this author witnessed a heavy invasion on the reefs near the Hon Gom Peninsula along the coast of central Vietnam. The reefs in this area were severely stressed during the war (bombing) and after (lime excavation, explosives fishery). In accordance with the author's observations, in 1981 the stars were practically absent there, but in 1984 their invasions at some reefs were terrific. Reefs were literally eaten up. Their density in many places was up to one per $1\,m^{-2}$. These observations seem to be evidencing, that the outbreaks in the above-mentioned sites were triggered by the destabilization of the reef communities there by man (Chesher 1969; Fishar 1969). This opinion was expressed by researchers who observed the first outbreaks of the 1960s (D. H. Kühlmann). In fact, they had been localized precisely in zones of intensive human activity moreover that the stars theirself are resistant to pollution, for example, in the GBR zone the acanthaster outbrake started first near a large turistic center and port of Cairns. The most severe attacks of *Acanthaster* in the 1960s–early 1970s occurred just in the places of active military construction works: at Okinawa, Saipan, Guam, Kwargelain, Majuro, Tahiti, Molokai, East Samoa. In the reef regions off Fiji the outbreaks coincided with the expansion of the tourist industry and with the mass collection of shells on reefs, especially of *Charonia tritonis* valued by run-of-the-mill tourists. On reefs of Palau outbreaks coincided with the construction of a large oil terminal complex. The reef ecosystems about all the regions listed were destabilized also by the development of agriculture with the employment of pesticides and fertilizers, by the intensive collection of shells and corals, and by the spear and explosious fishery. The activity of tourists and local people on the reefs greatly increased after the wide papularity of diving equipment: masks, flippers and scuba. All these kinds of human activity undermined the homeostasis in reef communities and thus promoted the population explosion of *A. planci* (Chesher 1969; Artonius 1971; Endean 1973; Nishihira and Yamazato 1974).

Being a highly specialized corallowore predator, *A. planci* has no competitors for food. Therefore its density on the reef should be controlled only by a low rate of propagation, connected with the low density of populations of this bisexual star and with grazing pressure at different stages of its development. The overcoming of these two barriers might lead to the possibility of its population explosion. According to Endean's hypothesis, in reefs subjected to modern human activity the number of fish and invertebrate predators feeding on spawns, larvae and the young of stars decreases. Among such predators are groupers, murexes, conuses, Charonia, Cassis – animals that are among those most hunted by tourists and sought after by the market. Special attention was given to Charonia by R. Endean, which he regards as a key species controlling the population density of adult stars (Endean and Stablum 1973). The decrease in grazing pressure results in the gradual building up of the number of young and adult stars. It increases further their successful multiplication. The appearance of a mass of larvae under conditions of decreased grazing pressure could cause a catastrophic growth in the number of this predator at some reefs. Accumulating on them during the periods of their multiplication, they are building up a front for the attack. Their hordes first gather in deep zones of the outer reef slope. After exterminating the corals there, many stars go up to the shallow part of the reef, leaving behind a devastated reef. Then the hordes of stars, moving rather fast by the bottom, migrate onto the neighboring reefs, thus creating new areas of infestation, which may not be directly connected with anthropogenic stress, such a secondary infestation might also come about by a transfer of a mass of star's larvae from places of invasion to healthy reefs by currents.

This ecological hypothesis offered by Kühlmann (1971) seems to this author to fit best the observed facts. The assumption of possible repetitions of invasions within a short time, especially in regions permanently stressed by man, which had been made on the basis of this hypothesis, was later proved. The subsequent invasions which occurred in the GBR zone in 1979–1983 prove the point that the *Acanthaster* outbreaks are a unique ecocatastrophe caused by an external cause (i.e., by man), rather than the consequence of natural periodic fluctuations in the star's numbers, which also occurred in the past. The opinions of the proponents of this latter aspect are divergent in their explanations of the cause of outbreaks. One of them sees it in their ability to migrate long distances and to form breeding aggregations. They suppose that the stars, thus accumulated in one place, graze out corals, which in turn also graze the star's larvae and eggs. As a result, the number of surviving star's larvae should increase, thus furthering their successful breeding. Repetitions of such cycles could gradually lead to an increase in the total density of their populations (Ormond and Campbell 1973, 1974). Other authors see as starting place for the outbreaks the reefs, damaged by cyclones (Dana et al. 1972). The cyclones, smashing all corals onto the shallows, force the stars to migrate down to the deeper zones of the slope for feeding, where corals remain safe. In this way they might form aggregations, which could form the start of the "army of invasion". Birkeland (1982) thinks that good nutrients conditions for phytoplankton, created by periodic rains near high islands, might improve the possibility of survival of the star's larvae. This could lead to an increase in the star's numbers. Vine (1970) also see the cause of outbreaks as a natural periodic phenomenon in the fluctuation of conditions for survival of the star's larvae. We may agree that all the above phenomena could cause the periodic fluctuations in the predator's population density, but only on condition that the pressure of grazers is lacking. But the main objection is that if these fluctuations are natural they ought to have been recorded before at some reefs, which actually is not the case. Many facts cannot be explained either by Kühlmann's hypothesis. For example, why did some reefs situated amid areas of the most severe star outbreaks remain safe? Moreover, in such uninfested zones some reefs heavily exploited by man also remained. Therefore, the problem of *Acanthaster* plagues is still one of the most intriguing and urgent in reef ecology.

As for the measures in the fight against the infestations, stars were collected by scuba divers. Off eastern Samoa and Okinawa divers collected hundred thousands of them. There were attempts to stop the moving front of stars by electrified wire – an electric barrier. A method was also developed for killing them by injection of copper-sulfate solution by divers (Chesher 1969; Sale et al. 1976). But all these measures appeared to be only palliative and extremely expensive. To kill one star on the reef costed 35 dollars (Johnson et al. 1990). Effective ones could be developed only after finding an adequate solution to the problem of the origin of *Acanthaster* outbreaks.

11.2 Destruction of Corals by Physical Stress

Among the factors of physical stress which damage corals, there could be mentioned wave action, decreases in salinity, siltation, exposure, and overheating during extra-low ebbs. Sometimes the action of these factors reach a catastrophic level, causing mass mortality among corals. Most often such catastrophic events are connected with the passing of tropical cyclones over the reefs – typhoons or hurricanes – which are accompanied by heavy rains causing the decreases in salinity and massive siltation of reefs. The rates of wind velocity in the zone of passage of an average strong tropical cyclone could be 100–250 km h^{-1}. In the open sea such a wind creates waves 8–10 m high. Near the reef edge such a storm will raise waves 4–8 m high, and in the lagoon up to 3 m (Highsmith et al. 1980). The waves induced by such a cyclone smash all the ramose corals, especially on the winddward side of the reef, where the depth of the zone of total destruction of corals may reach 20–25 m. The massive and firm foliose corals are destroyed by cyclones to a lesser degree (Stoddart 1985). Some of their large colonies on being thrown out onto the flat remain these as so-called nigger heads. The destruction of corals during the cyclone proceeds mostly under the action of masses of rubble moving to and fro at high speed. They smash the ramose corals and irreversibly damage the massive ones. The mass mortality of corals during the passing of cyclones occurs also on the windward sides of reefs. Corals perish there not only through large waves passing over the flat, but also under the impact of heavy siltation by stirred-up bottom sediments and by terrigenous suspended matter washed out by rains, which usually accompany such cyclones. The corals covered by sediments die off in 3–4 days. The rains cause decreases in salinity, apart from the cyclones themselves often causing extremely low ebbs.

The passage of an average-strength cyclone causes 50 to 80% mortality among corals in the upper reef zones. Some corals survive in patches situated in places shielded from the direct action of moving rubble masses. These patches become then the centers of rehabiliation of damaged coral communities (Graus et al. 1984). But extremely strong cyclones with winds of 300 km h^{-1}, as a rule, destroy corals on reefs almost totally. They induce the movement of millions of tons of rubble. The reefs may even change their configuration after the passage of this kind of cyclone. New keys and islands appear on them. Such cyclones occur in reef zones one or two times in a century (Maragos et al. 1973). After an extra-strong ccylone the reef benthic communities cannot completely recover for several tens of years. Often they do not recover at all, remaining overgrown by coralline algae and seaweeds (Woodley et al. 1981; Pearson 1981). For example, there could be mentioned the hurricane Hatty, which destroyed the reefs of British Honduras in 1961. Even after 11 years they showed no signs of recovery (Stoddard 1974). In 1980 the strongest hurricane in the century, Allen, destroyed the reefs off Jamaica (Woodley et al. 1981). The corals were smashed even in protected places. According to D. Stoddart, the now bare algal flats of many Indo-oceanic reefs, were not so long ago covered with corals, which were then extinguished by fierce typhoons and could no longer recover.

The recovery of coral communities, damaged by cyclones of average strength, starts with the recruitment of pieces of *Acropora* branches, most of which usually survive. For example, on the Belize reefs, damaged by the cyclone Greta with winds of 120–150 $km\,h^{-1}$, the ramose corals were destroyed down to a depth of 25 m. But about 40% of torn-away colonies of *Acropora palmata* survived and started to implant themselves and grow at new places. Most pieces of the branches of *A. palmata* and *A. cervicornis* also survived and recruiting in new places began to take place, thus giving start to the formation of new colonies (Highsmith et al. 1980). Because of the intensive growth of these newly settled pieces of ramose corals, the rate of calcification in these damaged reefs did not decrease but increased (Connell 1978). Periodic cyclones are an important factor in the formation of the zonal structure of reef bottom communities (Dollar 1982). Over 62 years, beginning with 1900, a total of 286 hurricanes passed over the Caribbean reefs. Periodic cyclones are an important factor in the evolution of reef corals. In perticular, this factor furthered in acroporids the ability of pieces of their colonies to survive, quickly settle and grow fast in new places (Highsmith et al. 1980).

In the recovery of coral communities after such catastrophic destruction an important role is played also by the recruitment of planulae and the growth of new coral colonies (Endean 1976). In 1975 we observed a mass colonization of the bare reef flat at the Funafuti atoll, from which the corals had been wiped off by the typhoon Baby in 1970, by young colonies of *Acropora hyacinthus*. The colonies had an even size of 5–7 cm and had grown up from the settled planulae. Their number attained ten per m^{-2}. The success of recolonization of damaged reef by corals depends greatly upon the amount of sites where their colonies survived, becoming sources of new planulae, and also upon the possibility of transfer from neighbouring, less destroyed reefs. The success of recruitment of new planulae depends upon the safety of the herbivorous fish populations after the cyclone, which clean the hard substrate for their settling. Within 10 years after the cyclone attack the colonies of *Acropora* attained a size of 50–70 cm. The colonies of massive corals could attain sizes of 20–40 cm only after 20 years. The total recovery of a coral community after the passage of a tropical cyclone takes 5

Table 11.1. Evaluations of rehabilitation time in coral communities destroyed during catastrophic events. (After Coles 1984)

Location	Cause of corals' mortality	Time of rehabilitation, years
Persian Gulf	Seasonal drops in temperature	4–5
Red Sea	Extra-low ebb	5–6
GBR	Typhoon	15–20
GBR	*Acanthaster* invasions	10–20
Reefs in Indian Ocean	Typhoon	20–25
Guam	*Acanthaster* invasions	20–30
British Honduras	Strong hurricane	75–100
Hawaii	Lava flows	20–50
Hawaii	Dredging and terrigenous siltation	15–30
Hawaii	Explosion excavations and siltation	8–64

to 25 years, depending on its strength, on the morphometry of reefs and the above-mentioned above conditions of planulae recruitment (Table 11.1). Thus, after the hurricane Dona, which had winds up to 200 km h^{-1}, the reefs off Florida recovered within 7 years (Shinn 1976). But there are also cases known when after such a cyclone of average strength reefs could not recover even after 10–15 years (Endean 1976).

Modern developments of anthroprogenic influence on reef ecosystems seriously undermine the rehabilitation ability of coral communities after catastrophic events, delaying so this process. It was, in perticular, demonstrated on the example of the Aquaba reef-flat coral communities killed by extra-low ebb. On reefs situated outside pollution influence, the coral communities recovered within 5–6 years, while on reefs subjected to pollution from the nearly port, they did not show even a sign of regeneration (Loya 1976c). Therefore, the opinion by Highsmith et al. (1980) about a probable enhancing of reef growth under cyclonic impacts seems to be premature (Pearson 1981; Porter et al. 1981; Rogers 1985).

The extra-low ebbs also result sometimes in mass mortality of corals in the reef-flat zone. It is caused by overheating of exposed corals or by drastic drops in salinity when it happens in rainy weather (Loya 1976c). During an extremely low ebb on reefs off Eilat over 80% of corals died. Mass mortality of corals on reefs, situated near continents or high islands, could be caused by heavy rains, which accompany the passing of tropical cyclones. They die through being affected by fresh water, by siltation and light deficiency, caused by elevated water turbidity (Goreau 1964b). Woodcutting and agricultural development in coastal areas greatly increase the terrigenous outflow from land to nearly reefs during cyclones, thus enhancing their damaging impact on them. For example the Stone Island reef (GBR) died off under the influence of massive siltation by terrigenous material washed out from fields by prolonged rains in the early 1930s, and has not recovered even after 35 years (Endean 1976). Mass mortality of corals happens also at

Table 11.2. Cases of mass bleaching and mortality of corals and other benthic animals during the period of El Niño in 1983. (After Lessios et al. 1983, 1984; Glynn 1984; Dahl 1985; Fisk and Done 1985; Harriott 1985; Oliver 1985; Jaap 1985)

Region	Phenomenon	Supposed cause
Eastern Pacific (Panama, Columbia)	Bleaching and death of 60–90% of corals	Overheating
Tokelau (French Polynesia)	Bleaching and death of corals on shallows	Long exposure and overheating
GBR near Townsville, Cairns, Lizard I.	Bleaching and death of corals (mainly young colonies of *Acropora* and *Pocillopora* down to 15 m depth	Exposure and overheating
Okinawa I.	Bleaching and death of corals and actinians down to 10 m	Overheating
Java, Pulau-Pari and Seribu I. (Indonesia)	Bleaching and death of corals and actinians down to 10 m	Causes are not clear
Bahamas	Bleaching of corals *Diploria* down to depth of 5–8 m	Causes are not clear
Florida	Bleaching and partial death of corals, gorgonians and *Millepora* down to depths of 14 m	Exposure, overheating
Caribbean reefs	Bleaching and partial death of corals and other anthozoans (25 species in total)	Overheating
Caribbean reefs	Mass mortality of urchins and asteroids	Causes not clear, possibly overheating

high-latitude reefs during severe winters or extra-hot summers (Shinn 1976; Rogers 1985). In some regions the mortality of corals has been caused by earthquakes and volcanic activity (Stoddart 1972, Grigg and Maragos 1974).

Recently, the phenomena of mass bleaching and subsequent mortality of corals were recorded on a global scale in vast areas of different reef regions (Table 11.2). In the eastern Pacific, it happened during the strong El Niño period of 1982–1983 at reefs off Panama, off Columbia, and off the Galapagos Islands (Glynn 1983, 1985, 1988; Glynn and D'Croz 1990). In the central Pacific, it was recorded on reefs of Tokelau, on Christmas Island and on some reefs of Polynesia and Indonesia (Brown and Suharsono 1990), and in the western Pacific on reefs of the central GBR (Brown, 1990). In the western Atlantic, mass bleaching, not only of scleractinian corals but also of hydrocorals, gorgonaceans and zoantharians, accompanied by mortality even of sponges and sea urchins occurred in the summer-autumn period of 1987. This phenomenon was observed on reefs of Jamaica, Puerto Rico, Florida, the Bahamas, and of San Croix (Jaap 1985, Roberts, 1987, Williams and Bunkley – Williams 1988). It is remarkable that during the first half of this century until the 1960s such phenomena had not been ever recorded in this area. Only casual accidents of mortality happened during that time on some reef shallows, caused by extra-low ebbs, so that their cause had been evident.

The phenomenon of bleaching of corals stressed by high temperature, oxygen deficiency, or by the presence of toxic pollutants manifests itself in the expulsion of zooxanthellae by their polyps. High temperature stimulates the activity of lysosomes in the cells of zooxanthellae, triggering the process of theirselfdigestion in polyps (Sucharsono and Brown 1989). The bleached corals then most often soon die. The percentage of mortality among bleached corals is especially high in corals sensitive to stress, like *Pocillopora*, and some species of *Acropora* and *Agaricia*. Over 60–90% of their bleached colonies subsequently die (Harriott 1985; Fisk and Done 1985; Cook et al. 1990). But some corals, like *Montastrea* and *Millepora*, may survive bleaching and later on may recover. There are also corals that do not bleach et all, like *Porites* or *Psammocora*. But because the first of their above-nentioned groups dominates in upper reef zones, the mass bleaching in coral communities leads to the death of up to a half of the coral populations in the upper 10–15 m zone.

The bleaching of corals was first recorded on Guam in 1962, on reefs of Florida in 1978, on Okinawa and on the Christmas Island in 1980 (Goreau 1964b; Jaap 1979; Yamazato 1981; Egana and Di Salvo 1982). These authors believe that the main cause of these events could be found in anthropogenic influence. In the first case it was the washing out of the soil, a consequence of military construction activity, and in two other cases pollution. The global destabilization of reef ecosystems took place in 1982–1983 and was accompanied by mass bleaching and mortality of scleractinian corals (Glynn, 1982; Williams and Williams 1990). Most resarchers see its cause in a general warming of the world ocean. It was brought about at that time by the strongest warming during the century of El Niño (Glynu 1985). The surface water temperature rose in the tropical eastern Pacific by 4–6 °C, and in other tropical regions by 2–3 °C. In the western and the central Pacific the sea level during the El Niño period sank by 50–70 cm. It resulted in long exposure of vast shallow areas during the low ebbs, moreover it happened in conditions of the above-mentioned general surface warming of oceanic waters (Glynn 1984; Oliver 1985; Fisk and Done 1985; Glynn et al. 1988).

From the viewpoint of the first approach, corals had he because adapted to the fluctuations of temperature. But actually this is not so. It appears that the usual summer temperatures of 29–31 °C in the tropics is the upper temperature limit for their normal growth. They can endure only short-term rises up to 34 °C. A long-term warming of the water up to 32–33 °C cause such stress that they start to expel their zooxanthellae and to excrete excess amounts of mucus. The mucus stimulates microbial growth in water and on the surface of corals weakened by the tmperature stress Blached corals lose gradually their tissues (Szamant and Gassman 1989). Finally, it leads to their death Ducklow and Mitchell 1979b; Lessios et al. 1984). At the time of water warming during the El Nino period many reef systems were also inhibited by anthropogenic influence and by the *Acanthaster* plagues, which could also weaken their endurance to temperature stress (Glynn 1984).

In the same El Niño period of 1982–1983 mass mortality of some fish, of the urchins *Diadema*, and of the reef sea stars was also recorded on reefs of the Caribbean basin and Florida (Dungan et al. 1982; Lessios et al. 1983). The authors who described these phenomena supposed that they had been caused by the joint impact of a number of factors and, above all ,by anthropogenic stress and overheating. An enhanced excretion of mucus by stressed and dying corals stimulates the development of bacteria, including the pathogenic ones, which cause the death of animals, also enfeebled by the stress caused by diseases. Corallovore fish and asteroids did not die during the periods of mass mortality of corals, but grazed the remaining ones, especially the young coral colonies (Glynn 1983). On the other hand, the mass mortality of the urchins *Diadema*, which graze on macrophytes, inhibits the recruitment of corals after their period of mortality (Lessios et al. 1984). Therefore, the regeneration of coral communities, destryed by the catastrophic events of bleaching and mass mortality, could be delayed and on some reefs it could even be stopped, resulting then in their final destruction (Glynn et al. 1988). The repetitions of these events on an even larger scale, which happened again in the western Atlantic in 1987, proves the contention, that we are witnessing the process of global destabilization of coral-reef ecosystems, which endangers their very existence (Glynn 1984). It happens because these mature and complex ecosystems, extremely sensitive to any kinds of external influence, are functioning even in natural conditions at their upper limits of endurance of physical stress. The corals, which compose the core their group, in the summer period exist at temperatares of only 2–3 °C below the lethal level. In zones of tropical cyclones they are subjected to periodical cathastrophic damage within a time-scale hardly permitting their complete recovery (i.e., 5–7 years). In this situation, the addition of anthropogenic stress causes their chronic global destabilization, which we are now witnessing (Bunkley-Williams and Williams 1990).

11.3 Coral Diseases

The exposure of corals to diseases had been first discovered by Antnnius in 1973. Later he described four kinds of these (Antonius 1981, 1984, 1989). Two of the diseases, "white bacteriosis" and the "pull-in of polyps" injure corals only under stress conditions, such as overheating or pollution (Lewin and Peters 1983; Antonius 1988; Chesher 1985; Kuhlmann 1988b; Edmunds 1991). Diseased corals pull in their polyps completely, abandoning activity.

After several days their tissues fall out from colonies in stripes and they die. Two other kinds of coral ills, the "white band" and the "black band" disease (WBD and BBD) could be found also on otherwise quite healthy reefs. Nevertheless, BBD might be stimulated by stress conditions (Antonius 1981, 1984, 1989). Outbreakes of these diseases were recorded on reefs where corals had been destroyed by cyclones, and then bored by sponges and clams. On the destabilized reefs of Jamaica and Curačao, WBD in-fected about 80% of corals (Rogers 1985; Peters 1988). Experiments showed that young colonies of corals, regenerated from fragments of branches, are most susceptive to diseases, and a significant part of them dies (Bak and Criens 1981).

BBD is caused by the blue-green filamentous alga *Phormidium corallicum* (Rützler and Santavy 1983). This pathogenic alga induces a diseases similar to BBD also in the gorgonian octocoral *Pseudopterogorgia* (Feingold 1988). The transfer of pieces of diseased coral to healthy ones can infect them. The infection agent of WBD is still unknown, but it is supposed to be a bacterium (Peters et al. 1983). It is inhibited by antibiotics, though this conclusion was put in doubt by Antonius (1984). Corals could become diseased by WBD without any external stress influence. It damages mostly the ramose corals acroporids and pocilloporids. The diseased corals mostly die. The disease usually starts at the base of a colony and then in the form of a white stripe extends upwards across the colony. Polyps in the damaged zone die and the tissues are eaten by microfauna (Gladfelter 1982). On this stripe often also BBD develops later on. This disease injures mostly massive corals, as faviids and *Platygyra*. According to Chesher (1985), acroporid corals at Tonga have been insured by the disease, induced by a blue-green filamentous algae resembling *Oscillatoria*.

Coral diseases are now under intensive investigation. It became evident also that they strike mostly corals enfeebled by stress, in particular by the activity of man of reefs (Segel and Ducklow 1982). Under the influence of siltation and pollution in zones of such an activity the corals increase their excretion of mucus, the excess of which stimulates the development of bacteria, which then infect the corals themselves (Mitchell and Chet 1975; Garrett and Ducklow 1975).

11.4 Anthropogenic Impacts

As mentioned above, reef ecosystems exist at their boundary of endurance of physical stress. But its periodic character has allowed them not only to survive, but also to flourish (Highsmith et al. 1980). Anthropogenic stress is much more dangerous for them because in most cases it is not only permanent but has also the tendency to increase with time (Kinsey 1988). In conditions of cumulative impact of physical and of anthropogenic stress the inhibition and destruction of reef systems could become irreversible (Mergner 1981). Therefore, the anthropogenic impacts as a cousequence of fisheries and of the industrial and recreational activity of man on reefs have become a main factor endangering their existence (Pearson 1981; Raymond 1981; Stoddart 1981).

The massive anthropogenic stress on reefs had its beginnings in the Second World War. During that time in the Pacific basin large contingents of people appeared, armed with a powerful technique. The construction of numerous military bases, ports and channels began (Banner 1974). The collection of shells and corals by the military and their families rapidly increased. A market of souvenirs was set up, which stimulated collection also among the aborigines. Because of the abundance of explosives and various antiseptic agents in areas of military activity, destructive and poisonous industrial fishing expanded (Dahl 1981). After the war, the development of economic and touristic activities proceeded in a rush in most of the reefs accessible to airplane or ship. Improved medical services for aborigines have led to the quick growth of local populations. This also resulted in a drastic increase in the number of people in reefal regions. The politics of governments in developing coastal and island states, whose main goal is to provide people with food and shelter, stimulated the exploitation of reef resources by activities such as fishery, extraction of lime and sand for the construction industry, and the collection of souvenirs for international trade. With the aid of western specialists the development of agriculture proceeded, using pesticides and fertilizers. In some coastal regions the mining of tin, copper, polymetallic ores, and oil began. In reef areas a mass of motor boats and fishery fleets with heavy trawls appeared. Tourists and the local people started to use widely diving equipment for collection of reef animals and for spear-fishing. The volume of various pollutants and hot water discharging on reefs from cities, touristic centers and power plants rapidly increased (Pecheaux 1989).

The negative consequences of such development in the man – reef relationship become more and more evident. On many reefs practically irreversible changes took place (Kühlmann 1988b). Anthropogenic stress enhanced greatly the processes of global destabilization of reef ecosystems. The evident endangerment of reefs by such developments impel governmental, public and scientific agencies to work out measures for the protection of reefs and their resources. In many countries corresponding natural reserves and national parks were organized, destined to preserve reef ecosystems in conditions of admissible use of their resources, because its complete cessation could not be afforded by either developed or developing countries (Stoddart 1981; Sudara 1981). In the following, various kinds of anthropogenic impacts on reefs are analyzed.

Discharge of various effluents to the reefs from coastal villages and hotels, as well as oceanic discharges from large cities, which are also driven to the reefs by oceanic currents, increase from year to year (Wood and Johannes 1975; Marshalek 1981). They not only exert a toxic effect on reef biota, but also foment some processes unusual for reef environment, such as eutrophication (Walker and Ormond 1982). The nutrients accumulating on the reef with wastewaters stimulate the growth of seaweeds, which inhibit and then replace the corals (Walker and Ormond 1982; Maragos et al. 1985; Rasmussen and Cuff 1990). In areas subjected to discharge the blooming of phytoplankton is furthered, including the toxic dinoflagellates and blue-green algae (Banner 1974; Bell and Gabric 1991). These blooms violate homeostatic conditions in reef ecosystems, thus destabilizing them. The discharge of man-made wastewaters led to sulfate reduction and H_2S formation, which inhibit the bottom fauna and undermine the benthic biofietering activity (Sorokin 1973c; Sorokin et al. 1982). The stimulation of bacterioplankton growth by the pollution of reef waters improves the trophic conditions for boring sponges, enhancing strongly their multiplication and activities. They weaken the resistance of infested corals to surf and are the cause of their mortaily (Rose and Risk 1985).

The waters discharged to reefs from the fields contain besides nutrient fertilizers also pesticides, polychlorphenols, insecticides (delderin, eudrine), and herbicides (high phenoxy acids). These substances accumulate in the tissues of corals and other reef fauna, being toxic for them in concentrations of $0.1\,mg\,l^{-1}$ (Glynn et al. 1984). In tissues of corals, which died in masses on reefs off the west coast of Panama during the 1982–1983 El Niño (cf. Sect. 7.3), pesticides were found (Glynn 1984). The local population has been using in many of the reef regions the insecticides as poisons in fishery (Endean 1976). Large amounts of detergents are dumped onto the reefs of the GBR region by the sugar industry. These substances kill corals at a concentration of 0.05%.

In reef areas numerous electricity and fresh-water producing plants were installed serving cities touristic centers, and military bases. They discharge warm waters with a high salinity, which kill corals at distances up to 1–2 km (Johannes 1972b; Jokiel and Coles 1974). These discharges are especially dangerous at high-latitude reefs, where corals are not adapted to high temperature (Nudebecker 1981). The wastewates often contain chloramine, used as an antifouling and antiseptic agent. It is poisonous to the bottom fauna at a concentration of about $1\,mg\,l^{-1}$. Especially sensitive to this poison are the larvae of bottom fauna (Best et al. 1981). Large amounts of pesticides and antiseptics are discharged on reefs situated close to sea ports (Loya 1976c). The industrial wastes and waste waters from mines bring heavy metals to the reefs, e.g., copper, mercuty, and tin, which accumulate, then in the tissues of corals and clams (Khristoforova et al. 1979; Fraizier et al. 1985; Brown 1986; Brown and Wells 1989). Corals stressed so die or, when they survive, they excrete a large quantitity of mucus. It stimulates microbial growth, which then causes the mortality of corals (Mitchell and Chet 1975). Often the mine waters contain much ammoniac, which is used

Table 11.3. Changes in structure of coral communities on the polluted reef *A* 5 km off the port of Eilat, and on the nonpolluted (control) reef *B* situated in the Gulf of Aquaba (Red Sea) after the mass mortality event of 1970, caused by extra-low ebb. (After Loya 1976a)

Parameters of community structure	Reef A	Reef B
Total number of species, 1973	23	39
Number of new species appearing since 1970	0	10
Number of species disappearing after 1970	19	4
Percentage of bottom areas recolonized by corals after 1970	23	86
Percentage of recolonized corals appearing after recruitment of planulae	2	42

for dissolving ores. These waters, being toxic for reef fauna, also stimulate the processes of eutrophication in reef waters (Carey 1981).

Oil pollution, especially long-term pollution, injures the reef benthic communities and undrmines their ability to regenerate after natural catastrophes, such as extra-low ebbs (Loya 1975, Table 11.3). On the reef shallows corals can be killed by oil films, which cover their colonies at low ebbs in the areas of pollution (Johannes 1972b). The presence of oil inhibits growth and the reproductive functions of polyps. Accumulating in coral tissues it causes pathological changes (Peters et al. 1981; Dodge 1982; Barnes and Knap 1989). The presence of oil provokes also the expusion of zooxanthellae by corals and enhances the production of mucus. In areas of permanent oil pollution in the Gulf of Aquaba the reduction in gonads in corals and their high mortality have been recorded (Rinkevich and Loya 1977, 1979c; Guzman et al. 1991). A dangerous feature of oil as a pollutant is that the oil slicks from the open ocean are driven by wind at long distances onto the shore, thus accumulating on reefs.

The oil pollution of reefs is rapidly increasing not only bacause of an increase in small fleet and tanker transportation, but also because of the development of transportation and oil extraction on the tropical shelf (Bak 1987). Around the oil-drilling platforms the corals soon die (Loya and Rinkevich 1980; Hudson et al. 1982). The drilling mud is also toxic for reef fauna (Thomson et al. 1980; Szamant-Froelich 1981). Addition of only 1 ml l^{-1} to water induced the expulsion of zooxanthellae by polyps, stopped photosynthesis in coral, and caused its subsequent death (Kendall et al. 1983).

The military construction and activities during and after the war appear to have been the first heavy blow from man to the reefs, actually starting the subsequent processes of their destabilization in the Pacific, and then in some other reef regions (Endean 1976). Many reefs of the Ryukyus, Marianas,

Palau, Solomons, Hawaii, French Polynesia and E. Samoa were seriously damaged. The military dredged reefs for passage through them, crushed them by ships and anchors, polluted them with various wastes and rubble, and bombed them with explosives to kill fish. Major destruction of reefs and the radioactive contamination of reef fauna for hundreds of square kilometers were the conseqences of nuclear tests on the atolls Bikini, Enivetok and Moruroa. Destruction was aggravated by the appearance of diving equipment, boat motors and air communication. Within a short time the collection of shells, corals and also of reef fish for aquaria became a kind of industry, promoted by the development of marketing and monetary systems in reef regions and countries. All these developments caused the rapid devastation of many reefs (Endean 1976).

The tourist boom of the last two decades became an importamt factor harmful to reefs (Sudara 1988; Bell and Gabric 1991) as it turned into a large-scale business. Thousands of hotels and recreation centers house been constructed for tourists. The wastes discharged from them are a main source of pollution on numerous reefs (Fraizier et al. 1985). Hordes of tourists wander on reef flats, trampling down everything that lives (Kay and Liddle 1984). They fish, collect shells and corals and buy them from local people. As a consequence, in regions of intensive tourism, the populations of many molluscs, gorgonians, asteroids, pencil urchins and corals seem to be diminishing. Among them were key species of the reef ecosystems, such as the large predatory gastropods *Melo*, *Charonia*, and *Cassis*. Their price during the tourist boom rose and they became objects of extensive collection. For example, the populations of *Charonia tritonis* abundant on reefs of the GBR at the beginning of the 1970s has been practically extinguished. Now this mollusc has disappeared from this area. The same happened in many other places. In 1970 on the market of Suva over 1000 species of *Charonia* were being sold at the same time. As mentioned above, it was just the extermination of this key species that most probably triggered the *Acanthaster* plague.

Trampling about on reefs tourists destroy the populations of ramose corals. For example, on the Wistari reef (GBR), which has been often visited by tourists, the bottom cover by these corals quickly decreased from 41 to 8% (Kay and Liddle 1984). In places of massive tourism reef suffer also from mechanical damage by anchors, anchor chains and motor propellers of numerous excursion boats and ships (Rogers 1985; Smith 1985). Tourists – amateur fishers employing modern equipment, including scuba apparatus and spear rifles destroyed on many reefs the populations of large predatory fish like groupers, sweetlips, and reef cods (Dahl 1985). Overfishing endangers also the populations of beautiful reef fish, which are caught for aquaria, like butterfly and angel fish (Lubbok and Polunin 1975). A world trade in corals has been set up, which undermines their populations in the Philippines, Thailand, Australia, Taiwan, and French Polynesia (Wells 1981; Gomez 1982).

Warious kinds of construction works significantly damaged reefs, among them the building of airstrips and dikes, the preparation of beaches, dredging and blasting in the construction of channels, passages and piers for boats. In places of construction corals are killed not only directly but also by siltation (Guilcher 1985). The same results from the extermination of forests and the extensive agriculture in coastal areas close to reefs (Johannes 1977b; Kühlmann 1985, 1988b). Most reef fish in such areas also die. Many of them being strictly terrestrial do not abandon their home ground even after being subjected to stress (Amesbury 1981). The building sites for construction works on many reefs were huge. For example, during construction of airstrips on Guam and Tahiti hundreds of hectares of beautiful reefs perished (Johannes 1970). Military construction altered the water exchange in the lagoon of the Palmira atoll so that the corals in it died. Hundreds of hectares of beautiful reefs perished also during the construction of a military chemical plant at the Johnston stoll. In many countries the reefs are endangered by excavation of sand and lime for construction works (Galzin 1981). It causes resuspension of sediments, increase in bacteria numbers by one order of magnitude, and decrease in oxygen content. The reef biocoenoses in areas of excavation were damaged, corals, especially the ramose species, died (Salvat et al. 1979; Najim 1981). Corals are also excavated as material for construction and for the production of lime (Brown and Wells 1989). Their massive colonies are sawed up for use in making lime plates for facings of buildings.

The level of anthropogenic stress and the state of the modern ecological situation in different reef regions we have tried to estimate relying on literature data (Brown 1986; Gomez 1981, 1982, Gomez et al. 1981) and our own experience (Table 11.4) The severest stress suffer reefs of S. E. Asia. The local people of this region have been using reef resources very extensively for a long time. They get lime from reefs, blast and excavate reef flat rocks, take the corals for construction, and collect shells and fish (Brown 1986). On the reefs of Hainan, where corals are taken for the souvenir trade and aquaria, in 1960 110 species of corals were recorded, but today only 20 are left. Destructive fishery is wide spread in this region and many important species of reef fauna are overfished. Many reefs here were inhibited and attacked by *Acanthaster*. But the reefs of S. E. Asia suffer most of all from siltation by terrigenous material coming from the fields (Soerigato 1981). The reefs of Thailand are stressed by waste waters from the tin mines, and by booming tourism and construction (Yap and Gomez 1983, 1984; Sudara 1988). The reefs of Polynesia and Micronesia suffer from the expansion of tourist business, desiructive fishery, and overfishing. More than half of the Tonga reefs are degrading. Mass mortality of corals on them proceeds, many species of fish and molluscs have disappeared (Chesher 1985). Reefs of New Caledonia are polluted from the nickel mines. On Hawaii arose the problem of wastewater discharge by crowds of tourists numbering millions. On Oahu Island these waters had been dumpted into Kaneohe bay, where

Table 11.4. Level of various types of anthropogenic stress in different reef regions as evaluated within a 4-grade scale

Types of stress	S.E. Asia	Micronesia	Polynesia and Melanesia	GBR, Australia	Hawaii	E. Pacific	Red Sea	W. Atlantic	Sum of grades in all regions
Destructive fisheries, employing explosives, heavy trawls, toxic substances	3	2	1	1	0	1	1	0	9
Overfishing of fish and other reef fauna, endangering their populations	3	4	4	1	4	2	1	4	23
Excavation of sand and lime for construction	4	1	3	0	0	0	2	3	13
Tourism	4	2	2	4	3	2	2	4	22
Collection of corals and shells	4	4	2	3	1	3	2	3	22
Discharge of industrial waste waters	2	1	2	1	2	2	1	2	13
Discharge of man-made wastewaters	3	2	3	1	2	3	1	3	20
Discharge of fertilizers and pesticides from fields	3	2	2	3	2	3	0	2	17
Pollution connected with construction and intensive agriculture	4	2	3	2	2	4	1	2	20
Sum of grades for all kinds of stress	30	20	22	16	15	20	10	23	

corals died and were replaced by colonial ascidians and zoantharians. In 1978 the dumping was stopped. In 1983 corals tolerant of pollution (*Montipora*, *Porites*) started to grow there again (Maragos et al. 1985).

On the Ryukyu Islands coral reefs were stressed in 1950 by military construction. Then they were attacked by *Acanthaster*, and then by the tourists. What remained of them were again attacked in the 1980s by sea stars. In many places on these islands reefs were destroyed by the cutting of wood in the coastal area and by water discharge from fertilized fields (Kühlmann 1988b). The local authorities did not undertake any measure against such developments (Muzik 1985). The same fate owaits the reefs off the Pacific coast of Panama. They are heavily stressed by siltation, effluents of pesticides, and by eutrophication (Glynn 1984). A dangerous situation is developing now on reefs of the western Atlantic also. Most of them are stressed by tourists, fishermen, and by total pollution of this region. Some recent reviews revealed their apparent degradation. Their regeneration after cyclones becomes more prolonged. Mass mcortality of corals and other benthic animals on them now is a usual phenomenon (Salvat 1987). Corals on many reefs are diseased. The fisheries in this region have been actually ruined. The formerly flourishing reefs off Venezuela, stressed by waste effluents and the tourist boom, were killed in only 10 years (Weiss and Goddart 1977). Colonies of dead acropores were overgrown by bryozoans and then collapsed. The situation in this region became still graver when the reefs of the Virgin Islands, Antiqua, Guadelupe, Martinique and the Dominican Republic had been damaged by the hurricanes David and Frederic in 1979, and the reefs of Jamaica and Caiman Island, by the hurricane Allen. The Atlantic reefs are now in a somewhat better position from the viewpoint of stress, compared with reefs of the Indo-Pacific because the *Acanthaster* is still absent there. However, the danger of its penetration in any case exists, especially if the proposed new Panama comal, without lock and filled with seawater, is constructed (Porter 1972).

The reefs of the GBR also suffer from significant anthropogenic stress (Kinsey 1991). One of its main factors is the development of agriculture and industry along the east coast of Australia, which produced much waste and terrigenic suspended matter, arriving with drainage waters to the GBR zone. Also, tourism is booming all along this zone to the south of Cooktown. The GBR reefs until recently had been preyed upon by collectors of molluscs and by Chinese (Taiwan) hunters for tridacnas and holothurians. Since 1960 many of them have been devastated by *Acanthaster*. Therefore, even if in this area most reefs are still preserving their natural communities, there is definite apprehension concerning their future.

The complete avoidance of external stress results in gradual recovery of the inhibited reef community (Coles 1984; Maragos et al. 1985; Fraizer et al. 1985). Even after the nuclear tests on Enivetok atoll the reef communities have regained in general their features in 13 years (Johannes 1970). More rapid regeneration of corals proceeds in places artificially "sowed" with pieces of branches of ramose corals (Harriott 1983; Harriott and Fisk 1988). Nevertheless, most researchers argue that the degradation of reef ecosystems is practically irreversible, because, in practice, to put an end to the human impact seems to be impossible (Weiss and Goddart 1977). In the near future, the tendency most probably will be the reverse. Moreover, as it was mentioned above, reefs might drown in future as a result of the expected rise in sea-level because of the "green house effect" (Smith and Buddemeyer 1986).

11.5 Reef Fishery

Among the resources of reefs most made use of by man are: (1) fish and edible invertebrates: molluscs, holothurians, crabs, shrimps, crayfish, polychaetes, urchins; (2) edible seaweeds; (3) shells and corals as souvenirs and for collection; (4) lime and sand for construction; (5) raw materials for production of drugs; and (6) recreational potential. Overexploitation of each of these resources destabilizes the reef ecosystem, undermines its ability to reproduce and thus leads to their loss. Therefore, the use of reef resources needs a rational approach and scientifically based regimentation.

Fishery had been for long time one of the important sources of food for people of tropical coasts (Marshall 1980). For aborigines of the islands, fishery supplies the main portion of needed proteins. But since ancient times reef fishery never has had a commercial purpose. All catch had been used mainly by the fishermen themselves (Stevenson and Marshall 1974). It corresponded not only to the old habits of the local people but also to the poor applicability of industrial methods in fishery (heavy trawls, large gill nets) in reef environments. Also, most reef fish are territorial and thus do not form large aggregations, profitable for industrial fishery (Marshall 1980). The total standing stock of reef fish is large, comprising about 9% of fish in the world ocean (S.V. Smith 1978). The biomass of fish on reefs is by 1–2 orders more than in pelagic intensively fished areas (cf. Table 6.7): 20–200 t ha^{-1} against 2–3 t (Carpenter 1977; Russ 1984). Nevertheless, the catch in reef zones does not largely exceed that in pelagic zones, comprising 5–50 kg ha^{-1} against 25 kg ha^{-1} in the fishery fields of the North Atlantic (Table 11.5). But on small intensively and variously fished reefs the catch could be kept up for a long time at a higher level – up to 100–200 kg ha^{-1} (Alcala 1979; Alcala and Luchavez 1981; Table 11.5). Such a high catch is achieved there by taking mostly middle-sized herbivorous, omnivorous and pelagic fishes, which represent the most numerous guilds. Among them there are the acanthurids, scarids, siganids, basistids, and caesionids. The curve of dependence of catch upon the intensity of fishing (number of

Table 11.5. Catch in different reef regions. (Data by Russ 1984; Munro and Williams 1985)

Place	Areas of reefs fished, km^{-2}	Objects fished	Catch per year	
			Total, tons	$kg\,ha^{-1}$
Puerto Rico	5 300	Fish and crustaceans	2 140	5
Jamaica	3 300	Fish and crustaceans	4 500	4.1
Tansania	12 160	Fish and various invertebrates	21 980	18
Mauritius	350	Fish and various invertebrates	1 645	47
Kora and Lakeba (Fiji)	8.4	Fish and various invertebrates	45	50
Philippines (total)	27 000	Fish and various invertebrates	130 000	49
Philippines (Apo I.)	0.75	Fish	9.6	128
Philippines (Sumillon I.)		Fish	11	183
East Samoa	3.0	Fish	63	212
Ifaluk atoll	6.0	Fish and invertebrates	3 006	51
Tarawa atoll	500	Fish	2 200	44
Kapingamarangi atoll	400	Fish	280	7
Ontong-Jawa atoll	122	Fish and invertebrates	79	6
Cuba	55 000	Fish	27 500	5
Bermuda	1 035	Fish	420	4

fishermen per unit of reef length) shows that it attains maximum, and then turns down. Further increase in fishing efforts results in a decrease in catch as a consequence of overfishing (Fig. 11.2). This level of overfishing is reached sooner, when large predators are caught: groopers, sweetlips, grunts, perch, coral cods, sharks, caranxes. But the catch of the middle-sized omnivores, benthovores, herbivores or planktovores might remain high even at a high level of fishery efforts (Munro and Williams 1985). The traditional regimentation of fishery by aborigines was based just upon the catch of small and medium-sized fish, and also of molluscs and crustaceans. It enabled them to have a permanently high catch without exhausting resources.

Traditional fishery habits were gradually replaced by marketing ones, which oriented fishermen towards plunder-fishery of large valuable predatory fish, which can be easily caught on line. In reef areas, where this commercial and amateur fishery had been rapidly developing, these kinds of fish were quickly overfished. They become rare and their average weight dropped 1.5–2 times. Overcatch is especially harmful to the hermaphrodite fish in which the females change sex only on becoming adult (Huntsman 1980). If because of overcatch they cannot survive till adult age, the population remains without males and its reproduction might stop.

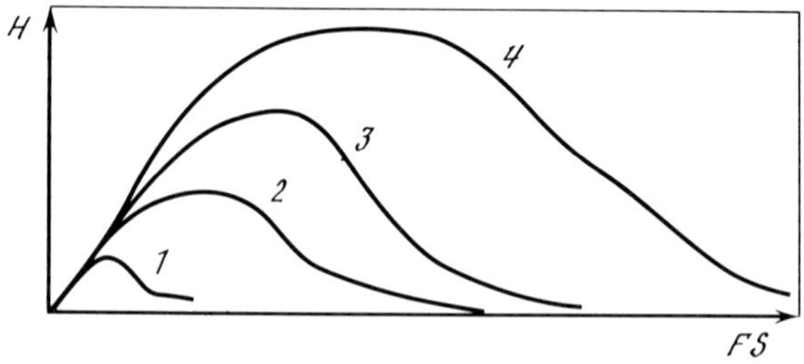

Fig. 11.2. Model of the dependence of fish catch (H) upon the intensity of fishing (FS); groups of fish: *1* large predators; *2* medium-sized predators; *3* omnivorous bottom fish; *4* small- to medium-sized planktovores and benthic phytophages. (After Munro and Williams 1985)

Nevertheless, there is an opinion that the overcatch of large predators could enhance total fish production in a given reef area because of a decrease in grazing pressure on other fish (Grigg et al. 1984; Munro and Williams 1985). It has been supposed that under these conditions the catch of middle-sized nonpredatory fish should increase, however, this opinion needs more proof.

Questions concerning fish production on reefs and optimal quotas of catch, still in dispute by scientists, have been solved also long ago in practice by aborigines. They appeared to be able to maintain the annual catch of reef fish permanently at a high level of 100–200 kg ha^{-1} for centuries, contradicting so some theoretical assumptions (Stevenson and Marshall 1974; Huntsman 1980). Indeed, in spite of a high gross primary production, the net autotrophic production of reef ecosystems is low or often even negative. But it is just the latter that could be taken without ruining it (Pichon and Morrissey 1985), although reef systems are moreover strictly limited in phosphorus. Thus taking out significant amounts of phosphorus with the catch could also undermine them. The truth proved by the practice of aborigines is that fish productivity in reef ecosystems is high indeed and exceeds greatly that of the temperate seas (Johannes 1978b; Alcala and Luchavez 1981; Ruddle and Johannes 1985). Regarding fishing of a wide spectrum of species, the permissible annual catch on reefs could be estimated to be within 50–150 kg ha^{-1}. The loss of nutrients with the catch for the reef ecosystem might be compensated by their input from the oceanic waters passing over the reef. Above, it was shown that rich reef communities also readily exploit the resources of the pelagic ecosystems in surrounding waters (cf. Sects. 2.2.2, 10.3). Therefore, the fishery potential of reef ecosystems is large, but to really obtain a high catch needs sensible regimentation of fishing and the rational use of stocks (Alcala 1979; Marshall 1980). The stocks of fish on reefs of the Indo-Pacific are higher compared with those in the West Atlantic because of the abundance in the former of populations of middle-sized benthovore, omnivore, and planktovore fish: acanthurids, pomacentrids, anthiids, and caesionids

(Munro and Williams 1985; Table 6.8). Presently, the annual catch of fish on reefs could be evaluated to be between 0.6–1 × 10^6 t, which comprises about 1.5% of the total world catch. Reef fishery provides 50 to 90% of necessary proteins for ca. 50–70 million people inhabiting tropical coasts.

Among the reef fish communities subjected to fishery only 1–3% of adult specimens has weights over 200 g. They comprise about 50% of the total biomass of fish on a reef. Among them up to 90% middle-sized fish: acanthurids, caesionids, mugilids, goat fish, and sardines (Table 6.8). The fish catch on reefs depends much upon the kinds of fishing gear and fishery techniques employed. In an optimal case they should be suitable for catching variously sized groups of fish, being at the same time the least destructive. In particular, successful fishery on reefs of the Philippines is ensured by the wide use of gill nets. The overcatch and degradation of fish resources at Tonga and the Marianas were caused by the use of destructive methods–toxic substance and explosives for fishery, and by the overcatch of large predators with line fishing. In areas of active commercial fishery the degradation of resources becomes habitual. For example, fish and crustacean resources of reefs off Florida and Guam were undermined by poorly regulated amateur fisheries. It is precisely amateur fishermen and sportsmen who violate fish stocks on reefs, eliminating mostly large predatory fish, which have a low growth rate and thus overcome overfishing very slowly. Significant damage to stocks of reef fish is made by mass catch with trawls of middle and average planktivore fish for bait by tuna fishermen. In many reef regions destructive methods of fishery are employed, e.g. explosives, ichthyocides and various other toxicants, which severely damage the whole reef communitiy (Campbell 1977). The same destructive impact is made by the booming amateur and sportive fishery in regions of mass tourism and near cities, where spear hunters and amateurs are literally devastating the reefs.

The negative consequences of these events are also obvious. For example, in the central GBR the number of objects of amateur fishing, like marlins, reef trout (*Plectroponus*), emperors (*Lutjanus sebae*), sweet-lips (*Lethrinus chrysostomus*), groupers (*Epipephelus*), as well as the average sized fish significantly decreased. Moreover, the amateur and sportive catch exceeds the commercial catch threefold. On the east coast of Queensland 75 000 motor boats of amateur fishermen were registered in 1979. The trade in reef fish for aquaria became a multimillion dollar business. Their export in the Philippines and Taiwan contzibutes a significant part to the total export of these countries. On the Philippines alone about 20 000 hundred weights of aquarium fish are exported annually. For fishing them besides gill nets and traps also cyanide and ichthyocides are employed which damage the reef fauna. For this business the most beautiful and rare species of reef fish are decimated, e.g. *Centropyge bispinosus*, *Pomacanthus imperator*, *Euxiphippos sextriatus*, and *Pygolites diacanthus*. This trade severely devastated their populations in numerous places in S.E. Asia (Lubbok and Polunin 1975; Albadejo and Corpuz 1981).

Among the important accompaying phenomena of reef fishery the siquaterra disease should be mentioned. This disease is caused by the consumption of poisonous fish (Banner 1976). It is especially widespread on the Marshalls, on Tuamotu and the Society Islands, where it injures 5 to 20% of local populations (Bagnis et al. 1988). On the atoll Hao (Tuamotu) 43% of the local people died due to siguaterra in 1969. The signs of siquaterra are fever reflexes, loss of reflexes, muscle pains and convulsions. Recovery takes several months. The initial source of the toxin seemed to be some toxic microalgae. Fish concentrate it in their bodies. Siguarotoxin has been isolated from fish meat. Its specific feature is thermostability. It survives even boiling and is present also in cooked fish. Its appearance in fish seems to be unpredictable. Many fish traded at certain periods might contain this toxin. But in some of them it happens to be present more often. Among these are the medium-sized and large perchiform fish (*Lutjanus bohar*, *Symphorus nematophorus*) and mackerels. Siguaterra seriously affects not only the health of people, but it harms also fishery, because the fishermen, afraid of this disease, often avoid some valuable and harmless fish too (Lewis et al. 1988).

Besides fish, a large quantity of invertebrates are also taken on reefs. The most sought after among them are the molluscs *Lambis*, *Strombus*, *Trochus*, large *Voluta*, *Cassis*, *Murex*, *Tridacna*, *Pinna*, *Pteria* – the pearl shell, *Cardium*, and *Pecten* (Salvat 1981b; Wells 1981). The meat of molluscs is used by local people as food and as fodder for pigs. The shells are utilized as lime material for construction or for the making of souvenirs, buttons and various adornments. The catch of such edible molluscs like Strombus (*S. luhuanus*) Trochus (*T. niloticus*), *Tridacna* and *Pteria* species in some reef regions has reached many thousand tonnes. Only from the Tuamotu atolls since 1810 about 100000 t of pearl shells alone have been exported. By the beginning of our century their stocks were devastated in this area. On many reefs also the rich populations of *Tridacna* and of the edible holothurians were exterminated (Dahl 1981; Salvat 1981a,b). The same happened to many common and even rare species of gastropods, especially cowries and conids, caught for selling to tourists and collectors. The catch of cowries alone on the Philippines reefs was estimated to be near 100 t. The total export of souvenir shells in 1970 was close to 4500 t, the total volume of their world trade amounting to about 20×10^3 t (Helsinga 1981; Wells 1981). The data on catch of holothurians were given above (cf. Sect. 5.5.5). The annual catch of the spiny lobster *Palinurus argus* on Florida reefs is valued at 10 million dollars. On the Pacific reefs the lobster *P. ornatus* is extensively fished.

The corals are taken from reefs for souvenirs and for the ornamentation of marine aquaria. The value of the annual trade of corals only from the Philippines in 1977 was about 3 million dollars (Johannes 1975, 1978b;

Gomez 1982). The damage to reefs by this trade was so great there that it was stopped in 1981 (Gomez 1982). On reefs of the Indo-Pacific the black coral *Anthipates dichotoma* and the pink coral *Corallum secunda* were also objects of vigorous trade. In places of intensive collection of these corals their resources were also undermined or even devastated (Grigg 1984). The regeneration of these slowly growing corals takes a long time. The black coral, for example, reaches market quality (diameter 2.5 cm) only after 20 years, and the pink coral after 34 years. Therefore, the governments introduced a quota for their catch. In Hawaii it was set with 6 t for black corals and 1 t for pink corals. The reef animals, especially the sessile ones, like gorgonians, zoantharians and sponges, appeared to be prospective raw materials for the pharmacentical industry. They contain antibacterial compounds, biostimulators and cancer-inhibiting agents (Jackson and Buss 1975; Ruggieri 1976; Angeles 1981; Allen et al. 1986; cf. Sect. 8.5). The catch of plexaurid gorgonians containing prostaglandins has become a significant source of hard currency for some developing countries.

11.6 Problems of Monitoring and Rational Use of Reef Resources

The permanent rise of anthropogenic stress caused a global destabilization of reef communities, and the death of corals in many regions. Such a trend of development leads to the devastation of reef resources exploited by man, endangering the economy of developing coastal countries of the tropical zone. It caused concern not only in the scientific community (Salvat 1978, 1987; Dahl 1981; Bradbury and Reichelt 1981; Stoddart 1981; Rogers 1985), but also in governmental and public agencies (Kelleher 1981; Leido 1981). Scientists supported by the governments carried out in the last few years investigations, appraising the actual state of reef ecosystems in many regions and monitored the tendency of recovery of damaged ones (Mergner 1981; Gomez 1982; Bakus 1982; Yap and Gomez 1983; Kay and Liddle 1984; Endean and Cameron 1985; Dahl 1985; Chesher 1985). The results of these studies make it possible to elaborate a strategy of undertakings for the preservation of reef resources. They should be based firstly upon a regulation of economic activities in reef areas, and secondly upon explaining to local people, to fishermen, and to tourists the endangered state of reef biota, and the possible degradation of reef resources, so that they look upon the reefs as treasures of their nations and of mankind. The experience gained so far in reef preservation shows that without solving the second problem all measures of forced regulations and punishment will show little effect.

In this connection experience gathered in traditional reef fishery in Polynesia and in some other reef regions should be made use of. Fish have been the main source of existence there providing 60–90% of necessary proteins (Johannes 1977a). Their populations exploited reef resources extremely carefully. Well-defined lots of reef and lagoon belonged to responsible families. Neighbors or visitors could not fish in them. The family took from its lot only the necessary amount of food. The head of the family and the fishermen kept strict count of the stock of fish and never permitted overcatch. Various taboos based on legends and religion controlled undesirable fishing in particular zones of the reefs in definite seasons. The chiefs of tribes looked after the stocks and watched over their use by their peoples (Johannes 1977b, 1978b). Due to such a regulation of fishery the reefs of Polynesia remained in a juvenile state for centuries despite a high density of the aboriginal populations. This state of affairs was maintained by subdivision of the reef area between responsible owners, and by the careful handling of resources by users (Johannes 1978b, 1981b). These attitudes fundamentally differed from the attitude of peoples in western civilizations and in most other regions of S.E. Asia. According to the latter the fish resources on marine shelves are not owned by anybody. So they could be used by everybody, free for all to catch as much as they could. As this coincided with market economy, it inevitably resulted in the devastation of natural fish resources. It happened also many times in the history of world fishery, that because of such attitudes of people and due to commercial relationships, all attempts to control fishery from the side of state yielded usually meagre results or needed immense efforts and expenditure (Parrish 1980).

At the beginning of the eighteenth century most countries and territories of coastal reef areas and islands had been colonized by western countries. The administration of the colonies by implanting christianity eradicated local religious traditions and habits, which had preserved reef resources. As a result, commercial fishery spread, replacing the traditional one. Such developments soon led in many reef regions to the degradation of reefs, overfishing of valuable species, and a drastic decrease in fish stocks, as it was mentioned above.

The task of the scientific community in these circumstances is to make a correct evaluation of the situation on modern reefs and to establish principles of more reasonable relationships between man and reef. These problems have been discussed at the International Symposia on Coral Reefs in Manila (1981), Tahiti (1985) and Townsville (1988). The active work of scientists enables governments to take measures regarding reef protection and the regulation of fishery on reefs. Laws have been enacted prohibiting trade in corals and shells and restricting construction activity in reef areas. Some states have organized national parks and reserves, which protect reef environments and nature. The most successful in this respect were Australia on the GBR, and the USA on reefs of the Virgin Islands. But many other states are still passively witnessing the imperilment of resources on their coastal reefs in spite of the alarming predictions of scientists (Japan,

Taiwan, Panama, USA on trust territories, Thailand, Vietnam, Tonga, France in Polynesia). Some governments have undertaken palliative measures, which however do not give results because of the opposition of people accustomed to take everything they can from the reefs (Indonesia, East Samoa). Therefore, the situation in most reef regions is still not one for optimism (Salvat 1981a, 1987).

As it has been mentioned above, the traditional rules of preservation of reef resources were abolished by the colonial administration to make place for the principles of free fishery, market economy and profit (Johannes 1982). But this did not happen everywhere. The traditional rules were preserved in Fiji, Palu, Yap, and in some regions of New Guinea and Indonesia. These traditional rules also included the main elements of scientifically based programs of fishery regulation, recommended for the temperate shelves. They propose the limitation of catch. Young fish that are caught have to be returned to the sea. Fishery is prohibited during the periods and in places of spawning. Definite lots of reef remain untouched for the recovery of the fish exploited populations. Overcatch is strictly prohibited. Aboriginal fishermen knew more about fish than scientists nowadays (Johannes 1981b). This author holds that the attempts of the administrations of developing countries to take measures for the protection of reefs and their resources should be based upon the centuries-old experience of aborigines in traditional fishery (Johannes 1982). The strategy of management of reef resources should be based also on ecological theory, recognizing man with his tendencies of social development as part of the ecosystem, i.e., as a large dominating predator (Bradbury and Reichelt 1981). Johannes (1978b) recommended to reestablish the traditional rules of fishery based on the distribution of the reef area between owner families. The families could make agreements with commercial fishermen. Other kinds of free commercial fishery should be stopped. The scientific recommendation on the regulation of fishery based upon the protection of particular overfished species did not produce the desired effect (Gomez 1982). Prohibition or the establishment of quotas, as a rule, does not work in practice, because they are not recognized by fishermen as necessary. For example, the government of Tonga appointed restricted zones for fishery and established quotas of catch. But the local fishermen have still exterminated *Tridacna* and overcaught large predatory fish. The aborigines use destructive methods of fishery. They crush corals to extract anything edible. As a result, the Tongan reefs are now seriously degraded (Chesher 1985; Dahl 1985).

In developing countries, such as French Polynesia, New Guinea, the Philippines, and Mauritius, reef resources are exploited rather extensively. But in these regions the local people depend less upon fish as a source of protein. Fishery is oriented there toward the catch of large predatory fish selling well on markets so as to earn money for importing foreign goods. The same possibility is offered by extensively developing tourism. In these countries measures regarding the preservation of nature and the resources of reefs have often been successful, especially in places where the traditional rules of fishery still exist, and where local fishermen are actively educated in this direction (Johannes 1981a; Alcala and Luchavez 1981; Cabanaban and White 1981). The developed countries direct the strategy of reef preservation and of fishery reglamentation toward conserving them as part of the wealth of nations and of mankind (Salvat 1981a,b; Bradbury and Reichelt 1981). This strategy has as its main goal the exploitation of the recreational potential of reefs, and as a secondary goal the use of their fish resources. Because of the evident harmful effect of all more or less intensive commercial fishery on the reefs and of their attractiveness to tourism, there are tendencies to prohibit them competely in natural reef reserves. In the USA an act was passed prescribing the preservation of reefs and regulating fishery on them. The same was done in Australia (Lilburn 1983). Quotas and licenses have been fixed in the GBR region, which was proclaimed National Park. A significant area of it was totally closed to fishing. For trawlers special fishery fields were reserved.

11.7 Protection, Monitoring and Rational Use of Resources; Artificial Reefs

The govermental agencies in many countries have organized national reserves or national parks in reef zones to preserve their nature and resources. Their main goal is to provide the use of recreational and fish resources of reefs without doing harm to their ecosystems. Among the most successful attempts in this field the national parks on the GBR (Kelleher and Dutton 1985), on the Sumillon and Apo Islands (Alcala and Luchavez 1981), and on the reefs of Florida and the Virgin Islands should be recognized as such (Dahl 1981; Bakus 1982). The administration of the GBR Marine Park, for example, first analized the available information on the resources of a given reserved reef area and on the ways they are used. Then the Zonal Plan of Park Area was drafted. In this plan the restricted zones were marked, where any human activity, except research, was prohibited. Such zones should serve as scientific reserves to which only scientists are admitted. Zones of tourism were also marked on the Plan, which could be used by tourists and for the scientific education of the local people, who had to develop a conscious relationship to the reef and its biota. In the Plan there were also places appointed where economic activity is permitted, such as regulated commercial and sportive fishery, and tourism. The success of such a Plan depends upon the conscious support of these measures by local fishermen and the public (Kelleher 1981).

This Zonal Plan was then realized by the GBR Marine Park Authority, which regulates all recreational, business, constructional and research activities on the entire territory of the GBR. Areas with limited access are carefully protected. Special permission is needed also for research. A large part of the territory of the Marine Park is closed to fishermen and tourists. All collections are prohibited. The administration of the Park has organized the monitoring of reefs. Counting of *Acanthaster*, and of some species of fish and turtles, was carried out. The tetritory of the Park (237 000 km^2) is periodically photographed from planes and satellites. Some parameters which characterize its current ecological state have been established, such as the distribution of phytoplankton and the areas of its "blooms". Using aerial microwave and infrared probes, pictures of temperature distribution, salinity, the water masses and oil pollution are periodically obtained for monitoring purposes (Jupp 1983). The Authority of the GBR Marine Park, presided by the well-known scientists, G. Kelleher and D. Kinzie, outstanding specialists in ecology and excellent managers, supports a wide program of scientific research destined to solve problems concerning the influence of tourism upon the reef ecosystem and to find ways how to accelerate the recovery of damaged reefs. It also promotes active educational work among tourists and local people, and has access to central television of Australia, edits various booklets, and organizes lectures and films (Kelleher 1983). In all places accessible to tourists stands have been posted with booklets which tell about the nature of reefs and about the necessary behavior of man on reefs to avoid damaging them.

The Marine Parks of Florida function with about the same efficiency: the Biscayan Park, the George Pennecamp Park, the Low Cay and the Dry Tortugas Parks, and the Marine Reserve Cay Largo. Their main goal is the conservation of the nature of coral reefs in the region, taking into account that the ecosystems of these high-latitude reefs, which exist under utmost physical stress, being situated in the zone of hurricanes and large seasonal fluctuations of water temperature, are extremely sensitive to additional (anthropogenic) stress. Within these Parks zones completely closed to tourists are marked. The Parks Authorities are carefully supervising and controlling movements of ships and boats as well as any kinds of activity and potential sources of pollution on their territories. Nevertheless, even such strict measures could not prevent the trend to reef degradation in the region. It was caused, most probably, by the overall pollution of the western Atlantic and overheating during the 1982–1983 El Niño (Jaap 1985).

Attempts to increase production of food on reefs tend now in the direction of development of aquaculture, introduction of new species, and the construction of artificial reefs. According to M. Doty (1981), there are now potential possibilities and also practical experience regarding the introduction of aquaculture on reefs. An objects, the algae agarophytes, *Euchemia*, *Gracillaria* and *Sargassum* have been tested. They were grown on ropes, stretched over the reef. The area of such a marine plantation should be fenced all around with a metallic net to keep out herbivorous fish (Lip and Porse 1981; Chiang 1981). The algae thus produced could be used for the fabrication of agar, as a fodder, for raising herbivorous fish in a complex aquaculture enterprise, and also as food for people, since the above algae are edible. On reefs of Palau the gastropods *Trochus* were successfully cultivated (Helsinga and Hillman 1981). The bivalve *Pinna* is cultivated on the reefs of Vietmam and on reefs of the Philippines *Tridacna* (Alcala 1981). The young of *Tridacna* are raised so as to revive their populations exterminated by illegal trade (Helsinga et al. 1984). On the Bahamas a large spider crab *Mithrax spinisissimus* was cultivated (Munro and Williams 1985). The raising of the noble rose and black corals on reefs is also planned (Grigg 1982), as well as the raising of large predatory fish. The fishermen of Thailand raise groupers and reef perches in cages, feeding them with small pelagic fish. Because they are territorial and are not grazed they could be raised even without cages (if the sharks are exterminated), being fed on the same small fish as found in places where they live.

The positive effects of the introduction of new species on reefs have been proved on Hawaii (Brock 1960; Oda and Parrish 1981). The fish fauna on reefs had been poor there, with large serranid, lutjanid and lethrinid fish and with edible molluscs, crabs and lobsters. Therefore, between 1920–1950 several tens of species had been introduced. These measures significantly increased the resources of reefs and catches increased. Among the species most successfully introduced, forming large stocks on Hawaiian reefs were the fish: *Lutjanus fulvus*, *L. kasmira*, *Cephalopholis guttatus*, the molluscs *Venerupis philippinarium*, *Cytherea merethrix*, the lobster *Procambarus clarkii*, and the crab *Seylla serrata*.

While the problem regarding the use of artificial reefs to improve fish productivity of natural reefs is still disputed, it has been proved that their construction might significantly enhance the fish stocks (Randall 1963; Ogawa 1973). For example, on an artificial reef constructed from hollow concrete blocks in the vicinity of a natural reef in Lameshur Bay (Virgin I.), after 1.5 years, the biomass of fish was about one order more than on the nearly reefs. On the artificial reefs, constructed from 500 spare tires in Florida, the composition and abundance of fish fauna were comparable to those in the natural neighboring patch reef. Within 1–2 years their surface became overgrown with periphyton, macrophyles and sessile reef fauna (Tsuda and Kami 1973). In tropical shelf areas the artificial reefs, even if they had been established on soft bottom far from natural reefs, soon become colonized by fish fauna, including typical reef fish like chaetodontids, apogonids, chromises, pomecentrids, caesionids, and sea perches from the genera *Haemulon*, *Lethrinus*, and *Luthjanus*. In the vicinity of such artificial reefs shoals of small pelagic fish appear.

Off the coast of the USA, either in tropical and temperate zones, now several thousands of artificial reefs exist, most of which testify to former oil-drilling platforms, now abandoned. Having numerous underwater constructions with vast areas, these platforms function as artificial reefs. In the Mexican Gulf alone they number over 3000. Each year there, and also along the Texas coast, about 40 such platforms are conserved (Hastings 1976; Mc Intosh 1981). Many artificial reefs have been constructed by clubs and associations of amateur fishermen from tires, old sunken ships, and gross remains of the construction industry. For their construction specially built concrete blocks were used (Woodhead et al. 1982). The catch in the vicinity of artificial reefs increases by up to one order sometimes. Amateur fishery along the USA coasts is now based mainly on the artificial reefs. Based on the number of these fishermen, running to several million, the total profit obtained by the USA from the use of artificial reefs was estimated at around 85 million dollars. A special company has recently been installed to work in this field: the "Artificial Reef Corporation". The center of development of artificial reefs is in Washington, and edits the bulletin *Reef briefs*. Its goal is to stimulate and act as consultant for artificial reef construction.

Artificial reefs are also constructed by the Japanese, who think that building them could be profitable even in the vicinity of natural reefs (Ogawa 1973; Stone et al. 1979; Nakamura 1985), however, an opposing point of view also exists (Munro and Williams 1985). The latter authors, summarizing the results of a special symposium, have concluded that the reliability of their construction in natural reef areas has still not been proved.

References[1]

Abdel Salam HA, Porter JM (1988) ACRC-6, p 1
Abe N (1938) Palao Trop Biol Stn Stud 1: 469–521
Abel EF (1970) Oecologia 4: 133–142
Abel KM, Drew EA (1985) PCRC-5 6, pp 21–26
Abel DI, Williams WT, Sammarco PF et al. (1983) Mar Ecol Prog Ser 12: 257–265
Abele LC (1979) In: Livingston R (ed) Ecological processes in marine and coastal systems. Plenum, New York, pp 265–287
Abele LC, Patton WK (1976) J Biogeogr 3: 35–47
Adey WH (1975) Atoll Res Bull 187: 1–67
Adey WH (1978) Science 202: 831–837
Adey WH (1983) Coral Reefs 1: 193–201
Adey WH, Macintyre IC (1973) Geol Soc Am Bull 84: 88–904
Adey WH, Vassar IM (1975) Phycologia 14: 55–69
Adlard RD, Lester R (1988) ACRC-6, p 1
Albadejo VD, Corpuz VT (1981) PCRS-4, 1, pp 75–81
Alberte RS, Cheng L, Lewin RA (1986) Mar Biol 90: 575–587
Alcala AC (1979) In: Pacific Sci Congr, Khabarovsk, Abstr 14, pp 16–17
Alcala AC (1981) PCRS-4, 2, pp 757–781
Alcala AC, Luchavez TF (1981) PCRS-4, 1, pp 69–74
Alcala AC, Alcala LC, Gomez ED et al. (1981) PCRS-4, 2, pp 215–220
Alexandersson T (1975) Nature (London) 254: 237–238
Alldredge AL, Hamner WM (1980) Estuary coast Mar Sci 10: 31–37
Alldredge AL, King JM (1977) Mar Biol 41: 317–333
Alldredge AL, King JM (1980) J Exp Mar Biol Ecol 44: 133–153
Allen MJ (1957) Biol Bull (Woods Hole) 113: 79–84
Allen TM, Sharma A, Dubin RE (1986) Bull Mar Sci 38: 4–8
Amede P, Pesando D (1985) PCRC-5, 5, pp 131–137
Amesbury SS (1981) PCRS-4, 1, pp 155–159
Anderson GR, Ehrlich AH, Ehrlich PR et al. (1981) Am Nat 117: 479–495
Andrews JC (1983a) Aust J Mar Freshwater Res 34: 821–834
Andrews JC (1983b) Aust J Mar Freshwater Res 34: 81–86
Andrews JC, Gentien P (1982) Mar Ecol Prog Ser 8: 257–269
Andrews JC, Muller HB (1983) Limnol Oceanogr 28: 215–227
Angeles L (1981) PCRS-4, 2, pp 759–760
Angot M (1968) Cah Orstom (Ser Oceanogr) 6: 3–31
Ansari ZA, Rivonker CY, Ramani P (1991) Coral Reefs 10: 127–131

---

[1] The names of Proceedings of International Coral Reef Symposia are abbreviated as: PCRS-1 (in Cochin), PCRS-2 (in Brisbane), PCRS-3 (in Miami) and PCRS-4 (in Manila). Subsequently, Symposia became the Congresses and they are abbreviated as: PCRC-5 (in Tahiti), PCRC-6 and ACRC-6 – Proceedings and Abstracts of the 6th International Congress on coral reefs (in Townsville). The Great Barrier Reef in titles is abbreviated GBR.

Antonius A (1971) Int Rev Ges Hydrobiol 56: 283–319
Antonius A (1981) PCRS-4, 2, pp 7–14
Antonius A (1984) Mar Ecol 6: 197–218
Antonius A (1988) ACRC-6, p 3
Antonius A (1989) In: Abstr Annu Meet Int Soc Reef Stud, Marseille, p 3
Appukutan KK (1972) PGRS-1, pp 379–389
Arx WS von (1948) Trans Am Geophys Un 29: 861–871
Atkinson M (1981a) PCRS-4, 1, pp 417–418
Atkinson M (1981b) Phosphate metabolism of coral reef flats. PhD Thesis, Univ Hawaii, 90 pp
Atkinson M (1983) In: Baker J (ed) Proc GBR Conf, Townsville, pp 271–273
Atkinson MJ (1987a) Limnol Oceanogr 32: 426–435
Atkinson MJ (1987b) Coral Reefs 6: 59–62
Atkinson MJ (1988) PCRC-6, 1, pp 57–66
Atkinson MJ, Grigg RW (1984) Coral Reefs 3: 13–22
Atkinson MJ, Smith SV, Stroup ED (1981) Limnol Oceanogr 26: 1074–1083
Atkinson MJ, Smith DF (1987) Limnol Oceanogr 32: 436–441
Atoda K (1951) J Morphol 89: 17–36
Atoda K (1953) Sci Rep Tohoku Univ Ser 4, 20: 105–121
Austin A, Austin S, Sale P (1980) Aust J Mar Freshwater Res 31: 163–174
Ayre DJ, Reising JM (1986) Mar Biol 90: 187–190
Babcock RC (1984) Coral Reefs 2: 187–195
Babcock RC (1985) PCRC-5, 4, pp 355–361
Babcock RC (1988) ACRC-6, p 4
Babcock RC, Bull GD, Harrison PL et al. (1986) Mar Biol 90: 379–394
Bablet JP (1985) PCRC-5, 4, pp 361–365
Bagnis R et al. (1988) ACRC-6, p 5
Bak RPM (1976) Neth J Sea Res 10: 285–337
Bak RPM (1977) In: Frost S (ed) Reefs and related carbonates. Geol Soc USA, Tulsa, pp 8–16
Bak RPM (1987) Mar Poll Bull 18: 534–539
Bak RPM (1989) In: Abstr Annu Meet Int Soc Reef Stud, Marseille, p 29
Bak RPM, Criens SR (1981) PCRS-4, 2, pp 221–226
Bak RPM, Elgershulzen JH (1976) Mar Biol 37: 105–113
Bak RPM, Engel MS (1979) Mar Biol 54: 341–352
Bak RPM, Luckhurst B (1980) Oecologia 47: 145–155
Bak RPM, Powell GDE (1988) ACRC-6, p 3
Bak RPM, van Eys G (1975) Oecologia 20: 111–115
Bak RPM, van Eys G (1980) Bull Mar Sci 30: 883–887
Bakus GJ (1964) Occ Pap Alan Handcock Found 27: 1–29
Bakus GJ (1967) Micronesica 3: 135–149
Bakus GJ (1968) Mar Biol 2: 23–32
Bakus GJ (1969) Int Rev Gen Exp Zool 4: 225–269
Bakus GJ (1973) In: Jones O, Endean R (eds) Biology and geology of coral reefs, vol 2: Biology 1. Academic Press, New York, pp 326–363
Bakus GJ (1981) Science 211: 497–499
Bakus GJ (1982) In: Ocean management. Elsevier, Amsterdam, pp 305–316
Banner AH (1974) PCRS-2, 2, pp 285–702
Banner AH (1976) In: Jones O, Endean R (eds) Biology and geology of coral reefs, vol 3. Academic Press, New York, pp 177–213
Bardach JE (1959) Limnol Oceanogr 4: 77–85
Bardach JE (1961) Science 133: 98–99
Barlow GW (1975) J Zool (London) 174: 175–194
Barnes DJ (1973) Bull Mar Sci 23: 280–298
Barnes DJ (1983) J Exp Mar Biol Ecol 66: 149–161
Barnes DJ, Crossland CJ (1978) Comp Biochem Physiol 59A: 133–138
Barnes DJ, Devereux MJ (1984) J Exp Mar Biol Ecol 79: 213–231

Barnes DJ, Taylor DL (1973) Helgoländer Wissensch Meersunters 24: 284–291
Barnes DJ, Caperon J, Cox TG et al. (1976) Search 7: 463–468
Barnes JH (1966) Aust J Nat Hist 15: 257–261
Barnes JH, Endean RA (1964) Med J Aust 1: 592–594
Barthel KW (1981) PCRS-2, 2, pp 649–659
Battey JF, Porter JW (1988) ACRC-6, p 6
Bayer FM (1961) In: Nijhoft M (ed) The shallow water Octocorallia of the West Indies. Nijhoft, The Hague, 373 pp
Bayer FM (1973) In: Broadman R (ed) Animal colonies. Dowden, Hutchinson Ross, Strondsburg, PA, pp 69–92
Bayer FM (1981) Proc Biol Soc Washington 94: 902–947
Bayer FM, Weinheimer AJ (1974) Stud Trop Oceanogr, Univ Miami 12, 165 p
Bell PRF, Gabric AG (1991) Search 22: 117–119
Benayahu Y (1985) PCRC-5, 6, pp 255–260
Benayahu Y (1989) In: Abstr 5th Int Conf Coelenterate biology, Southampton, p 11
Benayahu Y, Loya Y (1977) Helgoländer Wissausch Meeresunters 30: 362–382
Benayahu Y, Loya (1981) Bull Mar Sci 31: 514–521
Benayahu Y, Loya Y (1984) Biol Bull (Woods Hole) 166: 44–53
Benner R, Hodson RE (1985) Mar Ecol Prog Ser 23: 221–230
Bennett I (1984) GBR. Reader's Digest, Sydney, 384 pp
Benson AA, Muscatine L (1974) Limnol Oceanogr 19: 810–814
Benson AA, Patton JS, Abrahams A (1978) Atoll Res Bull 220: 33–35
Bergmann W (1962) Comp Biochem 3(A): 103–162
Bergmann W et al. (1956) J Org Chem 21: 721–728
Bernard JL (1976) Micronesica 12: 169–181
Bertold MA, Borowitzka MA, Mackay MA (1982) Phycologia 21: 327–335
Best BR, Bailey RD, Marsh JA et al. (1981) PCRS-4, 1, pp 169–172
Beyers RJ (1966) Arch Hydrobiol 62: 273–284
Binet D (1984) Mar Biol 82: 143–156
Birkeland C (1977) PCRS-3, 1, pp 15–22
Birkeland C (1982) Mar Biol 69: 175–185
Birkeland C, Neudecker S (1981) Copeia 198: 169–178
Birkeland C, Randall R (1981) PCRS-4, 1, pp 696–698
Birkeland C, Smalley D (1981) PCRS-4, 1, pp 437–445
Birkeland C, Rowley D, Randall RH (1981) PCRS-4, 2, pp 339–344
Bittaker HF, Iverson RL (1976) Mar Biol 37: 39–46
Black CC, Burris JE (1983) Mar Biol 75: 117–120
Black M (1933) Geol Mag 70: 455–466
Blanchot J, Charpy L, Le Borgne R (1989) Mar Biol 102: 329–339
Blank RJ, Trench RK (1985) PCRC-5, 4, pp 113–117
Bode L, Starck K (1983) In: Baker J (ed) Proc GBR Conf, Townsville, pp 409–414
Böhm L (1978) Mar Biol 47: 9–14
Bohnsack JA (1983) Coral Reefs 1: 223–228
Bonham K, Held EE (1963) Pac Sci 17: 305–314
Borowitzka MA (1977) Oceanogr Mar Biol Annu Rev 15: 189–223
Borowitzka MA (1979) Mar Biol 50: 399–347
Borowitzka MA (1983) In: Batnes D (ed) Perspectives in Coral Reefs, AIMS, Townsville, pp 16–21
Borowitzka MA, Larkum AW, Day R (1977) J Phycol 13: 16–23
Borowitzka MA, Day R, Larkum AW (1983) In: Baker J (ed) Proc GBR Conf, AIMS, Townsville, pp 287–292
Borowitzka MA et al. (1988) ACRC-6, p 10
Boshma H (1925a) Proc Am Acad Art Sci LX: 451–460
Boshma H (1925b) Biol Bull (Woods Hole) 49: 407–439
Boschma H (1956) Treatise in invertebrate paleontology: Coelenterata, Lawrence, pp 90–106
Bothwell AM (1981) PCRS-4, 2, pp 137–141

Bothwell AM (1983a) Abstr Annu Sci Meet Aust Coral Reef Soc, Brisbane, pp 8–11
Bothwell AM (1983b) Abstr Ann Sci Meet Aust Coral Reef Soc, Brisbane, pp 616–621
Bothwell AM (1984) Community organization and biology of scleractini an corals of Heron I PhD Thesis, Queensland Univ, Brisbane, 390 pp
Boto KG et al. (1984) Estuary Coast Shelf Sci 19: 321–329
Bottjer DJ (1980) J Paleontol 54: 1102–1107
Bouchon C (1981) Mar Ecol Prog Ser 4: 273–288
Bowden BF, Coll GJC (1983) In: Baber J (ed) Proc Great Barrier Reef Conf, Townsville, pp 281–287
Bowman TE, Lancaster LJ (1965) Limnol Oceanogr 10: 291–292
Bradbury RH (1977) PCRS-3, 1, pp 1–7
Bradbury RH, Reichelt R (1981) PCRS-4, 1, pp 219–223
Bradbury RH, Young PC (1981) Mar Ecol Prog Ser 5: 229–241
Brander KM, Mc Leod MA, Humphrey WF (1971) Symp Zool Soc London 28: 397–415
Brandon DE (1973) In: Jones O, Endean R (eds) Biology and geology of coral reefs, vol 1: Geology 1. Academic Press, New York, pp 187–232
Bray RD (1981) PCRS-4, 2, pp 619–629
Breder CM, Rosen DE (1966) Modes of reproduction in fishes. Nat Hist Press, New York, 941 pp
Bright T, Kramer TJ, Minnery GA (1984) Bull Mar Sci 34: 361–376
Britton EB (1971) J Entomol B 40: 83–91
Brock VE (1954) J Wildlife Manag 18: 284–308
Brock VE (1960) Int Rev Ges Hydrobiol 45: 463–480
Broecker WS, Takahashi T (1966) J Geophys Res 71: 1575–1602
Bromley RG (1978) Paleogeogr Paleoclimatol Paleoecol 23: 169–197
Brown BE (ed) (1986) Human damage to coral reefs. UNESCO Rep Mar Sci, Paris, 140, 176 pp
Brown BE (ed) (1990) Coral bleaching. Springer Berlin Heidelberg New York, 156 pp
Brown BE, Suharsono (1990) Coral Reefs 8: 163–170
Brown BE, Tudkope AW, Le Tissier MD et al. (1991) Coral Reefs 10: 211–215
Brown BE, Wells SM (1989) In: Proc 5th Int Conf Coelenterate biology, Southampton, p 14
Bruce AJ (1976) In: Jones O, Endean R (eds) Biology and geology of coral reefs, vol 3. Academic Press, New York, pp 37–94
Buckley RC (1982) Aust J Ecol 7: 103–106
Buckman NS, Ogden JC (1973) Ecology 54: 589–596
Buddemeyer RW, Kinzie RA (1976) Oceanogr Mar Biol Annu Rev 14: 183–225
Buddemeyer RW et al. (1974) J Exp Mar Biol Ecol 14: 179–200
Budin IN (1980) In: Preobrazhensky B (ed) Biology of coral reefs. Nauka, Moscow, pp 55–64
Budin IN, Preobrazhensky BV (1980) In: Preobrazhensky B (ed) Biology of coral reefs. Nauka, Moscow, pp 42–54
Bull GD (1982) Mar Ecol Prog Ser 7: 267–272
Bunkley-Williams D, Williams EHJ (1990) Nat Hist 4(90): 46–54
Bunt YS et al. (1975) Tech Rep Rosentiel School Mar Atmosph Sci Miami, 10 pp
Bunt JS, Boto KG, Boto G (1979) Marine Biol 52: 123–128
Burkholder PR (1973) In: Jones O, Endean R (eds) Biology and geology of coral reefs, vol 2: Biology 1. Academic Press, New York, pp 117–180
Burkholder PR, Burkholder LM (1960) Am J Bot 47: 866–874
Burkholder PR, Burkholder LM, Almodowar LR (1967) Bull Mar Sci 17: 1–15
Burns D, Andrews C, Craven D et al. (1984) Bull Mar Sci 34: 366–357
Burris KA, Knap AH (1989) Mar Poll Bull 20: 391–398
Burris RH (1976) Aust J Plant Physiol 3: 41–51
Burris RH (1983) Mar Biol 75: 151–155
Buzhinskaya GN, Obut AM, Potin VV (1980) In: Preobrazhenski B (ed) Biology of coral reefs. Nauka, Moscow, pp 225–266
Cabanaban AS, White AT (1981) PCRS-4, 1, pp 317–321

Cairns SD (1977) Guide to gorgonians. Sea grant field guides, N 6, Univ Press, Miami
Cairns SD, Stanley GD (1981) PCRS-4, 1, pp 611–617
Calvez Y, Salvat B (1980) Cah Micropaleontol Paris 4: 1–15
Cameron AM (1976) In: Jones O, Endean R (eds) Biology and geology of coral reefs, vol 3: Biology 2. Academic Press, New York, pp 155–176
Cameron AM (1977) PCRS-3, 1, pp 193–198
Cameron A, Endean R (1981) PCRS-4, 2, pp 593–596
Cameron A, Endean R (1985) PCRC-5, 5, pp 211–215
Cameron AM, Endean R et al. (1989) In: Abstr 5th Int Conf Coelenterate biology, Southampton, p 17
Campbell AC, Ormond R (1970) Biol Conserv 2: 246–251
Campbell DG (1977) PCRS-3, 2, p 594
Cantera JR (1989) In: Abstr Annu Meet Int Soc Reef studies, Marseille, p 43
Caperon JW, Harvey W, Steinhilper F (1976) Pac Sci 30: 317–327
Capone D (1977) PCRS-3, 1, pp 337–342
Capone DG et al. (1979) Limnol Oceanogr 24: 117–125
Carefoot TH (1981) PCRS-4, 2, pp 665–670
Carey J (1981) PCRS-4, 1, pp 137–146
Carpenter EJ, Mc Carthy JJ (1975) Limnol Oceanogr 20: 387–401
Carpenter KE (1977) Philip J Fish 17: 95–125
Carpenter KE, Miclat RI, Albadejo VD et al. (1981) PCRS-4, 2, pp 497–502
Cary LR (1931) Pap Dep Mar Biol, Carnegie Inst Washington 27: 53–98
Castro P (1976) Micronesica 12: 99–110
Chalker BE (1983) In: Barnes D (ed) Perspectives on coral reefs. AIMS, Townsville, pp 29–38
Chalker BE, Taylor DL (1978) Proc R Soc London Ser B 201: 179–189
Chalker BE, Dunlap WC, Oliver JK (1983) J Exp Mar Biol Ecol 73: 37–56
Chalker BE, Carr K, Gill E (1985) PCRC-5, 6, pp 167–172
Chalker BE et al. (1988) ACRC-6, p 15
Champalbert G (1981) PSRS-4, 1, pp 454
Chang SS, Prezelin BB, Trench RK (1983) Mar Biol 76: 219–231
Chappell J (1981) Int Assoc Hydrol Sci Publ 131, pp 411–430
Chappell J (1983) In: Barnes D (ed) Perspectives on coral reefs. AIMS, Townsville, pp 46–53
Chappell J, Polach HA (1976) Geol Soc Am Bull 87: 235–240
Chappell J, Veeh HM (1978) Bull Geol Soc Amer 89: 356–368
Chardy P, Chevillon C, Clavier J (1988) Coral Reefs 7: 69–76
Charpy L, Charpy-Rouband C (1988) PCRS-6, 2, pp 547–550
Chave EH, Eckert DB (1974) Pac Sci 28: 297–317
Chave KE, Smith SV, Ray KJ (1972) Mar Geol 12: 123–140
Cheney DP (1974) PCRS-2, 1, pp 591–595
Cheng L (ed) (1976) In: Marine insects. Elsevier, Amsterdam, pp 1–4
Cheng L (1985) PCRC-5, 5, pp 263–265
Chesher RH (1969) Science 165: 280–283
Chesher R (1985) PCRC-5, 4, pp 213–217
Chiang Y (1981) In: Proc 10th Int Seaweed Symp, Berlin, pp 569–580
Chidambaram K, Mukundan M (1944) Curr Sci 13: 263–265
Choat JH (1989) In: Abstr Annu Meet Int Soc Reef studies, Marseille, p 18
Chornesky EA (1983) Biol Bull 165: 569–581
Church JA, Andrews JC, Boland FM (1985) Cont Shelf Res 4: 515–531
Cieresko LS (1962) Trans NY Acad Sci 24: 502–503
Cieresko LS, Karns T (1973) In: Jones O, Endean R (eds) Biology and geology of coral reefs, vol 2; Biology 1. Academic Press, New York, pp 183–201
Cieresko LS et al. (1968) Comp Biochem Physiol 24: 899–910
Cintron G et al. (1980) In: Memoires Seminario estudà scientifico a impacto humano en el ecologia Manglaro. UNESCO, Montevideo, 405 pp
Claerebandt MR (1988) ACRC-6, Townsville, p 19

Clark AM (1974) Micronesica 12: 111–117
Clark AM (1976) In: Jones O, Endean R (eds) Biology and geology of Coral Reefs, vol 2: Biology 2. Academic Press, New York, pp 95–100
Clark KB, Busacca M (1978) J Molluscan Stud 44: 272–282
Clausen CD, Roth AA (1975) Mar Biol 33: 93–100
Cloud PE (1962) Prof Pap US Geol Surv 360: 1–138
Cloud PC, Semikhatov MA (1969) Am J Sci 267: 1017–1061
Clough BF (ed) (1982) Mangrove ecosystems in Australia. AIMS, ANU Press, Canberra, 302 pp
Coates AG, Oliver WA (1973) In: Broadman R (ed) Animal colonies. Dowden, Hutchinson & Ross, Stroudsburg, PA, pp 3–25
Coffroth MA (1984) Mar Ecol Prog Ser 17: 193–199
Coffroth MA (1985) PCRC-5, 4, pp 165–171
Coles SL (1980) Mar Ecol Prog Ser 2: 281–291
Coles SL (1984) Coral Reefs 3: 123–130
Coles SL (1985) PCRC-5, 4, pp 171–176
Coles SL, Jokiel PL (1978) Mar Biol 49: 187–195
Coles SL, Strathmann R (1973) Limnol Oceanogr 18: 673–678
Colgan MW (1981) PCRS-5, 2, pp 333–338
Colinvaux LH (1980) Adv Mar Biol 17: 1–35
Coll JC, Hanves GB, Liyanage N et al. (1977) Aust J Chem 30: 1305–1309
Coll JC, La Barre S, Sammarco PW et al. (1982) Mar Ecol Prog Ser 8: 271–278
Coll JC, Bowden BF, Tapiolas DM et al. (1985) Tetrahedron 41: 1085–1092
Collette BP, Earle SA (eds) (1972) Ecology of coral reef fishes. Nat Hist Mus Los Angeles City Soc Bull 14: 179
Collette BP, Talbot FH (1972) Bull Nat Hist Mus Los Angeles 14: 98–178
Colley F, Odum MT, Wilson RF (1962) Ecology 43: 9–29
Colley NJ, Trench RK (1985) Cell Tissue Res 239: 93–103
Conacher WJR, Lanzing WJ, Larcum AW (1979) Aust J Mar Freshwater Res 30: 387–400
Conand C (1981) Bull Mar Sci 31: 523–543
Conand C (1985) PCRC-5, 5, pp 291–295
Connell JH (1973) In: Jones O, Endean R (eds) Biology and geology of coral reefs, vol 2: Biology 1. Academic Press, New York, pp 205–245
Connell JH (1978) Science 199: 1302–1310
Connell JH (1983) In: Baker J (ed) Proc GBR Conf. AIMS, Townsville, pp 179–190
Connor JL, Adey WH (1977) Atoll Res Bull 211: 15–45
Cooke WJ (1976) In: Mackie O (ed) Coelenterate ecology and behavior. Plenum, New York, pp 281–288
Cooksey KE, Cooksey B (1972) Mar Biol 15: 289–292
Cope M (1981) PCRS-4, 2, pp 557–562
Copper P (1974) PCRS-2, 1, pp 365–380
Cox EF (1988) ACRC-6, p 22
Cresswell GR, Greig MA (1978) Aust J Mar Freshwater Res 29: 345–353
Crossland C (1938) Proc Zool Soc London Ser A 108: 513–523
Crossland C (1980) In: Schwemmler W (ed) Endocytobiology, endosymbiosis and cell biology, vol 1. De Gruyter, Berlin, pp 163–172
Crossland C (1981) PCRS-4, 1, pp 663–666
Crossland C (1983) In: Barnes D (ed) Perspectives on coral reefs. AIMS, Townsville, pp 56–68
Crossland CJ, Barnes DJ (1974) Mar Biol 28: 325–332
Crossland CJ, Barnes DJ (1977a) Mar Biol 40: 411–421
Crossland CJ, Barnes DJ (1977b) Mar Biol 43: 57–62
Crossland CJ, Barnes DJ (1983) Aust J Mar Freshwater Res 34: 835–844
Crossland CJ, Barnes DJ, Cox T (1980) Mar Biol 59: 181–187
Crossland CJ, Hatcher BG, Atkinson MJ et al. (1984) Mar Ecol Prog Ser 14: 159–163

Crossland CJ, Hatcher BG, Smith SV (1991) Coral Reefs 10: 55–64
Cuffey RJ (1973) Pac Geol 6: 25–50
Cuffey RJ (1974) PCRS-2, 2, pp 357–364
Cuffey RJ (1977) In: Frost S (ed) Reefs and related carbonates. Am Assoc Petrol Geol, Tulsa, Okl, pp 181–191
Cumings ER (1932) Bull Geol Soc Am 43: 331–352
Cushing DM (1971) FAO-JOC Rep N 72, Rome, pp 2–20
Cushman J, Todd R (1954) US Geol Surv Prof Pap 260-H: 319–384
Dahl AL (1973) Mar Biol 23: 239–249
Dahl AL (1974) PCRS-2, 1, pp 21–25
Dahl AL (1976) Micronesica 12: 43–47
Dahl AL (1981) Bull Mar Sci 31: 544–551
Dahl AL(1985) PCRC-5, 6, pp 509–512
Dahl AL et al. (1974) Atoll Res Bull 172: 7–36
Daly RA (1915) Proc Am Acad Art Sci 51: 155–251
Dana TF (1976) Bull Mar Sci 26: 1–13
Dana TF, Newman WA, Fager EW (1972) Pac Sci 26: 355–372
Darnell RM (1967) In: Lauff G (ed) Estuaries. Smithsonian Inst, Washington, pp 374–375
Darnell RM, Soniat TM (1979) In: Livingston R (ed) Ecology processes in coastal and marine systems. Plenum, New York, pp 487–525
Darwin C (1842) The structure and distribution of Coral Reefs. Smith Elder, London, 214 pp
Daumas R, Thomassin BA (1977) PCRS-3, 1, pp 517–523
Davies PJ (1977) PCRS-3, 2, pp 325–330
Davies PJ (1983) In: Barnes D (ed) Perspectives in Coral Reefs. AIMS, Townsville, pp 69–95
Davies PJ, Kinsey DW (1977) Mar Geology 24: M1–M11
Davies PJ, Marshall J (1979) Search 10: 276–279
Davies PS (1977) PCRS-3, 1, pp 391–396
Davies PS (1984) Coral Reefs, vol 2: 181–186
Davies PS (1989) In: Abstr Annu Meet Int Soc Reef studies, Marseille, p 51
Davies PS et al. (1985) PCRC-5, 3, pp 95–105
Davies PS (1980) Biol Bull 158: 187–194
Davis WP, Birdsong RS (1973) Helgoländer Wissensch Meersunters 24: 292–306
Dawydoff C (1936) Bull Soc Zool Fr 61: 461–484
Day JH (1967) Trop Polychaeta, Publ Brit Mus Nat Hist 656: 858 pp
Dayton PK, Hessler RR (1972) Deep Sea Res 19: 199–208
D'Elia CF (1977) Limnol Oceanogr 22: 301–315
D'Elia CF (1978) In: Stoddart D, Johannes R (eds) Coral reef research methods. UNESCO, Paris, pp 485–495
D'Elia CF, Webb KL (1977) PCRS-3, 1, pp 325–330
D'Elia CF, Webb KL, Porter JW (1981) Bull Mar Sci 31: 903–910
D'Elia EF, Domotor SL, Webb KL (1983) Mar Biol 75: 157–167
Demond J (1957) Pac Sci 11: 275–336
Dennison WC, Barnes DJ (1988) J Exp Mar Biol Ecol 115: 67–77
De Silva MV (1988) ACRC-6, p 24
Devaney DM (1974) Micronesica 10: 105–204
Devaney M, Reese E, Burch B, Helfrich P (eds) (1987) The natural history of the Enivetok atoll. US Dep Energy, Oakridge, 226 pp
Dinesen ZD (1977) PCRS-3, 1, pp 155–160
Dinesen ZD (1982) Mar Ecol Prog Ser 7: 117–123
Dinesen ZD (1983) Coral Reefs 1: 229–236
Di Salvo LH (1969) Pac Sci 23: 129
Di Salvo LH (1971a) In: Lenhoff H (ed) Experimental coelenterate biology. Univ Press, Honolulu, pp 129–139

Di Salvo LH (1971b) J Exp Mar Biol Ecol 7: 123–136
Di Salvo LH (1973) In: Jones O, Endean R (eds) Biology and ceology of coral reefs, vol 2: Biology 1. Academic Press, New York, pp 1–14
Di Salvo LH, Gundersen K (1971) Can J Microbiol 17: 1081–1090
Ditlev H (1978) Mar Biol 47: 29–39
Dodge RE (1981) PCRS-4, 2, pp 241–248
Dodge RE (1982) Mar Biol 71: 141–147
Dodge RE, Lang JC (1983) Linnol Oceanogr 28: 228–240
Dodge RE, Vaisnys JR (1977) J Marine Res 14: 32–35
Dodge RE, Aller RC, Thompson J (1974) Nature (London) 247: 574–577
Dodge RE, Jickells TD, Knap AH (1984) Mar Pollut Bull 15: 178–187
Dollar S (1982) Coral Reefs 1: 71–81
Domm SB, Domm AJ (1973) Pac Sci 27: 128–135
Done TJ (1981) PCRS-4, 2, pp 315–320
Done TJ (1982) Coral Reefs 1: 95–105
Done TJ (1983) In: Barnes D (ed) Perspectives on coral reefs. AIMS, Townsville, pp 107–139
Done TJ (1985) PCRC-5, 5, pp 315–319
Done TJ et al. (1981) PCRS-4, 1, pp 299–309
Doty MS (1974) PCRS-2, 1, pp 27–33
Doty MS (1981) The diversified farming of coral reefs. Arboreus Lectures, Univ Hawaii 11: 29 pp
Doty MS, Capurro RA (1956) World data center "A". Int geophys year, Washington Oceanogr Rep pt 1, vol 4, pp 1–625
Doty MS, Morrison JP (1954) Atoll Res Bull 35: 1–50
Drew EA (1972) J Exp Mar Biol Ecol 9: 71–75
Drew EA (1973) J Exp Mar Biol Ecol 13: 165–179
Drew EA, Abel KM (1983) In: Baker J (ed) Proc Great Barrier Reef Conf. AIMS, Townsville, pp 299–304
Drew-Ferrier M (1991) Coral Reefs 10: 183–187
Droop M (1963) In: Moss B (ed) Symbiotic associations. Univ Press Cambridge, pp 171–199
Dubinsky Z (ed) (1990) Coral reefs. Elsevier, Amsterdam, 450 pp
Dubinsky Z et al. (1984) Proc R Soc London (B) 222: 203–214
Ducklow HW, Mitchell R (1979a) Limnol Oceanogr 24: 715–725
Ducklow HW, Mitchell L (1979b) Microbiol Ecol 5: 215–223
Duguay LE, Taylor DL (1978) J Protozool 25: 356–361
Dumont J (1981) J Nat Hist 15: 623–637
Dungan ML, Miller TE, Thomson DA (1982) Science 216: 989–991
Durand D (1989) In: Abstr Annu Meet Int Soc Reef studies, Marseille, p 57
Dustan P (1975) Mar Biol 33: 101–107
Dustan P (1979) Bull Mar Sci 29: 79–95
Dustan P (1982) Mar Biol 68: 253–264
Earle SA (1972) Bull Nat Hist Mus Los Angeles 14: 18–35
Edmondson CH (1937) Occas Pap Bishop Mus Honolulu 13: 131–146
Edmunds M (1966) J Zool Soc London 46: 27–72
Edmunds PJ, Davies PS (1986) Mar Biol 92: 339–347
Egana AS and Di Salvo LH (1982) Pac Sci 36: 61–63
Ehrlich RR (1975) Annu Rev Ecol Syst 6: 21–247
Ehrlich PR et al. (1977) J Zool London 183: 213–228
Eldredge LG (1966) Micronesica 2: 163–198
Eldredge LG (1967) Micronesica 1: 161–261
Emery AR (1968) Limnol Oceanogr 13: 293–303
Emery AR (1973) Bull Mar Sci 23: 651–770
Emery KO (1956) Bull Geol Soc Am 67: 1505–1520
Emery KO, Tracey JT, Ladd MS (1954) US Geol Surv Pap 260-A: 1–150

Endean R (1957) Aust J Mar Freshwater Res 8: 233–273
Endean R (1971) J Mar Biol Assoc India 13: 1–13
Endean R (1973) In: Jones O, Endean R (eds) Biology and geology of coral reefs, vol 2: Biology 1. Academic Press, New York, pp 389–452
Endean R (1974) PCRS-2, 1, pp 563–576
Endean R (1976) In: Jones O, Endean R (eds) Biology and geology of coral reefs, vol 3: Biology 2. Academic Press, New York, pp 215–254
Endean R (1977) PCRS-3, 1, pp 186–191
Endean R (1982) Australia Great Barrier Reef. Univ Queensland, Brisbane, 348 pp
Endean R, Cameron AM (1985) PCRC-5, 5, pp 309–326
Endean R, Chesher RH (1973) Biol Conserv 5: 87–95
Endean R, Stablum W (1973) Atoll Res Bull 167: 1–62
Endean R, Stablum W (1975) Environ Conserv 2: 247–256
Entsch B, Sim RG, Hatcher BG (1983) Limnol Oceanogr 28: 465–476
Eppley RW (1981) In: Longhurst A (ed) Analysis of marine ecosystems. Academic Press, London, pp 343–357
Evseev GA (1981) PCRS-4, 2, pp 661–663
Evseev GA (1982) In: Krasnov E (ed) Biology of coral reefs. Acad Sci USSR, Vladivostok, pp 20–30
Evseev GA, Silina AV (1982) Biology of coral reefs. Acad Sci USSR, Vladivostok, pp 15–20
Fadallah YH (1983) Coral Reefs 2: 129–150
Fadallah YH, Karlson RH, Sebens KP (1984) Bull Mar Sci 35: 80–89
Fadeev VJ, Lukin VJ (1987) Proc 6th Symp or Modern and ancient reef systems. Vladivostok, pp 76–78
Fairbridge RV (1950) J Geol 58: 330–401
Falkowsky PG, Dubinsky Z (1981) Nature (London) 289: 172–174
Falkowsky PG, Dubinsky Z, Muscatine L (1984) Bioscience 34: 705–709
Fanning KA, Garder KL, Betzer PR (1982) Deep Sea Res 29: 953–965
Farran GP (1949) Sci Rep GBR Comm, Brisbane 1: 291–312
Farrant P (1985) PCRC-5, 4, pp 319–325
Fauchald K (1974) Micronesica 12: 163–167
Fauchald K, Jumars PA (1979) Oceanogr Mar Biol Annu Rev 17: 193–284
Faure G, Thomassin B, Vasseur P (1981) PCRS-4, 2, pp 293–301
Feder HM (1966) In: Mark H (ed) Symbiosis, vol 1. Academic Press, New York, pp 327–380
Feingold LS (1988) ACRC-6, p 29
Fenical WH, Harwell CD (1988) ACRC-6, p 30
Fenchel T (1969) Ophelia 6: 182–204
Ferraris JD (1982) In: Rützler K, Macintyre J (eds) The Atlantic Barrier Reef Belize. Smithsonian Inst Publ, Washington, pp 143–151
Finshelson L (1970) Nature (London) 227: 90–91
Finshelson L (1973) Mar Biol 19: 183–196
Finshelson L (1974) Mar Biol 26: 183–192
Fisher JL (1969) Science 165: 645–647
Fisk DA (1981) PCRS-4, 2, pp 21–26
Fisk DA, Done TJ (1985) PCRC-5, 6, pp 149–155
Fisk DA et al. (1988) ACRC-6, p 30
Fitt WK (1984) Mar Biol 81: 9–17
Fitt WK, Trench RK (1980) In: Schoenberg D (ed) Endocytobiology. De Gruyter, Berlin, pp 675–680
Fletcher M, Marshall KC (1982) Adv Microbiol Ecol 6: 199–235
Flood PG (1983) In: Proc Worksh North section of GBR. GBRMPA, Townsville, pp 270–278
Fontaine M, Momzikoff A, Taxit R et al. (1981) PCRS-4, 2, pp 75–80
Ford WL (1949) Trans Am Geophys Un 30: 46–52

Forsberg FR (1976) In: Jones O, Endean R (eds) Biology and geology of coral reefs, vol 4: Biology 2. Academic Press, New York, pp 256–280
Fortes MD (1981) PCRS-4, 2, pp 393–398
Foster AB (1979) J Exp Mar Biol Ecol 39: 25–54
Foster AB (1980) Bull Mar Sci 30: 678–709
Foster AB (1983) Coral Reefs 2: 14–27
Fraizier A, Fran D, Benente P (1985) PCRC-5, 6, pp 445–452
Frankel E (1977) PCRS-3, 1, pp 201–206
Franziskett L (1970) Int Rev Ges Hydrobiol 55: 1–12
Fricke HW, Hottinger L (1983) Mar Ecol Prog Ser 11: 113–117
Fricke HW, Meischner D (1985) Mar Biol 88: 175–185
Fricke HW, Schuhmacher H (1983) Mar Ecol 4: 163–194
Fricke HW, Vareshi E (1982) Mar Ecol Prog 7: 273–278
Frith CA (1981) PCRS-4, 1, pp 346–354
Frith CA (1983) In: Baker J (ed) Proc GBR Conf. AIMS, Townsville, pp 435–440
Fry WG (1970) Symp Zool Soc London 25: 3–22
Furnas MJ, Mitchell AW, Gilmartin M (1990) Coral Reefs 9: 1–10
Galkin SB (1982) In: Krasnov E (ed) Biology of coral reefs. Academic Press, Vladivostok, pp 31–56
Gallagher BS, Shimada KM, Gonzales FI (1971) Pac Sci 25: 191–205
Galzin R (1981) PCRS-4, 1, pp 115–121
Galzin R (1985) PCRC-5, 5, pp 451–455
Garrett P, Ducklow HW (1975) Nature (London) 253: 349–350
Garrett P, Smith DL, Wilson AO (1971) J Geol 79: 647–668
Gattuso JP (1985) PCRC-5, 6, pp 95–100
Geer LJK, Kjerfve B (1982) In: Rützler K, Macintyre J (eds) The Atlantic Barrier Reef Belize. Contrib Mar Sci, Smithsonian Inst, Washington 12: 321–330
Geister J (1977) PCRS-3, 1, pp 23–29
Geptner MV (1979) J Gen Biol (Moscow) 40: 544–553
Gerber MB, Gerber RP (1979) Mar Biol 52: 33–43
Gerber R, Marshall N (1974) PCRS-2, 1, pp 105–107
Gerhardt DJ (1984) Mar Ecol Prog Ser 19: 181–187
Gerhardt DJ (1985) PCRC-5, 6, pp 57–65
Gerhardt DJ (1986) Mar Ecol Prog Ser 31: 255–263
Gibbs PE (1978) Philos Trans R Soc London Ser B 284: 81–97
Gibson R (1982) Zool J Linn Soc 75: 269–296
Gill GA, Coates AG (1977) Lethaia 10: 119–134
Gillhurst SL (1985) PCRC-5, 5, pp 211–214
Gilmartin M, Relavante N (1974) J Exp Mar Biol Ecol 16: 181–204
Ginsburg RN (1956) Bull Am Assoc Petrol Geol 40: 2384–2427
Ginsburg RN (1983) In: Baker D (ed) Perspective on coral reefs. AIMS, Townsville, pp 148–152
Gladfelter EH (1988) ACRC-6, p 35
Gladfelter EH, Monahan RK, Gladfelter WB (1978) Bull Mar Sci 28: 728–734
Gladfelter W (1982) Bull Mar Sci 32: 639–643
Gladfelter WB, Ogden JC, Gladfelter EH (1980) Ecology 61: 1156–1168
Glenn C, Mc Manus J, Alino P (1981) PCRS-4, 2, pp 775–780
Gli-Turnes S, Corredor J (1981) PCRS-4, 2, pp 51–54
Glynn PW (1973a) Mar Biol 22: 1–23
Glynn PW (1973b) Mar Biol 20: 297–318
Glynn PW (1974) PCRS-2, 2, pp 183–198
Glynn PW (1976) Ecol Monogr 46: 431–456
Glynn PW (1977) J Mar Res 35: 567–585
Glynn PW (1981) PCRS-4, 2, pp 607–612
Glynn PW (1982) Coral Reefs 1: 89–94
Glynn PW (1983) Environ Conserv 10: 149–154
Glynn PW (1984) Environ Conserv 11: 133–146

Glynn PW (1985) PCRS-5, 4, pp 183–188
Glynn PW (1988) Annu Rev Ecol Syst 19: 309–345
Glynn PW, Stewart RH, Mc Closkey JE (1972) Geol Rundsch 61: 483–519
Glynn PW, Howard LC, Corcoran C (1984) Mar Pollut Bull 15: 370–374
Glynn PW et al. (1988) ACRC-6, p 36
Glynn PW, D'Croz L (1990) Coral Reefs 8: 81–91
Gohar HA (1940) Publ Mar Biol Stn Ghardaga 2: 25–118
Gohar HA (1948) Publ Mar Biol Stn Ghardaga 6: 3–33
Gohar HA, Soliman GN (1963) Publ Mar Biol Stn Ghardaga 2: 99–126
Goldberg WM (1973) Bull Mar Sci 23: 465–488
Goldman B, Talbot FH (1976) In: Jones O, Endean R (eds) Biology and geology of coral reefs, vol 3: Biology 2. Academic Press, New York, pp 125–154
Goldman B, Stroud G, Talbot F (1983) In: Baker J (ed) Proc GBR Conf. AIMS, Townsville, pp 203–207
Gomez ED (1981) Coral Reef Newslett 3: 28–32
Gomez ED (1982) In: Ocean management, vol 8. Elsevier, Amsterdam, pp 281–295
Gomez ED, Alcala AC, San Diego AC (1981) PCRS-4, 1, pp 275–285
Gonzales JC (1965) Rep Inst Mar Biol Puerto Rico 5: 1–49
Gordon DC (1971) Pac Sci 25: 222–227
Gordon DC, Fournier RY, Krasnick GJ (1971) Pac Sci 25: 228–233
Gordon MS, Kelley HM (1962) Ecology 43: 473–480
Goreau NJ, Goreau TJ, Hayes RM (1981) Bull Mar Sci 31: 424–435
Goreau TF (1959) Biol Bull 116: 59–75
Goreau TF (1961) In: Lenhoff H (ed) The biology of hydra. Univ Press, Miami, pp 269–285
Goreau TF (1963) Ann Acad Sci NY 109: 127–167
Goreau TF (1964a) Science 145: 383–386
Goreau TF (1964b) Bull Sea Fish Res Stn Haifa 35: 23–26
Goreau TF, Goreau NI (1960a) Science 131: 668–669
Goreau TF, Goreau NI (1960b) Biol Bull 119: 416–427
Goreau TF, Goreau NI (1973) Bull Mar Sci 23: 399–464
Goreau TF, Goreau NI, Yonge CM (1970) J Zool Lond 160: 154–166
Goreau TF, Hartman WD (1963) In: Sognaes R (ed) Mechanisms of hard tissues destruction. Smithsonian Inst Publ, Washington, pp 25–54
Goreau TF, Goreau NI, Yonge CM (1971) Biol Bull (Woods Hole) 141: 247–260
Goreau TF, Lang JC, Graham EA (1972) Bull Mar Sci 22: 113–152
Goreau TF, Goreau NI, Yonge CM (1973) J Zool London 169: 417–454
Goreau TF, Hartman WD (1966) Science 151: 343–344
Goreau TF, Philpott DE (1956) Exp Cell Res 10: 552–556
Goreau TF, Yonge CM (1968) Nature (London) 217: 421–423
Gosline WA (1971) Pac Sci 25: 282–287
Goswami SC (1983) Indian J Mar Res 12: 31–35
Gottfried M, Roman MR (1983) Mar Biol 72: 211–218
Grasshoff M (1976) Senckenberg Biol 57: 155–165
Grassle JF (1973) In: Jones O, Endean R (eds) Biology and geology of coral reefs, vol 2: Biology 1. Academic Press, New York, pp 247–269
Graus RR, Macintyre IG (1976) Science 193: 895–897
Graus RR, Macintyre IG, Merchenroder BE (1984) Coral Reefs 3: 59–68
Graus RR, Macintyre IG (1989) Coral Reefs 8: 9–18
Gravier C (1913) Rev Gen Sci 24: 15–21
Greenfield DW, Greenfield TA (1982) In: Rützler K, Macintyre J (eds) The Atlantic Barrier Reef Belize. Contrib Mar Sci, Smithsonian Inst, Washington 12: 499–507
Greenway M (1976) Aquat Bot 2: 117–126
Grelet Y (1985) PCRC-5, 5, pp 251–256
Gribb AB (1973) In: Jones O, Endean R (eds) Biology and geology of coral reefs, vol 2: Biology 1. Academic Press, New York, pp 47–75
Griffith DJ (1976) Aust J Plant Phys 3: 53–56

Griffiths DJ, Thinh LV (1987) Symbiosis 3: 87–109
Griffith PC, Cubit JD, Adey WH (1987) Limnol Oceanogr 32: 442–451
Grigg RW (1965) Pac Sci 19: 244–260
Grigg RW (1972) Limnol Oceanogr 17: 185–192
Grigg RW (1977) Ecology 58: 278–290
Grigg RW (1981a) Bull Mar Sci 31: 809–812
Grigg RW (1981b) PCRS-4, 1, pp 243–246
Grigg RW (1982) Coral Reefs 1: 29–34
Grigg RW (1983) Mar Ecol Prog Ser 11: 1–14
Grigg RW (1984) Mar Ecol 5: 57–73
Grigg RW, Maragos JE (1974) Ecology 55: 387–395
Grigg RW, Polovina JJ, Atkinson MJ (1984) Coral Reefs 3: 23–27
Guilcher A (1985) PCRC-5, 4, pp 207–210
Guilcher A (1988) Coral reef geomorphology. John Wiley & Sons, New York, 405 pp
Gulbin VV, Vinogragova KL, Nguen Van Chung (1987) Biology of the Sea (Vladivostok) 3: 56–65
Gundersen KR, Stroupe DB (1976) Univ Hawaii, Nat Res Tech Rep 12: 24 pp
Guzman HM, Jackson JB, Weil E (1991) Coral Reefs 10: 1–12
Gygi RA (1975) Ecol Geol Helv 68: 327–359
Hadfield MC (1976) Micronesica 12: 133–148
Hadfield MC, Kay EA, Gillete MU (1972) Mar Biol 11: 12–22
Haig J (1976) Micronesica 12: 183–185
Halldall P (1968) Biol Bull (Woods Hole) 134: 411–424
Hammond LS, Britles RA, Reichelt RE (1985) PCRC-5, 2, pp 285–297
Hamner WM, Carleton JH (1979) Limnol Oceanogr 24: 1–14
Hamner WM, Jones MS, Carleton JW (1988) Bull Mar Sci 42: 459–479
Hamner WM, Hauri IK (1981) Limnol Oceanogr 26: 1084–1112
Hansen JA, Alongi DM, Moriarty DJ (1987) Coral Reefs 6: 63–70
Hansen JA, Skilleter GA (1988) ACRC-6, p 41
Hanson RB (1977) J Exp Mar Biol Ecol 29: 107–119
Hanson RB, Gundersen K (1976) Appl Environ Microbiol 31: 942–948
Hargraves PE (1982) In: Rützler K, Macintyre J (eds) The Atlantic Barrier Reef Belize. Smithsonian Inst, Washington 12: 153–165
Harmelin-Vivien ML (1977) PCRS-3, 1, pp 289–295
Harmelin-Vivien ML (1981) Oceanol Acta 4: 365–376
Harmelin-Vivien ML (1989) Envir Biol Fish 25: 101–110
Harmelin-Vivien ML, Bouchon-Navaro Y (1981) PCRS-4, 2, pp 537–542
Harmelin-Vivien ML et al. (1988) ACRC-6, p 41
Harmelin-Vivien ML, Peyrot-Clausade M, Romano JC (1989) In: Abstr Annu Meet Int Soc Reef studies, Marseille, p 75
Harrigan JF (1972) Am Zool 12: 723–726
Harriott VJ (1982) Coral Reefs 2: 9–19
Harriott VJ (1983) In: Abstr Annu Sci Meet Aust Coral reef Soc, Brisbane, p 13
Harriott VJ (1985) Mar Ecol Prog Ser 21: 81–88
Harriott VJ, Fisk DA (1988) Austr J Mar Freshwater Res 39: 409–416, p 41
Harris LG (1975) Biol Bull 149: 539–550
Harrison JT (1983) In: Reaka M (ed) Ecology of deep and shallow coral reefs, vol 1. US Dep Commerce, Washington, pp 145–150
Harrison PL (1985) PCRC-5, 4, pp 337–342
Harrison PL, Babcock RC, Bull GD et al. (1984) Science 223: 1187–1189
Hartman WD (1977) Bull Am Assoc Petrol Geol Stud Geol 4: 127–134
Hartman WD, Goreau TF (1970) Symp Zool Soc London 25: 205–243
Hartmann G (1964) Kieler Meersforsch 20: 35–117
Hartog JC (1977) PCRS-3, 1, pp 463–469
Hashimoto Y, Fusetani (1972) In: Proc 7th Int Seaweed Symp Sec IV. Univ Press, Tokyo, pp 569–572

Hashimoto Y et al (1969) Bull Jpn Soc Fish 35: 1086–1096
Hastings RW (1976) Fish Bull 74: 387–402
Hatcher AI (1985) PCRC-5, 3, pp 407–413
Hatcher AI, Frith CA (1985) Coral Reefs 4:101–110
Hatcher AI, Hatcher BG (1981) PCRS-4, 1, pp 419–424
Hatcher BG (1981) PCRS-4, 2, pp 515–520
Hatcher BG (1983a) In: Baker J (ed) Proc GBR Conf. AIMS, Townsville, pp 317–327
Hatcher BG (1983b) In: Barnes D (ed) Perspectives on coral reefs. AIMS, Townsville, pp 164–172
Hatcher BG (1990) Trends in Ecol Evol 5: 149–155
Hatcher BG, Larkum AWD (1983) J Exp Mar Biol Ecol 69: 61–84
Hatcher BG, Rimmer DW (1985) PCRC-5, 6, pp 229–236
Heckel PH (1974) In: Laporte L (ed) Reefs in time and space. Am Assoc Petrol Geol, Tulsa, pp 90–154
Hedley RH (1964) Int Rev Gen Exp Zool 1: 1–45
Helfman GS (1978) Environ Biol Fish 3:129–149
Helsinga GA (1981) PCRS-4, 1, pp 39–45
Helsinga GA, Hillman A (1981) Aquaculture 22: 35–42
Helsinga GA, Perron FE, Orak O (1984) Aquaculture 39: 197–215
Henderson BS (1981) PCRS-4, 1, pp 678–686
Hentig RT von (1971) In: Miller J (ed) Scientists in the sea. Smithsonian Institute, Washington, pp 279–284
Herberts C (1972) Mar Biol 13: 127–136
Hernandez ML, Roberts MM, Ronse LJ (1977) PCRS-3, 2, pp 71–78
Herriott AB, Immerman FW (1979) Crustaceana 36: 166–172
Heyward AJ, Babcock RC (1986) Mar Biol 90: 191–195
Heyward AJ, Collins JD (1985) Coral Reefs 4: 35–40
Hiatt R, Strasburg D (1960) Ecol Monogr 30: 65–127
Hickel W (1972) In: Miller J (ed) Scientists in the sea. Smithsonian Institute, Washington, pp 275–279
Highsmith RC (1979a) J Exp Mar Biol Ecol 37: 105–125
Highsmith RC (1979b) J Exp Mar Biol Ecol 46: 177–196
Highsmith RC (1980) J Mar Biol Ecol 47: 55–67
Highsmith RC (1981) Int Rev Ges Hydrobiol 66:335–375
Highsmith RC (1982) Mar Ecol Prog Ser 7: 207–226
Highsmith RC, Riggs AC, D'Antonio MD (1980) Oecologia 46: 322–329
Hill D (1972) Archeocyatha. In: Teichert C (ed) Treatise on invertebrate paleontology, pt E, Lawrence, KS, 158 pp
Hill D, Stumm E (1956) Tabulata In: Moore R (ed) Treatise on invertebrate paleontology, pt F, Lawrence, KS, pp 444–479
Hillis-Colinvaux L (1974) PCRS-2, 1, pp 35–42
Hillis-Colinvaux L (1980) Adv Mar Biol 17: 1–327
Hirshfield H, Charmatz R, Nelson L (1968) J Protozool 15: 497–502
Hixon MA (1985) In: Reaka M (ed) Ecology of deep and shallow coral reefs, vol 1. US Dep Commerce, Washington, pp 27–33
Hobson ES (1972) Fish Bull US 70: 715–740
Hobson ES (1973) Helgoländer Wissensch Meersunters 24: 361–370
Hobson ES (1974) Fish Bull US 72: 915–1031
Hobson ES (1981) PCRS-4, 2, pp 562–567
Hobson ES, Chess JR (1978) Fish Bull US 76: 133–153
Hobson ES, Chess JR (1979) Fish Bull US 77: 275–280
Holt C von, Holt M von (1968) Comp Biochem Physiol 24: 83–92
Hopley D (1982) The geomorphology of Proc GBR. John Wiley & Sons, New York, 453 pp
Hopley D (1983) In: Barnes D (ed) Perspectives on coral reefs. AIMS, Townsville, pp 180–196

Hopley D, Kinsey DW (1988) In: Perman G (ed) Green house: planning for climatic change. CSIRO, Aust, pp 189–201
Hopley D, Mc Lean RF, Marshall J et al. (1978) Search 9: 323–325
Horridge GA (1956) Philos Trans R Soc London Ser B 240: 495–529
Hoskin CM (1966) J Sediment Petrol 36: 1058–1074
Hotchkiss FH (1982) In: Rützler K, Macintyre J (eds) The Atlantic Barrier Reef Belize. Smithsonian Inst, Washington, pp 261–280
Howard LA, Brown BE (1984) Oceanogr Mar Biol Ann Rev 22: 195–210
Hubbard DK (1985) PCRC-5, 4, pp 433–439
Hubbard DK, Sadd JL, Roberts HH (1981a) PCRS-4, 1, pp 399–404
Hubbard DK et al. (1981b) PCRS-4, 1, pp 382–387
Hubbard JA (1973) In: Broadman R (ed) Animal colonies. Dowden, Hutchinson & Ross, Strondsburg, PA, pp 31–42
Hubbard JA (1974) PCRS-2, 2, pp 107–126
Hubbard JA (1985) PCRC-5, 4, pp 433–439
Hubbard JA, Pocock GP (1972) Geol Rundsch 61: 598–626
Hudson JH (1977) PCRS-3, 2, pp 491–497
Hudson JH (1981) PCRS-4, 2, pp 233–240
Hudson JH, Shinn EA, Robbin DM (1982) Bull Mar Sci 32: 890–908
Huges RN (1983) Biol J Linn Soc London 20: 39–58
Hughes T (1983) In: Reaka M (ed) Ecology of deep and shallow coral reefs, vol 1. US Dep Commerce, Washington, pp 17–20
Hughes TP, Jackson JB (1980) Science 209: 713–715
Huges TP, Reed DC, Boyle MJ (1987) J Exp Mar Biol Ecol 113: 39–59
Huntsman GR (1980) In: Highman J (ed) Proc Gulf Carribb Fish Inst Curačao 32
Huston M (1985) Coral Reefs 4: 19–25
Hutchings PA (1974) PCRS-2, 1, pp 285–296
Hutchings PA (1981) Bull Mar Sci 31: 410–423
Hutchings PA (1985) PCRC-5, 5, pp 245–250
Hutchings PA, Howitt L (1988) ACRC-6, p 49
Hutchings PA, Kiene WE (1989) In: Abstr Annu Meet Soc Reef studies, Marseille, p 79
Hutchings PA, Weate PB (1978) Aust Zool 19: 315–319
Hyman LH (1940) The invertebrates: Protozoa through Ctenophora. Plenum, New York, 726 pp
Ikeda T, Hing Fay E, Hutchinson SA (1982) Aust J Mar Freshwater Res 33: 683–698
Isdale P (1983) In: Baker J (ed) GBR Conf. AIMS, Townsville, pp 327–330
Isdale PI (1984) Nature (London) 310: 578–579
Ivanovski AB (1975) Rugose corals. Nauka, Moscow, 121 pp
Jaap WC (1979) Bull Mar Sci 29: 414–422
Jaap WC (1985) PCRC-5, 6, pp 143–148
Jackson JB (1977) Am Nat 111: 743–767
Jackson JB (1984) J Exp Mar Biol Ecol 75: 37–57
Jackson J (1989) In: Abstr Annu Meet Int Soc Reef studies, Marseille, p 83
Jackson JB, Buss L (1975) Proc Natl Acad Sci USA 72: 5160–5163
Jamieson B (1977) Zool J Linn Soc 61: 329–349
Jaubert J (1977) PCRS-3, 1, pp 489–494
Jaubert J (1981) PCRS-4, 2, pp 425–429
Jaubert J, Vasseur P (1974) PCRS-2, 2, pp 127–142
Jeffrey SW (1968) Limnol Oceanogr 12: 189–195
Jeffrey SW, Haxo FT (1968) Biol Bull (Woods Hole) 135: 149–165
Johannes RE (1967) Limnol Oceanogr 12: 189–195
Johannes RE (1970) In: FAO Tech Conf Marine Pollution, Rome, pp 1–15
Johannes RE (1972a) Pollut Bull 3: 29–30
Johannes RE (1972b) Mem Ist Ital Idrobiol 29 (Suppl): 525–527
Johannes RE (1975) In: Wood E, Johannes R (eds) Oceanography, vol 12: Tropical marine pollution, Junk, The Haue, pp 13–51
Johannes RE (1977a) Micronesica 13: 121–127

Johannes RE (1977b) In: Coastal ecosystem management. John Wiley & Sons, New York, pp 593–595
Johannes RE (1978a) Environ Biol Fish 3: 65–84
Johannes RE (1978b) Annu Rev Ecol Syst 9: 349–369
Johannes RE (1980) Mar Ecol Prog Ser 3: 365–373
Johannes RE (1981a) Bull Mar Sci 31: 673–680
Johannes RE (1981b) Words of the lagoon. Univ Cal Press, Berkeley, 245 pp
Johannes RE (1982) Ambio 11: 258–261
Johannes RE, Gerber R (1974) PCRS-2, 1, pp 97–104
Johannes RE, Hearth CJ (1985) Estuar Coast Shelf Mar Sci 21: 789–800
Johannes RE, Tepley L (1974) PCRS-2, 1, pp 127–132
Johannes RE, Webb KL (1970) In: Hood D (ed) Organic matter in natural waters. Univ Alaska Press, Fairbanks, pp 257–273
Johannes RE, Wiebe WJ (1970) Limnol Oceanogr 15: 822–824
Johannes RE, Coles SL, Kuenzel NT (1970) Limnol Oceanogr 15: 579–586
Johannes RE, Coles SL, Gerber PR (1972) Bioscience 22: 541–543
Johannes RE, Wiebe WJ, Crossland CJ (1983a) Ecol Prog Ser 11: 105–111
Johannes RE, Wiebe WJ, Crossland CJ (1983b) Mar Ecol Prog Ser 12: 131–136
Johansen HW (1981) Coralline algae. CRC, Boca Raton, 240 pp
John BE (1974) PCRS-2, 2, pp 461–469
Johnson DB et al. (1990) Coral Reefs 9: 167–171
Johnson MW (1949) Trans Am Geophys Un 30: 238–244
Johnson MW (1954) Prof Pap US Geol Surv 200-E: 301–311
Johnston IS (1979) In: Johari O (ed) Scanning electron microscopy, pt 2. Chicago, pp 421–431
Jokiel PL (1978) J Exp Mar Biol Ecol 38: 87–87
Jokiel PL (1984) Coral Reefs 3: 113–116
Jokiel PL (1985) PCRC-5, 4, pp 307–312
Jokiel PL, Coles SL (1974) Pac Sci 28: 1–18
Jokiel PL, Coles SL (1990) Coral Reefs 8: 115–120
Jokiel PL, Morrissey JJ (1986) Mar Biol 91: 15–26
Jokiel PL, York RH (1982) Bull Mar Sci 32: 301–315
Jokiel PL, Maragos JE, Franzisket L (1978) In: Stoddart D (ed) Coral reef research methods. UNESCO, Paris, pp 529–541
Jones AR (1984) Coral Reefs 3: 101–111
Jones JA (1963) Bull Mar Sci Gulf Caribb 13: 282–307
Jones O, Endean R (eds) (1973–1977) Geology and biology of coral reefs, vols 1–4. Academic Press, New York
Jupp D (1983) In: Baker J (ed) Proc GBR Conf. AIMS, Townsville, pp 509–516
Kanazawa A, Teshima S, Ando T et al. (1976) Mar Biol 34: 53–57
Kanwisher JW, Wainwright SA (1967) Biol Bull (Woods Hole) 133: 378–390
Karlson RH (1980) Bull Mar Sci 30: 894–900
Karlson RH (1981) PCRS-4, 2, pp 699–704
Karlson RH (1983) Bull Mar Sci 33: 118–131
Karnaukhov VN, Yashin VA (1980) Spectral analysis of microplankton. Biol Cent Acad Sci USSR, Puschino, 54 pp
Kawaguti S (1937) Palao Trop Biol Stn Stud 1: 187–198
Kawaguti S (1941) Palao Trop Biol Stn Stud 2: 309–317
Kawaguti S (1944) Palao Trop Biol Stn Stud 62: 675–679
Kawaguti S (1953) Biol J Okayama Univ 1: 171–176
Kawaguti S (1964) Biol J Okayama Univ 15: 25–35
Kawaguti S (1971) In: Cheng T (ed) Aspects of biology of symbiosis. Univ Park Press, Baltimore, pp 265–275
Kawaguti S, Nakayama T (1973) Biol J Okayama Univ 16: 67–71
Kay AM, Liddle MJ (1984) Rep Great Barrier Reef. Marine park authorities. Griffith Univ, Brisbane, 180 pp
Kay EA (1971) Pac Sci 28: 247–255

Kay EA (1978) Atoll Res Bull 221: 161–169
Kay EA, Switzer MF (1974) Pac Sci 28: 275–295
Keller NB (1976) Trans Inst Oceanol Acad Sci USSR (Moscow) 99: 31–44
Kelleher GG (1981) PCRS-4, 1, pp 231–236
Kelleher GG (1983) In: Baker J (ed) Proc GBR Conf. AIMS, Townsville, pp 43–60
Kelleher GG, Dutton JM (1985) PCRC-5, 6, pp 525–530
Kellog RBB, Patton JS (1983) Mar Biol 75: 137–149
Kenchington RA (1977) Biol Conserv 11: 103–105
Kenchington RA (1984) Rep Mar Sci, UNESCO, Paris 21: 93–160
Kenchington RA (1987) Bull Mar Sci 41: 552–560
Kenchington RA, Pearson R (1981) PCRS-4, 2, pp 597–603
Kendall JJ, Powell EN, Connor SJ (1983) Bull Mar Sci 33: 336–352
Kendrick B, Risk MJ, Michael J (1982) Bull Mar Sci 32: 862–867
Kensley B (1982) In: Rützler K, Macintyre J (eds) Atlantic Barrier Reef Belize. Smithsonian Inst, Washington, pp 236–240
Khristoforova NK et al. (1979) Biology of the Sea (Vladivostok) 67–73
Kimmerer WJ, Walsh TW (1981) Micronesica 17: 161–179
Kimor B, Baras B, Goldansky N (1981) J Plankton Res 3: 445–458
Kimura S, Hashimoto Y, Yamazato K (1972) Toxicon 10: 611–617
King CA, Bergmann W (1942) J Org Chem 7: 427–440
Kinsey DW (1972) PCRS-1 1, pp 13–23
Kinsey DW (1977) PCRS-3, 2, pp 383–388
Kinsey DW (1978) Limnol Oceanogr 23: 989–991
Kinsey DW (1983) In: Baker D (ed) Perspectives on coral reefs. AIMS, Townsville, pp 209–218
Kinsey DW (1985a) PCRC-5, 4, pp 505–526
Kinsey DW (1985b) PCRC-5, 6, pp 223–228
Kinsey DW (1988) Galaxea 7: 113–128
Kinsey DW (1991) Search 22: 119–121
Kinsey DW, Davies PJ (1979a) Limnol Oceanogr 24: 935–940
Kinsey DW, Davies PJ (1979b) In: Trudinger P (ed) Biogeochemical cycling of mineral forming elements. Elsevier, Amsterdam, pp 131–162
Kinsey DW, Domm A (1974) PCRS-2, 1, pp 49–66
Kinsman DJ (1964) Nature (London) 202: 12801282
Kinzie RA (1970) PhD Thesis, Yale, 130 pp
Kinzie RA (1973) Bull Mar Sci 23: 93–155
Kinzie RA (1974a) In: Bayer F (ed) Prostaglandins from plexaura. Univ Press, Miami, pp 22–38
Kinzie RA (1974b) J Exp Mar Biol Ecol 15: 335–345
Kinzie RA, Araki G, Kia L (1977) Am Zool 17: 908
Kissling DI, Taylor GT (1977) PCRS. 1, pp 317–323
Kjerfve B, Dinnel SP (1983) Coral Reefs 1: 205–217
Kjerfve B, Rutzler K, Kjerspe GH (1982) In: Rützler K, Macintyre J (eds) The Atlantic Barrier Reef Belize. Smithsonian Inst Publ, Washington, pp 61–83
Kleeman KH (1960) J Mollusc Stud 46: 13–54
Klim DJ (1969) Hawaii Inst Geophys Tech Rep 16: 56 pp
Klumpp DW, Mc Kinnon AD (1988) ACRC-6, p 55
Klumpp DW et al. (1988) Mar Ecol Prog Ser 45: 95–108
Klumpp DW, Pulfrich A (1989) Coral Reefs 8: 135–144
Koblentz-Mishke OI, Fortus VN (1980) In: Ponomareva L (ed) Biological and geological studies of pacific islands. Nauka, Moscow, pp 74–84
Kobluk DR, Risk MJ (1977) J Mar Exp Biol Ecol 27: 107–115
Kohn AJ (1968) Ecology 49: 1046–1062
Kohn AJ, Helfrich P (1957) Limnol Oceanogr 2: 241–251
Kohn AJ, Levitien PJ (1976) Oecologia 25: 199–210
Kohn AJ, Lloyd MC (1973) Int Rev Ges Hydrobiol 58: 369–399

Kohn AJ, White JK (1977) Micronesica 13: 199–215
Kojis BJ (1986) Mar Biol 91: 291–309
Kojis BL, Quinn NJ (1980) Am Zool 20: 819–825
Kojis BL, Quinn NJ (1981) Bull Mar Sci 31: 558–573
Kojis BL, Quinn NJ (1982) Mar Ecol Prog Ser 8: 251–255
Kojis BL, Quinn NJ (1984) Coral Reefs 3: 165–172
Kojis BL, Quinn NJ (1985) PCRC-5, 4, pp 289–293
Kokke WC et al. (1984) J Biol Chem 259: 8168–8173
Kopylov AI (1985) In: Rep 6th cruise r/v "Alexander nesmeyanov". Oceanogr Commiss Acad Sci USSR, Moscow, pp 128–135
Kopylov AI, Orlova TP (1985) In: Rep 6th cruise r/v 'Alexander Nesmeyanov". Oceanogr Commiss Acad Sci USSR, Moscow, pp 115–128
Koroljuk IK, Mikhailova MV (1986) In: Sokolov B (ed) Phanerozoic reefs and corals. Nauka, Moscow, pp 134–142
Kott P (1982) Micronesica 18: 95–127
Kott et al. (1984) Bull Mar Sci 34: 308–312
Kramer A (1897) Über die Korallenriffe und die Planktonverteilung. Kiel, 163 pp
Krasnov EV (1976) In: 50 years of Soviet paleontology: on taxonomy of ancient organisms. Nauka, Leningrad, pp 36–48
Krasnov EV (1981) PCRS-4, Manila 2: 257–262
Krasnov EV, Preobrazhenski BV (1982) Paleontol J (Moscow) 2: 136–140
Krasnov EV et al. (1982) In: Biology of coral reefs. Far Eastern Center of Acad Sci USSR, Vladivostok, pp 35–48
Kropp RK, Birkeland C (1981) PCRS-4, 2, pp 627–632
Kubanin AA, Murachveri AM (1980) In: Preobrazhensky B (ed) Biology of coral reefs. Nauka, Moscow, pp 201–211
Kühlmann DHH (1971) Poseidon 109: 18–23
Kühlmann DHH (1974) Int Rev Ges Hydrobiol 59: 305–325
Kühlmann DHH (1975) Int Rev Ges Hydrobiol 60: 495–521
Kühlmann DHH (1982) Mar Ecol 3: 193–212
Kühlmann DHH (1983) Helgoländer Wissensch Meersunters 36: 183–204
Kühlmann DHH (1985) Das Lebende Riff, Academic, Leipzig Berlin, 420 pp
Kühlmann DHH (1988a) Mem Assoc Australas Paleontol 8: 433–438
Kühlmann DHH (1988b) Ambio 17: 13–22
Kühlmann DHH (1989) Mem Australas Paleontol 8: 833–838
Kuznetzov VG (1986) In: Sokolov B (ed) Phanerozoic reefs and corals. Nauka, Moscow, pp 110–123
La Barre S, Coll CJ (1982) Mar Biol 72: 119–124
Ladd HS (1969) In: Reef organisms through time, pt J. Symp N Am Paleontol Conv, pp 1273–1300
Ladd HS (1977) In: Jones O, Endean R (eds) Biolge and geology of coral reefs. vol 4: Geology 2. Academic Press, New York, pp 1–20
Lam RK (1974) J Geophys Res 79: 3073–3084
Land LS, Goreau TF (1970) J Sediment Petrol 40: 457–462
Landry MR, Maas LW, Fagernes VL et al. (1984) Mar Ecol Prog Ser 16: 127–133
Lang J (1973) Bull Mar Sci 23: 260–279
Lang J (1974) Am Sci 62: 272–281
Lang J (1975) J Mar Res 33: 223–231
Larcum AW (1983) In: Barnes D (ed) Perspectives on coral reefs. AIMS, Townsville, pp 221–227
Larcum AW, Cox GC (1988) ACRC-6, p 58
Lasker HR (1976) In: Mackie G (ed) Coelenterate ecology and behaviour. Plenum, New York, pp 101–111
Lasker HR (1977) PCRS-3, 1, pp 607–611
Lasker HR (1980) J Exp Mar Biol Ecol 47: 77–87
Lasker HR (1981) Biol Bull (Woods Hole) 160: 292–302

Lasker HR (1983) J Exp Mar Biol Ecol 72: 157–169
Lasker HR, Coffroth MA (1983) Mar Ecol Prog Ser 13: 21–18
Lasker HR, Coffroth MA (1985) PCRC-5, 4, pp 331–335
Lasker HR, Gottfried MD, Coffroth MA et al. (1983) Mar Biol 73: 73–78
Latypov Yu Y (1979) In: Abstr 14th Pac Sci Congr, pt G. Khabarovsk, p 20
Latypov Yu Y (1985) Mar Ecol Prog Ser 29: 261–270
Laubenfels MW de (1950) Trans Zool Soc London 27: 1–154
Le Campion-Alsumard T (1989) In: Annu Meet Int Soc Reef Studies, Marseille, p 91
Le Borgne R, Blanchot J, Charpy L (1989) Mar Biol 102: 341–353
Lecompte M (1956) In: Moore R (ed) Treatise on invertebrates paleontology, pt F, Coelenterata. Univ Kansas Press, Lawrence, KS, pp 107–144
Lee CC et al. (1975) Nat Hist Mus Los Angeles Sci Bull 20: 69–75
Lee JJ (1980) In: Levandovsky M (ed) Biochemistry and physiology of protozoa vol 3, 2nd edn. Academic Press, New York, pp 43–66
Lee JJ, Mc Enery ME, Kahn EG (1979a) Micropaleontology 25: 118–140
Lee JJ, Mc Enery ME, Shilo M (1979b) Nature (London) 280: 57–58
Legendre L, Demers S, Delesalle B et al. (1988) Mar Ecol Prog Ser 47: 153–160
Lehman JT, Porter JW (1973) Biol Bull (Woods Hole) 145: 140–149
Lehnhof HM (1968) In: Florkin M (ed) Chemical zoology, pt 2. Academic Press, New York, pp 157–221
Leido JJ (1981) PCRS-4, 1, pp 5–7
Leis JM (1981) PCRS-4, 2, pp 471–476
Leletkin VA, Zvalinsky VI (1981) PCRS-4, 2, pp 33–37
Leontjev OK, Belodedenko MV, Kosmynin VN (1979) In: New nomenclature in coral reef geomorphology. Preprint Fac Geograph Univ Moscow 2129: 17 pp
Lerch K, Cook C (1984) Bull Mar Sci 34: 477–483
Lessios H, Glynn P, Robertson D (1983) Science 220: 222–713
Lessios HA, Robertson DR, Cubit JD (1984) Science 226: 335–337
Leversee GJ (1976) Biol Bull (Woods Hole) 151: 344–351
Levi C, Levi P (1965) J Microsc 4: 151
Levi C, Levi P (1976) Ann Sci Nat Zool 18: 367–380
Levin VS (1979) In: Abstr 14th Pac Sci Congr, pt G. Khabarovsk, pp 6–7
Lewin RA (1977) Phycologia 16: 217–219
Lewin RA (1983) Bioscience 32: 254–260
Lewin RA (1984) Phycologia 23: 203–208
Lewin RA, Cheng L, Lafargue F (1980) Bull Mar Sci 30: 744–745
Lewin RA, Cheng L (1983) Micronesica 19: 165–170
Lewin RA, Pardy RL (1981) Phycologia 20: 109–118
Lewin RA, Peters EC (1983) Am Zool 23: 960–963
Lewis BJ et al. (1988) ACRC-6, p 62
Lewis JB (1973) Caribb J Sci 13: 207–209
Lewis JB (1974a) J Exp Mar Biol Ecol 15: 299–304
Lewis JB (1974b) PCRS-2, 2, pp 201–206
Lewis JB (1976) Mar Biol 36: 147–150
Lewis JB (1977a) Biol Rev 52: 305–347
Lewis JB (1977b) PCRS-3, 1, pp 406–408
Lewis JB (1978) J Zool (London) 186: 393–396
Lewis JB (1981a) PCRS-4, 2, pp 369–374
Lewis JB (1981b) In: Longhurst. Analysis of Marine Ecosystems Acad Press, NY, pp 127–158
Lewis JB (1982) J Zool (London) 196: 371–384
Lewis JB (1989) Coral Reefs 8: 99–107
Lewis JB, Post EE (1982) Comp Biochem Physiol 71A: 457–459
Lewis JB, Fish AG (1969) Carribb J Sci 9: 1–24
Lewis JB, Price WS (1975) J Zool (London) 176: 527–544
Lewis JB, Price WS (1976) J Zool (London) 178: 77–89

Liddell WD (1982) In: Lawrence J (ed) Echinoderms. Proc Int Symp Tampa, A Balekma, Rotterdam, pp 173–182
Lilburn B (1983) In: Baker J (ed) Proc GBR Conf. AIMS, Townsville, pp 525–530
Limbaugh C (1972) Bull Nat Hist Mus Los Angeles 14: 42–50
Limbaugh C, Pederson M, Chase FA (1961) Bull Mar Sci Gulf Caribb 11: 237–247
Linley EA, Kopp K (1986) Mar Biol 92: 457–464
Lip JR, Porse H (1981) In: Proc 10th Int Seaweed Symp. Berlin New York, pp 601–606
Littler MM (1972) Oceanogr Mar Biol Annu Rev 10: 311–347
Littler MM (1973a) Limnol Oceanogr 18: 946–952
Littler MM (1973b) J Exp Mar Biol Ecol 11: 103–120
Littler MM (1976) Micronesica 12: 27–41
Littler MM (1979) Aquatic Bot 7: 21–34
Littler MM, Doty MS (1975) J Ecol 63: 117–129
Littler MM, Littler DS, Titlyanov EA (1991) Coral Reefs 10: 199–209
Lobel PS, Ogden JS (1981) Mar Biol 64: 173–184
Logan A (1984) Coral Reefs 3: 131–138
Logan A, Tomascik T (1991) Coral Reefs 10: 155–160
Losey GS (1972) Copeia 4: 820–833
Lowenstam HA (1955) J Sediment Petrol 25: 270–272
Loya Y (1972) Mar Biol 13: 100–123
Loya Y (1975) Mar Biol 29: 177–185
Loya Y (1976a) Ecology 57: 278–289
Loya Y (1976b) In: Mackie D (ed) Coelenterate ecology and behavior. Plenum, New York, pp 89–100
Loya Y (1976c) Bull Mar Sci 26: 450–466
Loya Y (1989) In: Abstr Annu Meet Int Soc Coral Reef Stud, Marseille, p 95
Loya Y, Rinkevich B (1980) Mar Ecol Prog Ser 3: 167–180
Loya Y, Slobodkin L (1971) Symp Zool Soc London 21: 117–140
Lubbok HR, Polunin N (1975) In: Environemental conservation. Found Environ Conserv, Geneva, pp 229–232
Lucas JS (1973) Micronesica 9: 197–203
Lucas JS, Nash WJ (1985) PCRC-5, 5, pp 327–332
Lugo AE, Snedeaker SC (1974) Annu Rev Ecol Syst 5: 39–64
Lugo AE et al. (1975) In: Golley F (ed) Tropical ecological systems, pp 335–352
Lundington CA (1979) Aust J Mar Freshwater Res 30: 425–430
Macintyre JG (1976) J Sediment Petrol 48: 669–670
Macintyre JG, Smith SV (1974) PCRS-2, 2, pp 277–287
Macintyre JG, Towe KM (1976) Science 193: 701–702
Macintyre JG, Burke RB, Stuckenrath R (1977) Geology 5: 749–754
Mackenzie FT, Kulm LD, Cooley RL et al. (1965) J Sediment Petrol 35: 265–272
Macurda DB (1982) In: Rützler K, Macintyre J (eds) Atlantic Barrier Reef Belize. Smithsonian Inst Publ, Washington, pp 412–417
Macurda DB, Meyer DL (1974) Nature (London) 247: 394–396
Macurda DB, Meyer DL (1977) Stud Geol Am Assoc Petrol Geol 4: 231–237
Maes VO (1967) Proc Acad Nat Sci Philad 119: 93–217
Maestrini S, Pizarro MJ (1966) Ann Univ Madagascar Ser Sci Technol 4: 17–34
Magnus DB (1967) Stud Trop Oceanogr Miami 5: 635–663
Mague TH, Holm Hansen O (1975) Phycologia 14: 87–92
Mahoney BM (1981) Bull Mar Sci 31: 141–146
Maragos JE (1978) In: Stoddart D (ed) Coral reef research methods. UNESCO, Paris, pp 353–361
Maragos JE, Baines G, Beveridge PJ (1973) Science 181: 1161–1164
Maragos JE, Baines G, Beveridge PJ (1985) PCRC-5, 4, pp 189–193
Margalef R (1968) Perspectives in ecological theory. Univ Press, Chicago, 240 pp
Marian R (1979) Aspects of nitrogen metabolism in algal-invertebrate symbiosis. PhD Thesis, Univ Cal, Los Angeles, p 90

Mariscal RN (1970) Mar Biol 6: 58–65
Mariscal RN (1971) In: Lenhoff L (ed) Experimental coelenterate biology. Univ Hawaii Press, Honolulu, pp 100–118
Mariscal RN (1974) Z Zellforsch 147: 149–156
Mariscal RN, Bigger CH (1977) PCRS-3, 1, pp 127–133
Marphenin NN (1984) Zool J (Leningrad) 62: 5–13
Marsden JR (1962) Nature 193: 598–599
Marsden M (1983) In: Baker J (ed) Proc GBR Conf. AIMS, Townsville, pp 453–457
Marsh JA (1970) Ecology 51: 255–263
Marsh JA (1974) PCRS-2, 1, pp 139–145
Marsh JA (1977) PCRS-3, 1, pp 331–336
Marsh JA, Smith SV (1978) In: Stoddart D (ed) Coral reef research methods. UNESCO, Paris, pp 361–377
Marsh JA, Tsuda RT (1973) Atoll Res Bull 170: 1–16
Marsh JA, Ross RM, Zolan WJ (1981) PCRS-4, 1, pp 355–360
Marsh LM (1974) Micronesica 10: 66–107
Marsh LM, Marshall JJ (1983) Bull Mar Sci 33: 671–686
Marshall JF (1983) In: Barnes D (ed) Perspectives on coral reefs AIMS, Townsville pp 231–238
Marshall JF and Jackobson G (1985) Coral Reefs 4: 11–17
Marshall KC (1980) In: Berkley R (ed) Microbial adhesion to surfaces. Ellis Horwood, London, p 550
Marshall N (1965) Ecology 26: 343–344
Marshall N (1968) Mar Biol 2: 50–53
Marshall N (1972) PCRS-1, pp 59–65
Marshall N et al. (1975) Int Rev Ges Hydrobiol 60: 335–345
Marshall SM (1933) In: Sci Rep GBR Exped 2: 111–157
Marshall SM, Orr AP (1931) In: Sci Rep GBR Exped 1: 94–133
Mashansky VF, Lee CE, Titlyanov EA et al. (1980) In: Schwemmler W (ed) Endocytobiology, vol 1. Walter de Gruyter Berlin, pp 173–174
Marshalek DS (1981) PCRS-4, 1, pp 213
Mather P, Bennett I (1984) A coral reef handbook. Univ Press Brisbane, 143 pp
Maxwell WG (1968) Atlas of GBR. Elsevier Amsterdam, 258 pp
Maxwell WG (1973a) Ibid 1: Geology 1. Academic Press, New York, pp 299–345
Maxwell WG (1973b) In: Jones O, Endean R (eds) Biology and geology of coral reefs vol 1: Geology 1. Academic Press, New York, pp 233–272
Mc Closkey LR (1970) Int Rev Ges Hydrobiol 55: 13–81
Mc Closkey LR, Chester RH (1971) In: Miller J (ed) Scientists in the sea, pt 4. Smithsonian Inst Publ, Washington, pp 229–238
Mc Closkey LR, Muscatine L (1984) Proc R Soc London Ser B 222: 215–230
Mc Closkey LR, Wethey DS, Porter JW (1978) In: Stoddart D (ed) Coral reefs: research methods. UNESCO, Paris, pp 379–394
Mc Farland WN, Ogden JC (1985) In: Reaka M (ed) The ecology of deep and shallow coral reefs. US Dep Commerce, Washington, pp 37–51
Mc Intosh GS (1981) PCRS-4, 1, pp 99–103
Mc Kee ED, Chronic J, Leopold EB (1959) Bull Am Assoc Petrol Geol 43: 501–562
Mc Laughlin JJ, Zahl PA (1966) In: Henry S (ed) Symbiosis, vol 1. Academic Press, New York, pp 257–290
Mc Lean RE (1978) Phil Trans R Soc London Ser A 291: 167–186
Mc Michael DE (1974) PCRS-2, 1, pp 241–255
Mc William PS, Sale PF, Anderson DT (1981) J Exp Mar Biol Ecol 52: 185–203
Mergner H (1981) PCRS-4, 1, pp 193–207
Mergner H, Schummacher H (1974) Helgoländer Wissensch Meersunters 26: 238–358
Mergner H, Svoboda A (1977) Helgoländer Wissensch Meersunters 30: 383–399
Meyer DL (1973a) Mar Biol 22: 105–129
Meyer DL (1973b) Bull Mar Sci 23: 255–259

Meyer DL, Messing CG, Macurda DB (1978) Bull Mar Sci 28: 412–444
Meyer DL, Schultz ET, Helfman GS (1983) Science 220: 1047–1049
Meyer Reil LA, Bolter M, Dawson R (1981) Kieler Meersforsch Sonderh 5: 431–432
Michel A (1969) Cah Pac 13: 81–132
Michel A et al. (1972) Cah Off Rech Sci Tech Outre Mer Oceanogr 9: 375–402
Miller WJ et al. (1977) PCRS-3, 1, pp 349–352
Milliman JD (1967) J Sediment Petrol 37: 658–676
Milliman JD (1969) Atoll Res Bull 129: 26
Milliman JD (1974) Marine carbonates. Springer, Berlin, 375 pp
Milliman JD, Mahnken C (1972) Atoll Res Bull 129: 23–27
Mitchell R, Chet I (1975) Microbial Ecol 2: 227–233
Moiseev EV (1985) In: Rep 6th cruise r/v "Alexander Nesmeyanov". Oceanogr Commiss Acad Sci USSR, Moscow, pp 110–115
Montaggioni L (1977) Mem Bur Rech Geol Min Fr 89: 456–466
Montgomery RT, Gerodette T, Marshall LD (1977) PCRS-3, 1, pp 357–365
Moore CH, Shedd WW (1977) PCRS-3, 2, pp 499–506
Moore E, Sander F (1976) Estuar Coast Mar Sci 4: 589–607
Moore NH (1981) PCRS-4, 1, pp 361–367
Moore RE, Scheuer RJ (1971) Comp Biochem Physiol 39: 793–759
Moore RJ (1988) ACRS-6, p 72
Moran PJ (1986) Oceanogr Mar Biol Ann Rev 24: 379–480
Moran PJ, Bradbuary RH, Reichelt RE (1985) PCRC-5, 5, pp 321–326
Morelock J, Koenig KJ (1967) J Sediment Petrol 37: 1001–1005
Moriarty DJW (1979) Mar Biol 53: 193–200
Moriarty DJW (1982) Aust J Mar Freshwater Res 33: 255–263
Moriarty DJW, Mansen JA (1990) Aust J Mar Freshwater Res 41: 785–794
Moriarty DJW, Haywood AC (1982) Microbial Ecol 8: 1–14
Moriarty DJW, Pollard PC (1982) Mar Biol 72: 165–173
Moriarty DJW, Pollard PC, Hunt WG (1985a) Mar Biol 85: 293–300
Moriarty DJW, Pollard PC, Hunt WG (1985b) Mar Biol 85: 285–292
Morrissey J (1980) Aquat Bot 8: 91–139
Morrissey J (1985) PCRC-5, 5, pp 77–83
Morton J (1974) PCRS-2, 2, pp 31–53
Motoda S (1940) Trans Nat Hist Soc Sapporo 16: 100–104
Muller PH (1974) Limnol Oceanogr 19: 802–809
Munk W, Sargent MC (1948) Trans Am Geophys Un 19: 855–860
Munro JL, Gwyther J (1981) PCRS-4, 2, pp 633–638
Munro JL, Williams D (1985) PCRC-5, 4, pp 543–580
Murdock GR (1978) Bull Mar Sci 28: 354–362
Murphey RC, Kremer JN (1983) Bull Mar Sci 33: 152–164
Muscatine L (1967) Science 156: 516–519
Muscatine L (1971) In: Lenhoff H (ed) Experimental coelenterate biology. Univ Hawaii Press, Honolulu, pp 227–238
Muscatine L (1973) In: Jones O, Endean R (eds) Biol and geol of coral reefs v. 2: Biology 1. Academic Press, New York, pp 77–117
Muscatine L (1974) In: Muscatine L (ed) Coelenterate biology. Academic Press, New York, pp 359–395
Muscatine L (1980) In: Cooke C (ed) Cellular interactions in symbiosis and parasitism. Ohio Univ Press, Columbus, pp 229–242
Muscatine L (1983) In: Baker J (ed) Proc GBR Conf. AIMS, Townsville, pp 341–351
Muscatine L (1984) Proc R Soc London Ser B 222: 181–202
Muscatine L, Cernichari E (1969) Biol Bull (Woods Hole) 137: 506–523
Muscatine L, D'Elia CF (1978) Limnol Oceanogr 23: 724–734
Muscatine L, Mc Closkey LR, Marian RE (1981) Limnol Oceanogr 26: 601–611
Muscatine L, Masuda TM, Burnap R (1979) Bull Mar Sci 29: 572–575
Muscatine L, Pool RR, Cernichiari E (1972) Mar Biol 13: 298–308

Muscatine L, Pool RR, Trench R (1975) Trans Am Microsc 94: 450–469
Muscatine L, Porter JW (1977) Bioscience 27: 454–460
Muscatine L, Mc Closkey LR, Loya Y (1985) PCRC-5, 6, pp 119–124
Muzik KM (1982) In: Rützler K, Macintyre J (eds) Atlantic barrier reef belize. Smithsonian Inst Publ, Washington, pp 303–310
Muzik KM (1985) PCRC-5, 6, pp 483–492
Muzik KM, Wainwright S (1977) Bull Mar Sci 27: 308–337
Nair PVR, Pillai CS (1972) PCRS-1, pp 33–40
Najim O (1981) PCRS-4, 1, pp 122–127
Nakamura M (1985) Bull Mar Sci 37: 271–278
Nalivkin DV (1967) In: Proc Int Symp Devonian systems, vol 2. Alberta Soc Petral Ceol, Calgary, pp 331–332
Naumov DV, Vasilenko SV, Dolgolenko MA et al. (1980) In: Ponomareva L (ed) Biological and geological studies on pacific islands. Nauka, Moscow, pp 8–28
Nelson SG (1985) PCRC-5, 5, pp 65–70
Nelson SG, Tsutsui RN (1981) PCRS-4, 2, pp 503–507
Neudecker S (1979) Ecology 60: 666–672
Neumann AC (1965) Bull Mar Sci 15: 987–1035
Neumann AC (1966) Limnol Oceanogr 11: 92–108
Newcomb E, Pugh T (1975) Nature (London) 253: 533–534
Nguen Thuong Dao, Le Thi Ngoc Ahn (1972) Kuroshio III. In: Proc 3rd symp CSK, Bangkok, pp 280–288
Nishihira M (1981) PCRS-4, 1, p 722
Nishihira M, Yamazato K (1974) PCRS-2, 2, pp 577–590
Norris JN, Bucher KE (1982) In: Rützler K, Macintyre J (eds) Atlantic barrier reef belize. Smithsonian Inst Publ, Washington, pp 167–233
Novozhilov AB (1980) In: Preorazhensky B (ed) Biology of coral. reefs. Nauka, Moscow, pp 211–217
Nudebecker S (1981) PCRS-4, 1, pp 173–180
Oberdorfer JA, Buddemeyer RW (1985) PCRC-5, 3, pp 310–314
Oda DK, Parrish JD (1981) PCRS-4, 1, pp 59–68
Odinetz C, de Forges B (1985) PCRC-5, 5, pp 197–203
Odum HT (1956) Limnol Oceanogr 1: 102–107
Odum HT (1957) Limnol Oceanogr 2: 85–97
Odum HT, Burkholder PR, Ravero J (1959) Trans Inst Sci Puerto Rico 6: 157–175
Odum HT, Hoskin CM (1958) Publ Inst Mar Science, University of Texas, vol 5. pp 65–97.
Odum HT, Mc Connell WA, Abbet W (1958) Publ Inst Mar Sci Uviv Tex 5: 65–97
Odum HT, Odum E (1955) Ecol Monogr 25: 291–320
Odum HT, Odum EP (1956) Ecology 37: 385–386
Ogawa R (1973) Ocean Age 3: 21–30
Ogden JC (1976) Aquat Bot 2: 103–116
Ogden JC (1977) In: Frost S (ed) Reefs and related carbonates. Tulsa Okl, pp 281–288
Ogden JC, Buckman NS (1973) Ecology 54: 589–596
Ogden JC, Ebersole JP (1981) Mar Ecol Prog Ser 4: 97–103
Ogden JC, Erlich PR (1977) Mar Biol 40: 273–280
Ogden JC, Lobel PS (1978) Environ Biol Fish 3: 49–63
Ogden JC, Zieman JC (1977) PCRS-3, 1, pp 377–382
Ogden JC, Brown RA, Salesky N (1973) Science 182: 715–716
Ogden JC, Tighe S, Miller S (1980) Am Zool 20: 949–953
Ohlehorst SL (1982) J Exp Mar Biol Ecol 60: 1–15
Ohlehorst SL (1985) In: Reaka M (ed) Ecology of deep and shallow coral reefs, vol 3. US Dep Commerce, Washington, pp 1–15
Oliver J (1985) PCRS 5, 4, pp 201–266
Oliver JK, Chalker BE, Dunlap WC (1983) J Exp Mar Biol Ecol 73: 11–35
Olson RR, Porter JW (1985) PCRC-5, 5, pp 257–262
Olson RR (1987) Limnol Oceanogr 32: 895–904

Opresko DM (1973) Bull Mar Sci 23: 535–558
Orlova TP (1985) In: Report 6th cruise r/v "Alexander Nesmeyanov". Oceanogr Commiss Acad Sci USSR Moscow, pp 100–109
Ormond R, Campbell AC (1973) Nature (London) 246: 167–179
Ormond R, Campbell AC (1974) PCRS-2, vol 1, pp 577–590
Ormond R, Johnson M, Powell A (1989) In: Annu Meet Int Soc for Reef studies, Mazseille, p 111.
Orr AP (1933) In: Sci Rep GBR Exped London, vol 2, pp 37–86
Ostergaard JM (1955) Pac Sci 9: 110–136
Ott B, Lewis JB (1972) Can J Zool 50: 1651–1656
Ott B, Maurer L (1977) In: Keegan B (ed) Biology of benthic organisms. Pergamon, Oxford, pp 493–502
Paerl HW et al. (1981) In: Broughton W (ed) Biology of nitrogen fixation, vol 1. Univ Press Oxford, pp 193–240
Paerl MW, Levin RA, Cheng L (1984) Box Marina 27: 257–264
Paine RT (1966) Am Nat 100: 65–77
Paine RT (1974) Oecologia 15: 93–120
Palmieri V (1985) PCRC-5, 5, pp 221–226
Pang RK (1973) Bull Mar Sci 23: 227–244
Parker GM (1984) Biol Bull (Woods Hole) 167: 159–167
Parrish JD (1980) Pap U NESCO, Semon fishery resources, Madang, Papua New Guinea, pp 89–119
Pascal H, Vacelet E (1981) PCRS-4, 1, pp 669–677
Patterson GA (1964) Philos Trans R Soc London Ser B 247: 213–482
Patterson M (1984) Biol Bull 167: 613–629
Patton JS, Burris JE (1983) Mar Biol 75: 131–136
Patton J et al. (1977) Mar Biol 44: 235–247
Patton WK (1976) In: Jones O, Ehdean R (eds) Biol and geol of coral reefs vol 3: Biology 2. Academic Press, New York, pp 1–37
Pauri CE (1988) ACRC-6, p 79
Pearse VB, Muscatine L (1971) Biol Bull (Woods Hole) 141: 350–362
Pearson RG (1974) PCRS-2, 2, pp 207–215
Pearson RG (1981) Mar Ecol Prog Ser 4: 105–122
Pecheaux J-F (1989) In: Abstr Annu Meet Int Soc Reef studies, Marseille, p 121
Penhale PA, Capone DG (1981) Bull Mar Sci 31: 164–169
Penhale PA, Smith WO (1977) Limnol Oceanogr 22: 400–407
Pehnale PA, Thayer GW (1980) J Exp Mar Biol Ecol 42: 113–123
Peres JM, Pickard J (1969) Mar Biol 3: 227–232
Peters EC (1988) PCRS-6, 1, pp 205–209
Peters EC, Meyers PA, Yevich PP (1981) Mar Pollut Bull 12: 333–339
Peters EC, Oprandy JE, Yevich PP (1983) J Invertebr Pathol 41: 394–396
Peyrot-Clausade M (1977) PCRS-3, 1, pp 269–278
Peyrot-Clausade M (1981) PCRS-4, 2, pp 745–750
Peyrot-Clausade M, Hutchings P, Richard G (1989) In: Abstr Annu Meet Int Soc Reef studies, Marseille, p 127
Pichon ML (1964) Recl Trav Stn War Endoume 2 (Suppl): 81–203
Pichon M (1974) PCRS-2, 2, pp 173–181
Pichon M (1978) Atoll Res Bull 222: 1–447
Pichon M (1985) PCRC-5, 6, pp 173–177
Pichon M, Morrissey J (1981) Bull Mar Sci 31: 581–593
Pichon M, Morrissey J (1985) CR Acad Sci Paris Ser 3, 300: 99–101
Pickard GL (1983) In: Baker J (ed) Proc GBR Conf. AIMS, Townsville, pp 63–70
Pickard GL, Donguy JR, Menin C (1977) Review of physical oceanography of GBR and western Coral Sea. Monogr Ser 2. AIMS, Canberra, 134 pp
Pike JE (1974) In: Bayer F (ed) Prostaglandins from plexaura. Univ Press, Miami, pp 9–15
Pilson MEQ, Betzer SB (1973) Ecology 54: 581–588

Plante-Guny MR (1973) Cah ORSTOM (Oceanogr) 11: 317–348
Polovina JJ (1984) Coral Reefs 3: 1–11
Polunin N (1989) In: Abstr Annu Meet Soc Reef studies, Marseille, p 129
Pomeroy LR et al. (1974) PCRS-2, 1, pp 87–96
Pool DJ et al. (1975) In: Walsh G (ed) Proc symp biology and managament of mangroves, Honolulu, pp 213–217
Porter JW (1972) Ecology 53: 745–748
Porter JW (1973) Discovery 9: 3–12
Porter JW (1974a) Science 186: 543–545
Porter JW (1974b) PCRS-2, 1, pp 111–125
Porter JW (1976) Am Nat 110: 731–742
Porter JW (1987) In: Stoddart D (ed) Coral reef research methods. UNESCO, Paris, pp 500–514
Porter JW (1980) In: Falkovsky G (ed) Primary productivity in the sea. Plenum, New York, pp 403–410
Porter JW, Falkovsky PG, Dubinsky Z (1984) Proc R Soc London Ser B 222: 161–180
Porter JW, Porter KG (1977) Limnol Oceanogr 22: 553–556
Porter JW, Porter KG, Bataccalan Z (1977) PCRS-3, 1, pp 105–112
Porter JW, Porter KG, Bataccalan Z (1981) Nature (London) 294: 249–250
Potts DC (1977) J Mar Exp Biol Ecol 28: 207–216
Potts DC (1979) Atoll Res Bull 220: 55–74
Potts DC (1983) Bull Mar Sci 33: 619–632
Potts DC, Swart PK (1984) Limnol Oceanogr 29: 504–518
Potts DC, Done TJ, Isdale PJ (1985) Mar Ecol Prog Ser 23: 79–84
Potts M, Whitton BA (1977) Oecologia 27: 275–283
Prahl von H (1985) Bull ERFEN Lima 12: 21–25
Pratt EM (1906) Q J Microsc Sci 49: 327–362
Preobrazhensky BV (1975) Biology of the Sea (Vladivostok) 1: 23–30
Preobrazhensky BV (1986) Modern corals Nauka, Moscow, pp 243
Preobrazhensky BV, Ivanov NA (1982) In: Biology of coral reefs; communities of Australian waters. Acad Sci USSR, Vladivostok, pp 82–90
Preston EM, Preston JL (1975) Bull Mar Sci 25: 248–258
Primakova GA (1985) In: Rep 6th cruise r/v Alexander Nesmeyanov. Oceanogr Commiss Acad Sei USSR, Moscow
Propp MV (1981) Biology of the Sea (Vladivostok) 3: 55–62
Propp MV (1985) In: Rep 6th cruise of r/v "Alexander Nesmeyanov". Oceanogr Commiss Acad Sci USSR Moscow
Propp MV, Tarasov VG, Tchabargy II (1983) In: Biology of coral reefs. Acad Sci USSR Publ, divostok, pp 75–102
Purdy EG (1963) J Geol 71: 334–355
Purdy EG (1974) In: La Pote L (ed) Reefs in time and space, vol 18. Soc Econ Paleontol Mineral USA Spec Publ, pp 9–76
Quasim SZ (1970) Deep Sea Res 17: 655–660
Quasim SZ, Bhattathiri PM (1971) Hydrobiologia 31: 29–38
Quasim SZ, Sankaranarayanan VN (1970) Limnol Oceanogr 15: 574–578
Quasim SZ, Bhattathizi PB, Reddy CV (1972) Int Rev Ges Hydrobiol 57: 207–225
Randall JE (1961) Ecology 42: 812–814
Randall JE (1963) Caribb J Sci 3: 1–16
Randall JE (1967) Stud Trop Oceanogr (Miami) 5: 665–847
Randall JE (1972) Biotropica 4: 132–144
Randall JE (1974) PCRS-2, 1, pp 159–166
Randall RH (1981) PCRS-4, 2, pp 157–164
Rasmont R (1968) In: Florkin M (ed) Chemical zoology, vol 2. Academic Press, New York, pp 43–51
Rasmussen CE, Cuff C (1990) In: Proc 419 Pac Conf Mar Sci Technol, vol 1. Tokyo, pp 13–20

Raymond B (1981) PCRS-4, 1, pp 216–220
Reed JK (1985) PCRC-5, 6, pp 249–253
Reese ES (1975) Z Tierpsychol 31: 37–61
Reese ES (1977) PCRS-3, 1, pp 267–274
Reese ES (1981) Bull Mar Sci 31: 594–604
Reichelt RE (1979) Micronesica 15: 297–307
Reichelt RE (1982) Coral Reefs 1: 3–11
Reichelt RE, Bradbury RH, Moran PJ (1990) Coral Reefs, 9: 97–103
Reichelt RE, Kohn AJ (1985) PCRC-5, 5, pp 191–196
Reimer A (1970) In: Muscatine L (ed) Biology of coelenterates. Univ Press Hawaii, Honolulu, pp 209–217
Reimer A (1971) Pacific Sci 25: 512–520
Reimer A (1972) J Exp Mar Biol Ecol 10: 177–181
Reinboth R (1973) Helgoländer Wissensch Meersunters 24: 174–191
Reiswig HM (1973) Bull Mar Sci 23: 191–226
Reiswig HM (1974) J Exp Mar Biol Ecol 14: 231–249
Reiswig HM (1975) Can J Zool 53: 582–589
Relevante N, Gilmartin M (1982) J Plankton Res 4: 47–76
Renaud-Mornant J (1977) Cah Pacifique, 20: 1–6
Renaud-Mornant J, Salvat B, Boussy C (1971) Biotropica 3: 36–55.
Renon JP (1977) Ann Inst Oceanogr (Paris) 53: 213–326
Rhyne CF (1971) Atoll Res Bull 149: 41–65
Ricard G (1981) PCRS-4, 2, pp 637–642
Ricard G, Salvat B (1972) CR Acad Sci Paris Ser D 275: 1547–1550
Ricard M (1976) Cah Pac 19: 383–395
Ricard M (1977a) In: Simonsen B (ed) Proc 4th symp recent and fossile marine diatoms, Oslo, pp 389–405
Ricard M (1977b) Rev Algol 12: 141–336
Ricard M (1977c) PCRS-3, 1, pp 343–348
Ricard M (1981) PCRS-4, 1, pp 453
Ricard M, Delesalle B (1981) PCRS-4, 1, pp 425–429
Ricard M et al. (1979) J Soc Oceanistes (Paris) 62: 47–57
Rice ME (1976) Micronesica 12: 119–132
Rice ME, Macintyre JG (1982) In: Rützler K, Macintyre J (eds) The atlantic barrier reef belize. Simthsonican Inst Publ, Washington, pp 202–207
Richards GN, Meikle PJ, Yellowles D, (1983) In: Baker J (ed) Proc Conf. GBR (1983) AIMS, Townsville, Conf. pp 353–360
Richardson CA, Dustan D, Lang JS et al. (1979) Mar Biol 55: 181–186
Richman S, Loya Y, Slobodian LB (1975) Limnol Oceanogr 20: 918–923.
Richmond R (1981) PCRS-4, 2, pp 153–156
Richmond RH (1985) Mar Ecol Prog Ser 22: 181–185
Richmond RH, Jokiel PL (1984) Bull Mar Sci 34: 281–287
Riddle MJ (1988) Mar Ecol Prog Ser 47: 280–292
Rieper M (1978) Mar Biol 45: 337–346
Rinkevich B, Loya Y (1977) PCRS-3, 20, pp 585–591
Rinkevich B, Luya Y (1979a) Mar Ecol Prog Ser 1: 145–152
Rinkevich B, Loya Y (1979b) Mar Pollut Bull 10: 328–330
Rinkevich B, Loya Y (1979c) Mar Ecol Prog Ser 1: 133–144
Roberts HH (1974) PCRS-2, pp 497–512
Roberts HH (1977) Limnol Oceanogr 22: 152–156
Roberts HH, Murray SP, Suhaida JN (1975) J Mar Res 33: 233–259
Roberts HH, Wiseman WJ, Suchanek TH (1981) PCRS-4, 1, pp 459–465
Roberts HH, Rouse LJ, Walkez ND (1983) Caribb J Sci 19: 55–62
Roberts L (1987) Science, 238: 1128–1129
Robertson DR (1972) Science 177: 1007–1009
Robertson DR (1973) Z Tierpsychol 32: 319–324

Robertson DR (1982) Mar Ecol Prog Ser 7: 253–265
Robertson DR, Choat JH (1981) PGRS-4, 1, pp 217–225
Robertson DR, Reinboth R, Bruce RW (1982) Bull Mar Sci 32: 868–879
Robertson R (1970) Pac Sci 24: 45–54
Robertson R (1980) Pac Sci 34: 1–15
Robishaux DM (1985) PCRS-5, 5, pp 83–86
Robles CD, Cubit J (1981) Ecology 62: 1536–1547
Roffman B (1968) Comp Biochem Physiol 27: 405–418
Rogers CS (1979a) Limnol Oceanogr 24: 342–349
Rogers CS (1979b) J Exp Mar Biol Ecol 41: 269–288
Rogers CS (1985) PCRC-5, 6, pp 491–508
Rogers CS, Salesky NH (1981) J Exp Mar Biol Ecol 49: 179–187
Rogers CS, Fits MC, Gilnack M et al. (1984) Coral Reefs 3: 69–76
Romankevich EA (1979) In: Volkov I (ed) Chemistry of the ocean. Nauka, Moscow, pp, 266–277
Rose CS, Risk MJ (1985) Mar Ecol 6: 345–363
Rosen BR (1978) Philos Trans R Soc London Ser B 284: 115–122
Rosen BR (1981) In: Forey P (ed) The evoluting biosphere. Univ Press, Cambridge, pp 103–129
Rosen BR (1982) Bioscience 32: 519–525
Ross CA (1972) J Protozool 19: 181–192
Ross CA (1977) In: Frost S (ed) Reefs and related carbonates. Am Assoc Petrol Geol, Tulsa Okl, pp 217–230
Ross MA, Hodson G (1981) PCRS-4, 2, pp 281–286
Ross RM (1978) Z Tierpsychol 46: 71–83
Ross RM (1981) PCRS-4, 2, pp 575–580
Rostomov SA (1987) In: Abstr 6th Symp Fossile cnidaria and reefs, Acad Sci USSR, Veadivostou, pp 69–70
Roth AA, Clausen CD, Yahiki RY et al. (1982) Pac Sci 36: 65–81
Röttger R (1972) Mar Biol 15: 150–159
Rougerie F (1982) CR Acad Sci Paris Ser II 297: 909–912
Rougerie F (1985) PCRC-5, 6, pp 17–27
Roushdy HM, Hansen VK (1960) Nature (London) 188: 517–518
Roushdy HM, Hansen VK (1961) Nature (London) 190: 649–650
Rowe FW, Britless RA, Hoggett AK (1986) Zool J Linn Soc London 86: 197–277
Roy KJ, Smith SV (1971) Pac Sci 25: 234–348
Rublee PA, Lasker MR, Gottfried M et al. (1980) Bull Mar Sci 30: 888–893
Rublee MB, Zeeman C (1984) Bull Mar Sci 34: 335–346
Ruddle K, Johannes RE (1985) Traditional knowledge and management of coastal systems in Asia and Pacific. Pap Res TEA, Seeninai. UNESCO, Paris, 313 pp
Ruggieri GD (1976) Science 194: 491–497
Russ G (1984) Rep Mar Sci 27. UNESCO Paris, pp 75–90
Russell BC et al. (1974) PCRS-2, 1, pp 207–215
Russell ES (1934) In: Rep GBR London Exp, vol 2, pp 159–201
Rützler K (1975) Oecologia 19: 203–216
Rützler KJ et al. (1980) Mar Ecol 1: 65–71
Rützler KJ, Macintyre JG (eds) (1982) In: The Atlantic Barrie. Reef Belize Smithsonian Inst Publ, Washington, 380 pp
Rützler K, Santavy DL (1983) Mar Ecol 4: 301–319
Rylaarsdam KW (1983) Mar Ecol Prog Ser 13: 249–260
Sakai K, Nishihira M (1981) PCRS-4, 2, pp 763–766
Sale PF (1978a) Environ Biol Fish 3: 85–102
Sale PF (1978b) Mar Biol 49: 133–140
Sale PF (1980) Oceanogr Mar Biol Annu Rev 18: 367–421
Sale PF (1982) In: Pauly D (ed) Theory and management of tropical fishes Murphy, Manila, pp 241–253

Sale PF (1984) In: Strong D (ed) Ecological communities. Univ Press, Princeton, NJ, pp 479–489
Sale PF (1985) PCRC-5, 5, pp 381–386
Sale PF (1989) In: Abstr Meet Int Soc Coral reef studies, Marseille, p 145
Sale PF, MC William PS, Anderson DT (1976a) Mar Biol 34: 59–66
Sale PF, Potts DC, Frankel E (1976b) Search 7: 334–338
Sale PF, Doherty PJ, Eckert GJ (1984a) Oecologia 64: 191–198
Sale PF, Douglas WA, Doherty PJ (1984b) Coral Reefs 3: 91–99
Salvat B (1971) CR Acad Sci Paris Ser D 272: 211–14
Salvat B (1972) PCRS-1, pp 372–378
Salvat B (1978) Rev Int Parcs Nationaux 3: 1–4
Salvat B (1981a) PCRS-4, 1, pp 225–229
Salvat B (1981b) PCRS-4, 1, pp 503–509
Salvat B (1981c) PCRS-4, vol 1, p 331
Salvat B (ed) (1987) Human Impacts on coral reefs. Antenne Mus, Tahiti, 200 pp
Salvat B, Renand-Mornant J (1969) Cah Pac 13: 159–189
Salvat B, Vence-Peyre M (1981) PCRS-4, 2, pp 767–774
Salvat B et al. (1979) Abstr Pap 14th Pac Sci Congr (Suppl) Khabarovsk, pp 44–46
Sammarco PW (1980) J Exp Mar Biol Ecol 45: 245–272
Sammarco PW (1981) J Exp Mar Biol Ecol 61: 31–55
Sammarco PW (1982) J Exp Mar Biol Ecol 61: 31–55
Sammarco PW (1983) Baker J (ed) Proc GBR Conf AIMS, Townsville, pp 245–250
Sammarco PW (1985) PCRC-5, 4, pp 391–397
Sammarco PW, Crenshaw H (1984) Mar Biol 82: 167–180
Sammarco PW, Coll MC, Rabarre S (1983) Coral Reefs 1: 173–178
Sano M, Schimizu M, Nose I (1987) Mar Ecol Preg Ser 37: 191–199
Santavy DV (1985) PCRC-5, 5, pp 135–140
Sara M (1971) Mar Biol 11: 214–221
Sara M, Liaci L (1964) Nature (London) 203: 301–302
Sargent MC, Austin TS (1949) Trans Am Geophys Un 30: 245–249
Sargent MC, Austin TS (1954) Geol Surv US Dep Interior, 260-e: 291–301
Sato M (1985) Coral Reefs 4: 27–33
Scatterday J (1977) In: Frost S (ed) Reefs and related carbonates. Am Assoc Petrol Geol Tulsa, pp 155–169
Scheuer PJ (1969) Forlschr Chem Org Nat 27: 322–326
Scheuer PJ, Bagnis R (1985) PCRC-5, 4, pp 401–411
Schlanger D (1981) Soc Econ Paleontol Spec Publ 32: 209–226
Schlichter D (1982) Oecologia 53: 40–49
Schlichter D, Weber W, Fucke MW (1985) Mar Biol 89: 143–147
Schmaljohann R, Röttger R (1976) Naturwissenschaften 63: 486–487
Schmidt H (1972) Helgoländer Wissensch Meeresunters 23: 422–458
Schmitz K, Kremer BP (1977) Mar Biol 42: 305–313
Schoenberg DA (1980) In: Schwemmler W (ed) Endocytobiology, vol 1. Walter de Gruyter, Berlin, pp 155–163
Schoenberg DA, Trench RK (1980) Proc R Soc London (B) 207: 445–460
Schroeder JH (1985) PCRC-5, 5, pp 379–384
Schurhmacher H (1973) Helgoländer Wissensch Meeresunters 25: 461–472
Schuhmacher H (1979) Int Rev Ges Hydrobiol 640: 207–243
Schuhmacher H, Zibrorius H (1985) Coral Reefs 4: 1–9
Scoffin TP, Dixon JE (1983) Biol J Linn R Soc 20: 11–38
Scoffin TP, Stoddart DR (1978) Philos Trans R Soc London Ser B 284: 99–122
Scott BD, Dixon JE (1983) Biol J of Linnean Soc 20: 11–38
Scott BD, Jitts HR (1977) Mar Biol 41: 307–315
Scott G, Rotondo G (1983) Coral Reefs 1: 139–150
Sebens KP (1977) PCRS-3, 1, pp 397–404
Sebens KP, de Riemer K (1977) Mar Biol 43: 247–256

Sebens KP, Koehl MA (1984) Mar Biol 81: 255–271
Segel LA, Ducklow HW (1982) Bull Mar Sci 32: 919–935
Serene R (1972) PCRS-1, pp 419–429
Shakleton NJ, Opdyke NI (1973) Quatern Res 3, 39–55
Shapiro DJ (1977) PCRS-3, 1, pp 571–576
Shapiro DY (1979) Adv Stud Behav 10: 43–102
Shapiro DY (1981) Bull Mar Sci 31: 383–398
Shelley C (1985) PCRC-5, 5, pp 297–301
Sheppard AL (1984) Coral Reefs 3: 43–50
Sheppard CR (1979) Mar Ecol Prog Ser 1: 237–247
Sheppard CR (1980) Mar Ecol Prog Ser 2: 193–205
Sheppard CR (1981a) PCRS-4, 2, pp 363–369
Sheppard CR (1981b) J Nat Hist 15: 607–621
Sheppard CR (1982) Mar Ecol Prog Ser 7: 83–115
Sherr B, Sherr E (1983) Estuar Coast Shelf Sci 16: 1–7
Shinn EA (1963) J Sediment Petrol 33: 291–303
Shinn EA (1966) J Paleontol 40: 233–240
Shinn EA (1976) Environ Geol 1976 1: 241–254
Shirayama Y, Hovikoshi M (1982) J Oceanogr Soc Jpn 38: 193–207
Shirota H (1966) Plankton in waters off Vietnam. OTCa, Tokyo, 466 pp
Shlesinger Y, Loya Y (1985) Science 228: 1333–1335
Shornikov EJ (1980) In: Preobrazhemsky B, (ed) Biology of coral reefs by Acad Sci Publ, Moscow 1, pp 131–158
Short FT (1981) PhD Diss, Univ Alaska, 173 pp
Siebeck O (1981) Naturwiss enschaften 68: 426–428
Simkiss K (1964) Biol Rev Cambridge Philos Soc 39: 487–505
Sirenko BI (1980) In: Preobrazhensky B (ed) Biology of coral reefs. Nauka, Moscow, pp 87–112
Skyring GM (1985) PCRC-5 3, pp 421–425
Skyring GW, Chambers LA (1976) Aust J Mar Freshwater Res 27: 595–602
Slobodkin LB (1968) In: Lewontin R (ed) Population biology and evolution. Univ Press, Syracuse, pp 187–205
Smith CL (1978) Envlron Biol Fish 3: 109–178
Smith CL, Tyler JC (1972) In: Collette B, Rarle S (eds) Results of tektite program, vol 14. County Mus Sci Bull, Los Angeles, pp 125–176
Smith CL, Tyler JC (1973) Helgoländer Wissensch Meersunters 24: 264–275
Smith DC (1979) Proc R Soc London Ser B 204: 115–130
Smith DC (1980) In: Schwemmler W (ed) Endocytobiology. De Gruyter, Berlin, pp 317–332
Smith SR (1985) PCRC-5, 6, pp 497–508
Smith SV (1973) Limnol Oceanogr 18: 106–120
Smith SV (1974) PCRS-2, 1, pp 77–85
Smith SV (1977) PCRS-3, 2, pp 577–582
Smith SV (1978) Nature (London) 273: 225–226
Smith SV (1981) Limnol Oceanogr 26: 612–621
Smith SV (1983) In: Barnes D (ed) Perspectives on coral reefs. AIMS, Townsville, pp 240–244
Smith SV (1984) Limnol Oceanogr 29: 1149–1160
Smith SV, Buddemeyer RW (1986) Prediction of coral growth. Preprint Liver-more Lab Liver more USA
Smith SV, Harrison JT (1977) Science 197: 556–559
Smith SV, Jokiel PL (1978) Atoll Res Bull 221: 15–53
Smith SV, Kinsey DW (1976) Science 194: 937–939
Smith SV, Kinsey DW (1978) In: Stoddart D, Johannes R (eds) Coral reefs research methods. UNESCO, Paris, pp 469–484
Smith SV, Marsh JA (1973) Limnol Oceanogr, 18: 953–961

Smith SV, Pesret F (1974) Pacific Sci 28: 225–246
Smith SV, Roy KG, Schielsse HG et al. (1971) Pac Sci 25: 206–221
Smith SV et al. (1975) Atoll Res Bull 220: 1–11
Soerigato A (1981) PCRS-4, 1, pp 287–292
Sokolov BS (1955) Paleozoic tabulata from the European part of USSR. Gostoptekhisdat, Moscow, 525 pp
Sokolov BS (ed) (1986) Reefs of Phanerozoic Nanka, Moscow 232 pp
Sokolov BS (1987) Reefs and reef building corals. Nauka, Moscow, 292 pp
Soliman GW (1969) Am Zool 9: 887–894
Sorokin YuI (1968) The use of C^{14} in study of nutrition of aquatic animal. In: Rep Int Assoc Theor Appl Limnol 16. Springer, Berlin Heidelberg New York, 41 pp
Sorokin YuI (1970) Dokl Acad Sci USSR (Moscow) 192: 905–907
Sorokin YuI (1971a) J Gen Biol (Moscow), 32: 168–185
Sorokin YuI (1971b) Int Rev Ges Hydrobiol 56: 1–48
Sorokin YuI (1972) In: Dyrssen D (ed) The changing chemistry of oceans. John Wiley & Sons, London, pp 189–204
Sorokin YuI (1973a) Limnol Oceanogr 18: 380–385
Sorokin YuI (1973b) Nature (London) 242: 415–417
Sorokin YuI (1973c) In: Jones O, Endean R (eds) vol 2: Biology and Geology of Coral Refs Jones v 2: Biology 1. Academic Press, New York, pp 17–45
Sorokin YuI (1974) PCRS-2, 1, pp 3–10
Sorokin YuI (1977a) J Gen Biol (Moscow) 38: 185–197
Sorokin YuI (1977b) In: Vinogradov M (ed) Biology of the ocean, vol 2. Nauka, Moscow, pp 133–154
Sorokin YuI (1978a) In: Kinne O (ed) Marine ecology, vol 4. John Wiley & Sons, New York Longon, pp 501–616
Sorokin YuI (1978b) Arch Hydrobiol 83: 281–323
Sorokin YuI (1980) Hydrobiol J (Kiev) 16: 84–85
Sorokin YuI (1981a) PCRS-4, 2, pp 27–32
Sorokin YuI (1981b) PCRS-4 2: 443–447
Sorokin YuI (1981c) In: Longhurst A (ed) Analysis of marine ecosystems. Academic Press, New York, pp 293–342
Sorokin YuI (1984a) J Gen Biol 45: 203–216
Sorokin YuI (1984b) J Gen Biol 45: 813–828
Sorokin YuI (1985) Marine Ecol Prog Ser 27: 87–97
Sorokin YuI (1986a) Pol Arch Hydrobiol 33: 33–68
Sorokin YuI (1986b) Biology of the Sea (Vladivostok) 6: 13–14
Sorokin YuI (1989) J Gen Biol (Moscow) 50: 584–593
Sorokin YuI (1990a) Aust J Mar Freshwater Res 41: 775–784
Sorokin YuI (1990b) In: Dubinsky Z (ed) Coral reefs. Elsevier, Amsterdam, pp 401–410
Sorokin YuI (1990c) J Gen Biol (Moscow) 51: 61–71
Sorokin YuI (1991) Estuar Coast Sheef Sci 33: 259–280
Sorokin YuI, Kadota H (eds) (1972) Methods of assessment of microbial production and decomposition in fresh waters. John Whiley Sons, London, 112 pp
Sorokin YuI, Mamaeva TI (1980) In: Vinogradov M (ed) Pelagic ecosystems of Peruvian region. Nauka, Moscow, pp 104–115
Sorokin YuI, Tyapkin VS (1984) Biology of the Sea (Vladivostok) 4: 15–23
Sorolkin YuI, Tsretic AM (1972) Oceanology (Moscow) 12: 1047–1056
Sorokin YuI, Wyshkwartzev OI (1973) Aquaculture, 2: 141–148
Sorokin YuI, Sukanova JN, Konovalova GV et al. Pol Arch Hydrobiol 24 (Suppl): 145–162
Sorokin YuI, Tyapkin VS, Nguen Tak Ahn (1982) Biology of the Sea (Vladivostok) 6: 12–17
Sorokin YuI Tyapkin VS, Nguen Tak Ahn (1983) Biology of the Sea (Vladivostok) 3: 29–38
Soule DF, Soule JD (1976) Micronesica 12: 157–164

Soule JD, Soule DF (1977) PCRS-3, 1, pp 335–340
Sournia A (1968) Hydrobiologia 31: 545–560
Sournia A (1969) Mar Biol 3: 287–303
Sournia A (1976a) CR Acad Sci Paris Ser D 282: 553–555
Sournia A (1976b) Mar Biol 37: 29–32
Sournia A (1977) Ann Inst Oceanogr Paris 53: 47–74
Sournia A, Ricard M (1975a) Micronesica 11: 159–166
Sournia A, Ricard M (1975b) CR Acad Sci Paris Ser D 280: 741–743
Sournia A, Ricard M (1976a) Vie Milieu (2-B) 26: 243–279
Sournia A, Ricard M (1976b) J Exp War Biol Ecol 21: 129–140
Sournia A, Delesalle B, Ricard M (1981) Oceanol Acta 4: 112–119
Stanton RJ (1967) Bull Am Assoc Geol 51: 2463–2467
Starck L, Almodovar L, Krauss R (1969) J Phycol 5: 305–312
Starck WA, Schroeder KE (1965) Sea Frontier 11: 66–79
Stearn CW, Scaffin TP, Martindale W (1977) Bull Mar Sci 27: 479–510
Steele RD (1975) Biol Bull 149: 590–600
Stephens GG (1962) Biol Bull 23: 648–659
Stephens GC, Schinske RA (1961) Limnol Oceonogr 6: 175–181
Stevenson DK, Marshall N (1974) PCRS-2, 1, pp 147–156
Stiasny G (1940) Publ Mar Biol Stn Ghardaga 2: 121–131
Stimson JS (1978) Mar Biol 48: 173–184
Stoddart DR (1962) Nature (London) 196: 512–515
Stoddart DR (1969) Biol Rev 44: 433–498
Stoddart DR (1972) Nature (London) 239: 51–52
Stoddart DR (1974) PCRS-2, 2, vol 473–484
Stoddart DR (1980) J Biogeogr 7: 269–289
Stoddart DR (1981) PCRS-4, 1, pp 33–36
Stoddart DR (1985) PCRC-5, 3, pp 349–351
Stoddart DR, Johannes RE (eds) (1978) Coral reff research methods. UNESCO, Paris, 581 pp
Stoddart JA, Black R (1985) Mar Ecol Prog Ser 23: 153–164
Stone RB, Pratt MJ, Parker RO et al. (1979) Mar Rev Fish, 41: 1–11
Stubblefield CJ (1960) Q J Geol Soc London 116: 219–238
Suchanek TH (1983) In: Reaka M (ed) Ecology of deep and shallow coral reefs, vol 1. US Dep Commerce, Washington, pp 55–61
Sucharsono J, Brown BE (1989) In: Abstr Annu Meet Inst Soc Coral reef studies, Marseille, p 155
Sudara S (1981) PCRS-4, 2, pp 679–684
Sudara S (1988) ACRC-6, p 97
Sukhanova IN (1983) In: Vinogradov M (ed) Modern methods of quantifications of marine plankton. Nauka, Moscow, pp 97–106
Sumner R, Loya Y, Slobodlcin LB (1975) Limnol Oceanogr 20: 918–923
Suchanek TH, Green DJ (1981) FCRS-4, 2, pp 679–684
Sutton M (1983) In: Barnes D (ed) Perspectives on coral reefs. AIMS, Townsville, pp 248–252
Sutton DC, Clements KD (1988) ACRC-6, p 98
Swenney BM (1976) Biol Bull 151: 236–246
Sy JC, Herrera FS, McManus JW (1981) PCRS-4, 2, pp 263–269
Szamant A (1986) Coral Reefs 5: 43–54
Szamant A, Cassman NJ (1989) Coral Reefs 8: 217–224
Szamant-Froelich A (1981) J Exp Mar Biol Ecol 55: 133–144
Szamant-Froelich A (1983) In: Reaka M (ed) Ecology of deep and shallow coral reefs, vol 1. US Dep Commerce, Washington, pp 133–140
Szamant-Froelich A (1985) PCRC-5, 4, pp 399–402
Szamant-Froelich A, Pilson ME (1980) J Mar Exp Biol Ecol 48: 85–97
Szamant-Froelich A, Johnson V, Moehn T (1981) PCRS-4, 1, pp 163–168

Szamant-Froelich AL, Riggs L, Reutter M (1984) Am Zool 23: 961
Tait RJ (1972) J Geophys Res 77: 2077–2211
Talbot FH, Gilbert A (1981) PCRS-4, 2, pp 485–489
Talbot FH, Russel BC, Anderson GR (1978) Ecol Monogr 48: 425–440
Talbot FH, Talbot MS (1971) Endeaour 30: 38–42
Targett NM, Bishop SS, Connelloy MC et al. (1983) J Chem Ecol 9: 817–828
Taylor DL (1973a) Annu Rev Microbiol 27: 171–181
Taylor DL (1973b) Helgoländer Wissensch Meersunters 24: 276–296
Taylor DL (1977) PCRS-3, 1, pp 432–437
Taylor FJR (1980) In: Schoenberg D (ed) Endocytobiology, vol 2. De Gruiter, Berlin, pp 666–669
Taylor JD (1968) Philos Trans R Soc London Ser B 254: 129–206
Taylor JD (1976) Zool J Linn Soc 59: 155–193
Taylor JD (1978) J Exp Mar Biol Ecol 31: 83–107
Taylor JD (1984) J Exp Mar Biol Ecol 74: 273–290
Taylor JD, Lewis M (1970) J Nat Hist 4: 199–220
Taylor WR (1950) Plants of Bikiniatoll. Univ Mich Press, Ann Arbor, 227 pp
Tchabardgy II, Fam Van Huen (1988) In: Sorokin YuI (ed) Biology of coral reefs. Acad Sci USSR, Vladivostok, pp 81–99
Tchumakova NI, Gitelson II (1975) Luminose bacteria. Nauka, Moscow, 230 pp
Teas J (ed) (1983) Biology and Ecology of Mangroves. Junk Publ, The Hague, 188 pp
Theodor J (1963) Vie Milieu 14: 815–818
Theodor J (1967) Vie Milieu 18A: 291–301
Thom BC, Chappell J (1975) Search 6: 90–93
Thom BG, Chappell J (1978) Philos Trans R Soc London Ser A 291: 187–194
Thom BG, Chappell J, Veeh H (1978) Philos Trans R Soc London Ser A 291: 37–54
Thomas PA (1972) PCRS-1, pp 334–344
Thomassin BA (1974) PCRS-2, 1, pp 297–307
Thomassin BA (1976) Helgoländer Wissensch Meersunters 28: 51–65
Thomassin BA (1978) In: Stoddart D, Johannes R (eds) Coral reef research methods. UNESCO, Paris, pp 263–298
Thomassin BA, Cauwet G (1985) PCRC-5, 4, pp 377–382
Thomassin BA, Masse JP (1985) Tethys 11: 288–301
Thomassin BA, Vivier MM, Vitiello P (1976) J Exp Mar Biol Ecol 22: 31–53
Thomassin BA, Jouin C, Renand-Mornan J (1982) Tethys 10: 392–997
Thompson J, Skinn EA, Bright L (1980) In: Geyer R (ed) Marine environmental pollution. Elsevier, Amsterdam, pp 433–453
Thompson RD, Golding TJ (1982) J Geophys Res 86: 6517–6521
Thorniton EB, Lee TT, Blank KT (1976) Trans Am Geophys Un 79: 1–98
Thorp EM (1936) Calcareous deposits of Florida and Bahamas. Pap Dep Mar Biol Carnegie Inst Washington 29: 37–120
Thresher RE (1977) PCRS-3, 1, pp 551–553
Thresher RE (1984) Reproduction in reef fishes. TFH, Neptune City, 390 pp
Thresher RE, Moyer JT (1983) Anim Behav 31: 113–127
Thurber DJ (1965) Science 149: 55–58
Tiffon Y, Bouillon J (1975) J Exp Mar Biol Ecol 18: 255–269
Titlyanov EA (1981) PCRS-4, 2, pp 39–44
Titlyanov EA (1987) Biology of the Sea (Vladivostok) 3: 32–40
Titlyanov EA, Latypov YU (1991) Coral Reefs 10: 133–138
Titlyanov EA, Shaposhnikova MG, Zvalinsky VI (1980) Photosynthetica 14: 413–421
Titlyanov EA, Shaposhnikova EA, Zvalinsky VI (1981) Biology of the Sea (Vladivostok) 2: 23–31
Titlyanov EA, Shaposhnikova MG, Zvalinsky VI (1983a) In: Krashov E (ed) Biology of coral reefs. Acad Sci USSR, Vladivostok, pp 51–73
Titlyanov EA, Zvalinsky VI, Shaposhnikova MC (1983b) Biology of the Sea (Vladivostok) 3: 39–48

Titlyanov EA, Tchabazgy II, Zvalinksky VI (1985) In: Rep 6th cruise r/v "Alexander Nesmeyanov". Oceanogr Commiss Moscow pp 25–98
Titlyanov EA, Magomedov IM, Kolmakov PV et al. (1988a) Sorokin YuI (ed) Biology of coral reefs. Acad Sci USSR, Vladivostok, pp 19–45
Titlyanov EA, Kolmakov PV, Le Neguen Hieu (1988b) Sorokin Yu I (ed) Biology of coral reefs. Acad Sci USSR, Vladivostok, pp 52–271
Titlyanov EA, Latypov YuI, Ermak VD (1988c) In: Sorokin YuI (ed) Biology of coral reefs. photosynthesis of reef building corals. Acad Sci USSR, Vladivostok, pp 5–18
Tolbert NE, Osmond CB (1976) Aust J Plant Physiol 3: 1–8
Tranter DJ, George J (1972) PCRS-1, pp 239–256
Travers A, Travers M (1965) Ann Malagash Sci 2: 125–162
Trench RK (1971) Proc Roy Soc London, B, 177: 251–263
Trench RK (1974) Helgoländer Wiss Meersunters 26: 174–216
Trench RK (1979) Annu Rev Plant Physiol 30: 485–531
Trench RK, Blank RG (1987) J of Phycol 23: 469–481
Tresher RE (1984) In: Tropical fish hobbist, vol 7. TFN Publ, Neptune City, NJ, pp 343–388
Tribble GW et al. (1988) ACRC-6, p 6 (App)
Trudgill ST (1976) Geomorphology 26 (Suppl): 64–200
Trudgill ST (1983) In: Barnes D (ed) Perspectives on coral reefs. AIMS, Townsville, pp 256–260
Tsareva LA (1980) In: Preobrazhensky B (ed) Biology of coral reefs. Nauka, Moscow, pp 115–130
Tsuda RT (1974) PCRS-2, 1, pp 43–47
Tsuda RT (1977a) PCRS-3, 1, pp 371–376
Tsuda RT (1977b) Micronesica 13; 191–203
Tsuda RT (1979) In: Abstr 14th Pac Sci Congr, pt G, Khabarovsk, Tsuda RT, Kami HT (1973) J Phycol 9: 260–264
Tunnicliffe V (1978) In: Levi C (ed) Biologie des spongiaires. CNRS, Paris, pp 309–319
Tunnicliffe V (1981) Proc Natl Acad Sci USA 78: 2427–2431
Tunnicliffe V (1982) J Exp Mar Biol Ecol 64: 1–10
Tursch B, Tursch A (1982) Mar Biol 68: 321–332
Tursch B et al. (1978) In: Scheuer P (ed) Marine natural products. Academic Press, New York, pp 247–296
Vacelet J (1971) J Microsc 12: 363–380
Vacelet J (1979) In: Levi C (ed) Spongiaires. CNRS, Paris, pp 259–270
Vacelet J (1981) PCRS-4, 2, pp 713–719
Vacelet J, Donadey C (1977) J Exp Mar Biol Ecol 30: 301–314
Vacelet J, Vasseur zP (1977) PCRS-3, 1, pp 113–117
Vadas RL, Ogden JC, Taxis TM (1980) Am Zool 20: 811
Van den Hoek C (1969) Proc Kon Nederl Acad Wetensch Ser C, 72: 537–577
Van den Hoek C, Cortel-Breeman AM, Wanders JB (1978) Aquat Bot 5: 1–46
Vandermeulen JH, Davis N, Muscatine L (1972) Mar Biol 16: 185–195
Vandermeulen JH, Watabe N (1974) Marine Biol 23: 47–57
Vasseur P (1974) PCRS-2, 2, pp 143–159
Vasseur P (1977) PCRS-3, 1, pp 95–100
Vaugelas VJ de (1981) PCRS-4, 1, pp 411–417
Vaugelas VJ de, Naim O (1981) PCRS-4, 1, pp 405–410
Vaughan TW (1912) Studies of geology and of Madreporaria of the Bahamas. Yearb Carnegie Inst Washington 11: 230 pp
Vaughan TW (1919) Smithsonian Inst Washington Annu Rep, pp 189–258
Vaughan TW, Wells JM (1943) Bull Geol Soc Am Spec Pap 44: 363–450
Veeh H (1985) PCRC-5, 3, pp 463–466
Velimirov B, King J (1979) Mar Biol 50: 349–358
Venkataramanusiam K, Santhanam R (1989) In: Abstr 5th Int Conf Coelenterate biology, Southampton, p 97
Vereshi E, Fricke H (1986) Mar Biol 90: 395–401

Veron JEN (1985) PCRC-5, 4, pp 83–90
Veron JEN (1986) Corals of Australia the Indo-Pacific. Angus and Robertson, Sydney, 644 pp
Veron JEN, Hudson RC (1978) Philos Trans R Soc London Ser B 284: 3–21
Veron JEN, Pichon M (1976) Scleractinia of Eastern Australia, pt 1, vol 1. AIMS Monogr Ser. Aust Nat Univ Press, Canberra, pp 1–86
Veron JEN, Wallace CC (1984) Scleractinia of Eastern Australia, vol 6. AIMS Monogr Ser. Aust Nat Univ Press, Canberra, 481 pp
Veron JEN Pichon M, Wifsman-Best M (1977) Scleractinia of Eastern Australia, pt 2, vol 3. AIMS Monogr Ser. Aust Nat Univ Press, Canberra, pp 1–233
Verseveldt J (1974) Isr J Zool 23: 53–67
Verseveldt J (1977) Aust J Mar Freshwater Res 28: 171–240
Verseveldt J, Tursch A (1979) Zool Meded 54: 134–148
Villiers L (1988) ACRC-6, p 104
Vine PJ (1970) Nature (London) 228: 341–342
Vine PJ (1973) Atoll Res Bull 166: 10–15
Vine PJ (1974) Mar Biol 24: 131–136
Vinogradova KL, Kalugina-Gutnik AA (1985) In: Report of 6th cruise of r/v "Alexander Nesmeyanov". Oceanogr Commiss Acad Sci, USSR, Moscow, pp 180–210
Vittor BA, Johnson PG (1977) PCRS-3, 1, pp 163–168
Vivien M (1973) J Mar Biol Assoc India 15: 20–45
Vooren CM (1981) Aquat Bot 10: 143–154
Vosburgh F (1982) Proc R Soc London Ser B 214: 481–499
Voskoboynikov GM, Mashansky VE, Le Nguen Hieu (1983) Biology of the Sea (Vladivostok) 3: 54–60
Wafar M et al. (1985) PCRS-5, 6, pp 29–37
Wahbeh MI (1981) PCRS-4, 2, pp 453–459
Wahbeh MJ, Mahasnen AM (1988) ACRC-6, p 104
Wainwright SA, Koel MA (1976) In: "Coelenterate biology and behavior" NY Plenum Press, pp 5–21
Walker DI, Ormond RFG (1982) Mar Pollut Bull 13: 21–25
Walker DI, Ormond FG (1982) Mar Poll Poull 13: 21–25
Walker TA (1981) Aust J Mar Freshwater Res 32: 981–986
Walbran PD, Henderson RA, Faithful JW et al. (1989) Coral Reefs 8: 67–78
Wallace CC, Bull GD (1981) PCRS-4, 2, pp 345–350
Wallace RJ, Schaefersman SD (1977) In: Frost S (ed) Reefs and related carbonates. Am Assoc Petr Geol, pp 37–52
Walsh GE, Bowers RI (1977) In: Reef and shore fauna of Hawaii Univ. H of Hawaii Publ, pp 148–157
Walsh RJ et al. (1971) In: Rep Comm Crown of thorns problems. CSIRO, Melbourne, 45 pp
Walter C, Pasamonte JN, Tolane L (1981) PCRS-4, 1, pp 443–449
Wanders JB (1976a) Aquat Bot 2: 327–335
Wanders JB (1976b) Aquat Bot 2: 235–270
Wanders J (1977) Aquat Bot 3: 357–390
Warme JE (1977) In: Frost S (ed) Reefs and related carbonates. Amer Ass Petr Geol, pp 261–271
Webb KL, Wiebe WJ (1975) Can J Microobiol 21: 1427–1431
Webb KL, Wiebe WJ (1978) Mar Biol 47: 21–27
Webb KL, Du Panl WD, Wiebe W et al. (1975) Limnol Oceanogr 20: 198–210
Webb KL, Du Paul WD, D'Elia CF (1977) PCRS-3, 1, pp 409–415
Weinberg S, Weinberg F (1979) Bijdr Dierk 48: 127–140
Weinheimer AJ, Spraggins RL (1969) Tetrahedron Lett 59: 5185–5188
Weise W, Rheinheimer J (1978) Microbial Ecol 4: 175–178à
Weiss MP, Goddart DA (1977) In: Frost S (ed) Reefs and related carbonates. Tulsa, pp 111–124
Wellington GM (1980) Oecologia Assoc Petr Geol 47: 340–343

Wells JW (1954) US Geol Surv Prof Pap 260-1: 285–486
Wells JW (1956) In: Moore R (ed) Treatise on invertebrate paleontology pt F. Geol Soc Am, pp 329–443
Wells JW (1957) Geol Soc Am Mem 67: 609–631
Wells JW (1966) In Rees W (ed) The cnidaria and their evolution. Academic Press, London, pp 223–246
Wells JW (1971) In: Abstr Annu Meet Geol Soc Am, Washington, vol 3, pp 748
Wells SM (1981) PCRS-4, 1, pp 323–330
West N, Larcum A (1983) Proc Linn Soc 106: 213–223
Westlake DF (1963) Biol Rev 38: 385–425
Westneat MW, Reising JM (1988) Coral Reefs 7: 89–92
Westrum B, Meyers PA (1978) Bull Mar Sci Miami 28: 153–158
Wethey DS, Porter JW (1976) In: Mackie G (ed) Coelenterate biology and behavior. Plenum, New York, pp 59–66
Wickler W (1971) Verh Dtsch Zool Ges 64: 291–304
Wiebe WJ (1976) Micronesica 12: 23–26
Wiebe WJ (1985) PCRC-5, 3, pp 401–406
Wiebe WJ, Johannes RE, Webb KL (1975) Science 188: 257–259
Wiens HJ (1962) Atoll environment and ecology. Yale Univ. Press, New Haven, 535 pp
Wilkerson FP, Parker GM, Muscatine L (1983) Limnol Oceanogr 28: 1009–1014
Wilkinson CR (1978) Mar Biol 49: 161–176
Wilkinson CR (1980) In: Schwemmler W (ed) Endocytobiology, vol 1. De Gruyter, Berlin, pp 553–563
Wilkinson CR (1981) PCRS-4, 2, pp 705–718
Wilkinson CR (1983a) Science 219: 410–412
Wilkinson CR (1983b) In: Schwemmler W (ed) Endocytobiology, vol 2. De Gruyter Berlin, pp 993–1002
Wilkinson CR (1989) In: Abstr Annu Meet Int Soc Reef studies, Marseille, p 165
Wilkinson CR, Cheshik AC (1989) Coral Reefs 8: 127–134
Wilkinson CR, Fay P (1979) Nature (London) 279: 527–529
Wilkinson CR, Garrone R (1980) In: Smith D (ed) Nutrition in the lower Metazoa Pergamon, Oxford, pp 157–161
Wilkinson CR, Sammarco PW (1983) Mar Ecol Prog Ser 13: 15–19
Wilkinson CR, Trott LA (1985) PCRS-5, 5, pp 125–130
Wilkinson CR, Vacelet J (1979) J Exp Mar Biol Ecol 37: 91–104
Wilkinson CR, Novak M, Austin B et al. (1981) Microbial Ecol 7: 13–21
Wilkinson CR, Williams D, Sammareo PW (1984) Mar Biol 80: 255–62
Williams AH (1981) Ecology 62: 1107–1020
Williams DM (1982) Coral Reefs 1: 35–43
Williams DM (1983) Baker J (ed) Proc GBR Conf. AIMS, Townsville, pp 265–270
Williams DM, Hatcher AI (1983) Mar Ecol Prog Ser 10: 239–250
Williams DM, Wolanski E, Andrews JC (1984) Coral Reefs 3: 229–236
Williams EHJ, Williams J (1990) Atoll Res Bull 335: 1–71
Williams EH, Coenaga C, Vicente V et al. (1987) Science 238: 877–878
Williams EM, Bunkley-Williams L (1988) PCRS-6 vol 3, pp 313–318
Williams SL (1981) The uptake of NH_4^+ by caulerpa and its decomposition. PhD Diss, Univ Maryland, 77 pp
Willis BL, Babcock RC, Harrison PL et al. (1985) PCRC-5, 4, pp 310–316
Wilson BR, Marsh LM (1974) RCRS-2, 1, pp 621–630
Winn HE, Marshall JA, Hazlett B (1964) Copeia: 413–425
Wolanski E (1982) Limnol Oceanogr 32: 735–739
Wolanski E, Bennett AF (1983) Aust J Mar Freshwater Res 34: 23–47
Wolanski E, Jones M (1980) Aust J Mar Freshwater Res 31: 415–430
Wolanski E, Jupp DL, Pickard GL (1986) Oceanus 29: 83–89
Wolanski E, Pickard GL (1983a) Coral Reefs 2: 1–8
Wolanski E, Pickard GL (1983b) Aust J Mar Freshwater Res 34: 65–80

Wolanski E, Pickard GL (1985) Coral Reefs 4: 47–57
Wolanski E, van Senden D (1983) Aust J Mar Freshwater Res 34: 49–63
Womersly BS, Bailey A (1969) Philos Trans R Soc London Ser B 255: (B): 1–433
Wood EFJ (1960) In: Proc 9th Pac Sci Congr Honoluln, vol 7: Botany, pp 217–221
Woodhead PM, Parker JH, Duedall L (1982) Mar Fish Rev 44: 16–23
Woodley JD (1975) J Exp Mar Biol Ecol 18: 26–46
Woodley JD et al. (1981) Science 214: 749–755
Woscoboynikov GM et al. (1983) Biol of the Sea (Vladivostok) No 3 pp 54–60
Wyers SC (1985) PCRC-5, 4, pp 301–306
Wylie C (1988) ACRC-6, p 105
Yamaguchi M (1973) In: Jones O, Endean R (eds) Biol and geol of coral reefs, vol 2: Biology 1. Academic Press, New York, pp 369–386
Yamaguchi M (1977) Biol Conserv 11: 13–20
Yamazato K (1981) Sesoko Mar Sci Lab Tech Rep Okinawa 8: 9–18
Yamazato K, Sato M, Yamashira H (1981) PCRS-4, 2, pp 671–678
Yap HT, Gomez ED (1981) PCRS-4, 2, pp 207–213
Yap HT, Gomez ED (1983) In: Abstr 15th Int Pac Sci Congr, Dunedin p 81
Yap HT, Gomez ED (1984) Mar Biol 81: 209–215
Yonge CM (1930a) In: Sci Rep GBR Exped London, vol 1, pp 83–91
Yonge CM (1930b) In: Sci Rep GBR Exped London, vol 1, pp 14–57
Yonge CM (1940) In: Sci Rep GBR Exped, vol 1, pp 353–394
Yonge CM (1963) In: Russel F (ed) Advances in marine biology. Academic Press, New York, pp 209–260
Yonge CM (1973) Bull Mar Sci 23: 1–15
Yonge CM (1975) Sci 25: 96–105
Yonge CM, Nicholls AG (1930) In: Sci Rep GBR Exped, vol 1, pp 59–82
Yonge CM, Nicholls AG (1931) In: Sci Rep GBR Exped, Brit Mus Nut Hist Lond, vol 1, pp 135–178
Yoshioka PM, Yoshioka BB (1989) Coral Reefs 8: 145–152
Young SD (1973) Comp Biochem Phys 44A: 669–672
Zann LP, Bolton L (1985) Coral Reefs 4: 125–134
Zann L, Brodie J, Veikila V (1990) Coral Reefs 9: 135–144
Zharikov VV (1987) Abstr 6th Symp Modern and ancient reefs, Vladivostok, pp 28–30
Zhuravleva IT (1960) Acad Sci USSR Moscow, pp 210
Zhuravleva IT, Myagkova EI (1979) In: Environments and biota during the geological ages. Siberian Dep Acad Sci USSR Novosibirsk, pp 117–128
Zibrowius H (1980) Mem Inst Oceanogr Monaco 11: 1–284
Zibrowius H (1985) PCRS-5 5, pp 233–238
Zieman JC (1975) In: Estuarine research, vol 1. Academic Press, New York, pp 541–562
Zieman JC et al. (1979) In: Livingston R (ed) Ecological processes in coastal and marine ecosystems. Plenum New York, pp 21–33
Zlatarsky VN, Estalella MM (1980) Scleractinia of Cuba. Acad Sci Sofia, Bulg, 330 pp
Zmarzly DL, Holl and ND (1981) Mar Ecol Prog Ser 6: 229–230
Zohary T, Reiss Z, Hottinger L (1980) Helv Ecol Geol 73: 1071–1094
Zvalinski VI, Titlyanov EA, Leletkin VA et al. (1980) Photosynthetica 14: 421–430

Subject Index

Ecological Studies

Volumes published since 1990

Volume 80
Plant Biology of the Basin and Range (1990)
B. Osmond, G.M. Hidy, and L. Pitelka (Eds.)

Volume 81
Nitrogen in Terrestrial Ecosystems: Questions of Productivity, Vegetational Changes, and Ecosystem Stability (1991)
C.O. Tamm

Volume 82
Quantitative Methods in Landscape Ecology: The Analysis and Interpretation of Landscape Heterogeneity (1990)
M.G. Turner and R.H. Gardner (Eds.)

Volume 83
The Rivers of Florida (1990)
R.J. Livingston (Ed.)

Volume 84
Fire in the Tropical Biota: Ecosystem Processes and Global Challenges (1990)
J.G. Goldammer (Ed.)

Volume 85
The Mosaic-Cycle Concept of Ecosystems (1991)
H. Remmert (Ed.)

Volume 86
Ecological Heterogeneity (1991)
J. Kolasa and S.T.A. Pickett (Eds.)

Volume 87
Horses and Grasses: The Nutritional Ecology of Equids and Their Impact on the Camargue (1992)
P. Duncan

Volume 88
Pinnipeds and El Niño: Responses to Environmental Stress (1992)
F. Trillmich and K.A. Ono (Eds.)

Volume 89
Plantago: A Multidisciplinary Study (1992)
P.J.C. Kuiper and M. Bos (Eds.)

Volume 90
Biogeochemistry of a Subalpine Ecosystem: Loch Vale Watershed (1992)
J. Baron (Ed.)

Volume 91
Atmospheric Deposition and Forest Nutrient Cycling (1992)
D.W. Johnson and S.E. Lindberg (Eds.)

Volume 92
Landscape Boundaries: Consequences for Biotic Diversity and Ecological Flows (1992)
A.J. Hansen and F. di Castri (Eds.)

Volume 93
Fire in South African Mountain Fynbos: Ecosystem, Community, and Species Response at Swartboskloof (1992)
B.W. van Wilgen et al. (Eds.)

Volume 94
The Ecology of Aquatic Hyphomycetes (1992)
F. Bärlocher (Ed.)

Volume 95
Palms in Forest Ecosystems of Amazonia (1992)
F. Kahn and J.-J. DeGranville

Volume 96
Ecology and Decline of Red Spruce, in the Eastern United States (1992)
C. Eagar and M.B. Adams (Eds.)

Volume 97
The Response of Western Forests to Air Pollution (1992)
R.K. Olson, D. Binkley, and M. Böhm (Eds.)

Volume 98
Plankton Regulation Dynamics (1993)
N. Walz (Ed.)

Volume 99
Biodiversity and Ecosystem Function (1993)
E.-D. Schulze and H.A. Mooney (Eds.)

Volume 100
Ecophysiology of Photosynthesis (1994)
E.-D. Schulze and M.M. Caldwell (Eds.)

Ecological Studies

Volumes published since 1990

Volume 101
Effects of Land Use Change on Atmospheric CO_2 Concentrations: South and South East Asia as a Case Study (1993)
V.H. Dale (Ed.)

Volume 102
Coral Reef Ecology (1993)
Y.I. Sorokin

Volume 103
Rocky Shores: Exploitation in Chile and South Africa (1993)
W.R. Siegfried (Ed.)

Volume 104
Long-Term Experiments With Acid Rain in Norwegian Forest Ecosystems (1993)
G. Abrahamsen et al. (Eds.)

Volume 105
Microbial Ecology of Lake Plußsee (1993)
J. Overbeck and R.J. Chrost (Eds.)

Volume 106
Minimum Animal Populations (1994)
H. Remmert (Ed.)

Volume 107
The Role of Fire in Mediterranean-Type Ecosystems (1994)
J.M. Moreno and W.C. Oechel

Volume 108
Ecology and Biogeography of Mediterranean Ecosystems in Chile, California and Australia (1994)
M.T.K. Arroyo, P.H. Zedler, and M.D. Fox (Eds.)

Volume 109
Mediterranean-Type Ecosystems. The Function of Biodiversity (1995)
G.W. Davis and D.M. Richardson (Eds.)

Volume 110
Tropical Montane Cloud Forests (1995)
L.S. Hamilton, J.O. Juvik, and F.N. Scatena (Eds.)

Volume 111
Peatland Forestry. Ecology and Principles (1995)
E. Paavilainen and J. Päivänen

Volume 112
Tropical Forests: Management and Ecology (1995)
A.E. Lugo and C. Lowe (Eds.)

Volume 113
Arctic and Alpine Biodiversity. Patterns, Causes and Ecosystem Consequences (1995)
F.S. Chapin III and C. Körner (Eds.)

Volume 114
Crassulacean Acid Metabolism. Biochemistry, Ecophysiology and Evolution (1995)
K. Winter and J.A.C. Smith (Eds.)

Volume 115
Islands. Biological Diversity and Ecosystem Function (1995)
P.M. Vitousek, L.L. Loope, and H. Adsersen (Eds.)

Volume 116
High Latitude Rainforests and Associated Ecosystems of the West Coast of the Americas: Climate, Hydrology, Ecology and Conservation (1995)
R.G. Lawford, P. Alaback, and E.R. Fuentes (Eds.)

Volume 117
Anticipated Effects of a Changing Global Environment on Mediterranean-Type Ecosystems (1995)
J. Moreno and W.C. Oechel (Eds.)

Volume 118
Impact of Air Pollutants on Southern Pine Forests (1995)
S. Fox and R.A. Mickler (Eds.)

Volume 199
Freshwater Ecosystems of Alaska (1996)
A.M. Milner and M.W. Oswood (Eds.)

Volume 120
Landscape Function and Disturbance in Arctic Tundra (1996)
J.F. Reynolds and J.D. Tenhunen (Eds.)

Volume 121
Biodiversity and Savanna Ecosystem Processes. A Global Perspective (1996)
O.T. Solbrig, E. Medina, and J.F. Silva (Eds.)

Printing: Mercedesdruck, Berlin
Binding: Buchbinderei Lüderitz & Bauer, Berlin